T0314204

Solitons in Optical Fiber Systems

Solitons in Optical Fiber Systems

Mario F. S. Ferreira
University of Aveiro
Aviero, Portugal

This edition first published 2022

Registered Office
John Wiley & Sons, Inc., 111 River Street, Hoboken, NJ 07030, USA

Editorial Office
111 River Street, Hoboken, NJ 07030, USA

For details of our global editorial offices, customer services, and more information about Wiley products visit us at www.wiley.com.

Wiley also publishes its books in a variety of electronic formats and by print-on-demand. Some content that appears in standard print versions of this book may not be available in other formats.

Library of Congress Cataloging-in-Publication Data applied for:

Hardback 9781119506676

Cover image: © merrymoonmary/Getty Images
Cover design by Wiley

Set in 9.5/12.5pt STIXTwoText by Straive, Pondicherry, India

To the memory of my Parents
To Sara Maria, Sara Raquel, Mário Alberto, Maria do Céu, and Miguel Fernando

Contents

Preface

The concept of soliton was created by Norman Zabusky and Martin Kruskal in 1965 to refer to localized solutions of a particular class of models – the so-called integrable nonlinear equations. Such solutions are remarkable because they interact elastically and regain their forms after colliding with each other. Afterward, the soliton theory has been applied to numerous practical and fundamental problems in areas as diverse as hydrodynamics, plasma, nonlinear optics, molecular biology, field theory, and astrophysics. In the area of optics, integrable conventional solitons result from the single balance between nonlinearity and dispersion or diffraction.

Over the years, there have been attempts made in some parts of the mathematical community to keep the term "soliton" only for solutions of integrable equations. However, such attempts largely failed when physicists realized that solitary waves did exist in a wide range of nonintegrable and nonconservative systems. In practice, the detailed mathematical differences between solutions of integrable and nonintegrable models seem to be less important compared to the key soliton properties, such as localization, coherence, and stability that are observed in both cases. Solitary waves in nonlinear systems with nonlinear gain or loss mechanisms were thereafter referred to as dissipative solitons.

In contrast with the conventional conservative soliton, the characteristics of a dissipative soliton are predetermined by the equation parameters, rather than by the initial conditions. Moreover, a wide range of pulse profiles can be obtained by tuning such parameters. For example, a previously stationary soliton can evolve to a pulsating localized formation and even produce irregular and chaotic soliton dynamics. Actually, the field of mode-locked lasers constitutes an ideal playground for exploring nonlinear dissipative soliton dynamics. In an applied perspective, the dissipative soliton concept appears also highly useful in the development of compact, efficient, and reliable sources of ultrashort light pulses with unprecedented pulse energy and improved stability.

This book is intended to provide an introduction to the fascinating topic of optical solitons in fiber systems. Conventional fiber solitons are discussed first. Afterward, special attention is paid to dissipative soliton pulses, both stationary and pulsating, which appear in different fiber systems described by the complex Ginzburg–Landau equation, namely in mode-locked fiber lasers. Some recent experimental results in this area are reported. The peculiar phenomena occurring in highly nonlinear fibers and the role played by solitons in supercontinuum generation also deserve particular attention. The contents of this book make it of potential interest both to senior undergraduate and graduate students enrolled in MS and PhD degree programs, to engineers and technicians involved with the fiber optics industry, as well as to researchers working in the field of nonlinear fiber optics.

I am deeply grateful to the many students and colleagues with whom I have interacted over the years. All of them have contributed to this book either directly or indirectly. In particular, I thank Sofia Latas and Sílvia Rodrigues for providing many of the figures, as well as Margarida Facão, Armando Pinto, and Nelson Muga for several discussions concerning different parts of the text. Last, but not least, I thank my family for understanding why I needed to remain working during many of our weekends and holidays.

Mário F.S. Ferreira
Aveiro, Portugal
October 2021

List of Abbreviations

ASE	Amplified spontaneous emission
BER	Bit error rate
CGLE	Complex Ginzburg–Landau equation
CSHE	Complex Swift–Hohenberg equation
CP	Composite pulse
CW	Continuous wave
DCF	Dispersion-compensating fiber
DDF	Dispersion-decreasing fiber
DFT	Dispersive Fourier transform
DGD	Differential group delay
DM	Dispersion management
DSF	Dispersion-shifted fiber
DSR	Dissipative soliton resonance
EDF	Erbium-doped fiber
EDFA	Erbium-doped fiber amplifier
ESA	Excited-state absorption
FBG	Fiber Bragg grating
FOPA	Fiber-optical parametric amplifier
FRA	Fiber Raman amplifier
FWHM	Full width at half maximum (FWHM)
GNLSE	Generalized nonlinear Schrödinger equation
GVD	Group velocity dispersion
GVLVS	Group-velocity-locked vector solitons
HNLF	Highly nonlinear fibers
HOE	Higher-order effect
IRS	Intrapulse Raman scattering
KdV	Korteweg–de Vries
LAN	Local area network
LD	Laser diode
MDM	Mode-division multiplexing
MI	Modulation instability
MIR	Mid-infrared
MOF	Microstructured optical fiber
NALM	Nonlinear amplifying loop mirror

NOLM	Nonlinear optical loop mirror
NF	Noise figure
NIR	Near-infrared
NLSE	Nonlinear Schrödinger equation
NPR	Nonlinear polarization rotation
NSR	Nonsoliton radiation
NZDF	Nonzero dispersion fiber
OTDM	Optical time-division multiplexing
PCF	Photonic crystal fibers
PDM	Polarization-division multiplexing
PLVS	Polarization-locked vector solitons
PMD	Polarization-mode dispersion
PRVS	Polarization rotation vector solitons
PSP	Principal states of polarization
RZ	Return to zero
SA	Saturable absorber
SBS	Stimulated Brillouin scattering
SCG	Supercontinuum generation
SESAM	Semiconductor saturable absorber mirror
SMF	Single-mode fiber
SNR	Signal-to-noise ratio
SOP	State of polarization
SPM	Self-phase modulation
SRS	Stimulated Raman scattering
SSFS	Soliton self-frequency shift
SST	Self-steepening
TOD	Third-order dispersion
VHA	Very high amplitude
XPM	Cross-phase modulation
WDM	Wavelength-division multiplexing
WSC	Wavelength-selective coupler
ZDW	Zero-dispersion wavelength

1

Introduction

Solitary waves have been the subject of intense theoretical and experimental studies in many different fields, including hydrodynamics, nonlinear optics, plasma physics, and biology [1–5]. In fact, the history of solitons dates back to 1834, the year in which John Scott Russell observed that a heap of water in a canal propagated undistorted over several kilometers [6]. However, the term "soliton" was coined only in 1965 by Zabusky and Kruskal [7] to reflect the particle-like nature of solitary waves that remain intact even after mutual collisions. These waves correspond to localized solutions of integrable equations, such as the Korteweg-de Vries and the nonlinear Schrödinger equations. A rigorous mathematical theory was developed by Gardner, Greene, Kruskal, and Miura in 1967 [8], allowing the derivation of a general solution to these equations in terms of solitons and radiation waves. Since this pioneering work, scientists tend to use the term "soliton" to describe the modes of nonlinear partial differential equations that happen to be integrable by means of the "inverse scattering technique" [9, 10]. The integration of the nonlinear Schrödinger equation (NLSE) was demonstrated in 1971 by Zakharov and Shabat [9] using this technique.

Based on such restrictive definition, solitons were usually attributed, until the beginning of the 1990s, only to integrable systems. However, the soliton concept was subsequently broadened when many physicists understood the limitations of the theory. In fact, they observed that "solitary waves" do exist in a variety of systems, though they do not behave exactly as the classical theory predicts for "true" solitons.

The existence of solitons in optical fibers was suggested by Hasegawa and Tappert [11] in 1973. The formation of these solitons was shown to be the result of a balance between the negative (anomalous) group velocity dispersion (GVD) of the glass fiber, which occurs for wavelengths longer than 1.3 μm in a standard fiber, and the Kerr nonlinearity. The inclusion of both phenomena gives origin to the NLSE. The fiber solitons appeared as an ideal solution to the problem of pulse spreading caused by fiber dispersion. However, they could do nothing against the attenuation that any propagating pulse experiences due to the waveguide loss. In fact, at the time that paper was published, there were no practical methods to solve the fiber loss problem and the soliton idea remained nothing more than an elegant mathematical curiosity.

After many ingenious efforts, Mollenauer, Stolen, and Gordon [12] succeeded, in 1980, in observing the optical soliton in a fiber. The experimental results confirmed the predictions of the NLSE and proved that the properties of a pulse propagating in an optical fiber are described, to a remarkable degree, by this equation. However, few people expected some practical importance of solitons because there were erroneous beliefs that anything nonlinear would be too complicated.

Meanwhile, Hasegawa [13] made the imaginative proposal that solitons could be used in all-optical transmission systems based on optical amplifiers instead of regenerative repeaters, which

Solitons in Optical Fiber Systems, First Edition. Mario F. S. Ferreira.

were considered standard until 1990. In particular, he suggested using the Raman effect of transmission fiber itself for optical amplification. The idea was used in the first long-distance all-optical transmission experiment by Mollenauer and Smith [14] in 1988.

Two years before the experiment of Mollenauer and Smith, Gordon and Haus wrote a paper that is a milestone for the later development of the field of soliton transmission [15]. They predicted that the transmission of a signal made of optical solitons could not be extended over an unlimited distance when optical amplification is used. In fact, the amplifiers needed to compensate for the fiber loss generate amplified spontaneous emission (ASE), and this noise is, in part, incorporated by the soliton, whose mean frequency is then shifted. A frequency shift leads to a timing shift because pulses of different frequencies have different group velocities due to GVD. The arrival time of the soliton becomes then a random variable whose variance is proportional to the cube of the propagation distance. This is the so-called *Gordon–Haus effect.*

The concept of all-optical transmission using the distributed Raman amplification gave its position to erbium-doped fiber amplifiers (EDFAs) during the 1990s. Actually, this fiber amplifier is a few meters long and presents some remarkable properties. The first successful demonstration of soliton transmission using EDFAs was reported by Nakazawa, Kimura, and Suzuki in 1989 [16]. Facing this development, a number of theoretical and conceptual issues had to be solved. In fact, the soliton concept is associated with a lossless uniform fiber and hence one would expect that the actual loss of the fiber would have to be compensated by distributed gain. This was the reason for the use of distributed Raman gain in previous experiments. However, it was shown almost simultaneously by three research groups that lumped amplification is permissible provided that the amplifier spacing is not excessive [17–19]. This corresponds to the so-called average-soliton regime. The transmission of solitons at 2.5 Gb/s over transoceanic distances was achieved in 1991 within this regime [20]; the limit was set only by the Gordon–Haus effect.

At the end of 1991, two research groups suggested independently the use of frequency filters to extend the limit set by the Gordon–Haus effect [21, 22]. The validity of this proposal was confirmed experimentally by Mollenauer et al. [23]. The transmission distance was still limited in this experiment by the growth of narrow-band noise at the center frequency of the filters. In fact, in order to compensate for the filter-induced loss, some excess gain must be provided to the soliton. However, this excess gain amplifies also the small-amplitude waves coexistent with the soliton, resulting in a background instability that can affect significantly and even destroy the soliton itself.

The problem of background instability was beautifully solved by Mollenauer and coworkers, who developed the sliding guiding filter concept [24, 25]. In this scheme, the center frequency of the filters is displaced along the propagation distance. The soliton carrier frequency follows the filter center frequencies; the linear noise does not. The transmission line becomes then transparent to solitons and opaque to the noise. At the same time, when the sliding filters were proposed, Nakazawa and his group have demonstrated that practically unlimited distance of propagation can be achieved with a combination of optical amplifiers and amplitude modulators [26]. These modulators restored the timing of solitons that had strayed from the center of the bit interval, thus controlling the noise-induced timing jitter. However, this method is not immediately suitable for wavelength-division-multiplexed (WDM) systems.

Attempts to solve much of the intrinsic problems of optical soliton transmission by proper control of the fiber group dispersion have emerged in the latter half of the 1990s. Forysiak et al. [27] and Hasegawa et al. [28] have proposed adiabatic dispersion management by changing dispersion in proportion to the soliton power in order to reduce dispersive wave radiation and collision-induced frequency shift in WDM systems. In contrast, Suzuki et al. [29] used nonadiabatic periodic dispersion compensation to reduce the integrated dispersion and successfully transmit a 10 Gbit/s soliton signal over a cross-Pacific distance without soliton control. In 1996, Smith et al. [30] showed that a nonlinear soliton-like pulse can exist in a fiber having a periodic variation of the dispersion, even if the dispersion is almost zero. The nonlinear pulse that can propagate in such a system is usually called *dispersion-managed soliton* and presents several remarkable characteristics, namely an enhanced pulse energy, reduced Gordon–Haus timing jitter, longer collision lengths, and greater robustness to polarization-mode dispersion (PMD) [27–31]. Due to their characteristics, dispersion-managed solitons were considered the best option for use in new ultrahigh-speed multiplexed systems.

An alternative approach to avoid the background instability generated by the amplification of the linear waves in soliton fiber systems consists in the use of an amplifier having a nonlinear property of gain, or gain and saturable absorption in combination, such as the nonlinear loop mirror, or nonlinear polarization rotation in combination with a polarization-dependent loss element [32]. The key property of the nonlinearity in gain is to give an effective gain to the soliton and a suppression (or very small gain) to the noise. This method of nonlinear gain may be particularly useful for the transmission of solitons with sub-picosecond or femtosecond durations, where the gain bandwidth of amplifiers will not be wide enough for the sliding of the filter frequency to be allowed [33].

The pulses propagating in the presence of narrow-band filtering, linear, and nonlinear gain are termed *dissipative solitons* and they emerge as a result of a double balance: between nonlinearity and dispersion and also between gain and loss. Concerning the second condition, even the slightest imbalance will result in the solution either growing indefinitely, if gain prevails, or disappearing completely because of the dissipation. Such dissipative solitons have many unique properties that differ from those of their conservative counterparts. For example, except for very few cases [5], they form zero-parameter families and their properties are completely determined by the external parameters of the optical system. They can exist indefinitely in time, as long as these parameters stay constant. These features of dissipative solitons are highly desirable for several applications, such as in-line regeneration of optical data streams and generation of stable trains of laser pulses by mode-locked cavities.

The stability and other properties of a dissipative soliton are linked to the stability and characteristics of its attractor, in the terminology of nonlinear dynamics. For example, a stationary soliton is associated with a fixed point. By tuning the system parameters, we can transform such a fixed point into a limit cycle, which is an oscillatory attracting state. Thus, a previously stationary soliton becomes a pulsating soliton. Further changing the system parameters may result in irregular and chaotic dynamics.

Even if it is a stationary object, a dissipative soliton shows nontrivial energy flows with the environment and between different parts of the pulse. Hence, the dissipative soliton is an object that is far from equilibrium and presents characteristics similar to a living thing. In fact, we can consider animal species in nature as elaborate forms of dissipative solitons. An animal is a localized and persistent "structure," which has material and energy inputs and outputs and complicated internal dynamics. Moreover, it exists only for a certain range of parameters (pressure, temperature, humidity, etc.) and dies if the supply of energy is switched off.

Many nonequilibrium phenomena and dissipative systems, such as convection instabilities, binary fluid convection, and phase transitions, can be described by the complex Ginzburg–Landau equation (CGLE) [34–36]. In the field of nonlinear optics, the CGLE can describe various systems, namely optical parametric oscillators, free-electron laser oscillators, passively mode-locked fiber lasers, and all-optical transmission lines [32, 37–48]. In these systems, there are dispersive elements, linear and nonlinear gain, as well as losses. In some cases, the CGLE admits a multiplicity of solutions for the same range of system parameters. This reality again resembles the world of biology, where the number of species existing in the same environment is truly impressive.

Various techniques have been applied to investigate the soliton solutions of the CGLE. Exact analytical solutions were found [34, 47–51], but they can be presented explicitly only for certain relations between the parameters of the equation. Furthermore, so far, only stationary solutions of the CGLE are known in analytical form. The soliton perturbation theory can be used for small values of the parameters [42, 51]. Moreover, approximate expressions for some localized solutions can be derived for arbitrary values of the CGLE parameters by reducing this equation to finite-dimensional dynamical models. The reduced models can be obtained by applying the method of moments [52], or Lagrangian techniques [53–56].

To fully explore the CGLE, massive numerical simulations must be carried out. Different types of soliton solutions were obtained in this way, which can be divided into two main classes: localized fixed-shape solutions and localized pulsating solutions [56, 57].

One of the most striking forms of pulsating solitons is the erupting soliton, which has been found numerically for the first time in 2000 [58], where the soliton eruption manifests itself as a chaotic and quasi-periodic process when the dissipative system is in a metastable state. The soliton erupts abruptly into pieces in the temporal domain and after that recovers gradually its original state, which is similar to the exploding behavior and thus also called soliton explosion [58, 59]. The soliton explosion was experimentally observed for the first time by Cundiff et al. [60] in a solid-state Kerr-lens mode-locked Ti:sapphire laser. More recently, spectral and temporal signatures of soliton explosions were experimentally observed in an all-normal dispersion and polarization maintaining passively mode-locked Yb-doped fiber laser [61]. Some higher-order effects can have a significant impact on the behavior of erupting solitons, which can even be transformed into fixed-shape pulses [62–65].

Ultrashort optical pulses with extremely high energy are of great importance for a variety of applications. It has been found numerically that the energy of a CGLE dissipative soliton solution increases indefinitely when the equation parameters converge to a given region of the parameter space [66]. Such a set of parameters was called a dissipative soliton resonance (DSR) [67]. With the increase of pump power, the energy of a DSR pulse increases mainly due to the increase of the pulse width, while keeping the amplitude at a constant level [67, 68].

A different kind of high-energy ultrashort pulses is the very high amplitude (VHA) soliton solutions of the CGLE found in Refs [69, 70]. These VHA solutions occur when the nonlinear gain saturation effect tends to vanish. The increase in energy of such pulses is mainly due to the increase of the pulse amplitude, whereas the pulse width becomes narrower. VHA pulses with high energy were found mainly in the normal dispersion region [70], which is in agreement with the large majority of experimental results, reporting the observation of high-energy pulses in passively mode-locked lasers [71–73].

Starting in 1996, new types of fibers, known as tapered fibers, photonic crystal fibers, and microstructured fibers, have been developed [74–78]. Structural changes in such fibers affect their dispersive and nonlinear properties profoundly, including the soliton dynamics. Highly nonlinear fibers can be designed in order to provide more tight confinement of the propagating field to the fiber core. The efficiency of the nonlinear effects can be further increased using fibers made of materials with a nonlinear refractive index higher than that of the silica glass, namely lead silicate, tellurite, bismuth glasses, and chalcogenide glasses [79–82]. These kinds of fibers are commonly called highly nonlinear fibers.

New phenomena are observed in highly nonlinear fibers, like the supercontinuum generation (SCG), in which the optical spectrum of ultrashort pulses is broadened significantly over a short fiber length [83, 84]. A plethora of nonlinear mechanisms can explain such spectral broadening that may differ depending on waveguide dispersion profile, power, wavelength, and duration of pumping pulses. Soliton dynamical effects play a prominent role in SCG when pumping in the anomalous dispersion regime using femtosecond pulses. If the power of the pump pulses is high enough, they can evolve as higher-order solitons. These pulses begin experiencing spectral broadening and temporal compression that are typical of higher-order solitons [51]. However, because of perturbations such as higher-order dispersion or intra-pulse Raman scattering, the dynamics of such pulses departs from the behavior expected for ideal high-order solitons and the pulses break up. This process is known as soliton fission and determines the separation of each higher-order soliton pulse into several fundamental solitons.

Solitons also play a significant role in SCG when pumping with long-duration pulses or continuous wave pumping in the anomalous dispersion regime. In these cases, modulation instability becomes the dominant process in the initial propagation phase, leading to the splitting of the pumping wave into multiple fundamental solitons. The subsequent evolution of these solitons then leads to additional spectral broadening and supercontinuum formation through a variety of mechanisms.

This book is intended to provide an overview of the soliton phenomena and main applications in optical fiber systems. Chapter 2 presents some general concepts on linear and nonlinear waves and introduces some of the main equations describing the soliton evolution. Chapter 3 discusses the dispersive and nonlinear characteristics of optical fibers, including the derivation of the NLSE. Chapter 4 provides an overview of the main nonlinear effects in optical fibers, whereas Chapter 5 describes the basic properties of the NLSE soliton solutions. Chapter 6 discusses different optical amplification schemes, aiming to compensate for the fiber attenuation. The effects of different perturbations on soliton propagation, namely fiber losses, optical amplification, soliton interaction, and timing jitter are described in Chapter 7. Chapter 8 discusses various techniques aiming to guarantee stable soliton propagation in fiber transmission systems. Chapter 9 describes some higher-order linear and nonlinear effects affecting the propagation of ultrashort solitons, whereas Chapter 10 is dedicated to dispersion managed solitons. Chapter 11 discusses the effects of fiber birefringence both on conventional and dispersion-managed solitons. Chapter 12 describes the main properties of stationary dissipative solitons, whereas Chapter 13 is dedicated to different types of pulsating dissipative solitons. Chapter 14 discusses fundamental matters relating to soliton fiber lasers, and Chapter 15 provides an overview of several other applications of optical fiber solitons. Chapter 16 describes several examples and the main properties of highly nonlinear fibers, whereas Chapter 17 is dedicated to supercontinuum generation and the role played by optical solitons in this phenomenon.

References

1 Ablowitz, M.J., Clarkson, P.A., and Solitons, P.A. (1991). *Nonlinear Evolution Equations, and Inverse Scattering*. New York: Cambridge University Press.

2 Taylor, J.T. (ed.) (1992). *Optical Solitons – Theory and Experiment*. New York: Cambridge University Press.

3 Drazin, P.G. (1993). *Solitons: An Introduction*. New York: Cambridge University Press.

4 Gu, C.H. (1995). *Soliton Theory and its Applications*. New York: Springer.

5 Akhmediev, N. and Ankiewicz, A. (1997). *Solitons: Nonlinear Pulses and Beams*. London: Chapman and Hall.

6 Russell, J.S. (1844). Report on waves. In: *Report on the 14th Meeting of the British Association for the Advancement of Science*, York, September (ed. J. Murray) (London), 311–390.
7 Zabusky, N.J. and Kruskal, M.D. (1965). *Phys. Rev. Lett.* **15**: 240.
8 Gardner, C.S., Greene, J.M., Kruskal, M.D., and Miura, K.M. (1967). *Phys. Rev. Lett.* **19**: 109.
9 Zakharov, V.E. and Shabat, A.B. (1972). *Sov. Phys. JETP* **34**: 62; Zh. Eksp. Teor. Fiz. 61, 118 (1971).
10 Ablowitz, M.J. and Clarkson, P.A. (1991). *Solitons, Nonlinear Evolution Equations and Inverse Scattering*, London Mathematical Society Lecture Notes Series **149**. Cambridge: Cambridge University Press.
11 Hasegawa, A. and Tappert, F.D. (1973). *Appl. Phys. Lett.* **23**: 142.
12 Mollenauer, L.F., Stolen, R.H., and Gordon, J.P. (1980). *Phys. Rev. Lett.* **45**: 1095.
13 Hasegawa, A. (1983). *Opt. Lett.* **8**: 650.
14 Mollenauer, L.F. and Smith, K. (1988). *Opt. Lett.* **13**: 675.
15 Gordon, J.P. and Haus, H.A. (1986). *Opt. Lett.* **11**: 665.
16 Nakazawa, M., Kimura, Y., and Suzuki, K. (1989). *Electron. Lett.* **25**: 199.
17 Mollenauer, L.F., Evangelides, S.G., and Haus, H.A. (1991). *J. Lightwave Technol.* **9**: 194.
18 Hasegawa, A. and Kodama, Y. (1991). *Opt. Lett.* **16**: 1385.
19 Blow, K.J. and Doran, N.J. (1991). *IEEE Photon. Technol. Lett.* **3**: 369.
20 Mollenauer, L.F., Neubelt, M.J., Honer, M. et al. (1991). *Electron. Lett.* **27**: 2055.
21 Mecozzi, A., Moores, J.D., Haus, H.A., and Lai, Y. (1991). *Opt. Lett.* **16**: 1841.
22 Kodama, Y. and Hasegawa, A. (1991). *Opt. Lett.* **17**: 31.
23 Mollenauer, L.F., Lichtman, E., Harvey, G.T. et al. (1992). *Electron. Lett.* **28**: 792.
24 Mollenauer, L.F., Gordon, J.P., and Evangelides, S.G. (1992). *Opt. Lett.* **17**: 1575.
25 Mollenauer, L.F., Lichtman, E., Neubelt, M.J., and Harvey, G.T. (1993). *Electron. Lett.* **29**: 910.
26 Nakazawa, M., Kamada, Y., Kubota, H., and Suzuki, E. (1991). *Electron. Lett.* **27**: 1270.
27 Forysiak, W., Knox, F.M., and Doran, N.J. (1994). *Opt. Lett.* **19**: 174.
28 Hasegawa, A., Kumar, S., and Kodama, Y. (1996). *Opt. Lett.* **22**: 39.
29 Suzuki, M., Morita, I., Edagawa, N. et al. (1995). *Electron. Lett.* **31**: 2027.
30 Smith, N.J., Knox, F.M., Doran, N.J. et al. (1996). *Electron. Lett.* **32**: 54.
31 Hasegawa, A. (ed.) (1998). *New Trends in Optical Soliton Transmission Systems*. Dordrecht, Holland: Kluwer.
32 Matsumoto, M., Ikeda, H., Uda, T., and Hasegawa, A. (1995). *J. Lightwave Technol.* **13**: 658.
33 Ferreira, M.F. (1997). Ultrashort soliton stability in distributed fiber amplifiers with different pumping configurations. In: *Applications of Photonic Technology*, vol. **2** (ed. G. Lampropoulos and R. Lessard), 249. New York: Plenum Press.
34 Normand, C. and Pomeau, Y. (1977). *Rev. Mod. Phys.* **49**: 581.
35 Kolodner, P. (1991). *Phys. Rev. A* **44**: 6466.
36 Graham, R. (1975). *Fluctuations, Instabilities and Phase Transictions*. Berlin: Springer.
37 Staliunas, K. (1993). *Phys. Rev. A* **48**: 1573.
38 Jian, P., Torruellas, W., Haelterman, M. et al. (1999). *Opt. Lett.* **24**: 400.
39 Dunlop, A., Wright, E., and Firth, W. (1998). *Optics Commun.* **147**: 393.
40 Ng, C. and Bhattacharjee, A. (1999). *Phys. Rev. Lett.* **82**: 2665.
41 Akhmediev, N., Rodrigues, A., and Townes, G. (2001). *Opt. Commun.* **187**: 419.
42 Ferreira, M.F., Facão, M.V., and Latas, S.V. (2000). *Fiber Integrated Opt.* **19**: 31.
43 Ferreira, M.F., Facão, M.V., Latas, S.V., and Sousa, M.H. (2005). *Fiber Integrated Opt.* **24**: 287.
44 Ferreira, M.F., Facão, M.V., and Latas, S.V. (1999). *Phot. Optoelect.* **5**: 147.
45 Latas, S.V. and Ferreira, M.F. (1999). *SPIE Proc.* **3899**: 396.
46 Akhmediev, N., Ankiewicz, A., and Soto-Crespo, J. (1998). *J. Opt. Soc. Am. B* **15**: 515.

47 Akhmediev, N., Afanasjev, V., and Soto-Crespo, J. (1996). *Phys. Rev. E* **53**: 1190.
48 Soto-Crespo, J., Akhmediev, N., and Afanasjev, V. (1996). *J. Opt. Soc. Am. B* **13**: 1439.
49 Pereira, N. and Stenflo, L. (1977). *Phys. Fluids* **20**: 1733.
50 Nozaki, K. and Bekki, N. (1984). *Phys. Soc. Japan* **53**: 1581.
51 Ferreira, M.F. (2011). *Nonlinear Effects in Optical Fibers*. Hoboken, NJ: John Wiley & Sons.
52 Tsoy, E., Ankiewicz, A., and Akhmediev, N. (2006). *Phys. Rev. E.* **73**: 036621-1-10.
53 Manousakis, M., Papagiannis, P., Moshonas, N., and Hizanidis, K. (2001). *Opt. Commun.* **198**: 351.
54 Ankiewicz, A., Akhmediev, N., and Devine, N. (2007). *Optical Fiber Technol.* **13**: 91.
55 Mancas, S.C. and Choudhury, S.R. (2007). *Theor. Math. Phys.* **152**: 1160.
56 Mancas, S.C. and Choudhury, S.R. (2009). *Chaos, Solitons & Fractals* **40**: 91.
57 Chang, W., Ankiewicz, A., Akhmediev, N., and Soto-Crespo, J. (2007). *Phys. Rev. E.* **76**: 016607.
58 Soto-Crespo, J., Akhmediev, N., and Ankiewicz, A. (2000). *Phys. Rev. Lett.* **85**: 2937.
59 Akhmediev, N., Soto-Crespo, J., and Town, G. (2001). *Phys. Rev. E.* **63**: 056602.
60 Cundiff, S., Soto-Crespo, J., and Akhmediev, N. (2002). *Phys. Rev. Lett.* **88**: 073903.
61 Runge, A., Broderick, N., and Erkintalo, M. (2015). *Optica* **2**: 36.
62 Latas, S.C. and Ferreira, M.F. (2010). *Opt. Lett.* **35**: 1771.
63 Latas, S.C. and Ferreira, M.F. (2011). *Opt. Lett.* **36**: 3085.
64 Latas, S.C., Facão, M.V., and Ferreira, M.F. (2011). *Appl. Phys. B* **104**: 131.
65 Latas, S.C., Ferreira, M.F., and Facão, M.V. (2013). *Appl. Phys. B.* **116**: 279.
66 Akhmediev, N., Soto-Crespo, J., and Grelu, P. (2008). *Phys.Lett. A* **372**: 3124.
67 Chang, W., Ankiewicz, A., Soto-Crespo, J., and Akhmediev, N. (2008). *Phys. Rev. A* **78**: 023830.
68 Chang, W., Soto-Crespo, J., Ankiewicz, A., and Akhmediev, N. (2009). *Phys. Rev. A* **79**: 033840.
69 Latas, S.C., Ferreira, M.F.S., and Facão, M. (2017). *J. Opt. Soc. Am. B* **34**: 1033.
70 Latas, S.C. and Ferreira, M.F. (2019). *J. Opt. Soc. of Am. B* **36**: 3016.
71 Chong, A., Renninger, W., and Wise, F. (2007). *Opt. Lett.* **32**: 2408.
72 Zhao, L., Tang, D., and Wu, J. (2006). *Opt. Lett.* **31**: 1788.
73 Grelu, P. and Akhmediev, N. (2012). *Nat. Photonics* **6**: 84.
74 Knight, J., Birks, T., Russel, P., and Atkin, D. (1996). *Opt. Lett.* **21**: 1547.
75 Broeng, J., Mogilevstev, D., Barkou, S., and Bjarklev, A. (1999). *Opt. Fiber Technol.* **5**: 305.
76 Russell, P. (2003). *Science* **299**: 358.
77 Coen, S., Chau, A., Leonhardt, R. et al. (2002). *J. Opt. Soc. Am. B* **19**: 753.
78 Kudlinski, A., George, A., Knight, J. et al. (2006). *Opt. Express* **14**: 5715.
79 Asobe, M., Kanamori, T., and Kubodera, K. (1992). *IEEE Photon. Technol. Lett.* **4**: 362.
80 Kikuchi, K., Taira, K., and Sugimoto, N. (2002). *Electron. Lett.* **38**: 166.
81 Lee, J., Kikuchi, K., Nagashima, T. et al. (2005). *Opt. Express* **13**: 3144.
82 Ta'eed, V., Baker, N., Fu, L. et al. (2007). *Opt. Express* **15**: 9205.
83 Gorbach, A. and Skryabin, D. (2007). *Nat. Photonics* **1**: 653.
84 Kudlinski, A. and Mussot, A. (2008). *Opt. Lett.* **33**: 2407.

2

Waves Called Solitons

A wave is a time-evolution phenomenon that we generally model mathematically using partial differential equations (PDEs) which have a dependent variable $u(z,t)$, an independent variable time t, and one or more independent spatial variables z_n, where n is generally equal to 1, 2, or 3. Actually, there are many different kinds of wave phenomena, such as surface waves on the ocean, sound waves in the air or other media, and electromagnetic waves, of which visible light is a special case. In all these cases, waves are described by solutions to either linear or nonlinear PDEs. The purpose of this chapter is to introduce some basic concepts and provide an overview of various kinds of PDEs describing waves. In particular, we will focus on nonlinear equations and some of their wave solutions, called *solitons*.

2.1 Linear and Nonlinear Effects of a Wave

In its simplest form, the equation describing the one-dimensional motion of a linear wave with velocity v is given by:

$$\frac{\partial^2 u}{\partial t^2} - v^2 \frac{\partial^2 u}{\partial z^2} = 0 \tag{2.1}$$

where $u(z, t)$ is the wave amplitude, t is a time coordinate, and z is a spatial coordinate. The general solution of Eq. (2.1) can be written in the form:

$$u(z, t) = f(z - vt) + g(z + vt) \tag{2.2}$$

where f and g are arbitrary functions describing two distinct waves: the first propagating to the right and the second propagating to the left, respectively. The shape of these waves remains constant during propagation. This can be verified considering, for example, the wave described by function f and choosing a new coordinate system (ξ, τ):

$$\xi = z - vt \tag{2.3a}$$

$$\tau = t \tag{2.3b}$$

which moves with it. In such a case, assuming a given value of ξ, $f = f(\xi)$ remains constant as z and t change. As an example, we can consider a harmonic wave given by:

$$f(\xi) = A \cos k\xi \tag{2.4}$$

Solitons in Optical Fiber Systems, First Edition. Mario F. S. Ferreira.

where A is the wave amplitude and k is the *wave number*, which represents the periodicity in the ξ coordinate. The spatial period is the *wavelength*, $\lambda = 2\pi/k$. Using Eq. (2.3a), we can write Eq. (2.4) in terms of the stationary coordinate system (z, t):

$$f(z,t) = A\cos k(z - vt) = A\cos(kz - \omega t) \tag{2.5}$$

where

$$\omega = kv \tag{2.6}$$

is the angular frequency of the periodic motion for a given position z.

For an observer moving with velocity $v = dz/dt$, the phase $\phi = kz - \omega t$ of the wave given by Eq. (2.5) changes with time as:

$$\frac{d\phi}{dt} = kv - \omega \tag{2.7}$$

The required velocity to observe a constant phase, such that $d\phi/dt = 0$, is $v = \omega/k$, which is called *phase velocity*.

If, instead of one wave, we have two waves of slightly different frequencies, the resultant wave will be modulated. The required velocity for an observer to follow the modulation envelope is such that the rate of change of the phase difference between the two waves is zero. Such velocity is given by:

$$v = \frac{\omega_1 - \omega_2}{k_1 - k_2} \tag{2.8}$$

In the limit of a small frequency difference, this velocity corresponds to the so-called *group velocity*, v_g:

$$v_g = \frac{d\omega}{dk} \tag{2.9}$$

In general, the phase velocity can depend on the wavenumber k. If this is not the case, the wave profile remains invariant with time and we have:

$$\frac{\partial A}{\partial \tau} = 0 \tag{2.10}$$

In terms of the stationary coordinate system, Eq. (2.10) becomes:

$$\frac{\partial A}{\partial t} + v\frac{\partial A}{\partial z} = 0 \tag{2.11}$$

A linear dependence of the phase velocity on k is related with dissipation, while dependence on k^2 corresponds to the lowest-order dispersion. If this effect is relatively weak, it can be described as a small deviation from the dispersionless phase velocity v_0:

$$v = \frac{\omega}{k} = v_0 + b_d k^2 \tag{2.12}$$

where b_d is the coefficient accounting for the wave dispersion. In this case, the wave frequency is given by:

$$\omega = kv = v_0 k + b_d k^3 \tag{2.13}$$

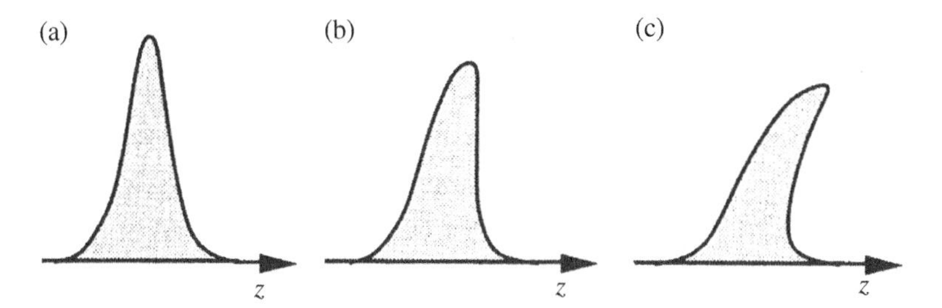

Figure 2.1 Formation of a shock wave due to nonlinearity. (a) The incident optical pulse with a Gaussian shape; (b) the optical pulse after propagating in a nonlinear medium showing the self-steepening effect; (c) optical shock wave formation due to self-steepening.

The term on k^3 corresponds to the third derivative with respect to ξ in the (ξ, τ) space. This means that, if we consider the dispersion effect, Eq. (2.10) becomes:

$$\frac{\partial A}{\partial \tau} + b_d \frac{\partial^3 A}{\partial \xi^3} = 0 \tag{2.14}$$

The wave phase velocity can depend also on the amplitude A, which corresponds to a nonlinear effect. If this effect is relatively weak, it can be described as a small deviation from the linear phase velocity v_0:

$$v = \frac{\omega}{k} = v_0 + b_{nl}A \tag{2.15}$$

where b_{nl} is the coefficient representing the nonlinear effect. As a consequence of this nonlinearity, the crest of the wave moves faster than the rest and its shape changes gradually during the propagation. The wave develops progressively a steepening front, giving origin to a *shock wave* (Figure 2.1).

The nonlinear term in Eq. (2.15) determines an additional contribution in Eq. (2.11), that is proportional to $b_{nl}A(\partial A/\partial z)$. As a consequence, when returning to the (ξ, τ) space, Eq. (2.10) becomes:

$$\frac{\partial A}{\partial \tau} + b_{nl}A \frac{\partial A}{\partial \xi} = 0 \tag{2.16}$$

Equation (2.16) shows that the wave height will depend both on time and space, even if we are moving with the linear phase velocity v_0.

2.2 Solitary Waves and Solitons

The first report on the observation of a solitary wave was made by the Scottish engineer John Scott Russell (1808–1882) in August 1834, when he saw a water wave about three kilometers long propagating without change in shape, on the Edinburgh–Glasgow Canal. He called it the "great wave of translation" and, then, reported his observations at the British Association in his 1844 paper "Report on Waves" [1]. The phenomenon was described by Russell as follows:

> *"I was observing the motion of a boat which was rapidly drawn along a narrow channel by a pair of horses, when the boat suddenly stopped – not so the mass of the water in the channel which it had put in motion; it accumulated round the prow of the vessel in a state of violent*

agitation, then suddenly leaving it behind, rolled forward with great velocity, assuming the form of a large solitary elevation, a rounded, smooth and well-defined heap of water, which continued its course along the channel apparently without change of form or diminution of speed. I followed it on horseback and overtook it still rolling on at a rate of some eight or nine miles an hour, preserving its original figure some thirty feet long and a foot to a foot and a half in height. Its height gradually diminished and after a chase of two miles I lost it in the windings of the channel. Such in the month of August 1834 was my first chance interview with that singular and beautiful phenomenon which I have called the Wave of Translation..."

These observations were followed by several laboratory experiments on the generation and propagation of such waves. Russell found that they are bell-shaped and have a velocity v that depends on the finite depth d of the channel and on the wave amplitude A above the free surface of the water in the form

$$v = \sqrt{g(d + A)} \tag{2.17}$$

where g is the acceleration due to gravity. Russell also observed that a given initial water elevation can give origin to one or more solitary waves.

Russell's predictions of solitary waves were strongly criticized by Airy [2] and Stokes [3]. They argued that such waves cannot propagate in a liquid medium without a change of form. However, Russell's predictions were later confirmed both by Joseph Boussinesq [4] and by Lord Rayleigh [5]. On the other hand, Korteweg and de Vries derived in 1895 an equation that aimed to describe the propagation of surface waves in water of finite depth. Such an equation can be obtained considering the combined effect of both dispersion and nonlinearity, as discussed in Section 2.1. Normalizing the quantities ξ^3 and A so that b_d and b_{nl} become unity, the resultant equation for the wave propagation becomes:

$$\frac{\partial A}{\partial \tau} + A\frac{\partial A}{\partial \xi} + \frac{\partial^3 A}{\partial \xi^3} = 0 \tag{2.18}$$

Equation (2.18) is known as the Korteweg–de Vries (KdV) equation and it has a solitary wave solution given by:

$$A(\xi, \tau) = 3w\mathrm{sech}^2\left[\frac{\sqrt{w}}{2}(\xi - w\tau)\right] \tag{2.19}$$

Actually, this solution represents the long-time evolution of a wave in which the nonlinear effect is counterbalanced by linear dispersion and corresponds to the solitary wave observed experimentally by Russell. This wave moves at a velocity $w + v_0$ in the stationary coordinate system. Its amplitude is proportional to the velocity w and its width is inversely proportional to the square root of the amplitude. Consequently, a taller (and thinner) soliton, initially behind a lower and wider one, will travel faster and eventually overtake and pass through the lower one.

After the work of Korteweg and de Vries, the solitary wave remained for a long time mainly as a simple curiosity in the area of nonlinear wave theory. The subject revived in 1955, when Fermi, Pasta, and Ulam (FPU) published a laboratory report on a numerical model of a discrete nonlinear mass–spring system [6]. The FPU model was described by a system of coupled nonlinear ordinary differential equations that, under certain approximations, can be transformed into the KdV equation. Such work inspired Martin Kruskal and Norman Zabusky, which solved numerically the KdV equation. They observed the formation of solitary waves that undergo nonlinear interaction, after which they emerge without any change of shape and amplitude, but with only a small change in their phases. Due to the particle-like behavior of such solitary waves, Kruskal and Zabusky called them for the first time "solitons" in a famous 1965 paper [7].

A significant contribution to the theory of soliton solutions of the KdV equation was given by Gardner et al. [8, 9] and Hirota [10, 11]. Using the so-called inverse scattering theory, Gardner et al. [8] demonstrated that a given input pulse can evolve to one or more solitons, together with a dispersive wave of small amplitude. The total number of solitons resulting from this process depends on the form of the initial pulse. These results are in good agreement with the experimental observations made by Russell.

By the same time, Zakharov [12] demonstrated that the temporal evolution of the envelope $\psi(\xi, \tau)$ of a wave train generated in deep water can be described by the following equation:

$$i\frac{\partial \psi}{\partial \tau} + D\frac{\partial^2 \psi}{\partial \xi^2} + Q|\psi|^2\psi = 0 \tag{2.20}$$

where D is the linear dispersion coefficient, Q is the nonlinear coefficient, and (ξ, τ) are the spatial and time coordinates in a frame of reference moving at the group velocity. Equation (2.20) is called the nonlinear Schrödinger equation (NLSE) because of its similarity to the Schrödinger equation of quantum mechanics:

$$i\hbar\frac{\partial \psi}{\partial \tau} + \frac{\hbar^2}{2m}\frac{\partial^2 \psi}{\partial \xi^2} - V\psi = 0 \tag{2.21}$$

Here, the potential V is proportional to the absolute square of the wave envelope and $\hbar = h/2\pi$, where h is Planck's constant.

When D and Q have the same sign, Eq. (2.20) has the following soliton solution, written in terms of the original space and time coordinates:

$$\psi(z, t) = A\text{sech}\left[\sqrt{\frac{Q}{2D}}A(z - v_g t)\right]\exp\left(i\frac{Q}{2}A^2 t\right) \tag{2.22}$$

The wave envelope propagates at the group velocity, whereas the phase velocity depends on the square of the amplitude.

Equation (2.20) was solved by Zakharov and Shabat [13] using the inverse scattering theory, showing that the evolution of an initial wave packet gives origin to a number of envelope solitons, together with some linear dispersive waves. These results were confirmed experimentally by Yuen and Lake [14].

In the following years, the NLSE was formulated for a wide variety of physical problems in fluid mechanics, plasma physics, and nonlinear optics. Similar to the KdV equation, the NLSE is actually regarded as a generic equation in nonlinear wave propagation.

2.3 Solitons in Optical Fibers

A medium exhibits *chromatic dispersion* if the propagation constant of a wave within it varies nonlinearly with frequency. In the case of an optical fiber, this phenomenon causes the broadening in time of a propagating signal, an effect known as *pulse dispersion*. This effect is characterized by the *group velocity dispersion* (GVD) parameter β_2, given by

$$\beta_2 = -\frac{dv_g/d\omega}{v_g^2} \tag{2.23}$$

where v_g represents the group velocity. For fused silica, it is known that β_2 becomes zero at the wavelength of 1.3 μm and is negative (positive) for a longer (shorter) wavelength.

In the presence of GVD, the information carried by different frequency components of an optical pulse propagate at different speeds. If $\beta_2 < 0$ (anomalous dispersion regime), higher frequency components of information travel faster, whereas for $\beta_2 > 0$ (normal dispersion regime) the opposite occurs. Consequently, in the presence of GVD, the spectrum of an optical pulse remains unchanged, but its time width increases during propagation.

The nonlinearity of an optical fiber originates from the dependence of the susceptibility on the electric field, which becomes important at high field strengths. This effect results in an intensity dependence of the refractive index of the form:

$$n = n_0 + n_2 I \tag{2.24}$$

In Eq. (2.24), n_0 is the linear refractive index and n_2 is the refractive index nonlinear coefficient, also known as the *Kerr coefficient*. In the case of fused silica, a value of $n_2 = 2.35 \times 10^{-20}\ \text{m}^2/\text{W}$ was measured at the main telecommunications window around 1.55 μm [15].

Early in 1973, Hasegawa and Tappert [16] were the first to point out that the propagation of an optical pulse in an ideal lossless optical fiber in the presence of chromatic dispersion and nonlinearity can be described by the following NLSE:

$$i\frac{\partial U}{\partial z} - \frac{1}{2}\beta_2\frac{\partial^2 U}{\partial \tau^2} + \gamma|U|^2 U = 0 \tag{2.25}$$

where U is the slowly varying complex amplitude of the field envelope z is the spatial coordinate along propagation, τ is the time coordinate in a system moving at the group velocity, and

$$\gamma = \frac{\omega_0 n_2}{cA_{eff}} \tag{2.26}$$

is the fiber nonlinear parameter, A_{eff} being the effective core area.

For $\beta_2 < 0$, that is, in the regime of anomalous dispersion, Eq. (2.25) admits an envelope-soliton solution given by:

$$U = U_0 \text{sech}\left[\sqrt{-\frac{\gamma}{\beta_2}}U_0\left(t - \frac{z}{v_g}\right)\right] \exp\left(i\frac{\gamma}{2}U_0{}^2 z\right) \tag{2.27}$$

The formation of solitons in optical fibers is the result of a balance between the negative (anomalous) GVD of the glass fiber, which occurs for wavelengths longer than 1.3 μm in a standard fiber, and the Kerr nonlinearity. The existence of fiber solitons was suggested for the first time by Hasegawa and Tappert [16], and it appeared soon as an ideal solution to the problem of pulse spreading caused by fiber dispersion.

After their first experimental observation by Mollenauer et al. [17], Hasegawa [18] suggested that solitons could be used in all-optical transmission systems based on optical amplifiers instead of regenerative repeaters, which were considered standard until 1990. In particular, he proposed using the Raman effect of transmission fiber itself for optical amplification. The distributed Raman amplification gave its position to erbium-doped fiber amplifiers (EDFAs) during the 1990s.

Meanwhile, Gordon and Haus [19] anticipated that the transmission of a signal made of optical solitons could not be extended over an unlimited distance when optical amplification is used. In fact, the amplifiers needed to compensate for the fiber loss generate amplified spontaneous emission (ASE), and this noise is, in part, incorporated by the soliton, whose mean frequency is then shifted. Due to GVD, the arrival time of the soliton becomes then a random variable, whose variance is proportional to the cube of the propagation distance. This is the so-called *Gordon–Haus effect.*

At the end of 1991, some research groups suggested the use of frequency filters to extend the limit set by the Gordon–Haus effect [20–22]. However, using this technique, the transmission distance was still limited by the growth of narrow-band noise at the center frequency of the filters. In order to solve this problem, Mollenauer and coworkers developed the sliding-guiding filter concept [23–25]. Other proposals to achieve stable soliton propagation in fiber systems made use of amplitude modulators [26] or nonlinear optical amplifiers [27–29].

A method of programming dispersion is called *dispersion management.* In linear systems, such a technique consists in introducing locally relatively large dispersion to make nonlinear effects relatively less important and to compensate for the accumulated dispersion at the end of the line so that the integrated dispersion becomes zero. Dispersion management was also found to be effective in soliton transmission systems since it can reduce significantly the timing jitter.

Smith et al. [30] proposed the use of a periodic map using both anomalous dispersion and normal dispersion fibers alternately. The nonlinear stationary pulse that propagates in such a fiber has a peak power larger than the soliton that propagates in a fiber with a constant dispersion equal to the average dispersion of the period map. Such a nonlinear stationary pulse is commonly called a *dispersion-managed soliton* and presents several remarkable characteristics [31, 32].

2.4 Dissipative Optical Solitons

As mentioned earlier, the background instability can be avoided through the use of optical amplifiers having a nonlinear property of gain, or gain and saturable absorption in combination, which provide an effective gain to the soliton and a suppression (or very small gain) to the noise. The pulse propagation in optical fibers where narrow-band filters and linear and nonlinear amplifiers are periodically inserted may be described by the following generalized NLSE [28, 29, 32–36]:

$$i\frac{\partial q}{\partial Z} + \frac{1}{2}\frac{\partial^2 q}{\partial T^2} + |q|^2 q = i\delta q + i\beta\frac{\partial^2 q}{\partial T^2} + i\varepsilon|q|^2 q + i\mu|q|^4 q \tag{2.28}$$

where Z is the normalized propagation distance, T is the retarded time, q is the normalized envelope of the electric field, β stands for spectral filtering ($\beta > 0$), δ is the linear gain or loss coefficient, ε accounts for nonlinear gain-absorption processes (for example, two-photon absorption), and μ represents a higher-order correction to the nonlinear gain absorption.

Equation (2.28) is known as the complex Ginzburg-Landau equation (CGLE) [32, 35, 36], so-called cubic for $\mu = 0$ and quintic for $\mu \neq 0$, and it reduces to the standard NLSE when the right-hand side is set to zero. The CGLE is rather general, as it includes dispersive and nonlinear effects, in both conservative and dissipative forms. It is known in many branches of physics and can describe several nonequilibrium phenomena, such as convection instabilities, binary fluid convection, and phase transitions [37–39]. In the field of nonlinear optics, the CGLE has been used to describe various systems, namely optical parametric oscillators, free-electron laser oscillators, spatial and temporal soliton lasers, and all-optical transmission lines [28, 32, 36, 40–45].

The soliton solutions of Eq. (2.28) are termed *dissipative solitons* and they differ significantly from conventional solitons. Actually, the conventional soliton concept implies a single balance between nonlinearity and dispersion, whereas a dissipative soliton arises as a result of a double balance: between nonlinearity and dispersion and also between gain and loss [32, 36]. The two balances

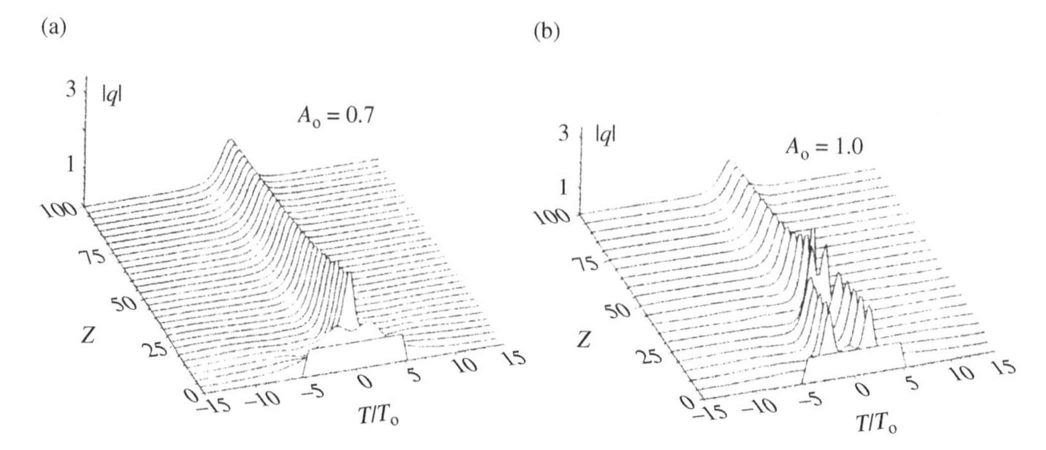

Figure 2.2 Formation of a fixed-amplitude soliton solution of the cubic CGLE starting from an initial pulse with a rectangular profile of amplitude $A_o = 0.7$ (a) and $A_o = 1.0$ (b), when $\delta = -0.003$, $\beta = 0.2$, $\varepsilon = 0.09$, and $\mu = 0$.

result in solutions in which the shape, amplitude, and width are all fixed and depend on the parameters of the equation. In particular, the balance between gain and loss, which should be exact in order to produce stationary localized solutions, plays a crucial role in the soliton dynamics. As a consequence, compared with the case of conventional (conservative) solitons, the impact of dissipative effects reveals unusual situations in which solitons can be found.

In general, Eq. (2.28) is non-integrable, and only particular exact solutions can be obtained. In the case of the cubic CGLE, exact solutions can be obtained using a special ansatz [46], Hirota bilinear method [47] or reduction to systems of linear PDEs [48]. Concerning the quintic CGLE, the existence of soliton-like solutions in the case $\varepsilon > 0$ has been demonstrated both analytically and numerically [29, 33, 34, 49]. Exact solutions of the quintic CGLE, including solitons, sinks, fronts, and sources, were obtained in [50], using Painlevé analysis and symbolic computations.

To fully explore the CGLE, numerical simulations must be carried out. Stable solitons can be found in this way taking as initial condition a pulse of somewhat arbitrary profile. In fact, such a profile appears to be of little importance. For example, Figure 2.2 illustrates the formation of a fixed amplitude soliton of the cubic CGLE starting from an initial pulse with a rectangular profile.

Different types of soliton solutions were obtained numerically from Eq. (2.28), which can be divided into two main classes: localized fixed-shape solutions and localized pulsating solutions [32, 51, 52]. Examples of localized fixed-shape solutions are the plain stationary pulses, the flat-top pulses, the composite pulses, and the moving pulses [33]. Among the localized pulsating solutions, we may refer to the plain pulsating and the creeping solitons, as well as the erupting solitons, which belong to the class of chaotic solutions [53–58]. Localized fixed-shape solutions and localized pulsating solutions of the CGLE will be discussed in Chapters 12 and 13, respectively.

References

1 Russell, J.S. (1844). Report on waves. In: *Report of 14th Meeting of the British Association for Advancement of Science*, York, September (ed. J. Murray) (London), 311–390.
2 Airy, G.B. (1845). *Tides and Waves*. B. Fellowes.
3 Stokes, C.G. (1847). *On the Theory of Oscillating Waves*, vol. **VIII**. Trans. Camb. Phil. Soc.

4 de Boussinesq, J. (1871). Théorie de l'intumescence liquid appeleé on the solitaire ou de translation se propageant dans un canal rectangulaire. *Comptes Rendus, Acad. Sci. (Paris)* **72**: 755.
5 Rayleig, L. (1876). *On waves. Phil. Mag.* **1**: 257.
6 Fermi, E., Pasta, J.R., and Ulam, S.M. (1955). Studies of Nonlinear Problems. *Los Alamos Sci. Lab. Rep., LA-1940.*
7 Zabusky, N. and Kruskal, M. (1965). *Phys. Rev. Lett.* **15**: 240.
8 Gardner, C., Greene, J., Kruskal, M., and Miura, R. (1967). *Phys. Rev. Lett.* **19**: 1095.
9 Gardner, C., Greene, J., Kruskal, M., and Miura, R. (1974). *Commun. Pure Appl. Math.* **27**: 97.
10 Hirota, R. (1971). *Phys. Rev. Lett.* **27**: 1192.
11 Hirota, R. (1973). *J. Math. Phys.* **14**: 810.
12 Zakharov, V.E. (1968). *J. Appl. Mech. Tech. Phys.* **9**: 86.
13 Zakharov, V.E. and Shabat, A.B. (1972). *Soviet Phys. JETP* **34**: 62.
14 Yuen, H.C. and Lake, B.M. (1975). *Phys. Fluids* **18**: 956.
15 Broderick, N.G.R., Monroe, T.M., Bennett, P.J., and Richardson, D.J. (1999). *Opt. Lett.* **24**: 1395.
16 Hasegawa, A. and Tappert, F. (1973). *Appl. Phys. Lett.* **23**: 142.
17 Mollenauer, L.F., Stolen, R.H., and Gordon, J.P. (1980). *Phys. Rev. Lett.* **45**: 1095.
18 Hasegawa, A. (1983). *Opt. Lett.* **8**: 650.
19 Gordon, J.P. and Haus, H.A. (1986). *Opt. Lett.* **11**: 665.
20 Mollenauer, L.F., Neubelt, M.J., Haner, M. et al. (1991). *Electron. Lett.* **27**: 2055.
21 Mecozzi, A., Moores, J.D., Haus, H.A., and Lai, Y. (1991). *Opt. Lett.* **16**: 1841.
22 Kodama, Y. and Hasegawa, A. (1992). *Opt. Lett.* **17**: 31.
23 Mollenauer, L.F., Gordon, J.P., and Evangelides, S.G. (1992). *Opt. Lett.* **17**: 1575.
24 Mollenauer, L.F., Lichtman, E., Neubelt, M.J., and Harvey, G.T. (1993). *Electron. Lett.* **29**: 910.
25 Kodama, Y. and Wabnitz, S. (1994). *Opt. Lett.* **19**: 162.
26 Nakazawa, M., Kamada, Y., Kubota, H., and Suzuki, E. (1991). *Electron. Lett.* **27**: 1270.
27 Kodama, Y., Romagnoli, M., and Wabnitz, S. (1992). *Electron. Lett.* **28**: 1981.
28 Matsumoto, M., Ikeda, H., Uda, T., and Hasegawa, A. (1995). *J. Lightwave Technol.* **13**: 658.
29 Ferreira, M.F., Facão, M.V., and Latas, S.V. (2000). *Fiber Integrat. Opt.* **19**: 31.
30 Smith, N.J., Knox, F.M., Doran, N.J. et al. (1996). *Electron. Lett.* **32**: 54.
31 Hasegawa, A. (ed.) (1998). *New Trends in Optical Soliton Transmission Systems.* Dordrecht, Holland: Kluwer.
32 Ferreira, M.F. (2011). *Nonlinear Effects in Optical Fibers.* Hoboken, NJ: John Wiley & Sons.
33 Akhmediev, N. and Ankiewicz, A. (1997). *Solitons: Nonlinear Pulses and Beams.* London: Chapman and Hall.
34 Akhmediev, N., Afanasjev, V., and Soto-Crespo, J. (1996). *Phys. Rev. E* **53**: 1190.
35 Ferreira, M.F., Facão, M.V., Latas, S.V., and Sousa, M.H. (2005). *Fiber Integrated Opt.* **24**: 287.
36 Akhmediev, N. and Ankiewicz, A. (2005). *Dissipative Solitons.* Berlin: Spinger.
37 Normand, C. and Pomeau, Y. (1977). *Rev. Mod. Phys.* **49**: 581.
38 Kolodner, P. (1991). *Phys. Rev. A* **44**: 6466.
39 Graham, R. (1975). *Fluctuations, Instabilities and Phase Transictions.* Berlin: Springer.
40 Staliunas, K. (1993). *Phys. Rev. A* **48**: 1573.
41 Jian, P., Torruellas, W., Haelterman, M. et al. (1999). *Opt. Lett.* **24**: 400.
42 Dunlop, A.M., Wright, E.M., and Firth, W.J. (1998). *Optics Commun.* **147**: 393.
43 Ng, C. and Bhattacharjee, A. (1999). *Phys. Rev. Lett.* **82**: 2665.
44 Akhmediev, N., Rodrigues, A., and Townes, G. (2001). *Opt. Commun.* **187**: 419.
45 Akhmediev, N., Ankiewicz, A., and Soto-Crespo, J. (2001). *J. Opt. Soc. Am. B* **15**: 515.
46 Pereira, N.R. and Stenflo, L. (1977). *Phys. Fluids* **20**: 1733.

47 Nozaki, K. and Bekki, N. (1984). *Phys. Soc. Japan* **53**: 1581.
48 Conte, R. and Musette, M. (1993). *Physica D* **69**: 1.
49 Thual, O. and Fauvre, S. (1988). *J. Phys.* **49**: 1829.
50 Marcq, P., Chaté, H., and Conte, R. (1994). *Physica D* **73**: 305.
51 Chang, W., Ankiewicz, A., Akhmediev, N., and Soto-Crespo, J. (2007). *Phys. Rev. E.* **76**: 016607.
52 Mancas, S. and Choudhury, S. (2009). *Chaos, Solitons Fractals* **40**: 91.
53 Soto-Crespo, J., Akhmediev, N., and Ankiewicz, A. (2000). *Phys. Rev. Lett.* **85**: 2937.
54 Akhmediev, N., Soto-Crespo, J., and Town, G. (2001). *Phys. Rev. E* **63**: 056602.
55 Akhmediev, N. and Soto-Crespo, J. (2003). *Phys. Lett A.* **317**: 287.
56 Akhmediev, N. and Soto-Crespo, J. (2004). *Phys. Rev. E.* **70**: 036613.
57 Soto-Crespo, J. and Akhmediev, N. (2005). *Phys. Rev. Lett.* **95**: 024101.
58 Soto-Crespo, J. and Akhmediev, N. (2005). *Math. Comp. Simulation* **69**: 526.

3

Fiber Dispersion and Nonlinearity

The suggestion of using optical fibers in communication systems has been made in 1966 by Kao and Hockam [1]. However, this suggestion was not feasible at that time since the most transparent glass available had losses of more than about 1000 dB/km. Meanwhile, following an impressive activity on the purification of fused silica, the fiber losses were reduced to 20 dB/km by 1970 [2] and to 0.2 dB/km near the 1550 nm spectral region by 1979 [3]. The availability of optical fibers with such low-loss transmission has led to a revolution in the field of optical communications.

Besides the fiber attenuation, there are two other main factors affecting the optical fiber communication systems: dispersion and nonlinearity. Dispersion refers to the phenomenon where different components of the signal travel at different velocities in the fiber and it became an important limiting factor as transmission systems evolved to longer distances and higher bit rates. Concerning the fiber nonlinearities, they can be readily observed due to two main reasons. On one hand, the field cross section is rather small, which results in relatively high field intensities. On the other hand, the optical fiber provides a long interaction length, which enhances significantly the efficiency of the nonlinear processes. Additional information concerning the topics discussed in this chapter can be found in several well-known books, namely those indicated by References [4–14].

3.1 Fiber Chromatic Dispersion

In digital optical fiber communication systems, information to be sent is first coded in the form of pulses, which are then transmitted through the fiber to the receiver, where the information is decoded. A pulse of light loses energy due to fiber attenuation. Silica glass, from which standard fibers are made, presents an attenuation below 1 dB/km for wavelengths in the range 0.8–1.8 μm. Outside this range, the attenuation increases rapidly. Figure 3.1 shows the typical spectral attenuation profile in a $GeO_2 - SiO_2$ fiber. The attenuation presents minima values around 1.3 and 1.55 μm, achieving in the last case a value of 0.2 dB/km. The two bands around these wavelengths correspond to the windows used actually in fiber communication systems.

Besides losing energy, a pulse also broadens in time as it propagates through the fiber; a phenomenon known as pulse dispersion. In an optical waveguide, this effect arises from two mechanisms: (i) refractive index variations with wavelength (material dispersion); and (ii) waveguide-related effects (waveguide dispersion).

Solitons in Optical Fiber Systems, First Edition. Mario F. S. Ferreira.

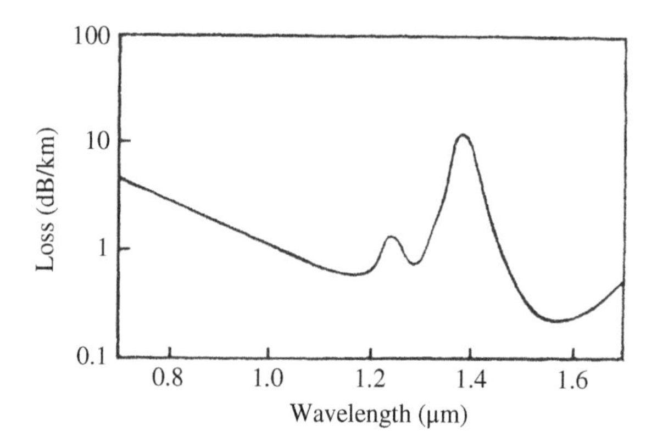

Figure 3.1 Typical spectral attenuation profile of a GeO_2-SiO_2 fiber.

Let us consider a pulse described by a function $\psi(0, t)$ at $z = 0$. This pulse can be represented as a superposition of harmonic waves.

$$\psi(0,t) = \int_{-\infty}^{\infty} A(\omega)e^{-i\omega t}d\omega \tag{3.1}$$

After propagation along a distance z, the spectral component of frequency ω experiments a phase variation $\beta(\omega)z$, where $\beta(\omega)$ is the propagation constant at this frequency. The total field is then given by

$$\psi(z,t) = \int_{-\infty}^{\infty} A(\omega)e^{i(\beta(\omega)z-\omega t)}d\omega \tag{3.2}$$

The propagation constant $\beta(\omega)$ is related with the refractive index $n(\omega)$ in the form

$$\beta(\omega) = \frac{\omega n(\omega)}{c} \tag{3.3}$$

We can expand $\beta(\omega)$ in a Taylor series about the carrier frequency ω_0:

$$\beta(\omega) = \beta_0 + \beta_1(\omega-\omega_0) + \frac{1}{2}\beta_2(\omega-\omega_0)^2 + \frac{1}{6}\beta_3(\omega-\omega_0)^3 + \cdots \tag{3.4a}$$

where $\beta_0 = \beta(\omega_0)$ and

$$\beta_j = \left(\frac{d^j\beta}{d\omega^j}\right)_{\omega=\omega_0} \qquad (j = 1, 2, \ldots) \tag{3.4b}$$

In particular, we have:

$$\beta_1 \equiv \frac{d\beta}{d\omega} = \frac{1}{c}\left[n + \omega\frac{dn}{d\omega}\right] = \frac{n_g}{c} = \frac{1}{v_g} \tag{3.5}$$

and

$$\beta_2 \equiv \frac{d^2\beta}{d\omega^2} = -\frac{dv_g/d\omega}{v_g^2} \tag{3.6}$$

In Eq. (3.5), $n_g = n + \omega(dn/d\omega)$ and $v_g = c/n_g$ represent the group refractive index and the group velocity, respectively. The parameter β_2 given by Eq. (3.6) characterizes the *group velocity dispersion* (GVD).

In the presence of GVD, the information carried by different frequency components of $\psi(0, t)$ propagates at different speeds and thus arrive at different times. The relative delay Δt_D of the arrival time of information at frequencies ω_1 and ω_2, after propagating a distance z, is given by

$$\Delta t_D = \frac{z}{v_g(\omega_1)} - \frac{z}{v_g(\omega_2)} = \frac{\frac{dv_g}{d\omega}(\omega_2 - \omega_1)z}{v_g^2} \tag{3.7}$$

Using Eq. (3.6), we have

$$\Delta t_D = -\beta_2(\omega_2 - \omega_1)z \tag{3.8}$$

Equation (3.8) shows that the difference of arrival time of information is proportional to the GVD, β_2, to the difference between the frequency components, $\omega_2 - \omega_1$, and to the propagation distance, z. We note that if $\beta_2 < 0$ (anomalous dispersion regime), higher frequency components of information arrive first, whereas for $\beta_2 > 0$ (normal dispersion regime), the reverse is true. If the information at different frequency components arrives at different time, the information may be lost. This problem becomes more serious when $\omega_2 - \omega_1$ is large.

In practice, the group dispersion is usually given by a group delay parameter, D, defined by the delay of arrival time, in ps unit, for two wavelength components separated by 1 nm, after propagating a distance of 1 km. From Eq. (3.8), the group delay may be described by:

$$D = \beta_2 c\left(\frac{2\pi}{\lambda_1} - \frac{2\pi}{\lambda_2}\right)z = -\beta_2 c\frac{2\pi}{\lambda^2}\Delta\lambda z \tag{3.9}$$

Thus, by taking $\Delta\lambda = 1$ nm and $z = 1$ km, D in the unit of ps/nm/km is related to β_2 through

$$D = -\frac{2\pi c\beta_2}{\lambda^2} \tag{3.10}$$

For a typical single-mode fiber, we have $D = 0$ (16 ps/(nm-km)) at $\lambda = 1.31$ μm (1.55 μm). This indicates that two wavelength components of a pulse separated by 1 nm arrive with a time delay on the order of a few ps when they propagate over a distance of 1 km in a typical fiber.

The fiber dispersion is given by the sum of the material dispersion, D_M, and waveguide dispersion, D_w. The material dispersion D_M is related with the rate of variation of the group index with wavelength, $D_M = c^{-1}(dn_g/d\lambda)$. In the case of fused silica, $dn_g/d\lambda = 0$ and consequently $D_M = 0$ at $\lambda = \lambda_{ZD} = 1.276$ μm, which is referred to as the material zero-dispersion wavelength.

The waveguide dispersion D_w depends on the fiber parameters such as the core radius and the index difference between the core and the cladding, and is negative in the range 0–1.6 μm. Figure 3.2 shows D_M, D_w, and their sum $D = D_M + D_w$ for a typical single-mode fiber. The main effect of waveguide dispersion is to shift λ_{ZD} so that the total dispersion is zero near 1.31 μm.

It is possible to design the fiber such that the zero-dispersion wavelength, λ_{ZD}, is shifted to the vicinity of 1.55 μm. Such fibers are referred to as *dispersion-shifted fibers* (DSFs). It is also possible to design the fiber such that the total dispersion D is relatively small over a wide wavelength range extending from 1.3 to 1.6 μm. Such fibers are called *dispersion-flattened fibers* (DFFs). Figure 3.3 illustrates the fiber dispersion as a function of wavelength for these kinds of fibers.

3.1.1 Gaussian Input Pulses

Let us consider a chirped Gaussian input pulse, whose electric field is given by:

$$E(0,t) = E_0 \exp\left\{-\frac{1+iC_0}{2}\left(\frac{t}{t_0}\right)^2\right\}\exp(-i\omega_0 t) \tag{3.11}$$

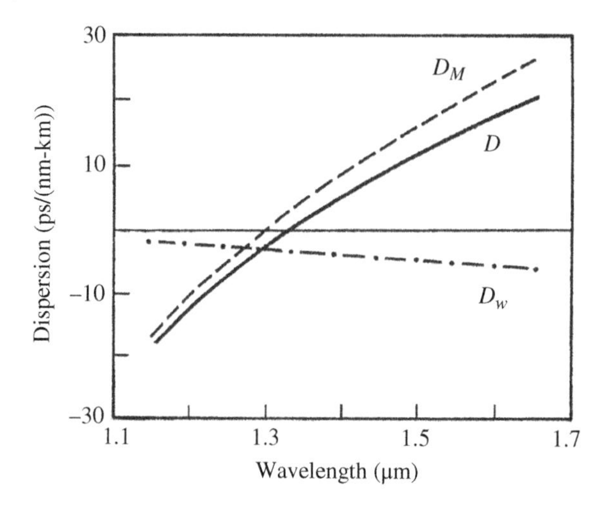

Figure 3.2 Relative contributions of material dispersion D_M and waveguide dispersion D_w, as well as total dispersion $D = D_M + D_w$ for a typical single-mode fiber.

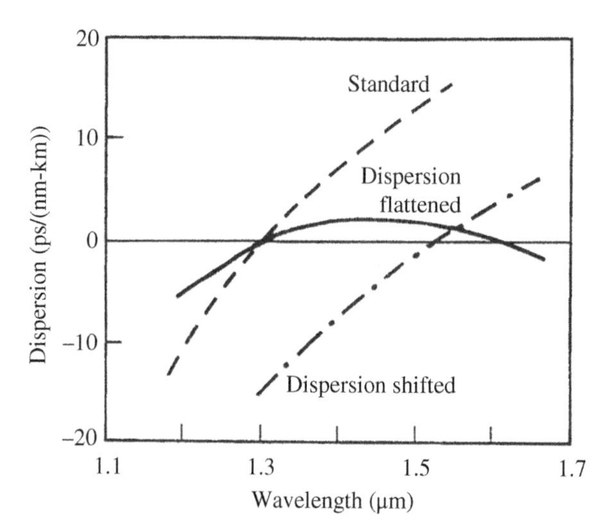

Figure 3.3 Fiber dispersion as a function of wavelength for several kinds of commercial fibers.

where E_0 is the peak amplitude, C_0 is the chirp coefficient, ω_0 is the carrier frequency, and the parameter t_0 represents the half-width at 1/e-intensity point. It is related to the full width at half maximum (FWHM) by the relation

$$t_{FWHM} = 2(\ln 2)^{1/2} t_0 \approx 1.665 t_0 \tag{3.12}$$

The amplitude spectrum of this pulse is obtained by calculating its Fourier transform:

$$E(0,\omega) = \frac{1}{2\pi} \int_{-\infty}^{\infty} E(0,t) \exp\{i(\omega - \omega_0)t\} dt = \frac{E_0 t_0}{\sqrt{2\pi(1 + iC_0)}} \exp\left\{ -\frac{t_0^2(\omega - \omega_o)^2}{2(1 + iC_0)} \right\} \tag{3.13}$$

Each spectral component of (3.13) is propagated over distance z in the medium with propagation constant $\beta(\omega)$ and acquires a phase variation $\beta(\omega)z$. Thus, the pulse at z is found through the inverse Fourier transform of $E(0, \omega)$ after propagation:

$$E(z,t) = \int_{-\infty}^{\infty} E(0,\omega) \exp\{i[\beta(\omega)z - \omega t]\}d\omega \tag{3.14}$$

Using the expansion of the propagation constant given by Eq. (3.4a), and retaining terms up to second-order in $(\omega - \omega_0)$, we obtain:

$$E(z,t) = \widetilde{E}_0 \exp\left\{-\frac{(1+iC)(t - z/v_g)^2}{2t_p^2(z)}\right\} \exp\{i[\beta_0 z - \omega_0 t]\} \tag{3.15}$$

where

$$\widetilde{E}_0 = \frac{E_0 t_0 [1 + i\sigma(1 - iC_0)]^{1/2}}{t_p} \tag{3.16}$$

$$\sigma = \frac{\beta_2 z}{t_o^2} \tag{3.17}$$

$$C = C_0 + \sigma\left(1 + C_0^2\right) \tag{3.18}$$

$$t_p^2 = t_0^2\left[(1 + C_0\sigma)^2 + \sigma^2\right] \tag{3.19}$$

In the case $C_0 = 0$, corresponding to an unchirped input Gaussian pulse, the optical field can be presented in the form:

$$E(z,t) = \frac{E_0}{(1+\sigma^2)^{1/4}} \exp\left\{-\frac{(t - z/v_g)^2}{2t_p^2(z)}\right\} \exp\{i[\beta_0 z - \phi(z,t)]\} \tag{3.20}$$

where

$$\phi(z,t) = \omega_o t + \kappa\left(t - \frac{z}{v_g}\right)^2 - \frac{1}{2}\text{tg}^{-1}(\sigma) \tag{3.21}$$

$$\kappa = \frac{\sigma}{2(1+\sigma^2)t_o^2}, \tag{3.22}$$

$$t_p^2(z) = t_o^2\left(1 + \sigma^2\right) \tag{3.23}$$

Equation (3.21) shows that the phase of the pulse varies with the square of time. As a consequence, the instantaneous frequency also varies with time and the pulse is said to be chirped. The instantaneous frequency is found by taking the time derivative of the phase:

$$\omega(t) = \omega_0 + 2\kappa\left(t - \frac{z}{v_g}\right) \tag{3.24}$$

This result shows that in the normal dispersion regime ($\kappa > 0$), the instantaneous frequency at a fixed position z increases with time (positive linear chirp). As a consequence, the leading edge of the pulse is red-shifted, whereas the trailing edge is blue-shifted. On the other hand, in the anomalous dispersion regime ($\kappa < 0$), the instantaneous frequency decreases with time (negative chirp). In this case, the leading edge of the pulse is blue-shifted and the trailing edge becomes red-shifted. Figure 3.4 illustrates the pulse chirping in the normal and anomalous dispersion regimes.

Equation (3.23) shows that the half-width of the pulse power $t_p(z)$ of an unchirped input pulse increases with the propagation distance by an amount that is governed by the GVD coefficient β_2.

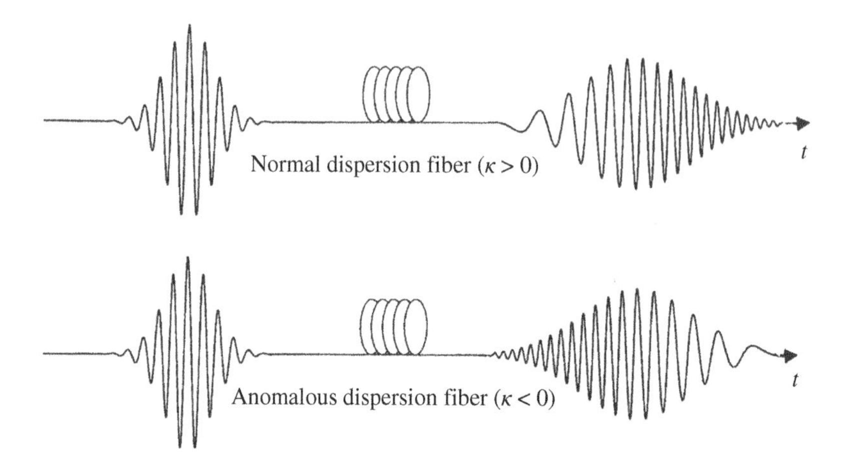

Figure 3.4 Illustration of the pulse chirp in the cases of anomalous and normal dispersion regimes.

The *dispersion length*, L_D, is defined as the distance over which a Gaussian pulse broadens to $\sqrt{2}$ times its initial width t_0. Considering Eqs. (3.17) and (3.23), it can be seen that

$$L_D = \frac{t_0^2}{|\beta_2|} \tag{3.25}$$

In the case of a chirped input pulse, Eqs. (3.17) and (3.19) show that the pulse broadens monotonously during propagation if C_0 and β_2 are of the same sign. However, if these parameters are of opposite sign, the pulse compresses until a given location z_{min}, after which it starts to broaden. This behavior is illustrated in Figure 3.5 for the case $\beta_2 < 0$.

The quantity $\left(1 + C^2\right)/t_p^2$ is related to the spectral width of the pulse and it can be verified from Eqs. (3.18) and (3.19) that it remains constant and equal to its initial value $\left(1 + C_0^2\right)/t_0^2$.

Besides the chromatic dispersion, another pulse-broadening mechanism in single-mode fibers is related with *polarization mode dispersion* (PMD) [15–18]. PMD is determined by the optical fiber birefringence and its main characteristics will be discussed in Chapter 11. Limitations due to PMD become particularly significant in high-speed long-haul optical fiber transmission systems [18–24].

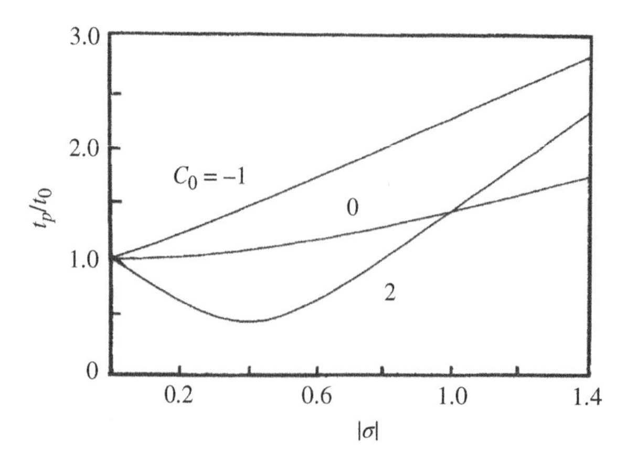

Figure 3.5 Evolution of the pulse width with the normalized distance $|\sigma|$, for three values of the chirp parameter C_0.

3.2 Fiber Nonlinearity

Nonlinear effects are attributed to the dependence of the susceptibility on the electric field, which becomes important at high field strengths. As a result, the total polarization vector **P** can be written in the frequency domain as a power series expansion in the electric field vector [25, 26]:

$$\mathbf{P}(r,\omega) = \varepsilon_0\left[\chi^{(1)}\cdot\mathbf{E} + \chi^{(2)} : \mathbf{EE} + \chi^{(3)} : \mathbf{EEE} + \cdots\right] = \mathbf{P}_L(r,\omega) + \mathbf{P}_{NL}(r,\omega) \tag{3.26}$$

where $\chi^{(j)}$ ($j = 1, 2, \ldots$) is the jth order susceptibility. To account for the light polarization effects, $\chi^{(j)}$ is a tensor of rank $j+1$. The linear susceptibility $\chi^{(1)}$ determines the linear part of the polarization, $\mathbf{P}_L$. On the other hand, terms of second and higher order in (3.26) determine the nonlinear polarization, $\mathbf{P}_{NL}$. Since SiO_2 is a symmetric molecule, the second-order susceptibility $\chi^{(2)}$ vanishes for silica glasses. As a consequence, virtually all nonlinear effects in optical fibers are determined by the third-order susceptibility, $\chi^{(3)}$. In the time domain, the form of the expansion is identical to (3.26) if the nonlinear response is assumed to be instantaneous.

3.2.1 The Nonlinear Refractive Index

Let us consider a plane optical wave propagating in a medium having inverse symmetry, with an electric field given by

$$E = \frac{1}{2}\left(\hat{E}(z)e^{-i\omega t} + c.c.\right) \tag{3.27}$$

where *c.c.* means the complex conjugate. The corresponding nonlinear third-order polarization is given by:

$$\begin{aligned} P_{NL}^{(3)} &= \varepsilon_0\chi^{(3)}EEE = \varepsilon_0\chi^{(3)}\left(\frac{\hat{E}e^{-i\omega t} + \hat{E}^{*}e^{i\omega t}}{2}\right)^3 \\ &= \frac{1}{8}\varepsilon_0\chi^{(3)}\left(\left(\hat{E}^3e^{-i3\omega t} + c.c.\right) + 3\left|\hat{E}\right|^2\left(\hat{E}e^{-i\omega t} + c.c.\right)\right) \end{aligned} \tag{3.28}$$

If the necessary phase-matching conditions for an efficient generation of the third harmonic generation are not provided, the first term in the right-hand side of Eq. (3.28) can be neglected and we have simply

$$P_{NL}^{(3)} \approx \frac{3}{4}\varepsilon_0\chi^{(3)}\left|\hat{E}\right|^2 E \tag{3.29}$$

Taking into account the result given by Eq. (3.29), the polarization of the material becomes

$$P = \varepsilon_0\left(\chi^{(1)} + \frac{3}{4}\chi^{(3)}\left|\hat{E}\right|^2\right)E \tag{3.30}$$

Since the intensity of the plane wave is given by

$$I = \frac{1}{2}c\varepsilon_0 n_0\left|\hat{E}\right|^2 \tag{3.31}$$

where n_0 is the linear refractive index, the polarization can be written as

$$P = \varepsilon_0\left(\chi^{(1)} + \frac{3}{2}\frac{\chi^{(3)}}{c\varepsilon_0 n_0}I\right)E \tag{3.32}$$

Neglecting the absorption by the medium, we have the following general relation between the polarization and refractive index:

$$P = \varepsilon_0 \left(n^2 - 1\right) E \tag{3.33}$$

Comparing Eqs. (3.32) and (3.33), the following result for the refractive index is obtained:

$$n = \sqrt{1 + \chi^{(1)} + \frac{3}{2} \frac{\chi^{(3)}}{c\varepsilon_0 n_0} I} \tag{3.34a}$$

The refractive index given by Eq. (3.34a) can be put in the form:

$$n = n_0 + n_2 I \tag{3.34b}$$

where

$$n_0 = \sqrt{1 + \chi^{(1)}} \tag{3.35}$$

is the linear refractive index and

$$n_2 = \frac{3}{4} \frac{\chi^{(3)}}{c\varepsilon_0 n_0^2} \tag{3.36}$$

is the refractive index nonlinear coefficient, also known as the *Kerr coefficient*. The dependence of the refractive index on the field intensity is known as the *nonlinear Kerr effect*.

Since the linear refractive index n_0 and the third-order susceptibility $\chi^{(3)}$ depend on the frequency, the Kerr coefficient given by Eq. (3.36) is frequency-dependent as well. In the case of fused silica, a value of $n_2 = 2.73 \times 10^{-20}\ \mathrm{m^2/W}$ was measured at a wavelength of 1.06 μm [27]. Several measurements indicate that the nonlinear index coefficient decreases with increasing wavelengths [28]. Considering the telecommunications window, a value $n_2 = 2.35 \times 10^{-20}\ \mathrm{m^2/W}$ was obtained for silica fibers [29]. Such value can be increased by doping the fiber core with germania [30].

Assuming a single-mode fiber with an effective mode area $A_{eff} = 50 \mu\mathrm{m}^2$ carrying an optical power $P = 100\ \mathrm{mW}$, the nonlinear part of the refractive index is

$$n_2 I = n_2 \frac{P}{A_{eff}} \approx 4.7 \times 10^{-11} \tag{3.37}$$

In spite of this very small value, the effects of the nonlinear component of the refractive index become significant due to the very long interaction lengths provided by the optical fibers.

3.2.2 Relevance of Nonlinear Effects in Fibers

The importance of a nonlinear process depends not only on the intensity, I, of the propagating wave but also on the effective interaction length, L_{eff}, along which that intensity remains sufficiently high. In the case of a Gaussian beam propagating in bulk media, the effective interaction length is [31].

$$L_{eff,bulk} \sim Z_R = \frac{\pi w_0^2}{\lambda} \tag{3.38}$$

where Z_R is the so-called *Rayleigh length* and w_0 is the spot size of the beam at the waist. The intensity of the focused beam can be increased by reducing the spot size w_0. However, a smaller spot size determines a faster divergence of the beam and a smaller Rayleigh range, as seen from Eq. (3.38). The product of the intensity by the effective length gives

$$\left(IL_{eff}\right)_{bulk} = \frac{P}{\pi w_0^2} Z_R = \frac{P}{\lambda} \tag{3.39}$$

In the case of an optical fiber, the field is guided and the power decreases only due to the attenuation, in the form

$$P(z) = P(0)e^{-\alpha z} \tag{3.40}$$

The product of the intensity and effective length becomes:

$$\left(IL_{eff}\right)_{fiber} = \int_0^L \frac{P(0)}{A_{eff}} e^{-\alpha z} dz = \frac{P(0)}{A_{eff}} \frac{1 - e^{-\alpha L}}{\alpha} = \frac{P(0)}{A_{eff}} L_{eff} \tag{3.41}$$

where

$$L_{eff} = \frac{1 - e^{-\alpha L}}{\alpha} \tag{3.42}$$

is the effective length of interaction. If the fiber real length, L, is sufficiently small, such that $\alpha L \ll 1$, we have $L_{eff} \approx L$. However, the difference between the real and the effective length increases with L. When $\alpha L \gg 1$, the effective length approaches a limiting value, given by

$$L_{eff}^{max} = \frac{1}{\alpha} \tag{3.43}$$

In the case of a fiber with an attenuation of 0.2 dB/km, the maximum effective length is $L_{eff}^{max} = 21.72$ km. Figure 3.6 shows the effective length of interaction against the real fiber length for an attenuation of 0.2 dB/km. In this case and considering the results given by Eqs. (3.39) and (3.41), it can be verified that the efficiency of a nonlinear process in a typical single-mode fiber can be enhanced by a factor ~10^9 relatively to a bulk media.

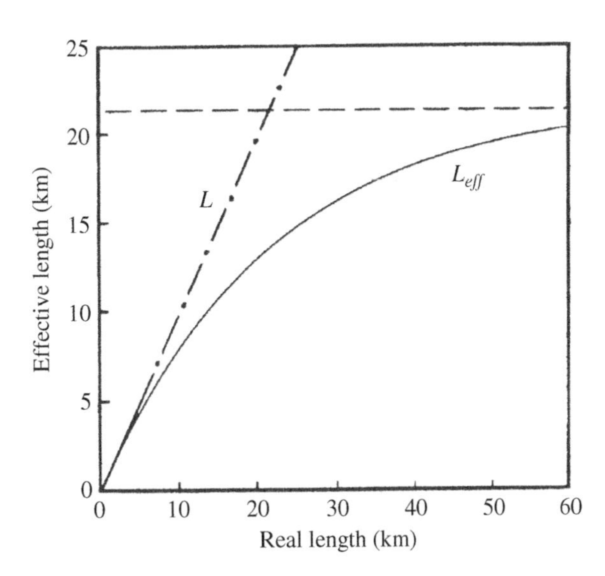

Figure 3.6 Effective length of interaction against the real fiber length for an attenuation of 0.2 dB/km.

In actual transmission systems, the fiber attenuation is compensated by periodically located optical amplifiers. Assuming that the spacing between consecutive amplifiers is L_A and that the power after each amplifier is equal to that at the fiber input, $P(0)$, the effective length of the transmission system becomes

$$L_{eff} = N\frac{1 - e^{-\alpha L_A}}{\alpha} \tag{3.44}$$

where $N = L/L_A$ is the number of sections constituting the transmission link. Then, the limiting value of the effective length is N times greater than in a non-amplified system:

$$L_{eff}^{\max} = \frac{N}{\alpha} \tag{3.45}$$

3.3 The Pulse Propagation Equation

The electric field associated with a bit stream propagating in a single-mode fiber along the z-direction can be written in the form

$$E = \frac{1}{2}\left(F(r,\phi)U(z,t)e^{i(\beta_0 z - \omega_0 t)} + c.c.\right) \tag{3.46}$$

where $U(z,t)$ describes the complex amplitude of the field envelope at a distance z inside the fiber, $F(r, \phi)$ gives the transverse distribution of the fundamental fiber mode, and β_0 is the mode propagation constant at the carrier frequency. The vector nature of the waves is neglected in this section, assuming that they are linearly polarized in the same direction.

The only quantity that changes during propagation is the amplitude $U(z,t)$. Since each frequency component of the optical field propagates with a slightly different propagation constant, it is useful to work in the spectral domain. The evolution of a specific spectral component $\tilde{U}(z,\omega)$ is given by

$$\tilde{U}(z,\omega) = \tilde{U}(0,\omega)\exp\left[i\beta(\omega)z - i\beta_0 z\right] \tag{3.47}$$

where $\tilde{U}(0,\omega)$ is the Fourier transform of the input signal $U(z = 0, t)$ and $\beta(\omega)$ is the propagation constant. Considering Eqs. (3.3) and (3.34), $\beta(\omega)$ can be written as the sum of a linear part, β_L, and a nonlinear part, β_{NL}:

$$\beta(\omega) = \beta_L(\omega) + \beta_{NL}(\omega_0) \tag{3.48}$$

with

$$\beta_L(\omega) = \frac{\tilde{n}(\omega)\omega}{c}, \quad \beta_{NL}(\omega_0) = n_2 I\frac{\omega_0}{c} \tag{3.49}$$

$\tilde{n}$ being the effective mode index. In Eq. (3.48), β_{NL} has been written as a frequency-independent quantity, considering that the spectrum of optical pulses used as bits in communication systems is usually much narrower than the carrier frequency ω_0.

Let us consider the expansion given by Eq. (3.4) for $\beta_L(\omega)$, retaining only terms up to second order in $(\omega - \omega_0)$. Substituting Eqs. (3.48) and (3.49) in Eq. (3.47), calculating the derivative $\partial\tilde{U}/\partial z$, and converting the resultant equation into the time domain by using the inverse Fourier transform and the relation

$$\Delta\omega \leftrightarrow i\frac{\partial}{\partial t} \tag{3.50}$$

we obtain [14]:

$$i\left(\frac{\partial}{\partial z}+\frac{1}{v_g}\frac{\partial}{\partial t}\right)U-\frac{1}{2}\beta_2\frac{\partial^2}{\partial t^2}U+\beta_{NL}U=0 \tag{3.51}$$

The amplitude U in Eq. (3.51) can be normalized such that its square gives the optical power, $P = |U|^2$. Making such normalization, we have $\beta_{NL} = \gamma|U|^2 = \gamma P$, where

$$\gamma=\frac{\omega_0 n_2}{cA_{eff}} \tag{3.52}$$

is known as the *fiber nonlinear parameter* and A_{eff} is the effective core area. Considering the symmetry of the single-mode waveguide, A_{eff} is given by:

$$A_{eff}=\frac{2\pi\left(\int\limits_0^\infty |F(r)|^2 r dr\right)^2}{\int\limits_0^\infty |F(r)|^4 r dr} \tag{3.53}$$

Assuming a moving frame propagating with the group velocity and using the new time variable

$$\tau=t-\frac{z}{v_g} \tag{3.54}$$

Equation (3.51) can be written in the form:

$$i\frac{\partial U}{\partial z}-\frac{1}{2}\beta_2\frac{\partial^2 U}{\partial \tau^2}+\gamma|U|^2U=0 \tag{3.55}$$

Equation (3.55) is usually called the *nonlinear Schrödinger equation* (NLSE) due to its similarity with the Schrödinger equation of quantum mechanics. The NLSE describes the propagation of pulses in optical fibers taking into account both the group-velocity dispersion, represented by the parameter β_2, and the fiber nonlinearity, represented by the parameter γ. Hasegawa and Tappert [32] were the first to point out that the propagation of an optical pulse in an ideal lossless optical fiber in the presence of both effects can be described by Eq. (3.55).

3.3.1 The Normalized NLSE

Considering Eq. (3.55), let us introduce a normalized amplitude Q given by

$$U(z,\tau)=\sqrt{P_0}Q(z,\tau) \tag{3.56}$$

where P_0 is the peak power of the incident pulse. $Q(z, \tau)$is found to satisfy the equation:

$$i\frac{\partial Q}{\partial z}-\frac{1}{2}\beta_2\frac{\partial^2 Q}{\partial \tau^2}+\gamma P_0|Q|^2Q=0 \tag{3.57}$$

Let us define a *nonlinear distance*, L_{NL}, as

$$L_{NL}=\frac{1}{\gamma P_0} \tag{3.58}$$

L_{NL} is the propagation distance required to produce a nonlinear phase change rotation of one radian at a power P_0. A *dispersion distance*, L_D, has been also defined in Eq. (3.25). These two characteristic distances provide the length scales over which nonlinear or dispersive effects become important for pulse evolution.

Using a distance Z normalized by the dispersion distance L_D and a time scale T normalized by t_0, Eq. (3.57) becomes

$$i\frac{\partial Q}{\partial Z} \pm \frac{1}{2}\frac{\partial^2 Q}{\partial T^2} + N^2|Q|^2 Q = 0 \tag{3.59}$$

where

$$N^2 = \frac{L_D}{L_{NL}} = \frac{\gamma P_0 t_0^2}{|\beta_2|} \tag{3.60}$$

In the second term of Eq. (3.59), the plus signal corresponds to the case of anomalous GVD (sgn $(\beta_2) = -1$), whereas the minus signal corresponds to normal GVD (sgn$(\beta_2) = +1$). The parameter N can be eliminated from Eq. (3.59) by introducing a new normalized amplitude $q = NQ$, with which the NLSE takes the following standard normalized form:

$$i\frac{\partial q}{\partial Z} + \frac{1}{2}\frac{\partial^2 q}{\partial T^2} + |q|^2 q = 0 \tag{3.61}$$

The case of anomalous GVD ($\beta_2 < 0$) was assumed in writing Eq. (3.61).

3.3.2 Propagation in the Absence of Dispersion and Nonlinearity

If there is no dispersion or fiber nonlinearity, Eq. (3.55) reduces to

$$\frac{\partial U(z,\tau)}{\partial z} = 0 \tag{3.62}$$

or

$$\frac{\partial U}{\partial z} + \frac{1}{v_g}\frac{\partial U}{\partial t} = 0 \tag{3.63}$$

Equation (3.62) has the solution

$$U = U_0(\tau) = U_0(t - z/v_g) \tag{3.64}$$

It is clear from this result that, in the absence of dispersion and nonlinearity, the pulse propagates without any distortion in shape with the group velocity v_g. The pulse energy also propagates with the group velocity since from Eq. (3.63) we get that

$$\frac{\partial |U|^2}{\partial z} + \frac{1}{v_g}\frac{\partial |U|^2}{\partial t} = 0 \tag{3.65}$$

3.3.3 Effect of Dispersion Only

Neglecting only the nonlinear term in Eq. (3.55), we have

$$i\frac{\partial U}{\partial z} - \frac{1}{2}\beta_2\frac{\partial^2 U}{\partial \tau^2} = 0 \tag{3.66}$$

Using the method of separation of variables, one can obtain the following general solution:

$$U(z,\tau) = \int_{-\infty}^{\infty} U(0,\omega) e^{i\left(\frac{1}{2}\beta_2\omega^2 z - \omega\tau\right)} d\omega \tag{3.67}$$

where $U(0, \omega)$ represents the frequency spectrum of the input pulse. The preceding equation is similar to Eq. (3.2), from which we have shown that the pulse becomes broadened and chirped due to dispersion.

Figure 3.7 illustrates the dependence of the GVD-induced pulse broadening on the pulse shape as obtained numerically from Equation (3.66) for the cases of a sech, a Gaussian, and two super-Gaussian pulses. The dashed curves in Figure 3.7 correspond to the input pulse profiles, whereas the full curves represent the pulse profiles after propagation over a distance $z = 3L_D$.

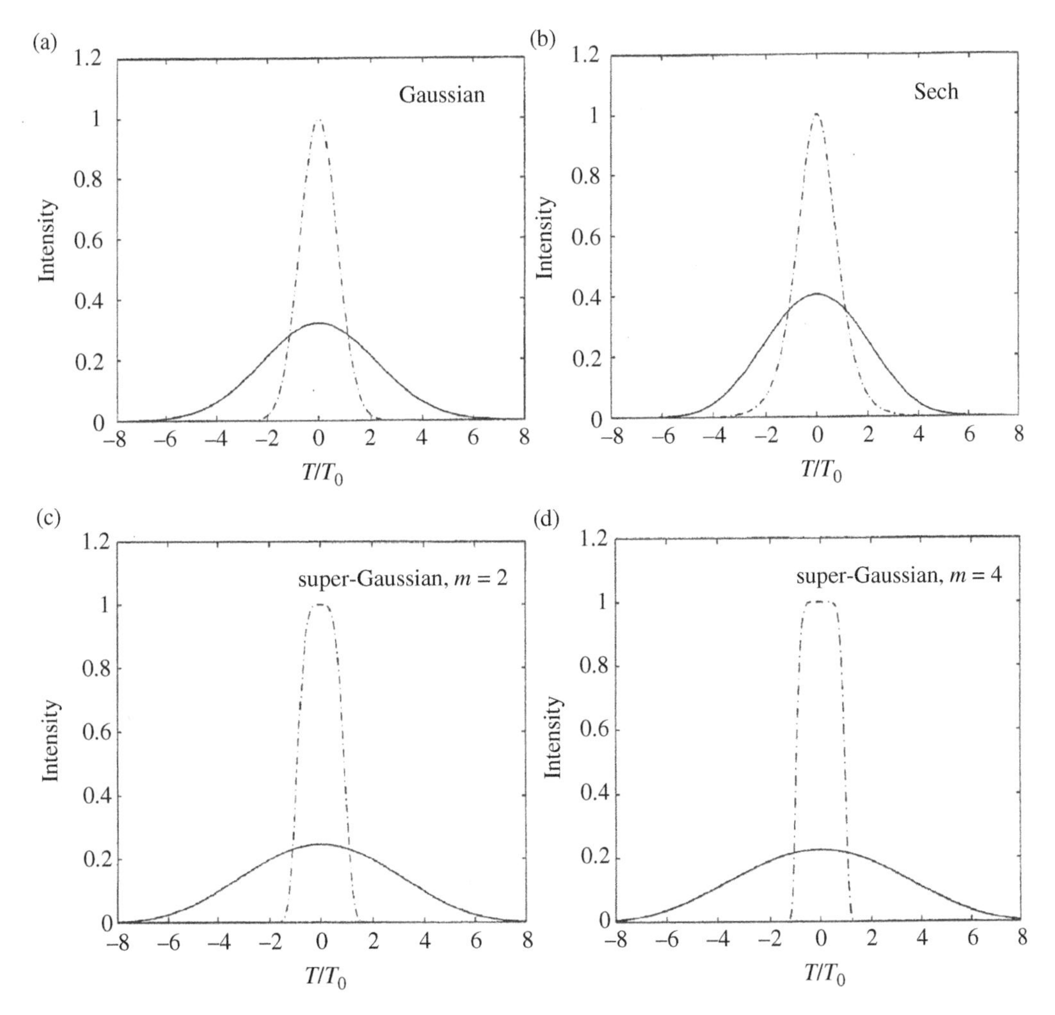

Figure 3.7 GVD-induced broadening for (a) a Gaussian pulse, (b) a sech pulse, and two super-Gaussian pulses with (c) $m = 2$ and (d) $m = 4$. The dashed curves correspond to the initial pulse profiles, whereas the full curves represent the pulse profiles after propagation over a distance $z = 3L_D$.

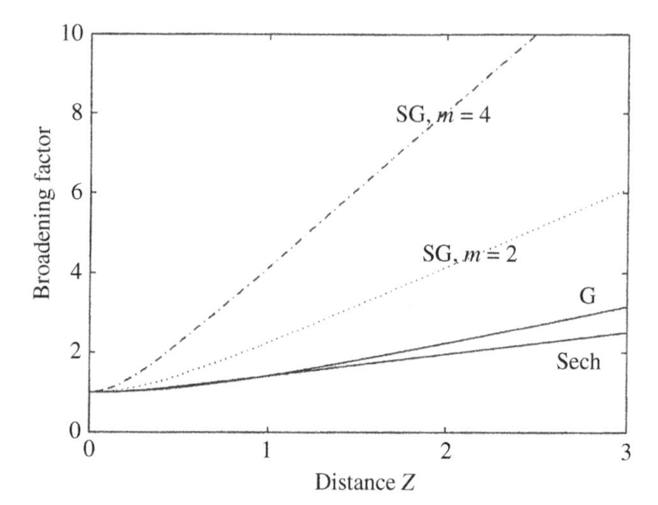

Figure 3.8 Broadening factor against the propagation distance for the same pulse profiles considered in Figure 3.7.

The amplitude of an input super-Gaussian pulse can be described in the form

$$U(0,\tau) = U_0 \exp\left\{ -\frac{1}{2}\left(\frac{\tau}{t_0}\right)^{2m} \right\} \tag{3.68}$$

where t_0 is the input pulse width and the parameter m controls the pulse shape. The Gaussian pulse corresponds to the case $m = 1$. Increasing the value of m, the pulse develops steeper leading and trailing edges, becoming nearly rectangular. As a consequence, its spectrum becomes larger, which determines a faster temporal broadening due to dispersion. This is illustrated in Figure 3.8, which shows the broadening factor against the propagation distance for the same cases considered in Figure 3.7.

3.3.4 Effect of Nonlinearity Only

Neglecting only the dispersion term in Equation (3.55), we obtain

$$i\frac{\partial U}{\partial z} + \gamma |U|^2 U = 0 \tag{3.69}$$

Multiplying this equation by U^* and its complex conjugate by U and subtracting the two equations, we obtain

$$\frac{\partial |U|^2}{\partial z} = 0 \tag{3.70}$$

which has the general solution

$$|U|^2 = f(\tau) = f(t - z/v_g) \tag{3.71}$$

This result shows that, in the presence of nonlinearity and neglecting the dispersion, the absolute square of the wave envelope maintains its shape during propagation. Based on this fact, the solution of Eq. (3.69) can be written in the form

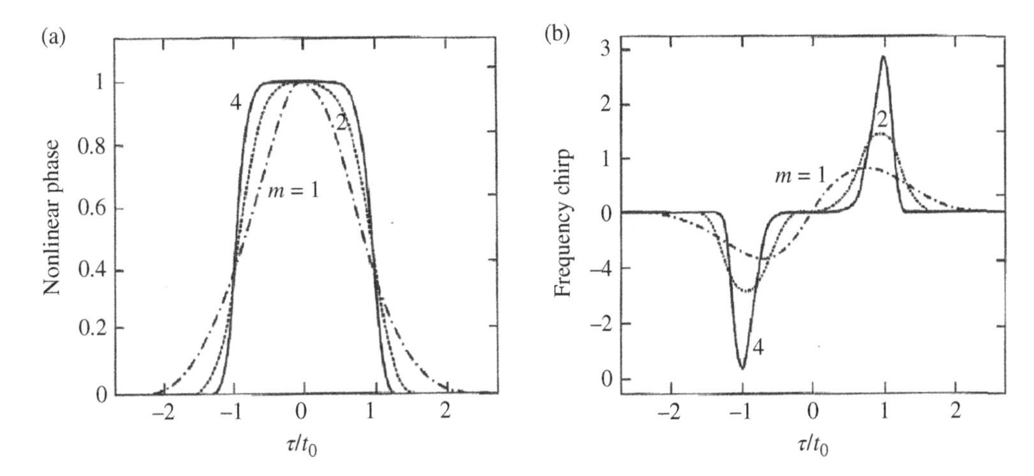

Figure 3.9 Temporal variations of (a) the nonlinear phase shift and (b) the nonlinear frequency chirp at a distance $z = L_{NL}$ for Gaussian (m = 1) and super-Gaussian (m = 2, 4) pulses.

$$U(z,\tau) = U_0(\tau)e^{i\phi_{NL}(z,\tau)} \tag{3.72}$$

where $U_0(\tau)$ and $\phi(z, \tau)$ are real functions. From Eqs. (3.69) and (3.72), we obtain

$$\phi_{NL}(z,\tau) = \gamma P(\tau)z \tag{3.73}$$

where $P(\tau) = |U(\tau)|^2$ is the pulse power. Eq. (3.73) indicates that the fiber nonlinearity leads to a phase modulation that is directly proportional to the pulse power and the distance of propagation. The time dependence of the nonlinear phase induces a frequency chirp given by

$$\Delta\omega_{NL} = -\frac{\partial\phi_{NL}}{\partial\tau} = -\gamma z\frac{\partial P}{\partial\tau} \tag{3.74}$$

The magnitude of the frequency chirp given by Eq. (7.74) depends clearly on the pulse profile. Figure 3.9 illustrates the time dependence of (a) the nonlinear phase and (b) the nonlinear frequency chirp for a Gaussian pulse and two super-Gaussian pulses, corresponding to $m = 2$ and $m = 4$ in Eq. (3.68), after propagation along a nonlinear length, $z = L_{NL}$. The magnitude of the nonlinear frequency chirp increases with m and corresponds to a redshift in the leading edge and a blueshift in the trailing edge of the pulse.

References

1 Kao, K.C. and Hocham, G.A. (1966). *Proc. IEE* **113**: 1151.
2 Kapron, F.P., Keck, D.B., and Maurer, R.D. (1970). *Appl. Phys. Lett.* **17**: 423.
3 Miya, T., Terunuma, Y., Hosaka, F., and Miyoshita, T. (1979). *Electron. Lett.* **15**: 106.
4 Snyder, A.W. and Love, J.D. (1983). *Optical Waveguide Theory*. Chapman Hall.
5 Palais, J.C. (1988). *Fiber Optic Communications*. Englewood Cliff: Prentice Hall.
6 Jeunhomme, L. (1990). *Single-Mode Fiber Optics*, 2e. New York: Marcel Dekker.
7 Marcuse, D. (1991). *Theory of Dielectric Optical Waveguides*, 2e. San Diego, CA: Academic Press.
8 Kaiser, A. (1991). *Optical Fibre Communications*, 2e. McGraw-Hill.
9 Senior, J.M. (1992). *Optical Fibre Communications: Principles and Practice*, 2e. Prentice Hall.

10 Gower, J. (1993). *Optical Communication Systems*, 2e. London: Prentice Hall.
11 Buck, J.A. (1995). *Fundamentals of Optical Fibers*. New York: Wiley.
12 Agrawal, G.P. (2002). *Fiber-Optic Communication Systems*, 3e. New York: Wiley.
13 Agrawal, G.P. (2007). *Nonlinear Fiber Optics*, 4e. Academic & Elsevier.
14 Ferreira, M.F. (2011). *Nonlinear Effects in Optical Fibers*. Hoboken, NJ: John Wiley & Sons.
15 Poole, C.D. and Nagel, J. (1997). Chapter 6 – Polarization effects in lightwave systems. In: *Optical Fiber Telecommunications*, vol. **IIIA** (ed. I.P. Kaminow). San Diego, CA: Academic Press.
16 H. Kogelnik and R. M. Jopson, and L. E. Nelson, Chapter15 – Polarization-Mode Dispersion. in *Optical Fiber Telecommunications*, vol. **IVB** (Academic Press, San Diego, CA, 2002).
17 Damask, J.N. (2005). *Polarization Optics in Telecommunications*. New York, NY: Springer.
18 Galtarrossa, A. and Menyuk, C.R. (ed.) (2005). *Polarization Mode Dispersion*. New York, NY: Springer.
19 Matera, F., Settembre, M., Tamburrini, M. et al. (1999). *J. Lightwave Technol.* **17**: 2225.
20 Kolltveit, E., Andrekson, P.A., Brentel, J. et al. (1999). *Electron. Lett.* **35**: 75.
21 Sunnerud, H., Karlsson, M., Xie, C., and Andrekson, P.A. (2002). *J. Lightwave Technol.* **20**: 2204.
22 Galtarrossa, A., Griggio, P., Palmieri, L., and Pizzinat, A. (2004). *J. Lightwave Technol.* **22**: 1127.
23 Chernyak, V., Chertkov, M., Gabitov, I. et al. (2004). *J. Lightwave Technol.* **22**: 1155.
24 Ning, G., Aditya, S., Shum, P. et al. (2006). *Opt. Commun.* **260**: 560.
25 Shen, Y.R. (1984). *Principles of Nonlinear Optics*. New York: Wiley.
26 Boyd, R.W. (2003). *Nonlinear Optics*, 2e. Boston: Academic Press.
27 Milam, D. and Weber, M.J. (1976). *J. Appl. Phys.* **47**: 2497.
28 Milam, D. (1998). *Appl. Opt.* **37**: 546.
29 Broderick, N.G.R., Monroe, T.M., Bennett, P.J., and Richardson, D.J. (1999). *Opt. Lett.* **24**: 1395.
30 Nakajima, K. and Ohashi, M. (2002). *IEEE Photon. Technol. Lett.* **14**: 492.
31 Yariv, A. (1977). *Introduction to Optical Electronics*. New York: Holt, Rinehart and Winston.
32 Hasegawa, A. and Tappert, F. (1973). *Appl. Phys. Lett.* **23**: 142.

4

Nonlinear Effects in Optical Fibers

The third-order nonlinear processes occurring in optical fibers can be classified into two general categories [1]. The first category arises from the modulation of the refractive index of silica by intensity changes in the signal (Kerr effect). This gives rise to nonlinearities such as self-phase modulation (SPM), whereby an optical signal alters its own phase; cross-phase modulation (XPM), where one signal affects the phases of all other optical signals and vice versa, and four-wave mixing (FWM), whereby signals with different frequencies interact to produce mixing sidebands. The second category of nonlinearities corresponds to stimulated scattering processes, such as stimulated Brillouin scattering (SBS) and stimulated Raman scattering (SRS), which are interactions between optical signals and acoustic or molecular vibrations in the fiber, respectively.

The aforementioned nonlinear effects impose several limitations on the performance of optical communications systems [2–4]. However, they also provide new possibilities for ultrafast all-optical signal processing, such as switching, wavelength conversion, amplification, and regeneration [1, 5].

4.1 Self-Phase Modulation

SPM is a nonlinear phenomenon affecting the phase of a propagating wave and which does not involve other waves. It arises because the refractive index of the fiber has an intensity-dependent component, as shown by Eq. (3.34). The nonlinear Shrödinger equation (NLSE) derived in Section 3.3 provides an adequate base for the theoretical description of SPM.

Pulse evolution in the presence of fiber chromatic dispersion, nonlinearity, and loss is governed by a generalized version of Eq. (3.55):

$$i\frac{\partial U}{\partial z} - \frac{1}{2}\beta_2\frac{\partial^2 U}{\partial \tau^2} + \gamma|U|^2U = -i\frac{\alpha}{2}U \tag{4.1}$$

The loss parameter α appearing in Eq. (4.1) not only reduces the signal power but also impacts the strength of the nonlinear effects. This can be seen by introducing a new amplitude $V(z, \tau)$ such that

$$U(z,\tau) = \exp(-\alpha z/2)V(z,\tau) \tag{4.2}$$

Substituting Eq. (4.2) into Eq. (4.1), the amplitude $V(z, \tau)$ is found to satisfy the following equation:

$$i\frac{\partial V}{\partial z} - \frac{1}{2}\beta_2\frac{\partial^2 V}{\partial \tau^2} + \gamma\exp(-\alpha z)\left|V^2\right|V = 0 \tag{4.3}$$

Solitons in Optical Fiber Systems, First Edition. Mario F. S. Ferreira.

Equation (4.3) shows that as the optical power $P(z, \tau) = |U(z, \tau|^2$ decreases exponentially during propagation, the nonlinear effects also become weaker, as expected.

Let us introduce a normalized amplitude Q given by

$$U(z,\tau) = \sqrt{P_0}\exp(-\alpha z/2)Q(z,\tau) \tag{4.4}$$

where P_0 is the peak power of the incident pulse. Using Eq. (4.4), Eq. (4.1) can be rewritten in the form

$$i\frac{\partial Q}{\partial z} - \frac{\beta_2}{2}\frac{\partial^2 Q}{\partial \tau^2} + \gamma P_0 \exp(-\alpha z)|Q|^2 Q = 0 \tag{4.5}$$

Neglecting the dispersive effects and following the discussion of Section 3.3.4, the field amplitude after propagation along a fiber of length L is given by

$$Q(L,\tau) = Q(0,\tau)\exp|\phi_{NL}(L,\tau)| \tag{4.6}$$

where the nonlinear phase is

$$\phi_{NL}(L,\tau) = \gamma P_0 L_{eff}|Q(0,\tau)|^2. \tag{4.7}$$

In Eq. (4.7), L_{eff} is the fiber-effective length, given by:

$$L_{eff} = \frac{1-\exp(-\alpha L)}{\alpha} \tag{4.8}$$

We observe from Eq. (4.7) that if the incident wave is a pulse with a given power temporal profile, the power variation within the pulse leads to its own phase modulation. Hence, this phenomenon is appropriately called *self-phase modulation*.

Let us consider the particular case of Gaussian pulse, with an electric field given by:

$$E(0,t) = E_0 \exp\left\{-\frac{1}{2}\left(\frac{t}{t_0}\right)^2\right\}\exp\{-i\omega_0 t\} \tag{4.9}$$

where E_0 is its peak amplitude. After propagating a distance z and neglecting the fiber dispersion, the pulse becomes:

$$E(z,t) = E_0 \exp\left\{-\frac{1}{2}\left(\frac{\tau}{t_0}\right)^2\right\}\exp\{i[\beta_0 + \gamma P(\tau)]z - i\omega_0 t\}. \tag{4.10}$$

where

$$P(\tau) = P_0 \exp\left[-\left(\frac{\tau}{t_0}\right)^2\right] \tag{4.11}$$

and

$$\tau = t - \frac{z}{v_g} \tag{4.12}$$

The instantaneous frequency within the pulse is given by:

$$\omega(t) = -\frac{\partial \phi}{\partial t} = \omega_0 - \gamma z\frac{\partial P}{\partial t} = \omega_0 + \gamma z\frac{2\tau}{t_0^2}P_0 e^{-\tau^2/t_0^2} \tag{4.13}$$

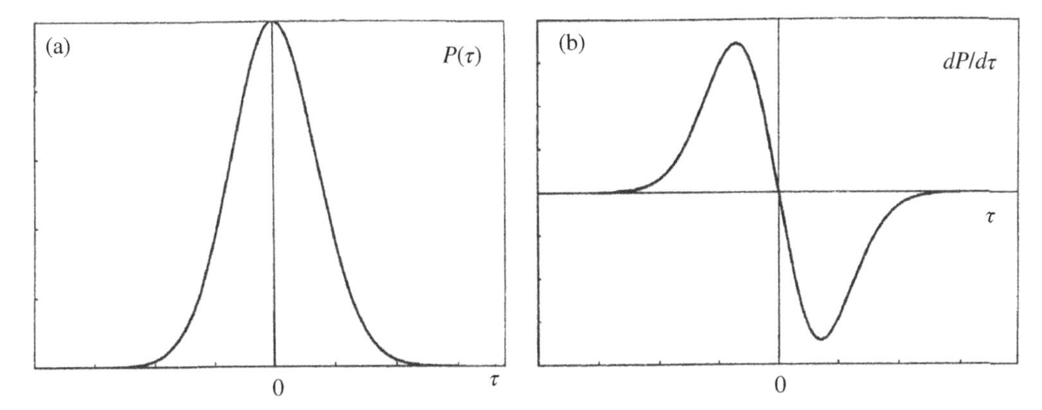

Figure 4.1 Temporal variation of (a) $P(\tau)$ and (b) $dP/d\tau$ for a Gaussian pulse.

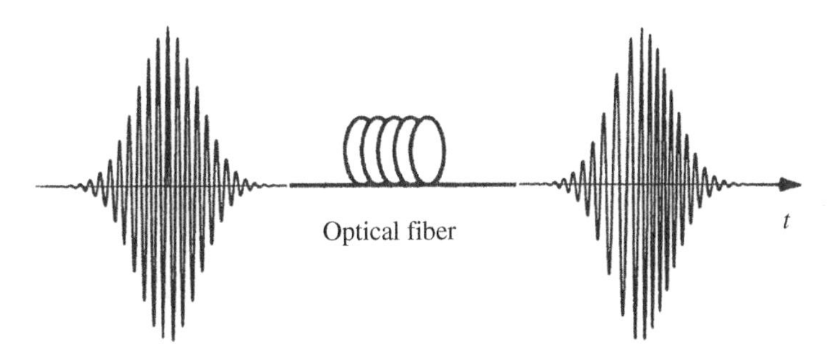

Figure 4.2 Unchirped input and chirped output pulse generated due to self-phase modulation.

Figure 4.1 shows the temporal variation of $P(\tau)$ and $dP/d\tau$ for a Gaussian pulse. In the leading edge of the pulse, where $dP/d\tau > 0$, the instantaneous frequency is downshifted from ω_0, whereas in the trailing edge, where $dP/d\tau < 0$, the instantaneous frequency is upshifted from ω_0. The frequency at the center of the pulse, ω_0, remains unchanged. Figure 4.2 illustrates the effect of SPM on an unchirped input pulse.

As a consequence of the SPM-induced frequency chirping, new frequency components are generated as the optical pulse propagates down the fiber, broadening the spectrum of the bit stream. The spectral broadening depends not only on the effective length but also on the temporal variation of the power of the input pulse. As a consequence, short pulses are much more affected by SPM than long pulses. The use of short pulses arises as a consequence of the increasing bit rates in modern optical transmission systems.

Figure 4.3 illustrates the SPM-induced spectral broadening experienced by an initially unchirped Gaussian, a sech-profile, and two super-Gaussian pulses, at a distance $z = 11L_{NL}$, where $L_{NL} = 1/(\gamma P_0)$ is the nonlinear length. The spectra develop multiple peaks and are similar for the Gaussian and sech pulses. In both cases, the two outermost peaks are the most intense, in accordance with the experimental observations [6]. By contrast, the spectra of the super-Gaussian pulses present one main peak and become larger as the parameter m in Eq. (3.68) increases.

In the case of a chirped input pulse, the SPM-induced spectral broadening can exhibit different behaviors. This is illustrated in Figure 4.4, which shows the propagation of an initially chirped Gaussian pulse, as given by Eq. (3 presenting a) a positive chirp parameter ($C_0 = 3$) and b) a negative

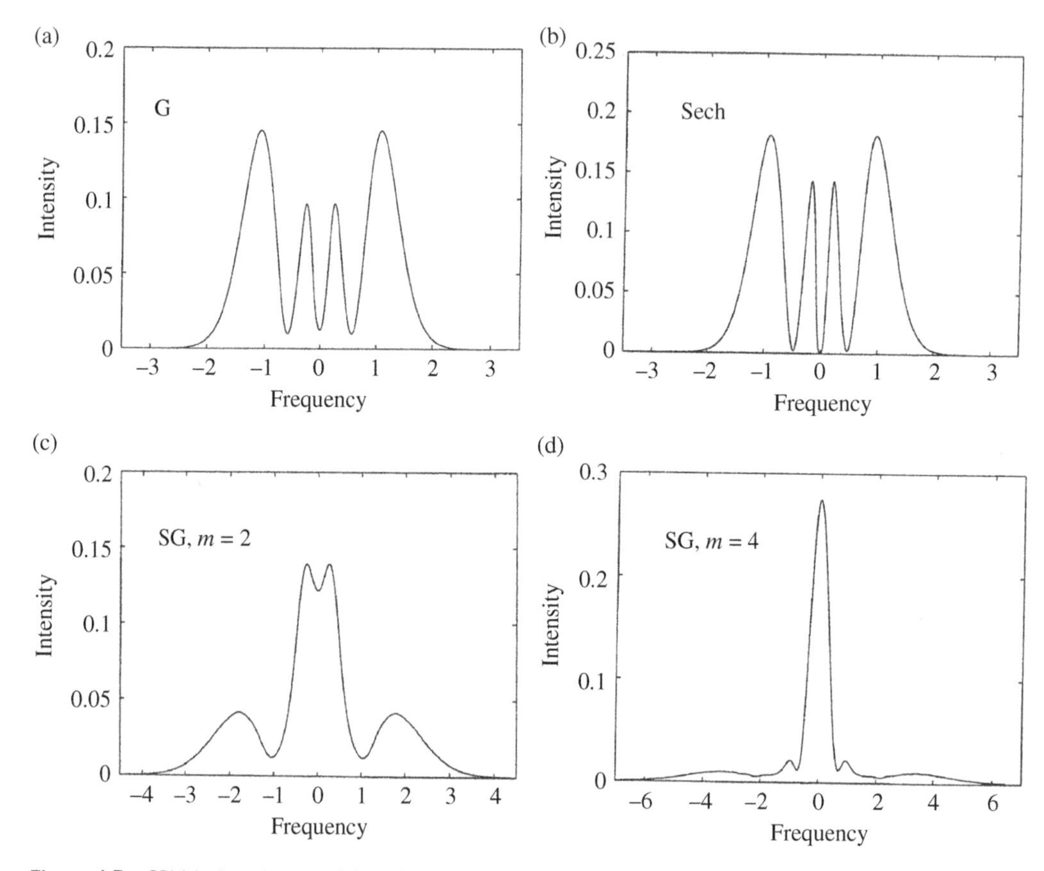

Figure 4.3 SPM-induced spectral broadening of initially unchirped (a) Gaussian, (b) sech-profile, and (c,d) super-Gaussian (m = 2 and m = 4) pulses at a distance $z = 11L_{NL}$.

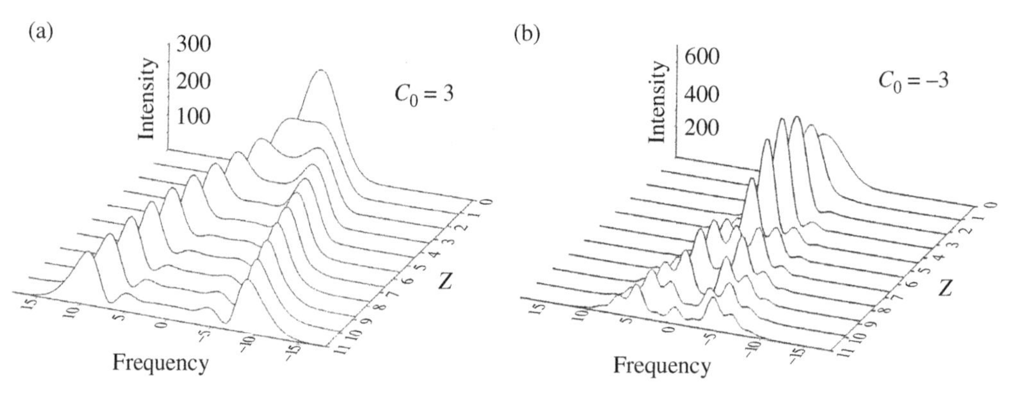

Figure 4.4 Spectral evolution of an initially chirped Gaussian pulse with (a) a positive chirp parameter ($C_0 = 3$) and (b) a negative chirp parameter ($C_0 = -3$).

chirp parameter ($C_0 = -3$). In the first case, the spectrum experiences a faster broadening in comparison with the unchirped case shown in Figure 4.3a. However, a rather different behavior is observed in the second case, in which the pulse spectrum experiences a compression in the first stage of propagation. After that, the spectrum starts to broaden.

In the presence of dispersion, the spectral broadening due to SPM determines two situations qualitatively different. As seen in Chapter 3, in the normal dispersion region (wavelength shorter than the zero-dispersion wavelength, λ_{ZD}), the chirping due to dispersion is to downshift the leading edge and to upshift the trailing edge of the pulse, which is a similar effect as that due to SPM. Thus, in this regime, the chirpings due to both effects add. As a consequence, the pulse experiences a more significant dispersion for high pulse powers than for low pulse powers. On the other hand, in the anomalous dispersion region, the chirping due to dispersion is opposite to that due to SPM. Consequently, nonlinearity- and dispersion-induced chirpings can partially or even completely cancel each other. When this cancellation is total, the pulse broadens neither in time nor in frequency domains and it is called a *soliton*.

4.1.1 Modulation Instability

A continuous wave in the anomalous dispersion regime is known to produce modulation instability (MI), a phenomenon in which the side-band component of the amplitude-modulated light grows exponentially [7]. To investigate analytically this phenomenon, we add a small perturbation u to the continuous wave (CW) solution of Eq. (3.55), such that

$$U = \left(\sqrt{P_0} + u\right) \exp\left(i\phi_{NL}\right) \tag{4.14}$$

where P_0 is the incident power and $\phi_{NL} = \gamma P_0 z$ is the nonlinear phase shift induced by SPM, neglecting the fiber attenuation. Substituting Eq. (4.14) into Eq. (3.55) under the assumption that the perturbation is sufficiently small we obtain the following differential equation for u:

$$i\frac{\partial u}{\partial z} = \frac{\beta_2}{2}\frac{\partial^2 u}{\partial \tau^2} - \gamma P_0(u + u^*) \tag{4.15}$$

Let us assume a general solution for $u(z, \tau)$ of the form

$$u(z,\tau) = a\exp\left[i(Kz - \Omega\tau)\right] + b\exp\left[-i(Kz - \Omega\tau)\right] \tag{4.16}$$

where Ω and K are the angular frequency and the wave number of the perturbation, respectively. By substituting Eq. (4.16) into Eq. (4.15), the following dispersion relation between K and Ω is obtained:

$$K^2 = \left(\frac{\beta_2\Omega}{2}\right)^2\left[\Omega^2 + \frac{4\gamma P_0}{\beta_2}\right] \tag{4.17}$$

Equation (4.17) shows that K becomes a pure imaginary number when $\beta_2 < 0$ (anomalous dispersion) and

$$\Omega < \Omega_c \equiv \left[\frac{4\gamma P_0}{|\beta_2|}\right]^{1/2} \tag{4.18}$$

In these circumstances, the CW solution is unstable, since the perturbation $u(z, \tau)$ grows exponentially with z. This phenomenon is called *modulation instability*.

The power gain of modulation instability is given from Eq. (4.17) as

$$g(\Omega) = |\beta_2\Omega|\left[\Omega_c^2 - \Omega^2\right]^{1/2} \tag{4.19}$$

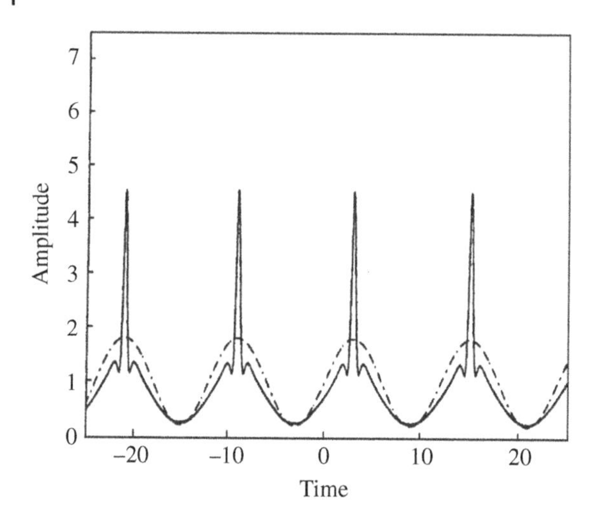

Figure 4.5 Computer simulation of induced modulational instability. The dash-dotted curve represents the modulated initial wave, whereas the full curve corresponds to the final profile after propagation along a normalized distance $Z = 1.27$.

The gain vanishes at $\Omega = 0$ and becomes maximum at $\Omega = \pm\Omega_c/\sqrt{2}$, with a peak value

$$g_{max} = \frac{1}{2}|\beta_2|\Omega_c^2 = 2\gamma P_0 \tag{4.20}$$

Even in the absence of other perturbations, modulation instability can be seeded by the broadband noise added by optical amplifiers. However, modulational instability can also be induced when a small modulation having an angular frequency in the vicinity of $\Omega_m = \Omega_c/\sqrt{2}$ is applied to an input signal. We call this an *induced modulational instability*. An ultrahigh-bit rate pulse train can be generated by using such induced modulational instability [7]. The repetition period of this pulse train is equal to the inverse of the given modulation frequency $\Omega_m/2\pi$.

Figure 4.5 shows the numerical evaluation of Eq. (3.61) at a distance $Z = 1.27$, considering an input signal $q(0, Z) = 1 + A_m\, sen\,(2\pi T/T_m)$, with $A_m = 0.8$ and $T_m = 12$. One can see that a soliton-like pulse train can be generated from this initial condition. A simple technique of generating induced modulational instability consists injecting into a fiber two continuous waves with different wavelengths [8].

Besides its fundamental role in the formation of optical solitons and in the generation of pulse trains [7, 9], SPM has been widely used to realize various optical signal processing functions, namely 2R optical regeneration [10–15], femtosecond pulse generation in fiber lasers [16–20], etc.

4.2 Cross-Phase Modulation

Waves with different wavelengths propagating in the same fiber can interact with each other since the refractive index that a wave experiences depends on the intensities of all other waves. Hence, a pulse at one wavelength has an influence on the phase of a pulse at another wavelength. This nonlinear phenomenon is known as XPM, and it can limit significantly the performance of multichannel optical communication systems [21–29].

Let us consider two channels propagating at the same time in the fiber. The two optical fields can be written in the form

$$E_j(r,t) = \frac{1}{2}F(r,\phi)U_j(z)\exp\left[i\left(\beta_j z - \omega_j t\right)\right] + c.c. \quad (j = 1, 2) \tag{4.21}$$

where $F(r, \phi)$ gives the transverse spatial distribution of the single mode supported by the fiber, $U_j(z)$ is the normalized slowly varying amplitude of the wave with frequency ω_j, and β_j is the propagation constant. The dispersive effects are taken into account by expanding the frequency-dependent propagation constant, given by Eq. (3.4). Retaining terms only up to the quadratic term in such expansion, the following coupled differential equations are obtained for the slowly varying amplitudes U_1 and U_2:

$$\frac{\partial U_1}{\partial z} + \frac{1}{v_{g1}}\frac{\partial U_1}{\partial t} + i\frac{\beta_{21}}{2}\frac{\partial^2 U_1}{\partial t^2} = i\gamma_1\left[\left(|U_1|^2 + 2|U_2|^2\right)U_1 + U_1^2 U_2^* \exp\{i(\Delta\beta z - \Delta\omega t)\}\right] \tag{4.22}$$

$$\frac{\partial U_2}{\partial z} + \frac{1}{v_{g2}}\frac{\partial U_2}{\partial t} + i\frac{\beta_{22}}{2}\frac{\partial^2 U_2}{\partial t^2} = i\gamma_2\left[\left(|U_2|^2 + 2|U_1|^2\right)U_2 + U_2^2 U_1^* \exp\{i(-\Delta\beta z + \Delta\omega t)\}\right] \tag{4.23}$$

where v_{gj} is the group velocity, β_{2j} is the group velocity dispersion (GVD) coefficient, $\Delta\omega = \omega_1 - \omega_2$, $\Delta\beta = \beta_1 - \beta_2$, and γ_j is the nonlinear parameter at frequency ω_j, defined as in Eq. (3.52):

$$\gamma_j = \frac{n_2\omega_j}{cA_{eff}} \tag{4.24}$$

The last terms in Eqs. (4.22) and (4.23) result from the phenomenon of FWM. However, these terms can be neglected if the phase-matching conditions for the occurrence of this phenomenon are not verified. Moreover, assuming that the pulse duration is long or the dispersion in the fiber is low, we can also neglect the linear terms in Eqs. (4.22) and (4.23), leading to the following simplified system of equations:

$$\frac{\partial U_1}{\partial z} = i\gamma_1(P_1 + 2P_2)U_1 \tag{4.25}$$

$$\frac{\partial U_2}{\partial z} = i\gamma_2(P_2 + 2P_1)U_2 \tag{4.26}$$

where $P_j = |U_j|^2$ is the power of the wave with frequency ω_j, $j = 1,2$. If, for simplicity, the powers are assumed to remain constant, the solution of Eq. (4.25) is given by

$$U_1(L) = U_1(0)\exp\{i\gamma_1(P_1 + 2P_2)L\} \tag{4.27}$$

A similar solution can be obtained from Eq. (4.26) for U_2. From Eq. (4.27), it is apparent that the phase of the signal at frequency ω_1 is modified not only due to its own power – which corresponds to the SPM effect – but also due to the power of the signal at frequency ω_2. This phenomenon is referred to as XPM and Eq. (4.27) shows that it is twice as effective as SPM.

The nonlinear phase shift resulting from the combination of SPM and XPM at the output of a fiber with length L is given by

$$\phi_{1NL} = \gamma_1(P_1 + 2P_2)L \tag{4.28}$$

The XPM-induced frequency shift $\Delta\omega_{1XPM}$ of the signal at frequency ω_1 is given by

$$\Delta\omega_{1XPM} = -2\gamma_1 L\frac{\partial P_2}{\partial t} \tag{4.29}$$

Equation (4.29) shows that the part of the signal at ω_1 that is affected by the leading edge of the signal at ω_2 will be downshifted in frequency, whereas the part overlapping with the trailing edge will be upshifted in frequency. This determines a spectral broadening of the signal at ω_1, that is twice the spectral broadening caused by SPM.

The XPM effect determines a mutual influence between two pulses only if they overlap at some extent. However, in the presence of finite dispersion, the two pulses with different wavelengths will move with different velocities and thus will walk off from each other. If the pulses enter the fiber separately, walk through each other, and again become separated, it is said that they experience a complete collision. In a lossless fiber, such collision is perfectly symmetric and no residual phase shift remains, since the pulses would have interacted equally with both the leading and the trailing edge of the other pulse. However, in case the pulses enter the fiber together, the result is a partial collision since each pulse will see only the trailing or the leading edge of the other pulse, which will lead to chirping. Moreover, in the case of a periodically amplified system, power variations also make complete collisions asymmetric, resulting in a net frequency shift that depends on the wavelength difference between the interacting pulses. Such frequency shifts lead to timing jitter in multichannel systems since their magnitude depends on the bit pattern as well as on channel wavelengths. The combination of amplitude and timing jitter degrades significantly the system performance [23].

Besides the limitations imposed by XPM in multichannel optical communication systems, this effect can be used to implement several optical signal processing functions, such as all-optical switching [9, 30–34], wavelength conversion [35–39], and full pulse regeneration [40–42].

4.3 Four-Wave Mixing

Four-wave mixing (FWM) is a parametric process in which four waves or photons interact with each other due to the third-order nonlinearity of the material. As a result, when several channels with frequencies ω_1, ..., ω_n are transmitted simultaneously over the same fiber, the intensity dependence of the refractive index not only leads to phase shifts within a channel, as discussed in previous sections, but also gives rise to signals at new frequencies such as $2\omega_i - \omega_j$ and $\omega_i + \omega_j - \omega_k$. In a wavelength-division multiplexing (WDM) system, if the various channels are equally spaced, all the new components generated within the bandwidth of the system fall at the original channel frequencies, giving rise to cross talk [43–48]. In contrast to SPM and XPM, which become more significant for high bit rate systems, the FWM does not depend on the bit rate. The efficiency of this phenomenon depends strongly on phase-matching conditions, as well as on the channel spacing, chromatic dispersion, and fiber length. The occurrence of the FWM phenomenon in optical fibers was observed for the first time in 1974 by Stolen et al. [49] using a 9-cm-long multimode fiber pumped by a doupled pulsed YAG laser at 532 nm.

In the quantum mechanical description, FWM occurs when photons from one or more waves are annihilated and new photons are created at different frequencies. In this process, the rules of conservation of energy and momentum must be fulfilled. The conservation of momentum leads to the phase-matching condition.

Considering the case in which two photons at frequencies ω_1 and ω_2 are annihilated with simultaneous creation of two photons at frequencies ω_3 and ω_4, the conservation of energy imposes the condition

$$\omega_1 + \omega_2 = \omega_3 + \omega_4 \tag{4.30}$$

On the other hand, the phase-matching is given by the condition $\Delta k = 0$, where

$$\Delta k = \beta_3 + \beta_4 - \beta_1 - \beta_2 = (n_3\omega_3 + n_4\omega_4 - n_1\omega_1 - n_2\omega_2)/c \tag{4.31}$$

The FWM efficiency is significantly higher in dispersion-shifted fibers than in standard fibers due to the low value of the dispersion in the first case. On the contrary, the FWM efficiency is effectively reduced in the case of dispersion-managed systems employing fibers with high dispersion, either normal or anomalous.

In practice, it is relatively easy to satisfy the phase-matching condition in the degenerate case $\omega_1 = \omega_2$. In this situation, a strong pump at $\omega_1 = \omega_2 \equiv \omega_p$ generates a low-frequency sideband at ω_3 and a high-frequency sideband at ω_4, when we assume $\omega_4 > \omega_3$. In analogy to Raman scattering, these sidebands are referred to as the Stokes and anti-Stokes bands, respectively, which are also often called the *signal* and *idler* bands. The frequency shift of the two sidebands is given by

$$\Omega_s = \omega_p - \omega_3 = \omega_4 - \omega_p \tag{4.32}$$

Assuming that the pulse durations are sufficiently long, we can neglect the linear term corresponding to the second-order dispersion in Eq. (3.55), which becomes

$$\frac{\partial U}{\partial z} = i\gamma |U|^2 U \tag{4.33}$$

In order to consider the FWM process and considering the nondegenerate case ($\omega_1 \neq \omega_2$), the amplitude U in Eq. (4.33) is assumed to be the result of the superposition of four waves:

$$U = U_1 e^{i(\beta_1 z - \omega_1 \tau)} + U_2 e^{i(\beta_2 z - \omega_2 \tau)} + U_3 e^{i(\beta_3 z - \omega_3 \tau)} + U_4 e^{i(\beta_4 z - \omega_4 \tau)} \tag{4.34}$$

In the following, we will consider that the four frequencies in Eq. (4.34) satisfy the relation given by Eq. (4.30).

Substituting Eq. (4.34) into Eq. (4.33) and considering Eq. (4.30), the following set of four coupled equations for the normalized amplitudes U_j is obtained:

$$\frac{\partial U_1}{\partial z} = i\gamma \left[\left(|U_1|^2 + 2\sum_{j \neq 1} |U_j|^2 \right) U_1 + 2U_3 U_4 U_2^* e^{i\Delta kz} \right] \tag{4.35}$$

$$\frac{\partial U_2}{\partial z} = i\gamma \left[\left(|U_2|^2 + 2\sum_{j \neq 2} |U_j|^2 \right) U_2 + 2U_3 U_4 U_1^* e^{i\Delta kz} \right] \tag{4.36}$$

$$\frac{\partial U_3}{\partial z} = i\gamma \left[\left(|U_3|^2 + 2\sum_{j \neq 3} |U_j|^2 \right) U_3 + 2U_1 U_2 U_4^* e^{-i\Delta kz} \right] \tag{4.37}$$

$$\frac{\partial U_4}{\partial z} = i\gamma \left[\left(|U_4|^2 + 2\sum_{j \neq 4} |U_j|^2 \right) U_4 + 2U_1 U_2 U_3^* e^{-i\Delta kz} \right] \tag{4.38}$$

where Δk is the wave-vector mismatch, given by Eq. (4.31), and γ is an averaged nonlinear parameter, given by Eq. (3.52), which ignores the small variation due to the slightly different frequencies of the involved waves. In deriving Eqs. (4.35)–(4.38), only nearly phase-matched terms were kept. Fiber loss may be included by adding the term $-(\alpha/2)U_j$, $j = 1,2,3,4$, to the right-hand side of each equation, respectively. The first term inside the brackets in Eqs. (4.35)–(4.38) describes the effect of SPM, whereas the second term is responsible for XPM. Since these terms can lead to a phase alteration only, the generation of new frequency components is provided by the remaining FWM terms. If only the two pump waves (with frequencies ω_1 and ω_2) are present at the fiber input, the signal and idler waves are generated from the wideband noise that is always present in communication systems. However, in some cases, the signal wave can already be present at the fiber input. If the phase-matching condition is satisfied, both the signal and the idler waves grow during propagation due to the optical power transferred from the two pump waves.

Let us assume that the pump waves are much more intense than the signal and idler waves and that they are not phase-matched. In such a case, the power transfer between these waves is very ineffective and we can assume that the pump waves remain undepleted during the FWM process. The solutions of Eqs. (4.35) and (4.36) are then given by:

$$U_j(z) = \sqrt{P_j} \exp\left[i\gamma\left(P_j + 2P_{3-j}\right)z\right],\ j = 1, 2 \tag{4.39}$$

where $P_j = |U_j(0)|^2$ are the input pump powers. In this approximation, the pump waves experience only a phase shift due to SPM and XPM. We will also consider the case in which the signal wave has already a finite value at the fiber input. However, the input signal power is assumed to be much less than the input pump power. In these circumstances, due to the phase mismatch between the involved waves, Eq. (4.37) has the following approximate solution:

$$U_3(z) = \sqrt{P_3} \exp\left[i\gamma 2(P_1 + P_2)z\right] \tag{4.40}$$

Substituting Eqs. (4.39) and (4.40) in Eq. (4.38), we obtain the following result for the power of the idler wave:

$$P_4(L) = 4\gamma^2 P_1 P_2 P_3 L^2 \left[\frac{\sin(\kappa L/2)}{\kappa L/2}\right]^2 \tag{4.41}$$

where the parameter

$$\kappa = \Delta k + \gamma(P_1 + P_2) \tag{4.42}$$

corresponds to the effective phase mismatch and $P_1 + P_2 = P_p$ is the total pump power.

If the phase-matching condition $\kappa = 0$ is fulfilled, Eq. (4.41) shows that the power of the idler increases quadratically with the fiber length, as represented in Figure 4.6 (full curve). However, in such a case, it is not reasonable to consider that the amplitudes of the pump and signal waves do remain constant along the fiber, as assumed in deriving Eq. (4.41). If the phases are not matched, the intensity of the idler waver shows a periodic evolution along the fiber, as described by Eq. (4.41) (dashed curve in Figure 4.6).

Let us consider the degenerate case $\omega_1 = \omega_2 \equiv \omega_p$, where ω_p denotes the pump frequency. In this case, the effective phase mismatch is given by

$$\kappa = \Delta k + 2\gamma P_p \tag{4.43}$$

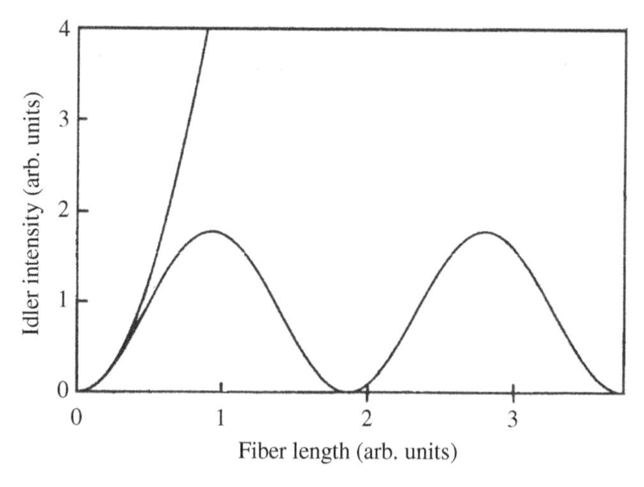

Figure 4.6 Idler against the fiber length for phase matching (full curve) and a phase mismatch between the waves (dashed curve).

where P_p is the pump power. Neglecting the phase mismatch contribution due to the waveguide dispersion, the linear phase mismatch, Δk, depends only on the material dispersion, as given by Eq. (4.31). Such phase mismatch can be expressed in terms of the frequency shift Ω_s given by Eq. (4.32) if we use an expansion of Δk about the pump frequency ω_p. By retaining up to terms quadratic in Ω_s in this expansion, one obtains

$$\Delta k \approx \beta_2 \Omega_s^2 \tag{4.44}$$

where $\beta_2 = d^2\beta/d\omega^2$ is the GVD coefficient at the pump frequency.

For a pump wavelength in the anomalous dispersion region ($\beta_2 < 0$), the negative value of Δk can be compensated by the fiber nonlinearity. For phase matching, one has

$$\kappa \approx \beta_2 \Omega_s^2 + 2\gamma P_p = 0 \tag{4.45}$$

From this, we obtain that the frequency shift is given by:

$$\Omega_s = \sqrt{2\gamma P_p/|\beta_2|} \tag{4.46}$$

We can notice that this frequency shift corresponds to the frequency for a maximum power gain of modulation instability, given by Eq. (4.19).

When $|\beta_2|$ is small, we must take into account the fourth-order term in the expansion of Δk, which becomes

$$\Delta k \approx \beta_2 \Omega_s^2 + \frac{\beta_4}{12} \Omega_s^4 \tag{4.47}$$

where $\beta_4 = d^4\beta/d\omega^4$. In this case, the phase-matching condition is given by

$$\kappa \approx \beta_2 \Omega_s^2 + \frac{\beta_4}{12} \Omega_s^4 + 2\gamma P_0 = 0 \tag{4.48}$$

We observe from Eq. (4.48) that phase matching can be achieved even for a pump wavelength in the normal GVD regime ($\beta_2 > 0$) if $\beta_4 < 0$. This condition can be easily realized in tapered and microstructured fibers [50, 51].

FWM can be used to realize fiber-optical parametric amplifiers (FOPAs), as seen in Chapter 5. A FOPA provides a high gain, a broad bandwidth, and may be tailored to operate at any wavelength [52–56]. The same effect has been explored to implement several optical signal-processing functions, such as all-optical switching [57–61], wavelength conversion [62–66], pulse regeneration [67–70], and optical phase conjugation [71].

4.4 Stimulated Raman Scattering

In experiments of light scattering in different media, we can observe the existence of a frequency-shifted field in addition to the incident field. The frequency shift of the scattered light is determined by the vibrational oscillations that occur between constituent atoms within the molecules of the material. A slight excitation of the molecular resonances due to the presence of the input field, referred to as the pump wave, results in spontaneous Raman scattering. As the scattered light increases in intensity, the scattering process becomes eventually stimulated. In this regime, the scattered field interacts coherently with the pump field to further excite the resonances, which significantly enhances the transfer of power between the two optical waves. Raman scattering can occur in all materials, but in silica glass the dominant Raman lines are due to the bending motion of the Si-O-Si bond.

The first experimental demonstration of stimulated Raman scattering in fibers was realized by Erich Ippen in early 1970, which constructed a CS_2-core CW fiber Raman laser [72]. Soon after, Stolen et al. observed for the first time the same effect in single-mode silica fibers and demonstrated its use in a Raman oscillator [73]. This work was followed by an amplifier experiment to directly measure the Raman gain in silica fibers [74].

Both downshift (Stokes) and upshifted (anti-Stokes) light can, in principle, be generated through the Raman scattering process. However, in the case of optical fibers, the Stokes radiation is generally much stronger than the anti-Stokes radiation due to two main reasons. On one hand, the anti-Stokes process is phase-mismatched, whereas the Stokes process is phase-matched, for collinear propagation. On the other hand, the anti-Stokes process involves the interaction of the pump light with previously excited molecular resonances, which is not a condition for the Stokes process. Due to these facts, only the Stokes wave will be considered in the following.

Considering the quantum mechanical picture, in the SRS process, one has simultaneously the absorption of a photon from the pump beam at frequency ω_p and the emission of a photon at the Stokes frequency, ω_S (Figure 4.7). The difference in energy is taken up by a high-energy phonon (molecular vibration) at frequency ω_r. Thus, SRS provides for energy gain at the Stokes frequency at the expense of the pump. This process is considered nonresonant because the upper state is a short-lived virtual state.

In the case of standard optical fibers, they are basically made of fused silica, which is an amorphous material. Several vibrational modes occur in the structure of amorphous silica, with resonance frequencies which overlap with each other and form broad frequency bands. As a consequence, the Raman scattering in optical fibers occurs over a relatively large frequency range.

The pump wave intensity (I_p) and the Stokes wave intensity (I_S) propagating along an optical fiber satisfy the following evolution equations [1]:

$$\frac{dI_S}{dz} = g_R I_S I_P - \alpha I_S \tag{4.49}$$

$$\frac{dI_P}{dz} = -\frac{\omega_P}{\omega_S} g_R I_S I_P - \alpha I_P \tag{4.50}$$

where α takes into account the fiber losses and g_R is the Raman gain coefficient.

Figure 4.8 shows the normalized Raman gain spectra for bulk fused silica, measured when the pump and signal light were either copolarized (full curve) or orthogonally polarized (dotted

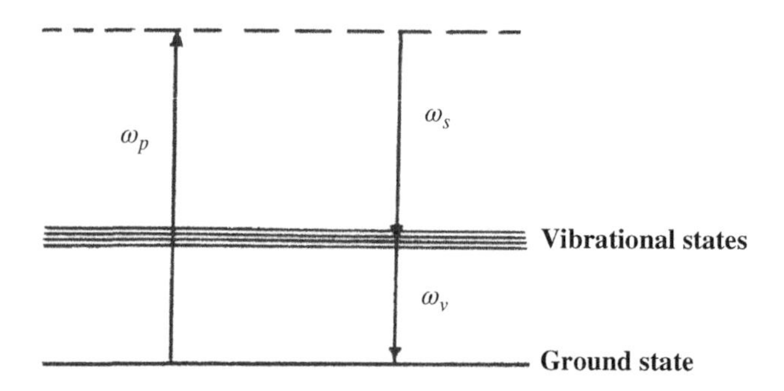

Figure 4.7 Energy-level illustration of the stimulated Raman scattering process.

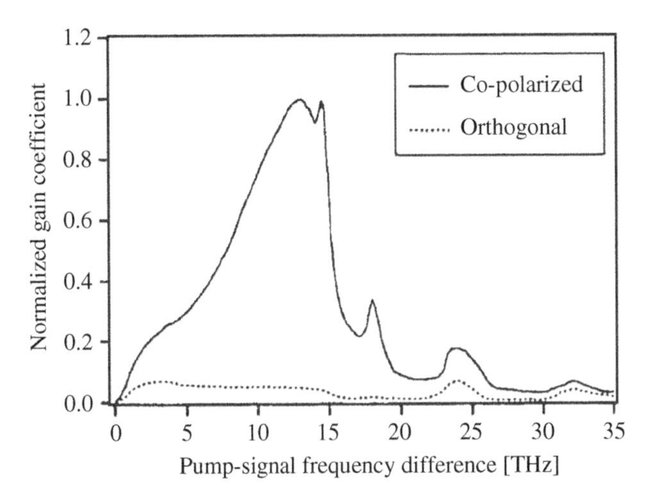

Figure 4.8 Normalized Raman gain coefficient for copolarized (full curve) and orthogonally polarized (dotted curve) pump and signal beams. *Source:* After Ref. [75].

curve) [75]. The most significant feature of the Raman gain in silica fibers is that g_R extends over a large frequency range (up to 40 THz). Optical signals whose bandwidths are of this order or less can be amplified using the Raman effect if a pump wave with the right wavelength is available. The Raman shift, corresponding to the location of the main peak in Figure 4.8, is close to 13 THz for silica fibers. The Raman gain depends on the relative state of polarization of the pump and signal fields. The copolarized gain is almost an order of magnitude larger than the orthogonally polarized gain near the peak of the gain curve. The peak value of the Raman gain coefficient for silica fibers is 9.4×10^{-14} m.W^{-1} for a pumping wavelength $\lambda_p = 1.0$ μm and varies as λ_p^{-1} [76]. The Raman gain coefficient can be enhanced by the use of oxide glasses as dopants for fiber fabrication.

When the input Stokes wave intensity is weak, such that $I_{S0} \ll I_{P0}$, the evolution of the Stokes wave intensity is given approximately by:

$$I_S(z) \approx I_{S0} \exp\left\{\frac{g_R I_{P0}[1 - \exp(-\alpha z]}{\alpha} - \alpha z\right\} \tag{4.51}$$

where $I_{p0} = I_p(z = 0)$ and $I_{s0} = I_s(z = 0)$ are the input pump power and the input Stokes power, respectively. In the absence of an input signal I_{S0}, the Stokes wave arises from spontaneous Raman scattering along the fiber. It was shown that this process is equivalent to injecting one fictitious photon per mode at the input end of the fiber [77]. The threshold for SRS is defined as the input pump power at which the output powers for pump and Stokes wave become equal. It is given approximately by [77]:

$$P_{p0}^{th} = 16 \frac{A_{eff}}{g_R L_{eff}} \tag{4.52}$$

where L_{eff} is the effective fiber length, as given by Eq. (4.8), and A_{eff} is the effective mode area. Eq. (4.52) was derived assuming that the polarization of the pump and Stokes waves is maintained along the fiber. However, in the case of standard single-mode fibers, due to the arbitrary distribution of the polarization states of both waves, the Raman threshold is increased by a factor of two. For example, the threshold for the SRS is $P_{P0}^{th} \approx 600$ mW at $\lambda_P = 1.55\ \mu m$ in long polarization-maintaining fibers, such that $L_{eff} \approx 22\ km$, considering an effective core area of $A_{eff} = 50\ \mu m^2$.

However, in standard single-mode fibers with similar characteristics, the threshold would be $P_{P0}^{th} \approx 1.2\ W$.

Because SRS has a relatively high threshold, it is not of concern for single-channel systems. However, in WDM systems, SRS can cause cross talk between channel signals whose wavelength separation falls within the Raman gain curve. Specifically, the long-wavelength signals are amplified by the short-wavelength signals, leading to power penalties for the latter signals. The shortest-wavelength signal is depleted most, as it acts as a pump for all other channels. The Raman-induced power transfer between two channels depends on the bit pattern, which leads to power fluctuations and determines additional receiver noise. The magnitude of these deleterious effects depends on several parameters, like the number of channels, their frequency spacing, and the power in each of them.

Besides the limitations that the SRS effect can impose on optical fiber communication systems, it can be used with advantage in several applications, namely to realize all-optical amplification, as will be seen in Chapter 5. The same effect plays a significant role in soliton dynamics and in the generation of the supercontinuum, as discussed in later chapters of this book.

In the SRS process, if the first-order Stokes wave is sufficiently strong, it can act as a pump source itself, resulting in the generation of higher-order Stokes waves. Figure 4.9 shows an example of a fiber output spectrum at a pump power of about 1 kW in which multiple Stokes orders occur [78]. Many researchers have devoted their attention to the study of cascaded SRS in optical fibers in the visible and near-infrared fields [79–82].

Compared with silica, chalcogenide glasses have a wider transmission window (from the visible up to the infrared region of 12 or 15 μm depending on the composition) and higher Raman gain coefficients ($4.3 - 5.7 \times 10^{-12}$ m/W for As_2S_3 at 1.5 μm, and $2 - 5 \times 10^{-11}$ m/W for As_2Se_3 at 1.5 μm) [83, 84]. As a result, optical fibers fabricated based on chalcogenide glasses are more

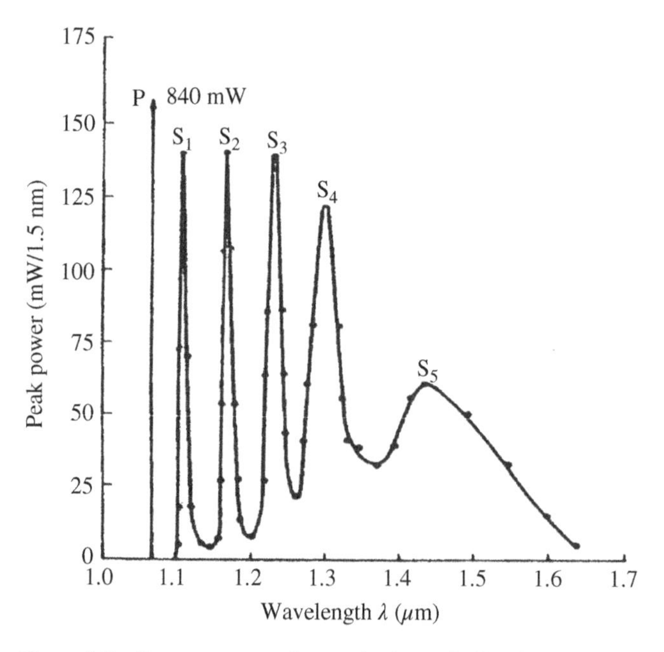

Figure 4.9 Output spectrum from a single-mode fiber Raman laser, showing multiple Stokes orders. The pump wavelength is $\lambda = 1.06$ μm. *Source:* After Ref. [78].

promising for generating cascaded SRS in the mid-infrared region [85–87]. This may be particularly useful for some applications, including light detection and ranging (LIDAR), gas sensing, and optical communications.

4.5 Stimulated Brillouin Scattering

Brillouin scattering is a phenomenon named after the French physicist Leon Brillouin, who investigated the scattering of light at acoustic waves during the 1920s. For low incident intensities, the scattered part of the field remains very weak. However, the process becomes stimulated and strong scattered fields are generated at high input intensities, which are readily available with lasers. Stimulated Brillouin scattering (SBS) in optical fibers was observed for the first time in 1972 [88], using a pulsed narrowband xenon laser operating at 535.3 nm.

The process of stimulated Brillouin scattering can be described as a classical three-wave interaction involving the incident (pump) wave with frequency ω_p, the Stokes wave with frequency ω_s, and an acoustic wave with frequency ω_a. The pump creates a pressure wave in the medium owing to electrostriction, which in turn causes a periodic modulation of the refractive index. This process is illustrated in Figure 4.10. Physically, each pump photon in the SBS process gives up its energy to create simultaneously a Stokes photon and an acoustic phonon.

The three waves involved in the SBS process must conserve both energy and momentum. The energy conservation requires that $\omega_p - \omega_s = 2\pi f_a$, where f_a is the linear frequency of the acoustic wave, which is about 11.1 GHz in standard fibers. The momentum conservation requires that the wave vectors of the three waves satisfy the condition

$$\boldsymbol{k}_a = \boldsymbol{k}_p - \boldsymbol{k}_s \tag{4.53}$$

In a single-mode fiber, optical waves can propagate only along the direction of the fiber axis. Since the acoustic wave velocity $v_a \approx 5.96$ km/s is by far smaller than the light velocity, we have

$$|\boldsymbol{k}_a| = 2\pi f_a/v_a > |\boldsymbol{k}_p| \approx |\boldsymbol{k}_s|. \tag{4.54}$$

Given this condition, the momentum conservation has the important consequence that the Brillouin effect occurs only if the Stokes and the pump waves propagate in opposite directions.

The response of the material to the interference of the pump and Stokes fields tends to increase the amplitude of the acoustic wave. Therefore, the beating of the pump wave with the acoustic wave tends to reinforce the Stokes wave, whereas the beating of the pump wave and the Stokes waves tends to reinforce the acoustic wave. This explains the appearance of the SBS process. In such

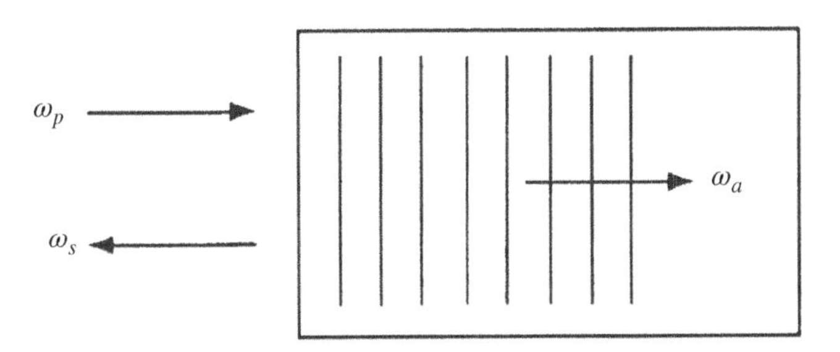

Figure 4.10 Schematic illustration of the stimulated Brillouin scattering process.

process, the pump wave intensity (I_p) and the Stokes wave intensity (I_S) propagating along an optical fiber satisfy the following evolution equations [1]:

$$\frac{dI_S}{dz} = g_R I_S I_P - \alpha I_S \tag{4.55}$$

$$\frac{dI_P}{dz} = -g_R I_S I_P - \alpha I_P \tag{4.56}$$

where α takes into account the fiber losses and g_B is the Brillouin gain coefficient, which for typical fibers is estimated to be about 2.5×10^{-11} $m.\ W^{-1}$ [1]. This value is two orders of magnitude larger than the Raman gain coefficient at λ_p = 1.55 μm.

The spectrum of the Brillouin gain is Lorentzian, given by [1]

$$g_B(\Delta\omega_S) = \frac{(\Gamma_B/2)^2}{(\Delta\omega_S)^2 + (\Gamma_B/2)^2} g_{B0} \tag{4.57}$$

where Γ_B is determined by the acoustic phonon lifetime. The spectrum of the Brillouin gain has a full width at half maximum (FWHM) $\Delta\nu_B = \Gamma_B/2\pi$, which for bulk silica is expected to be about 17 MHz at 1.5 μm. However, several experiments have shown much larger bandwidths for silica-based fibers, which in some cases can exceed 100 MHz [89, 90]. Such larger bandwidths are generally attributed both to the guided nature of acoustic modes and to inhomogeneities in fiber-core cross section along the fiber length.

Considering the case of an undepleted pump wave, the solutions of Eqs. (4.55) and (4.56) are given by:

$$I_p(z) = I_{p0} \exp(-\alpha z) \tag{4.58}$$

$$I_s(z) = I_{s0} \exp(\alpha z) \exp\left[-\frac{g_B I_{p0}(1-\exp(-\alpha z))}{\alpha}\right] \tag{4.59}$$

where $I_{p0} = I_p(z=0)$ and $I_{s0} = I_s(z=0)$ are the input pump power and the output Stokes power, respectively. Actually, a Stokes input signal at $z = L$ grows to produce an output signal at $z = 0$, given by

$$I_{s0} = I_s(L) \exp(-\alpha L) \exp\left[g_B I_{p0} L_{eff}\right], \tag{4.60}$$

where $L_{eff} = [1 - \exp(-\alpha L)]/\alpha$ is the effective fiber length.

In the absence of an input signal, the Stokes wave builds up from spontaneous scattering. The noise power provided by spontaneous Brillouin scattering is equivalent to injecting a fictitious photon per mode at a distance where the gain is equal to the fiber loss. The threshold pump power is defined as the input pump power that is equal to the backward output Stokes power, which gives the result [77]:

$$P_{p0}^{th} = 21 \frac{K_B A_{eff}}{g_{B0} L_{eff}} \tag{4.61}$$

where K_B is a factor varying between 1 and 2 that takes into account the polarization dependence of the pump and Stokes waves. If both waves propagate in a polarization-maintaining fiber and have the same polarization, we have $K_B = 1$. In the case of a nonpolarization-maintaining fiber, it is $K_B = 1.5$ [91]. Considering an effective area A_{eff} = 50 μm^2, an attenuation constant of α = 0.2 dB/km, a gain coefficient $g_{B0} = 2.5 \times 10^{-11}$ m/W, and $K_B = 1$ we obtain from Eq. (4.61) a threshold power $P_{p0}^{th} \approx 2$ mW. This value is around three orders of magnitude smaller than the threshold required for Raman scattering, which makes SBS the dominant nonlinear effect in some circumstances.

The value reported earlier for the Brillouin gain coefficient is valid only when the spectral width of the pump beam ($\Delta\nu_p$) is much narrower than the Brillouin linewidth ($\Delta\nu_B$). When this condition is not verified, the Brillouin gain coefficient is reduced and given by [1]:

$$\widetilde{g}_{B0} = \frac{\Delta\nu_B}{\Delta\nu_B + \Delta\nu_p} g_{B0} \tag{4.62}$$

SBS can affect seriously the performance of a transmission system since it determines the maximum power that can be launched into a system. Such maximum power can be of the order of some few mW, which limits the maximum signal-to-noise ratio (SNR) and the transmission distance that can be reached without amplification. These limitations are alleviated by increasing the SBS threshold, which can be achieved by broadening the spectrum of the transmitted light, according to Eq. (4.62).

The SBS process presents some new features if the transmission fiber is used also as an amplifying medium (e.g. as a distributed Raman or erbium-doped fiber amplifier), in which case, it is called an *active* fiber [92–98]. The equations describing the evolution of the signal (pump) power, P_p, first-order Stokes power, P_{S1}, and second-order Stokes power, P_{S2}, in an active fiber can be written in the form [96, 97]:

$$\frac{dP_p}{dz} = -\frac{g_B}{A_{eff}} P_p P_{S1} + \delta P_p \tag{4.63}$$

$$\frac{dP_{S1}}{dz} = -\frac{g_B}{A_{eff}} P_p P_{S1} + \frac{g_B}{A_{eff}} P_{S2} P_{S1} - \delta P_{S1} \tag{4.64}$$

$$\frac{dP_{S2}}{dz} = \frac{g_B}{A_{eff}} P_{S2} P_{S1} + \delta P_{S2} \tag{4.65}$$

where $\delta = g - \alpha$, g is the amplifier gain and α is the fiber loss.

The spatial dependence of the pump, first-order, and second-order Stokes powers is illustrated in Figure 4.11 against the normalized distance z/L, where L is the fiber length. The initial pump power is assumed to be $P_p(0) = 1$ mW, whereas the Stokes waves are seeded assuming initial noise powers

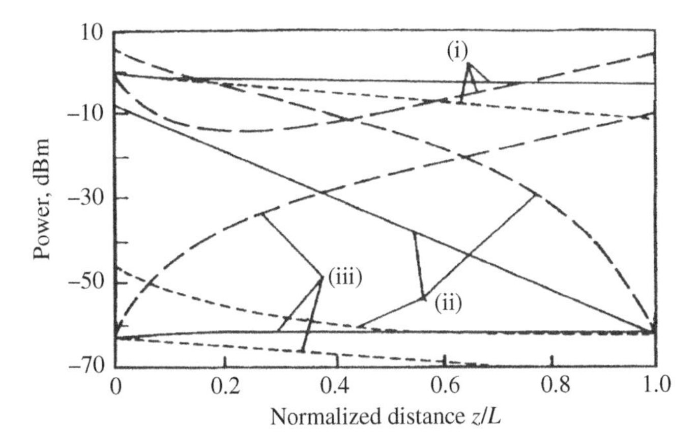

Figure 4.11 Pump (i), first-order (ii), and second-order (iii) Stokes powers against normalized fiber length for $\delta = -0.5$ (short-dashed curves), 0 (full curves), and 1.5 dB/km (long-dashed curves).

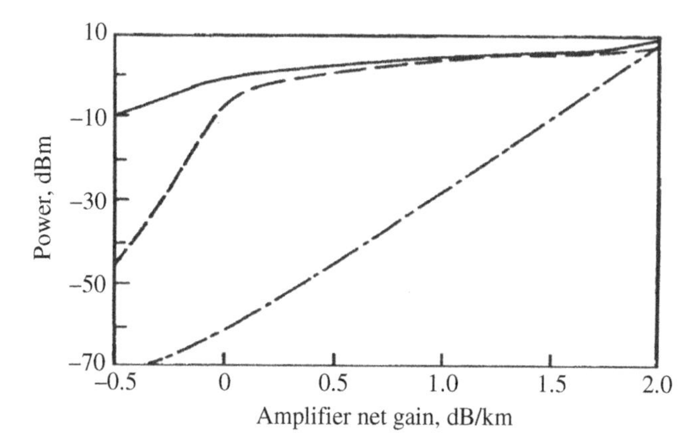

Figure 4.12 Pump (full curve), first-order (dashed curve), and second-order (dash-dotted curve) Stokes powers against fiber net gain for L = 20 km.

$P_{S1}(L) = P_{S2}(0) = 0.5$ nW. The fiber parameters are $L = 20$ km, $A_{eff} = 28.26\times10^{-12}$m^2, and $g_B = 2.2\times10^{-11}$ m/W, whereas the net gain/loss parameter is $\delta = -0.5$, 0, and 1.5 dB/km. In the case of a lossy fiber ($\delta = -0.5$ dB/km), the pump power decreases monotonically along the fiber, while the first-order Stokes power remains five orders of magnitude below the pump power. For a fiber with no net gain or loss ($\delta = 0$ dB/km), the pump power decreases slightly in the initial stage of the fiber and remains thereafter practically constant, whereas the first-order Stokes power increases to 0.19 mW and the second-order Stokes power remains negligible. Finally, when the fiber has a positive distributed gain ($\delta = 1.5$ dB/km), we observe that the pump power presents a non-monotonic evolution and increases after passing through a minimum value. The first-order Stokes power becomes similar to the pump power, whereas the output second-order Stokes power achieves a value of 0.13 mW.

Figure 4.12 shows that despite the fact that the output pump power increases monotonically with the net gain δ, this growth is slower for $\delta > 0$ dB/km. This is due to the nonlinear pump depletion effect determined by the first-order Stokes wave, which becomes of the same order of the pump for $\delta > 0$ dB/km. In this range, the second-order Stokes output power increases almost exponentially with the fiber gain by seven orders of magnitude. These results show that the presence of distributed gain in an optical fiber leads generally to the ready generation of multiple Stokes orders and to a significant variation of the signal pump power along the fiber.

Besides the SBS-induced limitations mentioned earlier, the Brillouin gain can also find some useful applications, namely for optical amplification [97, 98], lasing [99], channel selection in closely spaced wavelength-multiplexed network [100], optical phase conjugation [101], temperature and strain sensing [102, 103], all-optical slow-light control [104, 105], optical storage [106], etc.

References

1 Ferreira, M.F. (2011). *Nonlinear Effects in Optical Fibers*. Hoboken, NJ: John Wiley & Sons.
2 Chraplyvy, A. (1990). *J. Lightwave Technol.* **8**: 1548.
3 Mitra, P. and Stark, J. (2001). *Nature* **411**: 1027.
4 Wu, M. and Way, W. (2004). *J. Lightwave Technol.* **22**: 1483.

5 Ferreira, M.F. (2020). *Optical Signal Processing in Highly Nonlinear Fibers.* Abingdon, Oxon, UK: CRC Press.
6 Stolen, R.H. and Lin, C. (1978). *Phys. Rev. A* **17**: 1448.
7 Hasegawa, A. (1984). *Opt. Lett.* **9**: 288.
8 Tai, K., Tomita, A., Jewel, J.L., and Hasegawa, A. (1986). *Appl. Phys. Lett.* **49**: 236.
9 Igarashi, K. and Kikuchi, K. (2008). *IEEE J. Sel. Topics Quantum Electron.* **14**: 551.
10 Mamyshev P.V. (1998). All-optical data regeneration based on self-phase modulation effect. Presented at the European Conference on Optical Communications (ECOC '98), Madrid, Spain (20–24 September 1998), pp. 475–477.
11 Matsumoto, M. (2004). *J. Lightw. Technol* **23**: 1472.
12 Johannisson, P. and Karlsson, M. (2005). *IEEE Photon. Technol. Lett.* **17**: 2667.
13 Striegler, A.G. and Schmauss, B. (2006). *J. Lightwave Technol.* **24**: 2835.
14 Rochette, M., Libin, F., Ta'eed, V. et al. (2006). *IEEE J. Select. Topics Quantum Electron.* **12**: 736.
15 Fok, M.P. and Shu, C. (2008). *IEEE J. Sel. Topics Quantum Electron.* **14**: 587.
16 Liu, Z., Ziegler, Z.M., Wright, L.G., and Wise, F.W. (2017). *Optica* **4**: 649.
17 Liu, Z., Li, C., Zhang, Z. et al. (2016). *Opt. Express* **24**: 15328.
18 Liu, W., Chia, S.-H., Chung, H.-Y. et al. (2017). *Opt. Express* **25**: 6822.
19 Chung, H.-Y., Liu, W., Cao, Q. et al. (2017). *Opt. Express* **25**: 15760.
20 Chung, H., Liu, W., Cao, Q. et al. (2018). *Opt. Express* **26**: 3684.
21 Chraplyvy, A.R. and Stone, J. (1984). *Electron. Lett.* **20**: 996.
22 Wang, J. and Petermann, K. (1992). *J. Lightwave Technol.* **10**: 96.
23 Marcuse, D., Chraplyvy, A.R., and Tkach, R.W. (1994). *J. Lightwave Technol.* **12**: 885.
24 Chiang, T.K., Kagi, N., Marhic, M.E., and Kazovsky, L.G. (1996). *J. Lightwave Technol.* **14**: 249.
25 Belloti, G., Varani, M., Francia, C., and Bononi, A. (1998). *IEEE Photon Technol. Lett.* **10**: 1745.
26 Hui, R., Demarest, K.R., and Allen, C.T. (1999). *J. Lightwave Technol.* **17**: 1018.
27 Nelson, L.E., Jopson, R.M., Gnauck, A.H., and Chraplyvy, A.R. (1999). *IEEE Photon. Technol. Lett.* **11**: 907.
28 Betti, S. and Giaconi, M. (2001). *IEEE Photon. Technol. Lett.* **13**: 1304.
29 Thiele, H.J., Killey, R.I., and Bayvel, P. (2002). *Opt. Fiber Technol.* **8**: 71.
30 Blow, K.J., Doran, N.J., Nayar, B.K., and Nelson, B.P. (1990). *Opt. Lett.* **15**: 248.
31 Jino, M. and Matsumoto, T. (1991). *Electron. Lett.* **27**: 75.
32 Sharping, J.E., Fiorentino, M., Kumar, P., and Windeler, R.S. (2002). *IEEE Photon. Technol. Lett.* **14**: 77.
33 Lee, J.H., Tanemura, T., Nagashima, T. et al. (2005). *Opt. Lett.* **30**: 1267.
34 Salem, R., Lenihan, A.S., Carter, G.M., and Murphy, T.E. (2006). *IEEE Photon. Technol. Lett.* **18**: 2254.
35 McKinstrie, C., Radic, S., and Raymer, M. (2004). *Opt. Express* **12**: 5037.
36 Fu, L.B., Rochette, M., Ta'eed, V.G. et al. (2005). *Opt. Express* **13**: 7637.
37 Galili, M., Oxenlowe, L.K., Hansen, H.C.H. et al. (2008). *IEEE J. Sel. Top. Quantum Electron.* **14**: 573.
38 Fernández-Ruiz, M., Lei, L., Rochette, M., and Azaña, J. 1 (2015). *Opt. Express* **23**: 22847.
39 Watanabe, S., Kato, T., Tanimura, T. et al. (2019). *Opt. Express* **27**: 16767.
40 Rochette, M., Blows, J.L., and Eggleton, B.J. (2006). *Opt. Express* **14**: 6414.
41 Suzuki, J., Tanemura, T., Taira, K. et al. (2005). *IEEE Photon. Technol. Lett.* **17**: 423.
42 Daikoku, M., Yoshikane, N., Otani, T., and Tanaka, H. (2006). *J. Light. Technol.* **24**: 1142.
43 Shibata, N., Braun, R.P., and Waarts, R.G. (1987). *IEEE J. Quantum Electron.* **23**: 1205.
44 Inoue, K. (1992). *J. Lightwave Technol.* **10**: 1553.
45 Forghieri, F., Tkach, R.W., and Chraplyvy, A.R. (1995). *J. Lightwave Technol.* **13**: 889.

46 Taga, H. (1996). *J. Lightwave Technol.* **14**: 1287.
47 Suzuki, H., Ohteru, S., and Takachio, N. (1999). *IEEE Photon. Technol. Lett.* **11**: 1677.
48 Beti, S., Giaconi, M., and Nardini, M. (2003). *IEEE Photon. Technol. Lett.* **15**: 1079.
49 Stolen, R.H., Bjorkhholm, J.E., and Ashkin, A. (1974). *Appl. Phys. Lett.* **24**: 308.
50 Wadsworth, W.J., Joly, N., Knight, J.C. et al. (2004). *Opt. Express* **12**: 299.
51 Wong, G.K., Chen, A.Y., Murdoch, S.G. et al. (2005). *J. Opt. Soc. Am. B* **22**: 2505.
52 Karlsson, M. (1998). *J. Opt. Soc. Am. B* **15**: 2269.
53 Hansryd, J. and Andrekson, P.A. (2001). *IEEE Photon. Technol. Lett.* **13**: 194.
54 McKinstrie, C.J., Radic, S., and Chraplyvy, A.R. (2002). *IEEE J. Sel. Top. Quantum Electron.* **8**: 538.
55 Radic, S. and McKinstrie, C.J. (2003). *Opt. Fib. Technol.* **9**: 7.
56 Kalogerakis, G., Marhic, M.E., Wong, K.K., and Kazovsky, L.G. (2005). *J. Lightwave Technol.* **23**: 2945.
57 Morioka, T., Takara, H., Kawanishi, S. et al. (1996). *Electron. Lett.* **32**: 833.
58 Hansryd, J. and Andrekson, P.A. (2001). *IEEE Photon. Technol. Lett.* **13**: 732.
59 Sakamoto, T., Seo, K., Taira, K. et al. (2004). *IEEE Photon. Technol. Lett.* **16**: 563.
60 Tang, R., Lasri, J., Devgan, P.S. et al. (2005). *Opt. Express* **13**: 10483.
61 Cetina, J.P., Kumpera, A., Karlsson, M., and Andrekson, P.A. (2015). *Opt. Express* **23**: 33426.
62 Hansryd, J., Andrekson, P.A., Westlund, M. et al. (2002). *IEEE J. Sel. Top. Quantum Electron.* **8**: 506.
63 Chavez Boggio, J.M., Windmiller, J.R., Knutzen, M. et al. (2008). *Opt. Express* **16**: 5435.
64 M. Hirano, T. Nakanishi, T. Okuno, and M. Onishi, *IEEE J. Sel. Top. Quantum Electron.* **15**, 103 (2009).
65 Nodop, D., Jauregui, C., Schimpf, D. et al. (2009). *Opt. Lett.* **34**: 3499.
66 Víctor, J.F., Rancaño, F., Parmigiani, P. et al. (2014). *J. Lightwave Technol.* **32**: 3027.
67 Cappellini, G. and Trillo, S. (1991). *J. Opt. Soc. Am. B* **8**: 824.
68 Su, Y., Wang, L., Agrawal, A., and Kumar, P. (2000). *Electron. Lett.* **36**: 1103.
69 Matsumoto, M. (2005). *IEEE Photon. Technol. Lett.* **17**: 1055.
70 Matsumoto, M. (2012). *IEEE J. Select. Topics Quantum Electron.* **18**: 738.
71 Kuo, B.P.P., Myslivets, E., Wiberg, A.O.J. et al. (2011). *J. Lightwave Technol.* **29**: 516.
72 Ippen, E.P. (1970). *Appl. Phys. Lett.* **16**: 303.
73 Stolen, R.H., Ippen, E.P., and Tynes, A.R. (1972). *Appl. Phys. Lett.* **20**: 62.
74 Stolen, R.H. and Ippen, E.P. (1973). *Appl. Phys. Lett.* **22**: 276.
75 Bromage, J. (2004). *J. Lightwave Technol.* **22**: 79.
76 Buck, J.A. (2004). *Fundamental of Optical Fibers*, 2e. New York: Wiley.
77 Smith, R.G. (1972). *Appl. Opt.* **11**: 2489.
78 Cohen, L. and Lin, C. (1978). *IEEE J. Quantum Electron.* **14**: 855.
79 Pourbeyram, H., Agrawal, G.P., and Mafi, A. (2013). *Appl. Phys. Lett.* **102**: 201107.
80 Yin, K., Zhang, B., Yao, J. et al. (2016). *IEEE Photonics Technol. Lett.* **28**: 11107.
81 Zhang, L., Jiang, H., Yang, X. et al. (2016). *Opt. Lett.* **41**: 215.
82 Babin, S.A., Zlobina, E.A., Kablukov, S.I., and Podivilov, E.V. (2016). *Sci. Rep.* **6**: 22625.
83 Monro, T.M. (2011). *Opt. Lett.* **36**: 2351.
84 Tuniz, A., Brawley, G., Moss, D.J., and Eggleton, B.J. (2008). *Opt. Express* **16**: 18524.
85 Troles, J., Coulombier, Q., Canat, G. et al. (2010). *Opt. Express* **18**: 26647.
86 Gao, W., Cheng, T., Xue, X. et al. (2016). *Opt. Express* **24**: 3278.
87 Yao, J., Zhang, B., Yin, K. et al. (2016). *Opt. Express* **24**: 14717.
88 Ippen, E.P. and Stolen, R.H. (1972). *Appl. Phys. Lett.* **21**: 539.
89 Shiraki, K., Ohashi, M., and Tateda, M. (1996). *J. Lightwave Technol.* **14**: 50.
90 Yeniay, A., Delavaux, J.-M., and Toulouse, J. (2002). *J. Lightwave Technol.* **20**: 1425.

91 Van Deventer, M.O. and Boot, A.J. (1994). *J. Lightwave Technol.* **12**: 585.
92 Zhang, S.L. and O'Reilly, J.J. (1993). *IEEE Photon. Technol. Lett.* **5**: 537.
93 Pannel, C.N., St, P., Russel, J., and Newson, T.P. (1993). *J. Opt. Soc. Am. B* **10**: 684.
94 Ferreira, M.F. (1994). *Electron. Lett.* **30**: 40.
95 Ferreira, M.F. (1995). *J. Lightwave Technol.* **13**: 1692.
96 Kobyakov, A., Mehendale, M., Vasilyev, M. et al. (2002). *J. Lightwave Technol.* **20**: 1635.
97 Tkach, R.W. and Chraplyvy, A.R. (1989). *Opt. Quantum Electron.* **21**: S105.
98 Ferreira, M.F., Rocha, J.F., and Pinto, J.L. (1994). *Opt. Quantum Electron.* **26**: 35.
99 Stepanov, D.Y. and Cowle, G.J. (1997). *IEEE J. Sel. Top. Quantum Electron.* **3**: 1049.
100 Zadok, A., Eyal, A., and Tur, M. (2007). *Journal Ligtwave Technol.* **25**: 2168.
101 S. M. Massey and T. H. Russel, *IEEE J. Sel. Top. Quantum Electron.* **15**, 399 (2009).
102 Nikles, M., Thevenaz, L., and Robert, P.A. (1996). *Opt. Lett.* **21**: 738.
103 Hotate, K. and Tanaka, M. (2002). *IEEE Photon. Technol. Lett.* **14**: 179.
104 Song, K.Y., Herraez, M.G., and Thevenaz, L. (2005). *Opt. Express* **13**: 82.
105 Zhu, Z., Dawes, A.M., Gauthier, D.J. et al. (2007). *J. Lightwave Technol.* **25**: 201.
106 Zhu, Z., Gauthier, D.J., and Boyd, R.W. (2007). *Science* **318**: 1748.

5

Optical Amplification

In an early stage, the ultimate capacity limits of optic-optic communication systems were determined by the spectral bandwidth of the signal source and of the fundamental fiber parameters: loss and dispersion. However, in the mid-1980s, international development had reached a state at which not only dispersion-shifted fibers but also spectrally pure signal sources were available. In such circumstances, the remaining limitations for long-haul lightwave systems by that time were imposed by fiber loss. These limitations have traditionally been overcome by periodic regeneration of the optical signals at repeaters applying conversion to an intermediate electric signal. However, because of the complexity and high cost of such regenerators, the need for optical amplifiers soon became obvious.

Several means of obtaining optical amplification had been suggested since the 1970s, including direct use of the transmission fiber as the gain medium through nonlinear effects, semiconductor amplifiers, or doping optical waveguides with an active material (rare-earth ions) that could provide gain [1–4]. In this chapter, we provide a review of the main features of such optical fiber amplifiers.

5.1 General Concepts on Optical Amplifiers

An ideal lumped optical amplifier is a transparent box that provides gain and is also insensitive to the bit rate, modulation format, power, and wavelength of the signal passing through it. In practice, however, the optical gain depends not only on the wavelength (or frequency) of the incident signal but also on the electromagnetic field intensity at any point inside the amplifier. Details of wavelength and intensity dependence of the optical signal depend on the amplifying medium.

If the gain medium is modeled as a homogeneously broadened two-level system, the *gain coefficient* (i.e. the gain per unit length) can be written as [5]:

$$g(\nu, P) = \frac{g_0}{1 + \dfrac{(\nu - \nu_0)^2}{\Delta\nu_0^2} + \dfrac{P}{P_{sat}}} \tag{5.1}$$

where g_0 is the peak value of the gain coefficient determined by the pumping level of the amplifier, ν the optical frequency, ν_0 the atomic transition frequency, $\Delta\nu_0$ the 3-dB local gain bandwidth, P the optical power of the signal, and P_{sat} the saturation power, which depends on the gain medium parameters. The parameters $\Delta\nu_0$ and P_{sat} refer to the local gain. However, from the communication system point of view, it is more useful to use the related concepts of *amplifier bandwidth* and *amplifier saturation power*.

Solitons in Optical Fiber Systems, First Edition. Mario F. S. Ferreira.

The *amplifier gain G* is defined as

$$G = \frac{P_{\text{out}}}{P_{\text{in}}} \tag{5.2}$$

where P_{in} is the input power and P_{out} the output power of a continuous wave (CW) signal being amplified. The amplifier gain G may be found by using the relation

$$\frac{dP}{dz} = g(\nu, P)P \tag{5.3}$$

where $P(z)$ is the optical power at a distance z from the amplifier input end.

If the signal power obeys the condition $P << P_{sat}$ throughout the amplifier, the gain coefficient given by Eq. (5.1) can be considered independent of the signal power. In such a case, the amplifier is said to be operated in the unsaturated regime and the solution of Eq. (5.3) is an exponentially growing signal power, given by $P(z) = P_{\text{in}} \exp(gz)$. For an amplifier length L, we then find the linear amplifier gain, given by:

$$G(\nu) = \exp(gL) = \exp\left[\frac{g_0 L}{1 + (\nu - \nu_0)^2/\Delta\nu_0^2}\right] \tag{5.4}$$

Both the amplifier gain $G(\nu)$ and the gain coefficient $g(\nu)$ are maximum when $\nu = \nu_0$. However, $G(\nu)$ decreases much faster than $g(\nu)$ with the signal detuning $\nu - \nu_0$ because of the exponential dependence of G on g. As a consequence, the *amplifier bandwidth* $\Delta\nu_A$ is generally much smaller than the gain bandwidth $\Delta\nu_0$. This is illustrated in Figure 5.1.

An important limitation of an optical amplifier is related with the power dependence of the gain coefficient given by Eq. (5.1). This property is known as *gain saturation* and it appears when the signal power ratio P/P_{sat} is non-negligible. Since the gain coefficient is reduced when the signal power P becomes comparable to the saturation power P_{sat}, the amplifier gain G will also decrease. The *output saturation power* P_{out}^s is defined as the output power for which the amplifier gain G at $\nu = \nu_0$ is reduced by a factor of 2 from its unsaturated value, given by Eq. (5.4). Figure 5.2

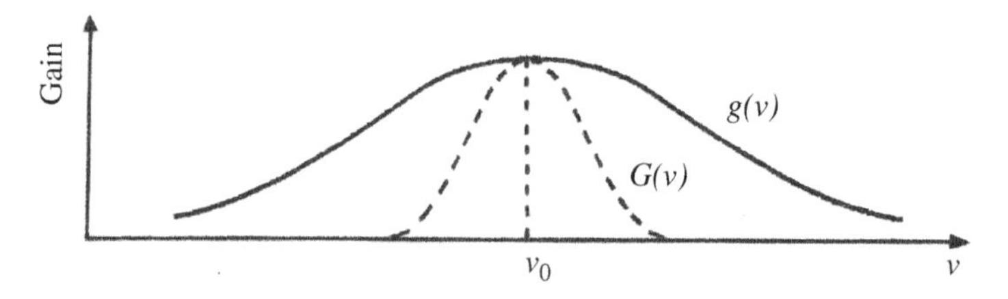

Figure 5.1 Gain coefficient profile $g(\nu)$ and the corresponding amplifier gain spectrum $G(\nu)$.

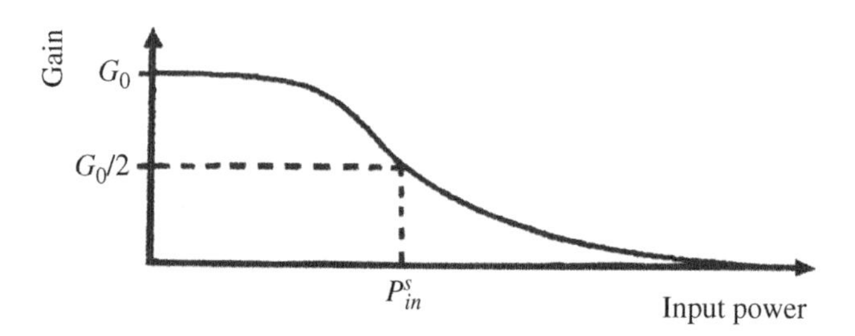

Figure 5.2 Saturated amplifier gain as a function of the input power.

illustrates this concept. In practice, P^{s}_{out} is found to be smaller than P_{sat} by about 30%. Gain saturation can be seen as a serious limitation, particularly for multichannel communication systems.

Besides the bandwidth and gain saturation limitations, the performance of practical optical amplifiers is also limited by their noise characteristics. Actually, optical amplifiers always add spontaneously emitted photons to the signal during the amplification process. These photons are amplified besides the signal photons so that, at the amplifier output, an amplified spontaneous emission (ASE) noise is present. Since spontaneous emission always takes place, ASE noise is unavoidable.

The ASE determines the degradation of the so-called *signal-to-noise ratio* (*SNR*). The *SNR* degradation is commonly characterized by the amplifier *noise figure* (*NF*), which is defined as the *SNR* ratio between input and output:

$$NF = \frac{SNR_{\mathrm{in}}}{SNR_{\mathrm{out}}} \tag{5.5}$$

The SNR is usually referred to as the electrical power generated when the optical signal is converted to electrical current by using a photodetector. Considering a low-noise amplifier, the signal power impinging the photodetector is by far larger than the optical noise power and the shot noise power. As a consequence, the electrical noise due to the signal-*ASE* beating is the dominant contribution and the amplifier NF defined by Eq. (5.5) becomes [5, 6]:

$$NF = 2n_{sp}\frac{G-1}{G} \approx 2n_{sp} \tag{5.6}$$

where n_{sp} is the *spontaneous emission factor* and the approximation holds when the gain is much higher than one. In the case of an ideal amplifier, $n_{sp} = 1$ and Eq. (5.6) shows that the *SNR* is degraded by 3 dB. For most practical amplifiers, NF can be as large as 6–8 dB.

The relative importance of the different limiting factors discussed above depends on the actual amplifier application. There are four main configurations concerning the application of lumped optical amplifiers [1, 4]. The first configuration is to place the amplifier immediately following the laser transmitter to act as a *power amplifier* or *booster*. The main purpose of such amplifiers is to boost the signal power, which can increase the transmission distance by 100 km or more. The second configuration is to place the amplifier in-line and perhaps incorporated at one or more places along the transmission path, replacing the electronic regenerators. Such *in-line amplifier* corrects for periodic signal attenuation and may exist in a cascaded form. The third configuration consists of using the amplifier immediately before the receiver. So, it functions as a *preamplifier*. The purpose of such an amplifier is to improve the receiver sensitivity. The fourth application of lumped optical amplifiers consists of using them for compensating the distribution losses in local area networks (LANs).

5.2 Erbium-Doped Fiber Amplifiers

Many different rare-earth elements, such as erbium, holmium, neodymium, samarium, thulium, and ytterbium, can be used as dopants to realize fiber amplifiers operating at different wavelengths in the range 0.5–3.5 μm. The properties of such amplifiers, namely the operating wavelength and the gain bandwidth, are determined by the dopants rather than by the silica fiber. Among them, erbium-doped fiber amplifiers (EDFAs) have attracted the most attention because they operate in the wavelength region near 1.55 μm, the main window of nowadays' communication systems

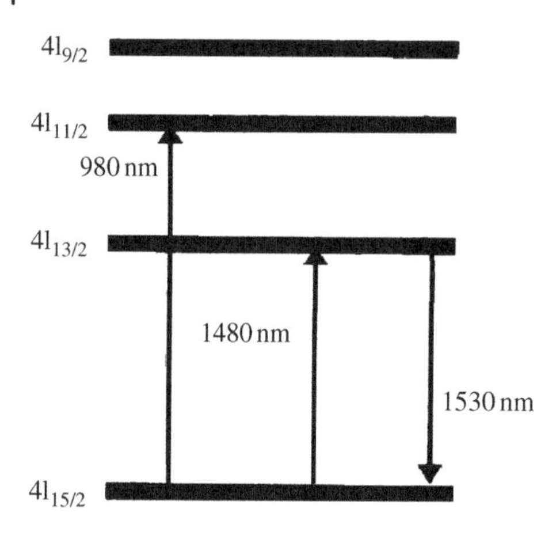

Figure 5.3 Schematic of energy levels of erbium ions in silica fibers.

[7–10]. The spreading in the use of EDFAs is due to their intriguing characteristics such as high gain [11–15], high saturation output power, polarization-independent gain [11], crosstalk absence [14], low NF [15], and low insertion loss.

Erbium-doped fibers are made by incorporating erbium ions into the glass matrix that forms the fiber core. Figure 5.3 illustrates the relevant energy levels of erbium ions in silica fibers. Due to the amorphous nature of silica, each energy level of the erbium ion broadens into a band.

In general, erbium ions act as a three-level system in the amplification of 1550-nm-wavelength light. The three levels involved are the ground state, $4I_{15/2}$, the first excited state, $4I_{13/2}$, and one of the other higher excited states. Figure 5.3 illustrates two possible pumping schemes. In the case of 1480-nm pumping, the upper portion of the $4I_{13/2}$ state is used as a pumping level, and the Er ions are operated as a quasi-three-level system. A gain in the Er-doped fiber (EDF) occurs when an inverted population is achieved between the $4I_{13/2}$ and $4I_{15/2}$ states.

When an erbium ion is excited to higher energy levels, it can decay to lower energy levels either radiatively or nonradiatively. Nonradiative decay involves the creation of phonons, whereas, in radiative decay, photons are emitted. The emission of these photons can assume two forms: spontaneous or stimulated. In the stimulated emission process, the emitted photons are in phase with an incident photon carrying energy equal to the difference in energy of the excited and ground states. This is the process that allows signal amplification to occur and, therefore, is the desired property of the fiber amplifier.

The shape of the gain spectrum is affected considerably by the amorphous nature of silica and by the presence of other co-dopants within the fiber core such as germania and alumina [16–18]. Figure 5.4 shows the gain spectrum of an EDF for various levels of population inversion [9]. Without any inversion (*bottom curve*), the fiber absorbs light. As the inversion increases, gain first appears at the long-wavelength side. At the highest inversion shown (*top curve*), the gain has spread across the entire band. The gain spectrum is quite broad but it is not uniform over the entire bandwidth.

5.2.1 Two-Level Model

Considering the rapid transfer of the pumped population to the first excited state, we can assume that the highest energy level remains practically unpopulated. In this case, the rate equations for the population densities in the ground and excited states can be written in the form

$$\frac{\partial N_1}{\partial t} = \left(\sigma_p^e N_2 - \sigma_p^a N_1\right)\phi_p + \left(\sigma_s^e N_2 - \sigma_s^a N_1\right)\phi_s + \frac{N_2}{T_1} \tag{5.7}$$

$$\frac{\partial N_2}{\partial t} = \left(\sigma_p^a N_1 - \sigma_p^e N_2\right)\phi_p + \left(\sigma_s^a N_1 - \sigma_s^e N_2\right)\phi_s - \frac{N_2}{T_1} \tag{5.8}$$

where T_1 is the spontaneous lifetime of the excited state, σ_j^a and σ_j^e are the absorption, and emission cross sections at the frequency ν_j ($j = s, p$), $\phi_j = P_j/(A_j h\nu_j)$ is the photon flux for the pump and signal waves, P_j being the optical power and A_j the area of the fiber mode cross section. Neglecting the

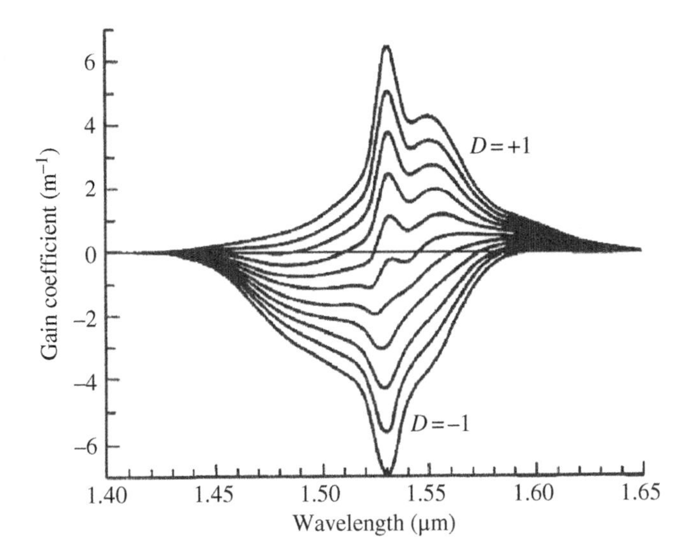

Figure 5.4 Gain spectrum of an Er-doped fiber for various levels of population inversion. *Source:* After Ref. [9].

spontaneous emission effect, the evolution equations for the optical power of the pump and signal waves are given by

$$\frac{\partial P_s}{\partial z} = \Gamma_s(\sigma_s^e N_2 - \sigma_s^a N_1)P_s - \alpha_s P_s \tag{5.9}$$

$$s\frac{\partial P_p}{\partial z} = \Gamma_p\left(\sigma_p^e N_2 - \sigma_p^a N_1\right)P_p - \alpha_p P_p \tag{5.10}$$

where α_j ($j = s, p$) is the fiber loss and Γ_j is the confinement factor at the signal and pump frequencies. The factor $s = \pm 1$ in Eq. (5.10) depends on the direction of propagation of the pump wave. The effect of fiber loss can be neglected in the case of lumped amplifiers since the fiber length is short enough in such a case.

Equations (5.7)–(5.10) can be solved analytically and the powers P_s and P_p at the fiber output obtained after assuming some reasonable approximations. In particular, when the signal power inside the amplifiers is relatively low, the ϕ_s terms in Eqs. (5.7) and (5.8) can be neglected. This corresponds to the so-called small-signal regime, in which the gain coefficient $g(z) = \sigma_s^e N_2 - \sigma_s^a N_1$ does not depend on the signal power. In this case, Eq. (5.9) can be easily integrated, providing the following result for the total gain G of an EDFA of length L:

$$G = \exp\left[\Gamma_s \int_0^L [g(z) - \alpha_s]dz\right] \tag{5.11}$$

Figure 5.5 shows the dependence of the small-signal gain at 1550 nm on the pump power and amplifier length. We observe from Figure 5.5a that, for a given amplifier length, the amplifier gain begins increasing exponentially with the pump power, but such increase becomes much slower when the pump power is above a given value. On the other hand, considering a fixed pump power, Figure 5.5b shows that the amplifier gain is maximum at an optimum amplifier length L, decreasing for longer values of L.

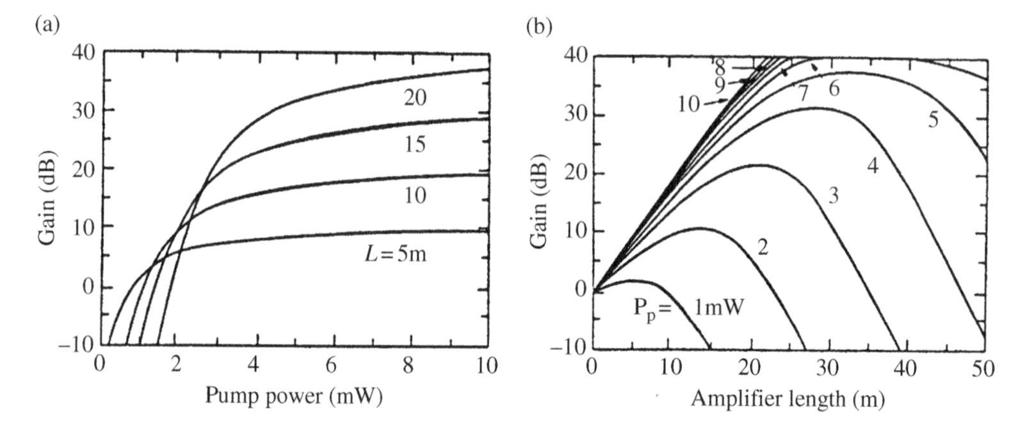

Figure 5.5 Small-signal gain as a function of (a) pump power and (b) amplifier length for an EDFA pumped at 1480 nm. *Source:* After Ref. [18].

A typical EDFA configuration consists of an EDF, a pumping light source, and a wavelength selective coupler (WSC). Polarization-independent optical isolators are also generally used in order to reject any unwanted scattered light. Pump light is input by way of a WSC, which can be configured for forward, backward, or bidirectional pumping. A better NF can be obtained by the copropagating pump scheme as compared to that of the counterpropagating pump. However, the counterpropagating scheme is suitable as a booster application since a higher conversion efficiency is obtained with this scheme. Figure 5.6 illustrates the general configuration of an EDFA with bidirectional pumping. A bandpass filter is usually added to the output path to eliminate the ASE light.

Depending on the intended use for the amplifier, the EDFA can operate in different regimes. In the small-signal or linear regime, low input signal levels (<1 μW) are amplified with negligible gain saturation. Using an optimum amplifier length, typical EDFA gains of 25–35 dB can be achieved in this regime. In the saturation regime, the input signal level is high enough to cause a measurable reduction in the net gain. This is typical of power amplifier applications, in which input signal levels are relatively high.

Semiconductor pump sources are generally used to pump EDFAs in the 800-, 980-, and 1480-nm absorption bands. It has been noted, however, that in the 800-nm band, the occurrence of the excited state absorption (ESA) phenomenon becomes significant. In this case, once an ion is excited through the absorption of a pump photon, it relaxes on a metastable state, where, instead of causing

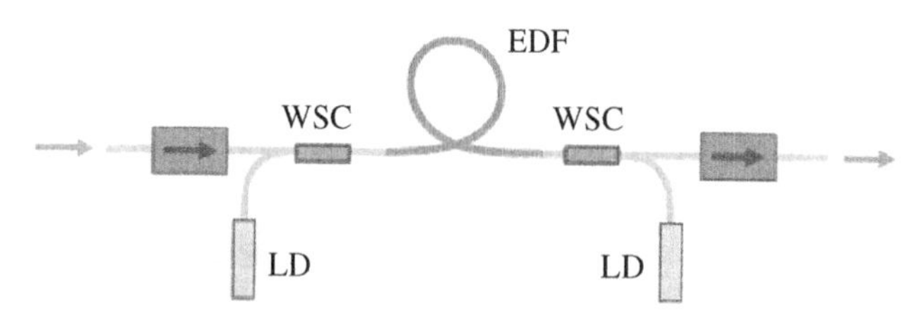

Figure 5.6 General erbium-doped fiber amplifier (EDFA) configuration showing bidirectional pumping. EDF: erbium-doped fiber; WSC: wavelength selective coupler; LD: laser diode.

stimulated emission by decaying toward the ground-state, another pump photon is absorbed so as to make the ion jump to a different excited state. This phenomenon strongly reduces the pumping efficiency at 800 nm.

The candidates for pump wavelength are, therefore, currently restricted to either 980 or 1480 nm, and gains in excess of 30 dB can be obtained in both cases. The gain efficiency of a rare-earth-doped fiber is defined as the ratio of the maximum small signal gain to the input pump power, using the optimized fiber length. EDFA efficiencies are typically on the order of 10 dB/mW for pumping at 980 nm [19]. Moreover, since an NF very close to 3 dB can be achieved in this case, the 980-nm pumping solution is also the preferred choice in cases where very low NFs are required [20]. For pumping at 1480 nm, efficiencies are about half the values obtainable at 980 nm, and require about twice the fiber length [9]. NFs down to 5 dB are generally obtainable at this pump wavelength [20].

Depending on the fiber parameters, the 3-dB bandwidth of a standard EDFA in a small-signal amplification regime ranges from 25 to 44 nm, which enables its use to amplify several channels simultaneously, as required for dense wavelength-division multiplexing (DWDM). The same feature could allow its use also in the L-band of the fiber transmission window. However, the nonuniform gain spectrum of EDFA causes problems when multiple EDFAs are cascaded in the system. Actually, if several channels are located on the relatively flat shoulder region of the gain spectrum, then the gain differential after a single amplifier will be within a few decibels. However, when a cascade of EDFAs is used to periodically compensate for losses, the differential in gain and resultant optical signal-to-noise ratio (OSNR) can become quite severe. This feature could limit the usable bandwidth of EDFAs and, hence, the amount of data transmission by the system. Several approaches have been used to mitigate this limitation, namely by using gain-equalizing filters, long-period fiber grating, chirped fiber Bragg grating, and so on [21–23].

Extensive research has been done during recent years in order to enhance the performance of EDFAs. This has been done using various glass host and co-dopant materials, such as fluorozirconate, chalcogenides, bismuth, aluminum, phosphorus, lanthanum, etc. [24–26]. Very recently, an EDFA co-doped with zirconia-yttria–alumina-barium (ZYAB) with a length of only 1 m has been demonstrated with a flat gain band of 25 dB with a variation of less than 3 dB, a width of 40 nm, and an NF less than 4.2 dB [27]. Such ZYAB-EDFA was shown to be suitable for optical amplification in C + L band [27].

5.3 Fiber Raman Amplifiers

As discussed in Chapter 4, stimulated Raman scattering (SRS) is likely to occur in fibers, producing appreciable amplification for downshifted signals when adequate pump power levels are used. Raman amplification in optical fibers was demonstrated early in the 1970s by Stolen and Ippen [28]. The benefits from Raman amplification were elucidated by many research papers in the mid-1980s [29–35]. However, due principally to its poor pumping efficiency and the scarcity of high power pumps at appropriate wavelengths, much of that work was overtaken by EDFAs by the late 1980s. Then, in the mid-1990s, the development of suitable high-power pumps and the availability of higher Raman gain fibers motivated a renewed interest in Raman amplification.

The most important feature of fiber Raman amplifiers (FRAs) is their capability to provide gain at any signal wavelength, as opposed to fiber amplifiers based on the doped ions in the fibers. The position of the gain bandwidth within the wavelength domain can be adjusted simply by tuning the pump wavelength. Thus, Raman amplification potentially can be achieved in every region of the transmission window of the optical fiber.

As seen in Chapter 4, the Raman gain in silica fibers extends over a frequency range of about 40 THz. Optical signals whose bandwidths are of this order or less can be amplified using the Raman effect if a pump wave with the right wavelength is available. The Raman gain depends on the relative state of polarization of the pump and signal fields. The peak value of the Raman gain decreases with increasing the pump wavelength and it is about 6×10^{-14} m/W in the wavelength region around 1.5 μm.

Oxide glasses used as dopants of pure silica in fiber manufacture exhibit Raman gain coefficients that can be significantly higher than that of silica. The most important dopant in communications fiber is GeO_2, whose Raman gain coefficient is about 8.2 times that for pure SiO_2 glass. The Raman gain coefficient increases above that of pure SiO_2 in proportion to the concentration GeO_2 [36]. Besides, doping the core of a silica fiber with GeO_2 also raises the relative index difference, Δ, between the core and cladding, which increases the effective waveguiding and, in turn, reduces the effective cross-sectional area A_{eff} [37]. As a consequence, the rate of stimulated Raman scattering is increased. However, doping with GeO_2 also increases the fiber losses and this effect must be considered together with the improvement in amplifier gain. Higher Raman gains have been observed in several non-silica materials and structures, including fluoride [38, 39], tellurite [40–43], chalcogenide [44, 45], and photonic crystal fibers [46, 47], and have put it into practice [48–51].

The interaction between the pump and signal waves is governed by the following coupled equations [52]:

$$\frac{dP_s}{dz} = -\alpha_s P_s + \frac{g_R}{A_{eff}} P_p P_s \tag{5.12}$$

$$\frac{dP_p}{dz} = -\alpha_p P_p - \frac{\omega_p}{\omega_s} \frac{g_R}{A_{eff}} P_s P_p \tag{5.13}$$

where A_{eff} is the effective cross-sectional area of the fiber mode, α_s and α_p are absorption coefficients which account for the fiber loss at the signal and pump frequencies, respectively, and the signal wave is considered to be copropagating with the pump wave. Assuming that $\alpha_s = \alpha_p = \alpha$ (which is a reasonable approximation around the 1.5 μm wavelength region in a low loss fiber), the following approximate solutions of Eqs. (5.12) and (5.13) can be written when $P_p(0) >> P_s(0)$:

$$P_p(z) = \frac{P_p(0)\exp(-\alpha z)}{1 + F(z)} \tag{5.14}$$

$$P_s(z) = \frac{P_s(0)\exp(\Gamma(1 - e^{-\alpha z}) - \alpha z)}{1 + F(z)} \tag{5.15}$$

where

$$\Gamma = \frac{g_R P_p(0)}{\alpha A_{eff}} \tag{5.16}$$

and

$$F(z) = \frac{\omega_p P_s(0)}{\omega_s P_p(0)} \exp(\Gamma(1 - e^{-\alpha z})) \tag{5.17}$$

The dimensionless parameter $F(z)$ accounts for the effects of pump depletion and the attendant saturation of the gain seen by the signal.

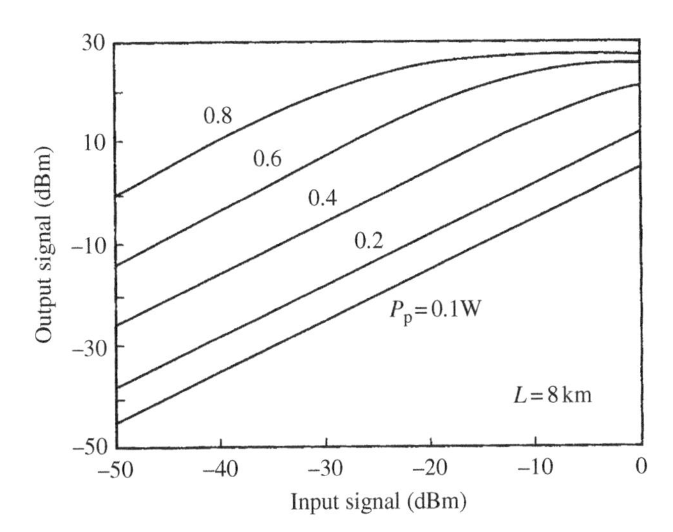

Figure 5.7 Transfer characteristics of a fiber Raman amplifier with length L = 8 km for several values of the input pump power. The amplifier parameters are: g_R = 6.7 × 10^{-14} m.W^{-1}, α = 0.2 dB.km^{-1}, A_{eff} = 30 × 10^{-12} m^2, λ_p = 1.46 μm and λ_S = 1.55 μm.

Figure 5.7 shows the output signal power $P_s(L)$ versus the input signal power $P_s(0)$ for a Raman amplifier with length $L = 8$ km and several values of the input pump power. Typical values $g_R = 6.7 \times 10^{-14}$ m.W^{-1}, $\alpha = 0.2$ dB.km^{-1}, $A_{eff} = 30 \times 10^{-12}$ m^2, $\lambda_p = 1.46$ μm, and $\lambda_s = 1.55$ μm were assumed. The transfer characteristics are linear for $P_p(0) = 0.1$, 0.2, and 0.4 W, but the effects of pump depletion become discernible for input signal levels $P_s(0) > -15$ dBm when $P_p(0) = 0.6$ W. For a pump power $P_p(0) = 0.8$ W, the linear behavior of the Raman amplifier cannot be observed even for signal power levels as low as −25 dBm.

Since, in the absence of Raman amplification, the signal power at the amplifier output would be $P_s(L) = P_s(0) \exp(-\alpha L)$, the amplifier gain is given by

$$G_R = \frac{P_s(L)}{P_s(0)\exp(-\alpha L)} = \frac{\exp\left(g_R P_p(0) L_{eff}/A_{eff}\right)}{1 + F(L)} \tag{5.18}$$

where L_{eff} is the effective interaction length, given by Eq. (4.8). The amplifier gain is seen to be a function of the input signal power (i.e. a saturation nonlinearity) through the term $F(L)$.

When $F(L) >> 1$, we have from Eq. (5.15) that

$$P_s(L) = \frac{\omega_s}{\omega_p} P_p(0) \exp(-\alpha L) \tag{5.19}$$

In this case, the output signal reaches the pump level irrespective of the input signal level. This implies that any spontaneous Raman scattered in the fiber will be amplified up to power levels comparable to that of the pump, which must be avoided in practice.

On the other hand, when $F(L) << 1$, we have from Eq. (5.15) that

$$P_s(L) = P_s(0) \exp\left(g_R L_{eff} P_p(0)/A_{eff} - \alpha L\right) \tag{5.20}$$

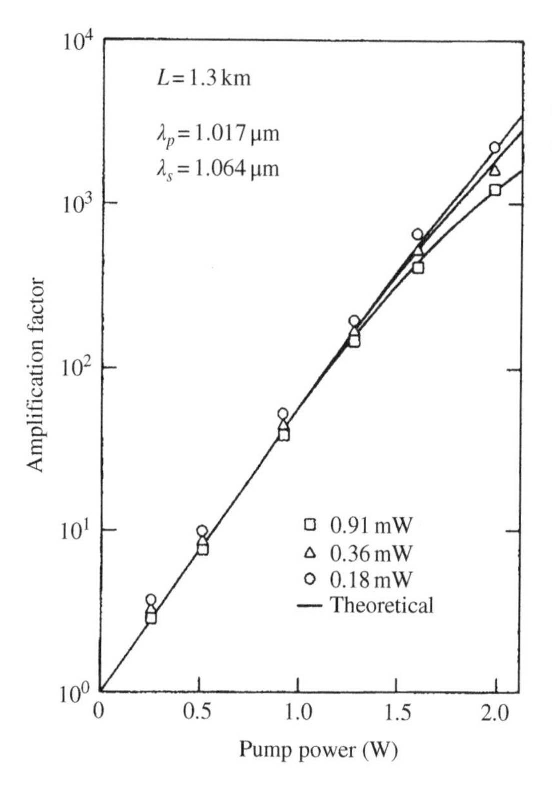

Figure 5.8 Experimentally observed variation of amplifier gain with the pump power operating at 1.064 μm by using a 1.017 μm pump for several values of the input signal power for a 1.3-km FRA. *Source:* After Ref. [53].

In this case, the amplifier gain is

$$G_R = \frac{P_s(L)}{P_s(0)\exp(-\alpha L)} = \exp\left(g_R L_{eff} P_0 / A_{eff}\right) \tag{5.21}$$

From Eq. (5.21), the Raman gain in decibels is expected to increase linearly with the pump input power. This fact was confirmed experimentally, as illustrated in Figure 5.8 [53]. The beginning of saturation of the amplifier gain can be observed in Figure 5.8 for high values of the pump and signal powers.

Figure 5.9 shows schematically a fiber Raman amplifier. The pump and the signal beams are injected into the fiber through a WDM fiber coupler. The case illustrated in Figure 5.9 shows the two beams co-propagating inside the fiber, but the counter-propagating configuration is also possible. In fact, the SRS can occur in both directions, forward and backward. It was confirmed experimentally that the Raman gain is almost the same in the two cases.

Considering its broad bandwidth (>5 THz), FRAs can be used to amplify several channels simultaneously in a multichannel communication system. The same characteristic also makes such

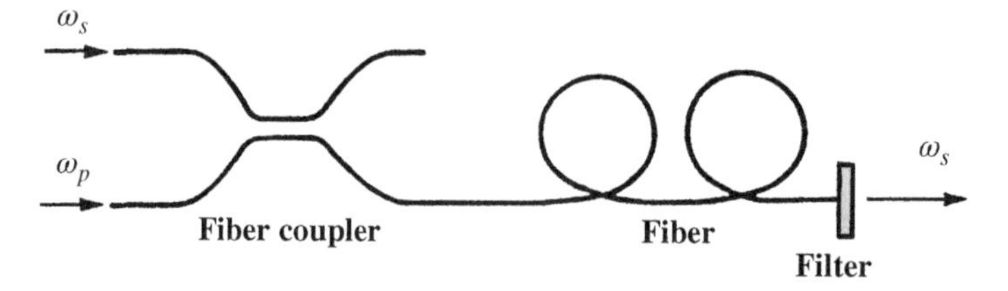

Figure 5.9 Schematic of a fiber Raman amplifier.

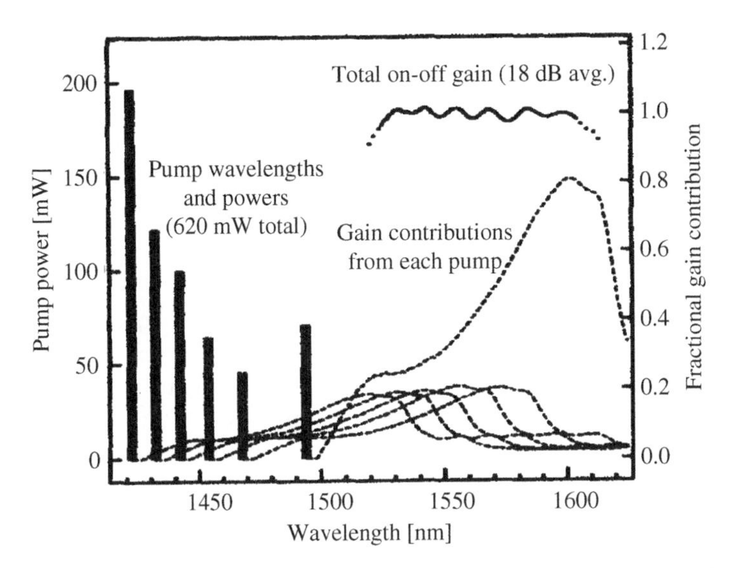

Figure 5.10 Numerical example of a broadband Raman gain obtained using a broad pump spectrum to pump a nonzero dispersion fiber (NZDF). Bars show the pump wavelengths and input powers. The solid line shows the total small signal on-off gain. Dashed lines show the fractional gain contribution from each pump wavelength. *Source:* After Ref. [56].

amplifiers suitable for the amplification of short optical pulses. Moreover, the Raman gain spectrum can be shaped by combining multiple pump wavelengths. In practice, CW power at several wavelengths from a set of high-power semiconductor lasers located at the pump stations [54]. The wavelengths of pump lasers should be in the vicinity of 1450 nm for amplifying optical signals in the 1550 nm spectral region. These wavelengths and pump-power levels are chosen to provide a uniform gain over the entire C band. Fiber Raman amplifiers with gain bandwidths greater than 100 nm were demonstrated [55].

Figure 5.10 shows a numerical example of a broadband Raman gain obtained using a broad pump spectrum to pump a nonzero dispersion fiber (NZDF). The short wavelength pumps amplify the longer wavelengths, and so more power is typically needed at the shortest wavelengths; this is indicated by the height of the bars in Figure 5.10.

When using a broad pump spectrum, an important issue is an interaction between the pumps, which affects the noise properties of the amplifier. Problems arise particularly due to four-wave mixing (FWM) between the pumps since it can create light at new frequencies within the signal band. This new light can interfere with the signal channels, producing beat noise [56].

Another problem can arise from stimulated Brillouin scattering (SBS). Since the intrinsic gain coefficient for stimulated SBS is two orders of magnitude larger than that for SRS, in some cases, the SBS may occur at lower pump powers and may affect significantly the SRS process [57–59]. In particular, the Raman gain may not only be reduced but also become unstable. However, the SBS can be suppressed if a pump laser with a linewidth broader than the Brillouin bandwidth is employed.

The dominant noise light in fiber Raman amplification is due to the amplified spontaneously Raman-scattered light. In fact, a part of the pump energy is spontaneously converted into Stokes radiation extending over the entire bandwidth of the Raman gain spectrum and is amplified together with the signal. The output thus consists not only of the desired signal but also of background noise extending over a wide frequency range (~10 THz or more).

It has been shown that, for both forward and backward pumping, the noise light power is equivalent to a hypothetical injection of a single-photon per unit frequency at the fiber input end for forward pumping, and at some distance away from the fiber output for backward pumping. Assuming that $G_R >> 1$ and $\alpha L << 1$, the noise light powers for forward pumping, $P_{fasp}(L)$, and backward pumping, $P_{basp}(L)$, are approximately given by [60, 61]

$$P_{fasp}(L) \approx h\nu_s \Delta\nu_R (G_R - 1) \exp(-\alpha L) \tag{5.22}$$

$$P_{basp}(L) \approx h\nu_s \Delta\nu_R (G_R - 1)/\ln G_R \tag{5.23}$$

where $\Delta\nu_R$ is the Raman gain bandwidth, h is Planck's constant and ν_s is the signal frequency. We can observe from Eqs. (5.22) and (5.23) that the noise light power in the case of forward pumping decreases as the fiber length is increased. However, in the case of backward pumping, it depends mainly on the Raman gain, being nearly independent of fiber length and loss.

Raman amplifiers can be realized considering two main options. One is the lumped Raman amplifier, in which all the pump power is confined to a relatively short fiber element that is inserted into the transmission line to provide gain. The primary use of the lumped amplifier is to open new wavelength bands between about 1280 and 1530 nm, a wavelength range that is inaccessible by EDFAs. The other option is the distributed Raman amplifier (DRA), which utilizes the transmission fiber itself as the Raman gain medium to obtain amplification. This option has the merit of reducing the overall excursion experienced by the signal power. Consequently, nonlinear effects are reduced at higher signal levels, whereas the SNR remains relatively high at lower signal levels.

Much work has been carried out during the last years in order to extend the operating wavelength of fiber lasers and amplifiers toward the longer mid-infrared wavelength region [62–64], driven by a large number of promising applications including light detection and ranging (LIDAR), gas sensing, and optical communication. Thulium-doped fiber lasers and amplifiers operate efficiently between 1.8 and 2.1 μm wavelength regions [65–68], whereas holmium-doped fiber lasers and amplifiers have been demonstrated to be efficient at >2.1 μm wavelength [69–71]. Raman fiber lasers or amplifiers constitute a promising alternative for operation at >2.1 μm wavelength. Actually, the SRS has permitted the development of a wide variety of fiber lasers and amplifiers at wavelengths for which there are no rare-earth-doped gain media available [72–74]. The principle is to wavelength convert the output of a rare-earth-doped fiber laser or amplifier to the required output wavelength by using the first-order or cascaded Raman Stokes shifts [75]. Cascaded Raman wavelength shifting up to the fourth order ranging from 2.1 to 2.4 μm in a low-loss chalcogenide fiber has been demonstrated [76].

In recent years, few-mode DRAs (FM-DRAs) have attracted increasing attention [77–80] since they are an attractive amplification solution for long-haul mode-division multiplexing (MDM) transmission based on few-mode fibers (FMFs) [81–83]. Actually, compared with few-mode erbium-doped fiber amplifiers (FM-EDFAs), FM-DRAs have some fundamental advantages, such as simple configuration, broad gain bandwidth, flexible operation window, and low NF [84].

5.4 Fiber Parametric Amplifiers

Fiber-optical parametric amplifiers (FOPAs) are based on the four-wave mixing effect discussed in Chapter 4. A FOPA may be tailored to operate at any wavelength and offers a wide gain bandwidth [85–90]. Moreover, it can operate with a very low NF [91].

Using the result given by Eq. (4.40) for the pump fields and the transformation

$$V_j = U_j \exp\{-2i\gamma(P_1 + P_2)z\},\ (j = 3, 4) \tag{5.24}$$

we obtain from Eqs. (4.37) and (4.38) the following coupled equations governing the evolution of the signal and idler waves:

$$\frac{dV_3}{dz} = 2i\gamma\sqrt{P_1P_2}\exp(-i\kappa z)V_4^* \tag{5.25}$$

$$\frac{dV_4^*}{dz} = -2i\gamma\sqrt{P_1P_2}\exp(i\kappa z)V_3 \tag{5.26}$$

where the effective phase mismatch κ is given by Eq. (4.42). If only the signal and the pumps are launched at $z = 0$, i.e. assuming that $V_4^*(0) = 0$, Eqs. (5.25) and (5.26) have the following solutions:

$$V_3(z) = V_3(0)(\cosh(gz) + (i\kappa/2g)\sinh(gz))\exp(-j\kappa z/2) \tag{5.27}$$

$$V_4^*(z) = -i(\gamma/g)\sqrt{P_1P_2}V_3(0)\sinh(gz)\exp(j\kappa z/2) \tag{5.28}$$

where g is the parametric gain coefficient, given by

$$g = \sqrt{4\gamma^2P_1P_2 - \left(\frac{\kappa}{2}\right)^2} \tag{5.29}$$

The maximum gain occurs for perfect phase matching ($\kappa = 0$) and is given by

$$g_{max} = 2\gamma\sqrt{P_1P_2} \tag{5.30}$$

From Eq. (5.27), we can write the following result for the signal power, $P_3 = |V_3|^2$:

$$P_3(z) = P_3(0)\left[1 + \left(1 + \frac{\kappa^2}{4g^2}\right)\sinh^2(gz)\right] = P_3(0)\left[1 + \frac{4\gamma^2P_1P_2}{g^2}\sinh^2(gz)\right] \tag{5.31}$$

The idler power $P_4 = |V_4|^2$ can be obtained by noting from Eqs. (5.27) and (5.28) that $P_3(z) - P_4(z) = constant = P_3(0)$. The result is:

$$P_4(z) = P_3(0)\frac{4\gamma^2P_1P_2}{g^2}\sinh^2(gz) \tag{5.32}$$

The unsaturated single-pass gain of a FOPA of length L becomes,

$$G_P = \frac{P_3(L)}{P_3(0)} = 1 + \frac{4\gamma^2P_1P_2}{g^2}\sinh^2(gL) \tag{5.33}$$

According to Eq. (5.29), amplification ($g > 0$) occurs only in the case

$$|\kappa| < 4\gamma\sqrt{P_1P_2} \tag{5.34}$$

Since $|\kappa|$ must be small, this also means good phase matching. For $|\kappa| > 4\gamma\sqrt{P_1P_2}$, the parametric gain becomes imaginary and there is no longer amplification, but rather a periodical power variation of the signal and idler waves.

In the case of a single pump, the effective phase mismatch is given by Eq. (4.43) and the parametric gain coefficient becomes:

$$g^2 = \left[(\gamma P_p)^2 - (\kappa/2)^2\right] = -\Delta k\left(\frac{\Delta k}{4} + \gamma P_p\right) \tag{5.35}$$

where P_p is the pump power. The maximum value of the parametric gain occurs when $\Delta k = -2\gamma P_p$ and is given by

$$g_{\max} = \gamma P_p \tag{5.36}$$

The unsaturated single-pass gain" may be written as

$$G_p = \frac{P_3(L)}{P_3(0)} = 1 + \left[\frac{\gamma P_p}{g}\sinh(gL)\right]^2 = 1 + (\gamma P_p L)^2\left[1 + \frac{gL^2}{6} + \frac{gL^4}{120} + \ldots\right]^2 \tag{5.37}$$

For signal wavelengths close to the pump wavelength, $\Delta k \approx 0$ and

$$G_p \approx 1 + (\gamma P_p L)^2 \tag{5.38}$$

In this case of perfect phase matching ($\kappa = 0$) and assuming that $\gamma P_p L \gg 1$, the unsaturated single-pass gain becomes

$$G_p \approx \sinh^2(gL) \approx \frac{1}{4}\exp(2\gamma P_p L) \tag{5.39}$$

A very simple expression for the FOPA peak gain may be obtained if Eq. (5.39) is rewritten in decibel units

$$G_p(\text{dB}) = 10\log_{10}\left[\frac{1}{4}\exp(2\gamma P_p L)\right] = P_p L S_P - 6 \tag{5.40}$$

where

$$S_P = 10\log_{10}(e^2)\gamma \approx 8.7\gamma \tag{5.41}$$

is the parametric gain slope in [dB/W/km].

Figure 5.11 shows the small signal gain against the pump power provided by a parametric amplifier constituted by a 500-m-long highly nonlinear fiber (HNLF) with $\gamma = 11\ \text{W}^{-1}/\text{km}$ [87, 89]. The gain slope in the undepleted pump region is 100 dB/W/km. The input signal power was −19.5 dBm and the maximum achieved output signal power was 21 dBm, limited by signal-induced stimulated Brillouin scattering. Figure 5.12 shows the amplifier gain versus the wavelength difference between pump and signal for the same fiber of Figure 5.11, considering a pump power of 1.4 W. The region

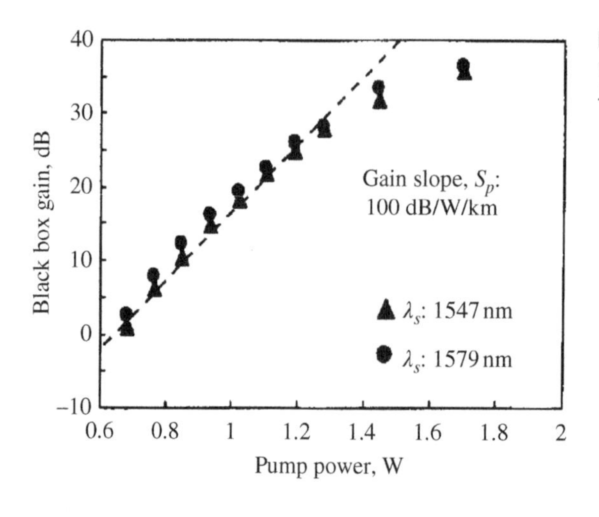

Figure 5.11 Small signal parametric gain using a 500 m long HNLF with $\gamma = 11\ \text{W}^{-1}/\text{km}$. *Source:* After Ref. [89].

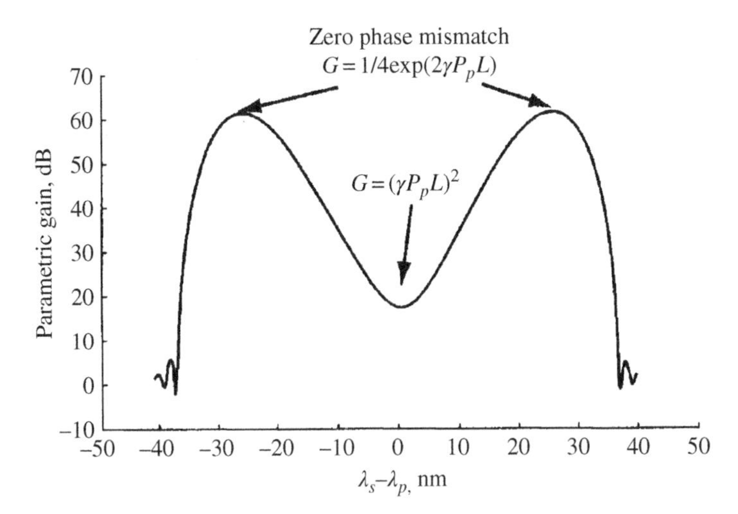

Figure 5.12 Calculated amplifier gain versus the wavelength difference between pump and signal for the same fiber of Figure 5.11, considering a pump power of 1.4 W. *Source:* After Ref. [89].

with exponential gain (corresponding to perfect phase matching) and the region with quadratic gain (corresponding to $\lambda_3 \approx \lambda_1$) are marked in the figure.

Considering the approximation given by Eq. (4.44) for the linear phase mismatch, the FOPA bandwidth $\Delta\Omega$ can be obtained from the maximum effective phase mismatch,

$$\kappa_m = \beta_2(\Omega_s + \Delta\Omega)^2 + 2\gamma P_p \tag{5.42}$$

with $\Delta\Omega \ll \Omega_s$ and Ω_s given by Eq. (4.46). This maximum value of the effective phase mismatch occurs when the parametric gain in Eq. (5.35) vanishes, which gives $\kappa_m = 2\gamma P_p$. In these circumstances, the FOPA bandwidth is approximately given by

$$\Delta\Omega \approx \sqrt{\frac{\gamma P_p}{2|\beta_2|}} \tag{5.43}$$

Equation (5.43) shows that the FOPA bandwidth can be increased by increasing the nonlinear parameter γ and reducing $|\beta_2|$. This is the reason why modern FOPAs use generally HNLFs and the pump wavelengths are chosen near the zero-dispersion wavelength (ZDW) of the fiber. Moreover, Eq. (5.37) shows that, for a fixed value of the FOPA gain, the value of $\gamma P_p L$ must be kept constant. In such a case, the amplifier bandwidth will increase with decreasing L. Using an HNLF provides the possibility of simultaneously decreasing L and increasing the nonlinear parameter γ. For example, a value $\gamma P_p L = 10$ can be achieved with a pump power of 1 W and an HNLF with $L = 1$ km and $\gamma = 10\ \text{W}^{-1}$/km. The bandwidth will be increased 20 times using, for instance, a fiber length of 50 m, a pump power of 5 W and a nonlinear parameter $\gamma = 40\ \text{W}^{-1}$/km. Such a high value of γ can be achieved using some types of HNLFs.

As the FOPA bandwidth is proportional to the square root of the pump power, it could be, in principle, arbitrarily increased if enough optical power is available. However, such high values of the pump power are limited in practice by stimulated Brillouin scattering (SBS). SBS shows a reduced threshold in the case of HNLFs due to the small effective core area of this type of fibers.

The limitations imposed by the SBS process can be partially circumvented if the pump power is distributed between two pumps, instead of being concentrated in one pump. Moreover, a two-pump FOPA offers additional degrees of design freedom which makes it fundamentally different from the conventional one-pump device [92–97].

The polarization sensitivity is an important issue of any fiber communication device. This is indeed the main problem of one-pump FOPA, which presents a high degree of polarization sensitivity. In the case of a two-pump FOPA, the gain is maximized when the pump and signal are copolarized along the entire interaction length. Any deviation from such copolarized state determines a significant reduction of the parametric gain, which can become negligible for a signal polarization orthogonal to two copolarized pumps [98]. A simple and elegant solution to this problem can be achieved using orthogonally multiplexed pumps [98–101]. This configuration results in the near polarization invariance of the parametric process. However, such an improvement is achieved at the expense of a significant reduction of the parametric gain, compared with the copolarized two-pump scheme [94]. In a 2017 experiment, polarization-division multiplexed (PDM) DWDM data transmission was demonstrated for the first time in a range of systems incorporating a net-gain polarization-insensitive (PI) FOPA for loss compensation. The PI-FOPA comprised a modified diversity-loop architecture to achieve 15 dB net-gain, and up to 2.3 THz (~18 nm) bandwidth [102, 103].

5.5 Lumped versus Distributed Amplification

Optical amplifiers can be divided into two categories: lumped and distributed amplifiers. In the first case, the fiber losses are compensated using some few meters-long special fibers. These include both EDFAs and lumped Raman fiber amplifiers, as described in Sections 5.2 and 5.3. In contrast, a distributed amplifier uses the transmission fiber itself for signal amplification. In this case, pump lasers are included periodically along the transmission link in order to inject optical power at suitable wavelengths.

When using a lumped amplification scheme, the length of the fiber amplifier is usually much shorter than the amplifier spacing, L_A. After propagating this distance, the average signal power is reduced from its initial value P_0 to $P(L_A) = P_0 \exp(-\alpha L_A)$. For example, this means a reduction of 20 dB for a standard 100-km-long fiber, presenting a minimum attenuation of 0.2 dB/km. In order to compensate for this loss, the lumped amplifier gain must be $G = \exp(\alpha L_A)$. In this case, there is a significant variation of the signal power, which shows a sawtooth-like profile. This behavior is illustrated by the dotted curve in Figure 5.13.

Backward distributed Raman amplification can be used to assist the lumped amplifier in order to reduce the overall excursion of the signal power. This scheme provides two main advantages. On one hand, the top signal level is reduced, thus minimizing the impact of nonlinear effects. On the other hand, as the signal does not dip down so profoundly, the signal-to-noise ratio remains higher than when using only lumped amplifiers. Actually, a further reduction of the signal power excursion can be achieved using distributed Raman amplification with a bidirectional pumping scheme, as illustrated by the dashed curve in Figure 5.13.

The propagation of an optical pulse along a lossy fiber in the presence of optical amplification can be described by the following generalized nonlinear Schrödinger (NLS) equation:

$$i\frac{\partial U}{\partial z} - \frac{1}{2}\beta_2\frac{\partial^2 U}{\partial \tau^2} + \gamma|U|^2 U = i\frac{1}{2}[g(z) - \alpha]U \tag{5.44}$$

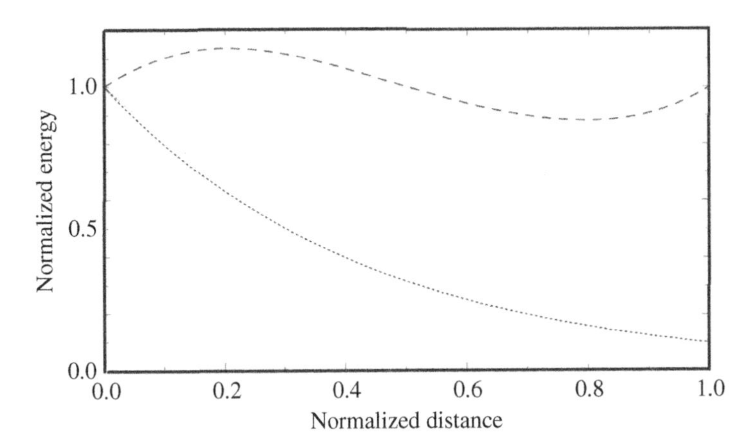

Figure 5.13 Normalized signal energy against the normalized distance, z/L_A. Comparison is made between a purely lumped amplified system (dotted curve) and a system using distributed amplification with bidirectional pumping (dashed curve).

where g(z) is the gain coefficient. We can write the general solution of Eq. (5.44) in the form:

$$U(z,\tau) = \sqrt{p(z)}V(z,\tau) \tag{5.45}$$

where $p(z)$ describes the variations of the time-averaged power along the link length due to fiber losses and optical amplification. Substituting Eq. (5.45) into (5.44), it is found that $p(z)$ obeys the equation:

$$\frac{dp}{dz} = [g(z) - \alpha]p \tag{5.46}$$

whereas $V(z, \tau)$ satisfies the equation

$$i\frac{\partial V}{\partial z} - \frac{1}{2}\beta_2\frac{\partial^2 V}{\partial \tau^2} + \gamma p(z)|V|^2 V = 0 \tag{5.47}$$

An ideal situation is achieved by designing the gain so that g(z) is constant and equal to α for all z. In such a case, the average power of the optical signal would remain constant along the fiber link. In practice, the gain coefficient cannot be made constant and the fiber losses cannot be compensated everywhere locally since the pump power also suffers from fiber loss. However, the fiber losses over a distance L_A, corresponding to the pump-station spacing, can be fully compensated if the following condition is satisfied:

$$\int_0^{L_A} g(z)dz = \alpha L_A \tag{5.48}$$

This is the condition that any distributed amplification scheme aims to satisfy.

Assuming a bidirectional pumping scheme, the gain coefficient g(z) can be approximated as:

$$g(z) = g_0\left\{ \exp\left(-\alpha_p z\right) + \exp\left[-\alpha_p(L_A - z)\right]\right\} \tag{5.49}$$

where α_p is the fiber loss at the pump wavelength. The gain constant g_0 is related to the pump power injected at both ends. Using Eq. (5.49), the integration of Eq. (5.46) gives the following result for $p(z)$:

$$\ln\left(p(z)\right) = \alpha_0 L_A \left\{ \frac{\sinh\left[\alpha_p(z - L_A/2)\right] + \sinh\left(\alpha_p L_A/2\right)}{2\sinh\left(\alpha_p L_A/2\right)} - \frac{z}{L_A} \right\} \tag{5.50}$$

Equation (5.50) shows that the range of energy variations in a distributed amplification scheme using bidirectional pumping increases with L_A. However, it remains much smaller than that occurring in the lumped-amplification scheme, as illustrated by Fig. 5.

One possibility to achieve distributed amplification is by lightly doping the transmission fiber itself. It has been shown that, in some circumstances, the performance of such "active fibers" can be limited by the occurrence of some nonlinear effects, namely the stimulated Brilloun scattering, as discussed in Section 4.5. In such cases, the presence of distributed gain leads to the ready generation of multiple Stokes orders and to a significant variation of the signal power along the fiber [104, 105].

5.6 Parabolic Pulses

A type of optical pulses with a parabolic profile in both the time and frequency domains have generated considerable interest since their first experimental demonstration [106]. Such pulses propagate in nonlinear optical fibers with normal GVD [107] and in optical fiber amplifiers with constant and distributed gain functions [108–110]. Parabolic pulses are generated asymptotically in the fiber amplifier independent of the shape or the noise properties of the input pulse and propagate self-similarly subject to exponential scaling of amplitude and temporal width. Moreover, they possess a linear chirp which leads to highly efficient pulse compression to the sub-100-fs domain [111].

Pulse evolution in a fiber amplifier in the presence of dispersion, nonlinearity, and gain is governed by the generalized NLS equation given by Eq. (5.44). In writing this equation, we have neglected gain saturation and assumed that the propagating pulse has a spectral bandwidth less than the amplifier bandwidth. The evolution of the pulse energy

$$E(z) = \int_{-\infty}^{\infty} |U(z,\tau)|^2 d\tau \tag{5.51}$$

along the fiber amplifier is given by

$$E(z) = E_0 \exp\left(g' z\right) \tag{5.52}$$

where $E_0 = E(z = 0)$ and $g' = g - \alpha > 0$ is the distributed net gain coefficient, which we assume to be constant.

In the normal dispersion regime ($\beta_2 > 0$), Eq. (5.44) has an asymptotic ($z \rightarrow \infty$) analytical solution given by [109, 112]:

$$U(z,\tau) = U_p(z)\left[1 - \frac{\tau^2}{t_p^2}\right]^{1/2} \exp\left[i\psi_p(z,\tau)\right], \text{ for } |\tau| \le t_p(z) \tag{5.53a}$$

$$U(z,\tau) = 0, \text{ for } |\tau| > t_p(z) \tag{5.53b}$$

The asymptotic pulse amplitude U_p, width t_p, and phase ψ_p are given by:

$$U_p(z) = \frac{1}{2}\left(\frac{g' E_0}{\sqrt{\gamma\beta_2/2}}\right)^{1/3} \exp\left(g' z/3\right) \tag{5.54}$$

$$t_p(z) = \frac{6(\gamma\beta_2/2)^{1/2} U_p(z)}{g'} \tag{5.55}$$

$$\psi_p(z,\tau) = \psi_0 + \frac{3\gamma U_p^2(z)}{2g'} - \frac{g'}{6\beta_2}\tau^2 \tag{5.56}$$

where ψ_0 is an arbitrary constant. Equation (5.53a) represents a parabolic pulse, not only near the pulse center but also well toward the wings up to the point where the amplitude goes to zero. Such parabolic pulse has a linear chirp across its temporal profile, given by:

$$\delta\omega(\tau) = -\frac{\partial\psi_p}{\partial\tau} = \frac{g'}{3\beta_2}\tau, \ |\tau| \le t_p(z) \tag{5.57}$$

Equation (5.55) shows that the pulse width $t_p(z)$ scales linearly with the amplitude $U_p(z)$, which indicates the self-similar nature of this solution [113]. Such asymptotic scaling is related with the exponential growth of the pulse amplitude and width. We also observe from Eqs. (5.53a)–(5.56) that the asymptotic solution is determined only by the energy of the initial pulse, and not by its specific shape or width. These characteristics have been experimentally verified [114].

The theoretical prediction and experimental demonstration of self-similar propagation in a fiber amplifier were reported for the first time in 2000 [108]. Since then, the generation of parabolic pulses has been demonstrated in various fiber-based amplifiers providing the amplification of picosecond or femtosecond pulses in the normal dispersion regime. Both rare-earth-doped amplifiers [108, 115–117] and Raman amplifiers [118] have been used in such demonstrations.

In a 2004 experiment, a 5.3-km-long Raman amplifier was used to amplify 7-ps pulses launched with 2.16-pJ energy. Experimental results agreed well with numerical predictions [119]. Figure 5.14 shows the simulation result illustrating the nonlinear transformation of the initial pulse toward a parabolic pulse during propagation through the 5.3-km Raman fiber amplifier. The solid lines represent the experimental input and output pulse profiles. Figure 5.15 shows the intensity and chirp

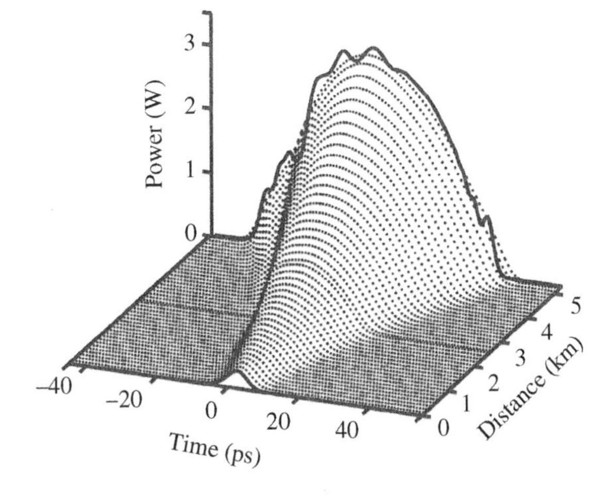

Figure 5.14 Simulation (dashed lines) of parabolic pulse evolution over 5.3 km of a Raman fiber amplifier. FROG retrievals of the intensity profiles of input and output pulses (solid lines). *Source:* After Ref. [119].

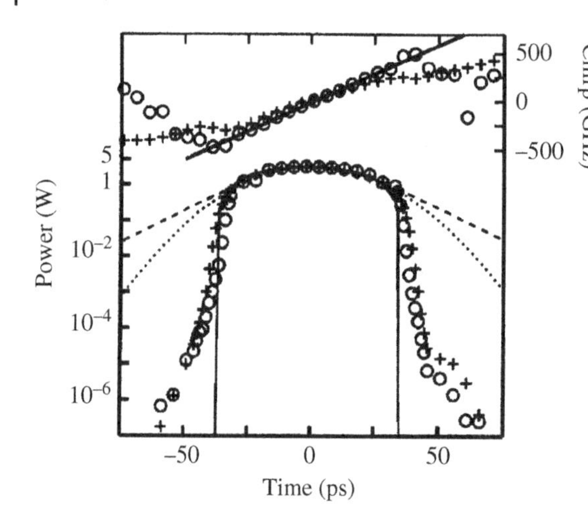

Figure 5.15 Intensity and chirp profiles of the output pulse after 5.3 km propagation along a Raman fiber amplifier. Results from FROG trace retrieval (circles) compared with numerical simulations (crosses). Linear and parabolic fit (solid lines) of, respectively, chirp and intensity profiles. The dashed and dotted lines correspond to Gaussian and $sech^2$ fits, respectively. *Source:* After Ref. [119].

profiles of the output pulse, which are compared to least-squares parabolic and linear fits shown as solid lines. The linearly chirped parabolic nature of pulses at the Raman amplifier output is well demonstrated by the good fits which can be observed. In contrast, Figure 5.15 also shows the comparatively poor fits obtained when using a Gaussian (dotted line) or a sech^2 (dashed line) pulse profile.

Parabolic pulses are of fundamental interest as they represent a particular class of solutions of the NLS equation with gain which is associated with self-similar evolution of the pulse intensity and that resists the deleterious effects of optical wave breaking. This property allows the scaling of fiber amplifiers and lasers to higher power regimes [115, 116]. Moreover, their linear chirp facilitates efficient temporal compression [108, 115, 116]. In a 2002 experiment, amplified parabolic pulses with a 17-W average power were obtained by launching 180-fs pulses with 10-mW average power in a 9-m-long ytterbium-doped fiber amplifier (YDFA) [115]. Pulses broadened to 5.6 ps but could be compressed down to 80 fs because of their linear chirp. By 2005, the highest average power realized with this approach was 131 W [120]. Other experimental studies have exploited the remarkable properties of parabolic pulses to realize optical pulse synthesis [121], 10-GHz telecom multiwavelength sources [122], and optical regeneration of telecom signals [123].

References

1 Agrawal, G.P. (2002). *Fiber-Optic Communication Systems*, 3e. New York: Wiley.

2 Ferreira, M.F. (2004). Basic concepts of optical amplifiers. In: *Encyclopedia of Modern Optics* (ed. B. Guenther, A. Miller, L. Bayvel and J. Midwinter), 171–174. Elsevier Physics/Academic Press.

3 Ferreira, M.F. (2004). Raman, brillouin and parametric amplifiers. In: *Encyclopedia of Modern Optics* (ed. B. Guenther, A. Miller, L. Bayvel and J. Midwinter), 297–308. Elsevier Physics/Academic Press.

4 Ferreira, M.F. (2015). Optical amplifiers. In: *Encyclopedia of Optical and Photonic Engineering*, 2e (ed. C. Hoffman and R. Driggers). CRC Press.

5 Siegman, A.E. (1986). *Lasers*. Mill Valley, CA: University Science Books.

6 Yariv, A. (1990). *Opt. Lett.* **15**: 1064.

7 Digonnet, M.F. (ed.) (1993). *Rare-Earth Doped Fiber Lasers and Amplifiers*. New York: Marcel Dekker.

8 Bjarklev, A. (1993). *Optical Fiber Amplifiers: Design and System Applications*. Boston: Artech House.

9 Desurvire, E. (1994). *Erbium-Doped Fiber Amplifiers*. New York: Wiley.
10 Agrawal, G.P. (2001). *Applications of Nonlinear Fiber Optics*. San Diego, CA: Academic Press.
11 Foschini, G. and Habbab, I. (1995). *J. Lightwave Technol.* **13**: 507.
12 Otani, T., Goto, K., Kawazawa, T. et al. (1997). *J. Lightwave Technol.* **15**: 737.
13 Giles, C., Desurvire, E., Talman, J. et al. (1989). *J. Lightwave Technol.* **7**: 651.
14 Desurvire, E., Giles, C., and Simpson, J. (1989). *J. Lightwave Technol.* **7**: 2095.
15 Lumholt, O., Povlsen, J.H., Schusler, K. et al. (1993). *J. Lightwave Technol.* **11**: 1344.
16 Miniscalco, W. (1991). *J. Lightwave Technol.* **9**: 234.
17 Zyskind, J., Desurvire, E., Sulhoff, J., and DiGiovanni, D. (1990). *IEEE Photon. Technol. Lett.* **2**: 869.
18 Giles, C. and Desurvire, E. (1991). *J. Lightwave Technol.* **9**: 271.
19 Shimizu, M., Yamada, M., Horiguchi, H. et al. (1990). *Electron. Lett* **26**: 1641.
20 Yamada, M., Shimizu, M., Okayasu, M. et al. (1990). *IEEE Photon. Technol. Lett.* **2**: 205.
21 Pal, M., Bandyopadhyay, S., Biswas, P. et al. (2007). *Opt. Quantum Electron.* **39**: 1231.
22 Vengsarkar, A.M., Bergano, N.S., Davidson, C.R. et al. (1996). *Opt. Lett.* **21**: 336.
23 Kim, H.S., Yun, S.H., Kim, H.K. et al. (1998). *IEEE Photonics Technol. Lett.* **10**: 790.
24 Wang, B., Pub, G., Osnato, R., and Palsdottir, B. (2004). *Proc. SPIE* **5280**: 162.
25 Yamada, M., Kanamori, T., Terunuma, Y. et al. (1996). *IEEE Photonics Technol. Lett.* **8**: 882.
26 Jha, A., Shen, S., and Naftaly, M. (2000). *Phys. Rev. B* **62**: 6215.
27 Duarte, J., Paul, M.C., Das, S. et al. (2019). *Opt. Mat. Express* **9**: 2652.
28 Stolen, R.H. and Ippen, E.P. (1973). *Appl. Phys. Lett.* **22**: 276.
29 Ikeda, M. (1981). *Opt. Commun.* **39**: 148.
30 Chraplyvy, A.R., Stone, J., and Burrus, C.A. (1983). *Opt. Lett.* **8**: 415.
31 Nakazawa, M. (1985). *Appl. Phys. Lett.* **46**: 628.
32 Nakazawa, M., Nakashima, T., and Seikai, S. (1985). *J. Opt. Soc. Am. B* **2**: 215.
33 Dakss, M.L. and Melman, P. (1985). *J. Lightwave Technol.* **3**: 806.
34 Olsson, N.A. and Hegarty, J. (1986). *J. Lightwave Technol.* **4**: 391.
35 Aoki, Y., Kishida, S., and Washio, K. (1986). *Appl. Opt.* **25**: 1056.
36 Nakashima, T., Seikai, S., and Nakazawa, M. (1985). *Opt. Lett.* **10**: 420.
37 Shibata, N., Horigudhi, M., and Edahiro, T. (1981). *J. Non.Cryst. Solids* **45**: 115.
38 Cheng, T.L., Gao, W.Q., Xue, X.J. et al. (2017). *Opt. Fiber Technol.* **36**: 245.
39 Fortin, V., Bernier, M., Faucher, D. et al. (2012). *Opt. Express* **20**: 19412.
40 Mori, A., Masuda, H., Shikano, K., and Shimizu, M. (2003). *J. Lightwave Technol.* **21**: 1300.
41 Zhu, G., Geng, L., Zhu, X. et al. (2015). *Opt. Express* **23**: 7559.
42 Liao, M., Yan, X., Gao, W. et al. (2011). *Opt. Express* **19**: 15389.
43 Cheng, T.L., Gao, W.Q., Xue, X.J. et al. (2016). *Opt. Mater. Express* **6**: 3438.
44 Jackson, S.D. and Anzueto-Sánchez, G. (2006). *Appl. Phys. Lett.* **88**: 221106.
45 Bernier, M., Fortin, V., El-Amraoui, M. et al. (2014). *Opt. Lett.* **39**: 2052.
46 Benabid, F., Knight, J.C., Antonopoulos, G. et al. (2002). *Science* **298**: 399.
47 Coen, S., Chau, A.H.L., Leonhardt, R. et al. (2002). *J. Opt. Soc. Am. B* **19**: 753.
48 Qin, G.S., Jose, R., and Ohishi, Y. (2007). *J. Appl. Phys.* **101**: 093109.
49 Shi, J.D., Feng, X., Horak, P. et al. (2011). *J. Lightwave Technol.* **29**: 3461.
50 Vieira, J., Trines, R.M.G.M., Alves, E.P. et al. (2016). *Nat. Commun.* **7**: 10371.
51 Robles, F.E., Zhou, K.C., Fischer, M.C., and Warren, W.S. (2017). *Optica* **4**: 243.
52 Ferreira, M. (2011). *Nonlinear Effects in Optical Fibers*. Hoboken, NJ: John Wiley & Sons.
53 Tomita, A. (1983). *Opt. Lett.* **8**: 412.
54 Namiki, S. and Emori, Y. (2001). *IEEE J. Sel. Topics Quantum Electron.* **7**: 3.
55 Fukai, C., Nakajima, K., Zhou, J. et al. (2004). *Opt. Lett.* **29**: 545.

56 Bromage, J. (2004). *J. Lightwave Technol.* **22**: 79.
57 Foley, B., Dakss, M.L., Davies, R.W., and Melman, P. (1989). *J. Lightwave Technol.* **7**: 2024.
58 Ferreira, M.F., Rocha, J.F., and Pinto, J.L. (1991). *Electron. Lett.* **27**: 1576.
59 Hamidi, S., Simeonidou, D., Siddiqui, A.S., and Chaleon, T. (1992). *Electron. Lett.* **28**: 1768.
60 Aoki, Y. (1989). *Opt. Quantum Electron.* **21**: S89.
61 Aoki, Y. (1988). *IEEE J. Lightwave Technol.* **6**: 1225.
62 Jackson, S.D. (2012). *Nat. Photonics* **6**: 423.
63 Rudy, C.W., Marandi, A., Vodopyanov, K.L., and Byer, R.L. (2013). *Opt. Lett.* **38**: 2865.
64 Liu, K., Liu, J., Shi, H. et al. (2014). *Opt. Express* **22**: 24384.
65 Geng, J., Wang, Q., Luo, T. et al. (2009). *Opt. Lett.* **34**: 3493.
66 Fang, Q., Shi, W., Kieu, K. et al. (2012). *Opt. Express* **20**: 16410.
67 Wang, X., Zhou, P., Wang, X. et al. (2013). *Opt. Express* **21**: 32386.
68 Liu, J., Xu, J., Liu, K. et al. (2013). *Opt. Lett.* **38**: 4150.
69 Jackson, S.D., Sabella, A., Hemming, A. et al. (2007). *Opt. Lett.* **32**: 241.
70 Hemming, A., Simakov, N., Davidson, A., et al. *CLEO, OSA Technical Digest* (online) (Optical Society of America, 2013), paper CW1M.1 (2013).
71 Simakov, N., Hemming, A., Clarkson, W.A. et al. (2013). *Opt. Express* **21**: 28415.
72 Feng, Y., Taylor, L.R., and Calia, D.B. (2009). *Opt. Express* **17**: 23678.
73 Zhang, L., Jiang, H., Cui, S., and Feng, Y. (2014). *Opt. Lett.* **39**: 1933.
74 Zhang, H., Xiao, H., Zhou, P. et al. (2014). *Opt. Express* **22**: 10248.
75 Dianov, E.M., Bufetov, I.A., Mashinsky, V.M. et al. (2004). *Quantum Electron.* **34**: 695.
76 Duhant, M., Renard, W., Canat, G. et al. (2011). *Opt. Lett.* **36**: 2859.
77 Antonelli, C., Mecozzi, A., and Shtaif, M. (2013). *Opt. Lett.* **38**: 1188.
78 Zhou, J. (2014). *Opt. Express* **22**: 21393.
79 Christensen, E.N., Koefoed, J.G., Friis, S.M.M. et al. (2016). *Sci. Rep.* **6**: 34693.
80 Esmaeelpour, M., Ryf, R., Fontaine, N.K. et al. (2016). *J. Lightwave Technol.* **34**: 1864.
81 Richardson, D.J., Fini, J.M., and Nelson, L.E. (2013). *Nat. Photonics* **7**: 354.
82 van Uden, R.G.H., Amezcua Correa, R., Antonio Lopez, E. et al. (2014). *Nat. Photonics* **8**: 865.
83 Li, G., Bai, N., Zhao, N., and Xia, C. (2014). *Adv. Opt. Photonics* **6**: 413.
84 Islam, M.N. (2002). *IEEE J. Sel. Top. Quantum Electron.* **8**: 548.
85 Marhic, M.E., Kagi, N., Chiang, T.-K., and Kazovsky, L.G. (1996). *Opt. Lett.* **21**: 573.
86 Karlsson, M. (1998). *J. Opt. Soc. Am. B* **15**: 2269.
87 Hansryd, J. and Andrekson, P.A. (2001). *IEEE Photon. Technol. Lett.* **13**: 194.
88 Westlund, M., Hansryd, J., Andrekson, P.A., and Knudsen, S.N. (2002). *Electron. Lett.* **38**: 85.
89 Hansryd, J., Andrekson, P.A., Westlund, M. et al. (2002). *IEEE J. Sel. Top. Quantum Electron.* **8**: 506.
90 Gordienko, V., Stephens, M.F.C., El-Taher, A.E., and Doran, N.J. (2017). *Opt. Express* **25**: 4810.
91 Tong, Z., Lundstrom, C., Andrekson, P.A. et al. (2012). *IEEE J. Sel. Topics Quantum Electron.* **18**: 1016.
92 Marhic, M.E., Park, Y., Yang, F.S., and Kazovsky, L.G. (1996). *Opt. Lett.* **21**: 1354.
93 Wong, K.K.Y., Marhic, M.E., Uesaka, K., and Kazovsky, L.G. (2002). *IEEE Photon. Technol. Lett.* **14**: 911.
94 McKinstrie, C.J. and Radic, S. (2002). *Opt. Lett.* **27**: 1138.
95 McKinstrie, C.J., Radic, S., and Chraplyvy, A.R. (2002). *IEEE J. Sel. Top. Quantum Electron.* **8**: 538.
96 Radic, S. and McKinstrie, C.J. (2003). *Opt. Fib. Technol.* **9**: 7.
97 Radic, S., Mckinstrie, C.J., Chraplyvy, A.R. et al. (2002). *IEEE Photon. Technol. Lett.* **14**: 1406.
98 Stolen, R.H. and Bjorkholm, J.E. (1982). *IEEE J. Quantum Electron.* **18**: 1062.
99 Inoue, K. (1994). *J. Lightwave Technol.* **12**: 1916.

100 Jopson, R.M. and Tench, R.E. (1993). *Electron. Lett.* **29**: 2216.

101 Leclerc, O., Lavigne, B., Balmefrezol, E. et al. (2003). *J. Lightwave Technol.* **21**: 2779.

102 Stephens, M.F., Tan, M., Gordienko, V. et al. (2017). *Opt. Express* **25**: 24312.

103 Stephens, M.F., Gordienko, V., and Doran, N.J. (2017). *Opt. Express* **25**: 10597.

104 Ferreira, M.F. (1994). *Electron. Lett.* **30**: 40.

105 Ferreira, M.F. (1995). *J. Lightwave Technol.* **13**: 1692.

106 Tamura, K. and Nakazawa, M. (1996). *Opt. Lett.* **21**: 68.

107 Anderson, D., Desaix, M., Karlsson, M. et al. (1993). *J. Opt. Soc. Am. B* **10**: 1185.

108 Fermann, M.E., Kruglov, V.I., Thomson, B.C. et al. (2000). *Phys. Rev. Lett.* **84**: 6010.

109 Kruglov, V.I., Peacock, A.C., Dudley, J.M., and Harvey, J.D. (2000). *Opt. Lett.* **25**: 1753.

110 Kruglov, V.I., Peacock, A.C., Harvey, J.D., and Dudley, J.M. (2002). *J. Opt. Soc. Am. B* **19**: 461.

111 Ruehl, A., Marcinkevicius, A., Fermann, M.E., and Hartl, I. (2010). *Opt. Lett.* **35**: 3015.

112 Finot, C., Parmigiani, F., Petropoulos, P., and Richardson, D.J. (2006). *Opt. Express* **14**: 3161.

113 Barenblatt, G.I. (1996). *Scaling, Self-Similarity, and Intermediate Asymptotics*. Cambridge, UK: Cambridge University Press.

114 Finot, C., Millot, G., and Dudley, J.M. (2004). *Opt. Lett.* **29**: 2533.

115 Limpert, J.P., Schreiber, T., Clausnitzer, T. et al. (2002). *Opt. Express* **10**: 628.

116 Malinowski, A., Piper, A., Price, J.H.V. et al. (2004). *Opt. Lett.* **29**: 2073.

117 Billet, C., Dudley, J.M., Joly, N., and Knight, J.C. (2005). *Opt. Express* **13**: 3236.

118 Finot, C., Millot, G., Billet, C., and Dudley, J.M. (2003). *Opt. Express* **11**: 1547.

119 Finot, C., Millot, G., Pitois, S. et al. (2004). *IEEE J. Sel. Topics Quantum Electron.* **10**: 1211.

120 Röser, F., Rothhard, J., Ortac, B. et al. (2005). *Opt. Lett.* **30**: 2754.

121 Finot, C. and Millot, G. (2004). *Opt. Express* **12**: 5104.

122 Ozeki, Y., Takushima, Y., Aiso, K. et al. (2004). *Electron. Lett.* **40**: 1103.

123 Finot, C., Pitois, S., and Millot, G. (2005). *Opt. Lett.* **30**: 1776.

6

Solitons in Optical Fibers

As seen in Chapter 3, if the operating wavelength is in the anomalous dispersion regime, the group velocity dispersion (GVD) and the self-phase modulation (SPM) effects can compensate each other and the propagating pulse would broaden neither in the time domain nor in the frequency domain. In such case, the pulse is called "soliton," a term that has been used for the first time in 1965 [1]. This name reflects the particle-like nature of these kind of pulses, which remain intact even after mutual collisions. The existence of solitons in optical fibers was suggested for the first time in 1973 by Hasegawa and Tappert [2], whereas their first experimental observation was reported only in 1980 by Mollenauer et al. [3]. The experimental results confirmed that the properties of a pulse propagating in an optical fiber are described, to a remarkable degree, by the nonlinear Schrödinger equation (NLSE).

The nonlinear Schrödinger equation is a nonlinear partial differential equation that happens to be integrable by means of the "inverse scattering technique" [4, 5], as demonstrated by Zakharov and Shabat in 1971 [6]. Details about this method can be found in many texts [7–12].

6.1 The Fundamental Soliton Solution

The nonlinear Schrödinger equation, as given by Eq. (3.57), has pulse-like solutions called bright solitons, which have the potential of many technological applications. In the following, we find a simple analytical solution, corresponding to the so-called fundamental soliton. Other analytical soliton solutions can be found by solving the NLSE using the inverse scattering method [4–6].

Let us consider a pulse solution in the form

$$Q(z, \tau) = S(\tau)e^{i\phi(z)} \tag{6.1}$$

where the envelope function $S(\tau)$ is assumed to be a real function of τ. Substituting Eq. (6.1) into Eq. (3.57) and rearranging, we obtain

$$\frac{d\phi}{dz} = \gamma P_0 S^2(\tau) - \frac{1}{2}\beta_2 \frac{d^2S/d\tau^2}{S(\tau)} \tag{6.2}$$

Since the left-hand side of Eq. (6.2) depends only on z and the right-hand side depends only on τ, we can set each side equal to a constant, C_1. The equation resulting from the left-hand side has the solution

$$\phi(z) = C_1 z + \phi_0 \tag{6.3}$$

Solitons in Optical Fiber Systems, First Edition. Mario F. S. Ferreira.

where ϕ_0 is a constant of integration, which we will assume to be equal to zero in the following. Concerning the right-hand side of Eq. (6.2), it gives the equation

$$\frac{d^2S}{d\tau^2} - \frac{2\gamma}{\beta_2}P_0S^3(\tau) + \frac{2C_1}{\beta_2}S(\tau) = 0 \tag{6.4}$$

We can integrate Eq. (6.4) to obtain

$$\left(\frac{dS}{d\tau}\right)^2 = \frac{\gamma}{\beta_2}P_0S^4(\tau) - \frac{2C_1}{\beta_2}S^2(\tau) + K \tag{6.5}$$

where K is a constant of integration. However, this constant is zero, as for a localized solution, we must have

$$\lim_{\tau\to\pm\infty} S(\tau) = \lim_{\tau\to\pm\infty}\frac{dS}{d\tau} = 0 \tag{6.6}$$

Moreover, considering that the envelope function presents a maximum such that $S(\tau) = 1$ and $\frac{dS}{d\tau} = 0$, we obtain:

$$C_1 = \frac{\gamma P_0}{2} \tag{6.7}$$

Using this result, we can rewrite Eq. (6.5) in the form

$$\frac{dS}{d\tau} = \alpha S\sqrt{1 - S^2} \tag{6.8}$$

where

$$\alpha = \sqrt{\gamma P_0/|\beta_2|} \tag{6.9}$$

Equation (6.8) can be easily integrated, giving the result

$$S(\tau) = \text{sech}(\alpha\tau) \tag{6.10}$$

Thus, the soliton solution given by Eq. (6.1) becomes

$$Q(z, \tau) = \text{sech}[\alpha\tau]\exp\left\{i\frac{\gamma P_0}{2}z\right\} \tag{6.11}$$

Considering the normalizations used to derive Eq. (3.61), the above result can be written in the form

$$q(Z, T) = \eta\text{sech}[\eta T]\exp\left\{i\frac{1}{2}\eta^2 Z\right\} \tag{6.12}$$

where

$$\eta = \left(\frac{\gamma P_0 t_0^2}{|\beta_2|}\right)^{1/2} \tag{6.13}$$

is simultaneously the soliton amplitude and the inverse of pulse width. In real units, the soliton width changes as t_0/η, i.e., it scales inversely with the soliton amplitude. This inverse relationship between the amplitude and the width of a soliton is the most crucial property of NLSE solitons.

The canonical form of the fundamental soliton can be obtained from Eq. (6.12), considering $\eta = 1$ and is given by:

$$q(Z, T) = \text{sech}(T) \exp(iZ/2) \tag{6.14}$$

It can be shown that, given a solution $q(Z, T)$, the nonlinear Schrödinger equation is satisfied by another function

$$q'(Z, T) = q(T + \kappa Z, Z) \exp\left[-i\left(\kappa T + \frac{1}{2}\kappa^2 Z\right)\right] \tag{6.15}$$

which corresponds to the Galilean invariance of the NLSE. Introducing a phase constant σ and a time position T_0, the solitary wave solution can be written in the more general form:

$$q(Z, T) = \eta \text{sech}[\eta(T + \kappa Z - T_0)] \exp\left(-i\kappa T + \frac{i}{2}(\eta^2 - \kappa^2)Z + i\sigma\right) \tag{6.16}$$

The solitary wave solution given by Eq. (6.16) is characterized by four parameters: the amplitude η (also the inverse of pulse width), the frequency κ (also the pulse speed), the time position T_0, and the phase σ.

6.2 Higher-Order Solitons

The nonlinear Schrödinger equation given by Eq. (3.61) is a completely integrable equation that can be solved exactly using the inverse scattering method [4, 5], as shown by Zakharov and Shabat [6]. The main goal of this method is to find an appropriate scattering problem whose potential is $q(Z, T)$. The initial scattering data is determined by the initial field $q(0, T)$. The field $q(Z, T)$ after propagation along a distance Z is obtained from the evolved scattering data by solving a linear integral equation.

For the NLSE given by Eq. (3.61), the suitable scattering problem becomes [11, 13]:

$$i\frac{\partial \psi_1}{\partial T} + q(Z, T)\psi_2 = \varsigma\psi_1 \tag{6.17}$$

$$i\frac{\partial \psi_2}{\partial T} + q^*(Z, T)\psi_1 = -\varsigma\psi_2 \tag{6.18}$$

where ψ_1 and ψ_2 are the amplitudes of the two waves scattered by the potential $q(Z, T)$, and ς is the Z-independent eigenvalue. In general, we can have N eigenvalues, which may be written as

$$\varsigma_n = \frac{1}{2}(\kappa_n + i\eta_n), \quad \text{for } n = 1, 2, \ldots, N \tag{6.19}$$

In the case $N = 1$, the fundamental soliton solution given by Eq. (6.16) is obtained. The amplitude and speed of the soliton are given by the imaginary and real parts, respectively, of the eigenvalue (6.19).

In general, if the κ_n in Eq. (6.19) are all distinct, the N-soliton solutions that arise from the initial wave form are asymptotically given in the form of N separated solitons:

$$q(Z, T) = \sum_{j=1}^{N} \eta_j \text{sech}\left[\eta_j(T + \kappa_j Z - T_{0j})\right] \exp\left[-i\kappa_j T + i\frac{1}{2}\left(\eta_j^2 - \kappa_j^2\right)Z - i\sigma_j\right] \tag{6.20}$$

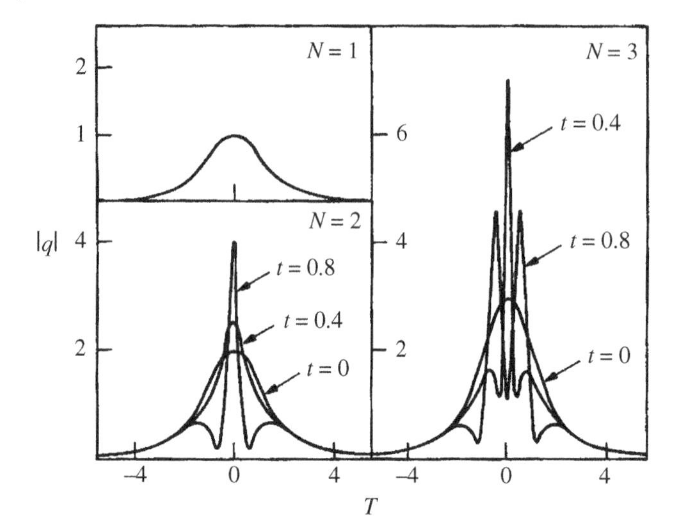

Figure 6.1 Fundamental and higher-order solitons that arise from an input pulse shape $q(0, T) = N\,\mathrm{sech}(T)$. *Source:* After Ref. [13].

If the input pulse shape is symmetrical, it can be shown that the eigenvalues of Eqs. (6.17) and (6.18) are purely imaginary. For example, considering an input pulse shape given by

$$q(0,T) = N\mathrm{sech}(T) \tag{6.21}$$

the corresponding eigenvalues are given by

$$\varsigma_n = \frac{i}{2}\eta_n = i\left(N - n + \frac{1}{2}\right), n = 1, 2, \ldots, N \tag{6.22}$$

Since $\kappa_n = 0$, all the output solitons propagate at exactly the same speed. This type of solution is called the *bound-soliton solution* and its shape evolves periodically during the propagation due to the phase interference among the constituting solitons. Figure 6.1 shows the profiles of the fundamental, second- and third-order solitons at different positions. Compared with the fundamental soliton, the shape of higher-order solitons will change during propagation, returning to their initial shape after a certain distance.

For the second-order soliton ($N = 2$), the eigenvalues are $\eta_1 = 1/2$ and $\eta_2 = 3/2$ and the field distribution is given by [13]:

$$q(Z,T) = \frac{4[\cosh(3T) + 3\exp(4iZ)\cosh(T)]\exp(iZ/2)}{[\cosh(4T) + 4\cosh(2T) + 3\cos(4Z)]} \tag{6.23}$$

Figure 6.2 provides a perspective of the second-order soliton evolution. We observe that such evolution is indeed periodic, and that, in the initial part, the pulse contracts in the time domain and increases its peak power.

The period of oscillation of the higher-order solitons resulting from the initial pulse given by Eq. (6.21) is

$$Z_0 = \frac{\pi}{2} \tag{6.24}$$

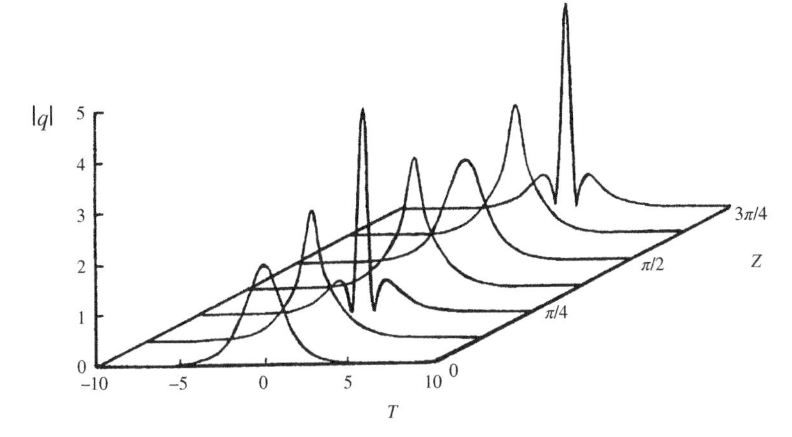

Figure 6.2 Evolution of the second-order soliton.

which can be written in terms of real units as

$$z_0 = \frac{\pi}{2} L_D = \frac{\pi}{2} \frac{t_0^2}{|\beta_2|} \tag{6.25}$$

The length z_0 is usually called the *soliton period.* Considering a pulse with a temporal width $t_0 = 10$ ps propagating in a dispersion-shifted fiber with $\beta_2 = -1$ ps^2/km in the C-band ($\lambda \approx$ 1550 nm), we have a propagation period $z_0 \approx 150$ km.

The periodic evolution of higher-order solitons can be understood as follows. In the initial part of its evolution, the pulse contracts in the time domain and increases its peak power. As seen in Chapter 4, SPM generates a frequency chirp such that the leading edge of the pulse is red-shifted, while its trailing-edge is blue-shifted from the central frequency, resulting in a broad spectrum in the frequency domain. In the absence of GVD, the pulse shape would have remained unchanged. However, in the presence of anomalous GVD, the red-shifted components in the leading edge of the pulse move slower, whereas the blue-shifted components in the trailing edge are faster. As a consequence, the pulse spectrum is compressed and the pulse reaches again its original width in the time domain. After that, everything starts again. It is this mutual interaction between the SPM and GVD effects that explains the periodic evolution for higher-order solitons.

The first experimental observation of solitons in optical fibers was realized by Mollenauer et al. in 1980 [3]. In this experiment, a mode-locked color-center laser was used to obtain 7-ps pulses near 1550 nm, a wavelength close to the fiber loss minimum. Considering the fiber dispersion parameter $D = 16$ ps/nm-km, the soliton period should be about 1260 m, or very nearly twice the 700-m length of the used fiber. Figure 6.3 shows the measured autocorrelation traces and the pulse spectra at certain critical peak pulse powers at the fiber input. The autocorrelation trace and the spectrum of the input pulses are also shown for comparison. From left to right, as the input power increases, we see first a significant dispersive broadening, followed by return to the original pulse width at the power (~1.2 W) for the fundamental soliton. As the power is increased, the pulses exhibit dramatic changes in their shape and develop a multipeak structure. At 4.2x the previous power (5 W), we observe the narrowing expected at the half soliton period of the second-order soliton, as illustrated in Figure 6.2. At the power levels of 11.4 W (9.5x) and 22.5 W (18x), we observe the autocorrelation patterns of the expected two- and three-fold splitting at the half soliton period of the third- and fourth-order solitons.

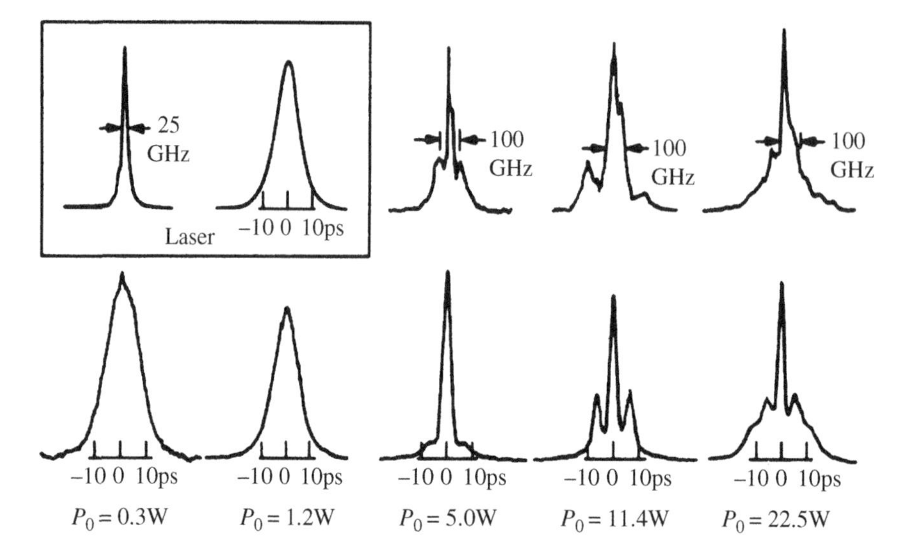

Figure 6.3 Autocorrelation traces (lower row) and corresponding power spectra (upper row) at the fiber output for several values of the input peak power. The autocorrelation trace and the spectrum of the input pulses are shown inside the rectangular box. *Source:* After Ref. [3].

6.3 Soliton Units

The full width at half maximum (FWHM) of the canonical single soliton solution given by Eq. (6.14) is $\Delta t = 2\cosh^{-1}(\sqrt{2}) \approx 1.763$. Consequently, the relation between the pulse width t_0 and its FWHM t_{FWHM} is given by

$$t_0 = \frac{t_{FWHM}}{1.763} \tag{6.26}$$

Using this relation, the dispersion length given by Eq. (3.25) can be rewritten in the form

$$L_D = \frac{1}{(1.763)^2}\frac{t_{FWHM}^2}{|\beta_2|} = \frac{1}{(1.763)^2}\frac{2\pi c}{\lambda^2}\frac{t_{FWHM}^2}{D} \tag{6.27}$$

where D is the dispersion parameter, as defined in Eq. (3.10).

According to Eq. (3.60), the peak power of the fundamental soliton ($N = 1$) in an optical fiber is given by

$$P_0 = \frac{|\beta_2|}{\gamma t_0^2} = \left(\frac{1.763}{2\pi}\right)^2 \frac{A_{eff}\lambda^3}{n_2 c}\frac{D}{t_{FWHM}^2} \tag{6.28}$$

where Eq. (3.52) has been used. Considering pulses with a temporal width $t_0 = 6$ ps and using typical parameter values $\beta_2 = -1$ ps^2/km and $\gamma = 3$ W^{-1}/km for dispersion-shifted fibers, we obtain P_0~10 mW in the C-band. The required power to launch the Nth-order soliton is N^2 times that for the fundamental one. Moreover, higher-order solitons compress periodically (see Figure 6.2), resulting in soliton chirping and spectral broadening. In contrast, fundamental solitons preserve their shape during propagation. This fact, together with the lower power required for their generation, makes fundamental solitons the preferred option in soliton communication systems.

The soliton pulse energy is given by

$$E = P_0 \int_{-\infty}^{\infty} \mathrm{sech}^2(\tau/t_0)d\tau = 2P_0 t_0 = \frac{2}{1.763} P_0 t_{FWHM} \tag{6.29}$$

Considering the typical values assumed above, we have $E \sim 110$ fJ in the C-band.

6.4 Dark Solitons

In the normal dispersion regime ($\beta_2 > 0$), the normalized nonlinear Schrödinger equation becomes

$$i\frac{\partial q}{\partial Z} - \frac{1}{2}\frac{\partial^2 q}{\partial T^2} + |q|^2 q = 0 \tag{6.30}$$

Equation (6.30) differs from Eq. (3.61) only in the sign of its dispersion term. The soliton solution of Eq. (6.30) can be written in the form [14]:

$$q(Z, T) = \sqrt{\rho} e^{i\sigma} \tag{6.31}$$

where

$$\rho = \rho_0 \left[1 - a^2 \mathrm{sech}^2\left(\sqrt{\rho_0} aT\right)\right], \quad a^2 \leq 1 \tag{6.32}$$

$$\sigma = \left[\rho_0 \left(1 - a^2\right)\right]^{1/2} T + \tan^{-1}\left[\frac{1}{\sqrt{1 - a^2}} \tanh\left(\sqrt{\rho_0} aT\right)\right] - \frac{\rho_0 (3 - a^2)}{2} Z \tag{6.33}$$

The parameter a is known as depth of modulation. In fact, this kind of solitons appears in the section where light waves are absent, and thus they are called *dark solitons*. Pulse-like solitons discussed in previous sections are called *bright* to make the distinction clear. When $a = 1$, we have a *black soliton*, given by

$$q = \sqrt{\rho_0} \tanh\left(\sqrt{\rho_0} T\right) \tag{6.34}$$

Thus, an input pulse with a "tanh" profile, exhibiting an intensity "hole" at the center, would propagate unchanged in the normal-dispersion region of optical fibers. The two types of dark solitons are illustrated in Figure 6.4.

Dark solitons were first observed by Emplit et al. [15] and Kröbel et al. [16] independently by transmitting a lightwave in the normal dispersion region of a fiber. In Ref. [15], 26-ps input pulses (at 595 nm) with a 5-ps-wide central hole were launched along a 52-m fiber. In Ref. [16], 100-ps pulses (at 532 nm) with a 0.3-ps-wide central hole were launched at the input of a 10-m long fiber. After these experiments, considerable attention has been paid to dark solitons [17–29]. In one experiment [20], two optical pulses propagated in the normal-GVD region of the fiber, with a relative time delay between them. The two pulses broaden, become chirped, and acquire a nearly rectangular shape during propagation. As these chirped pulses merge into each other, they interfere, producing a train of isolated dark solitons at the fiber output. In another experiment [23], nonlinear conversion of a beat signal propagating along a dispersion-decreasing fiber in the normal-dispersion regime was used to generate a train of dark solitons. Some numerical results have shown that dark solitons are more stable and are less affected than bright solitons in the presence of several perturbations, demonstrating their potential application for optical communication systems [30].

(a)

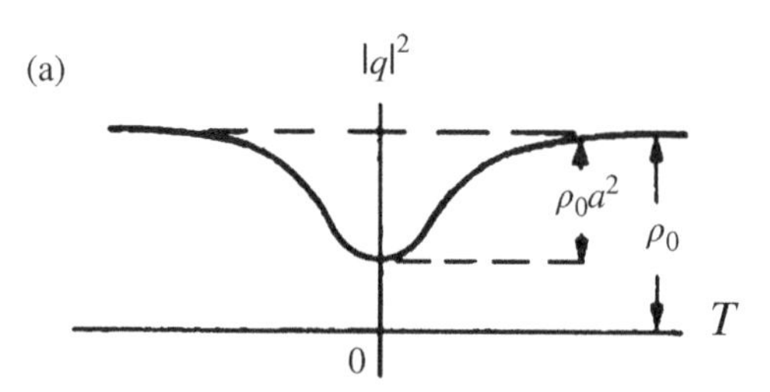

(b)

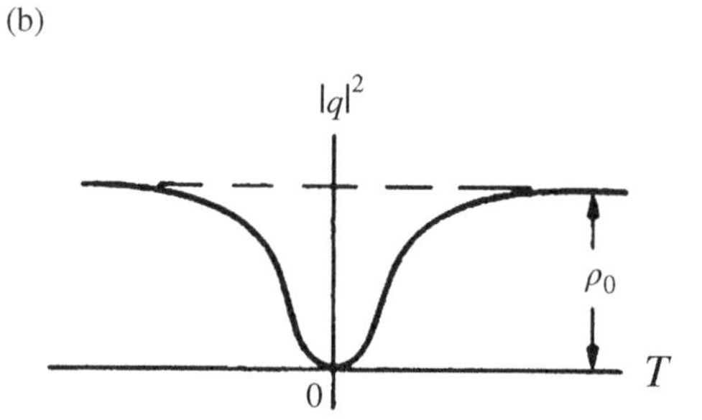

Figure 6.4 Schematic representation of dark solitons with (a) $a < 1$ and (b) $a = 1$.

6.5 Bistable Solitons

The nonlinear Schrödinger equation given by Eq. (3.61) was obtained considering the Kerr nonlinearity, i.e., a linear dependence of the refractive index relatively to the mode intensity, as given by Eq. (3.34b). However, such linear dependence is no longer verified when the nonlinear response of the material begins to saturate, as happens at sufficiently high intensities. In this case, Eq. (3.34b) should be replaced by

$$n = n_0 + n_2 f(I) \tag{6.35}$$

where $f(I)$ is some function of the mode intensity I, with $f(0) = 0$. Considering this more general form for the refractive index, the NLSE can be generalized and becomes

$$i\frac{\partial q}{\partial Z} + \frac{1}{2}\frac{\partial^2 q}{\partial T^2} + f\left(|q|^2\right)q = 0 \tag{6.36}$$

The stationary solutions of Eq. (6.36) have non-varying intensity profiles and can be written as

$$q(Z, T) = A(T)\exp(ikZ) \tag{6.37}$$

where $A(T)$ is the real amplitude and k is an (unknown) propagation constant of the soliton. By substituting Eq. (6.37) into Eq. (6.36), we find the following equation for $A(T)$:

$$\frac{d^2A}{dT^2} + 2A\left[f\left(A^2\right) - k\right] = 0 \tag{6.38}$$

The soliton solution must satisfy the condition $A \to 0$ as $|T| \to \infty$ in order for the soliton energy

$$E = \int_{-\infty}^{+\infty} A^2 dT \tag{6.39}$$

to be limited. This provides for the first integral of Eq. (6.38) in the form

$$\left(\frac{dA}{dT}\right)^2 = 4\int_0^A A\left[k - f\left(A^2\right)\right]dA \tag{6.40}$$

The integration of Eq. (6.40) provides the soliton amplitude profile $A(T)$ for each particular k and $f(A^2)$. The respective integrals can be analytically evaluated only for some particular functions $f(A^2)$. Using Eq. (6.40), the soliton energy E is shown to depend on the parameter k as

$$E(k) = \frac{1}{2}\int_0^{P_m(k)} [k - F(P)]^{-1/2} dP \tag{6.41}$$

where $P = A^2$ and

$$F(P) = \frac{1}{P}\int_0^P f(P)dP \tag{6.42}$$

In Eq. (6.41), $P_m(k)$ is a peak power of the soliton; it is defined as a minimal positive root of the equation $F(P) = k$.

For some functions $f(P)$, Eq. (6.41) can have more than one solution, each having the same energy E but different values of k and P_m. In general, only two solutions correspond to stable solitons, which are known as *bistable solitons*. Actually, for a given value of the pulse energy, this kind of solitons propagate in two different stable states and can be made to switch from one state to another [31]. Bistable solitons were discovered in 1985 by Kaplan [32], and they have been studied extensively since then. In fact, they are considered to be very important for some nonlinear optical applications, namely, the compression of optical pulses, optical switching, and bistability [31–41].

6.6 XPM-Paired Solitons

Several studies have shown that the coupled NLSEs given by Eqs. (4.22) and (4.23) have solitary wave solutions that maintain their shape during propagation through the XPM interaction [42–44]. Since such solutions always occur in pairs, they are referred to as XPM-paired solitons. Such solitons can have the same group velocity ($v_{g1} = v_{g2} = v_g$) if their wavelengths are chosen appropriately on opposite sides of the fiber zero-dispersion wavelength.

Assuming that the phase matching condition for the FWM process does not occur, the last terms in Eqs. (4.22) and (4.23) can be neglected. Under these conditions, let us look for soliton solutions of these equations in the form

$$U_j(z,t) = V_j\left(t - z/v_{ge}\right)\exp\left[i\left(K_j z - \Omega_j t + \phi_j\right)\right],\quad j = 1,2 \tag{6.43}$$

where v_{ge} is the effective group velocity of the soliton pair, V_j describes the soliton shape, K_j and Ω_j represent changes in the propagation constant and frequency of the two solitons, whereas ϕ_j is the phase. The solution for the soliton shape has the form [44]:

$$V_1\left(t - z/v_{ge}\right) = V_{10}\left[1 - a^2\text{sech}^2\left(B\left(t - z/v_{ge}\right)\right)\right]^{1/2} \tag{6.44}$$

$$V_2(t - z/v_{ge}) = V_{20}\mathrm{sech}(B(t - z/v_{ge})) \tag{6.45}$$

where

$$B^2 = \frac{3\gamma_1\gamma_2}{2\gamma_1\beta_{22} - 4\gamma_2\beta_{21}} V_{20}^2 \tag{6.46}$$

$$a^2 = \frac{2\gamma_1\beta_{22} - \gamma_2\beta_{21}}{\gamma_1\beta_{22} - 2\gamma_2\beta_{21}} \frac{V_{20}^2}{V_{10}^2} \tag{6.47}$$

and the effective velocity of the soliton pair, v_{ge}, is given by

$$v_{ge}^{-1} = v_g^{-1} + \beta_{21}\Omega_1 = v_g^{-1} + \beta_{22}\Omega_2 \tag{6.48}$$

Equations (6.44) and (6.45) correspond to a XPM-coupled bright-gray soliton pair, which have the same width but different amplitudes. The parameter a indicates the depth of the intensity dip of the gray soliton. If $a = 1, \beta_{21} < 0$, and $\beta_{22} > 0$, a bright-dark soliton pair is obtained [42]. An interesting feature of such solution is that the bright soliton propagates in the normal dispersion regime, whereas the dark soliton propagates in the anomalous dispersion regime. This behavior is opposite to what would normally be expected and is due solely to the effect of XPM. In fact, as seen in Section 4.2, XPM is twice as strong as SPM and induces a mutual frequency shift between the two pulses. The leading edge of the bright pulse is up-shifted in frequency, while its trailing edge is down-shifted in frequency by XPM. The reverse occurs for the dark pulse. In the normal (anomalous) dispersion regime, the group velocity decreases (increases) when the frequency increases; therefore, the leading edges of the pulses are retarded and the trailing edges are advanced. As a consequence, XPM counteracts the temporal spreading of the pulses induced by SPM and GVD, and the final result is that the coupled pulses propagate with their shape undistorted.

Besides the above-mentioned soliton-pair solutions, Eqs. (4.22) and (4.23) can also support pairs with two bright or two dark solitons [45], as well as periodic solutions representing two pulse trains [46, 47]. XPM-coupled soliton pairs can be generalized to multicomponent solitons, constituted by multiple pulses at different carrier frequencies. Actually, such multicomponent solitons can occur in multichannel communication systems [48].

6.7 Optical Similaritons

Similaritons are self-similar waves, whose envelope maintains its overall shape but its parameters such as amplitude, width, and chirp evolve with propagation inside nonlinear media [49]. As seen in Section 5.6, parabolic-shape similaritons form asymptotically in the normal-dispersion region of a homogeneous fiber amplifier. Here, we consider the more general case of a fiber amplifier with nonuniform characteristics.

Let us assume that the GVD parameter β_2, the nonlinear parameter γ, and the net gain coefficient g' are nonuniform along the fiber length. In this case, Eq. (5.44) must be replaced by the following generic inhomogeneous NLSE:

$$i\frac{\partial U}{\partial z} - \frac{1}{2}\beta_2(z)\frac{\partial^2 U}{\partial \tau^2} + \gamma(z)|U|^2 U = i\frac{g'(z)}{2}U \tag{6.49}$$

The model given by Eq. (6.49) can describe different physical systems in nonlinear optics and condensed matter physics, and it is known to have similariton solutions [50, 51], provided a certain relation exists among the parameters $\beta_2(z)$, $\gamma(z)$, and $g'(z)$.

Let us consider the self-similar transformation:

$$U(z,\tau) = u(z)\Psi\left[\zeta, \frac{\tau - \tau_c(z)}{t_p(z)}\right] \exp\left[i\phi(z,\tau\right] \tag{6.50}$$

where $\zeta(z)$ is an effective propagation distance, $u(z)$, $t_p(z)$, and $\tau_c(z)$ represent the amplitude, width, and center position of the similariton, respectively. In order to have a linear chirp, we assume that the phase front of the similariton in Eq. (6.50) is parabolic, given by:

$$\phi(z,\tau) = \frac{C(z)}{2t_0^2}\tau^2 + b(z)\tau + d(z) \tag{6.51}$$

where $C(z)$ is the chirp parameter, t_0 is the input pulse width, $b(z)$ corresponds to a frequency shift, and $d(z)$ is independent of τ. Actually, chirp-free similaritons with $C = 0$ also exist.

Substituting Eqs. (6.50) and (6.51) into Eq. (6.49), we obtain a set of differential equations for the evolution of the pulse parameters, such that Ψ obeys the standard homogeneous NLSE:

$$i\frac{\partial\Psi}{\partial\zeta} \pm \frac{1}{2}\frac{\partial^2\Psi}{\partial\chi^2} + |\Psi|^2\Psi = 0 \tag{6.52}$$

In Eq. (6.52), the upper and lower signs in the second term correspond to the cases of anomalous and normal dispersion, respectively, while the similarity variable $\chi(z, \tau)$ is given by:

$$\chi(z,\tau) = \frac{\tau - \tau_c(z)}{t_p(z)} \tag{6.53}$$

Solving the differential equations for the evolution of pulse parameters, we obtain the following expressions for the effective propagation distance, amplitude, width, and position of the pulse:

$$\zeta(z) = \frac{D(z)}{1 - C_0 D(z)} \tag{6.54}$$

$$u(z) = \frac{1}{t_p(z)}\sqrt{\frac{|\beta_2(z)|}{\gamma(z)}} \tag{6.55}$$

$$t_p(z) = t_0[1 - C_0 D(z)] \tag{6.56}$$

$$\tau_c(z) = \tau_{c0} - (C_0\tau_{c0} + b_0)D(z) \tag{6.57}$$

where

$$D(z) = \frac{1}{t_0^2}\int_0^z \beta_2(z)dz \tag{6.58}$$

corresponds to the accumulated dispersion over the distance z.

The parameters related to the phase are given by:

$$C(z) = \frac{C_0}{1 - C_0 D(z)}, \quad b(z) = \frac{b_0}{1 - C_0 D(z)}, \quad d(z) = \frac{b_0^2 D(z)/2}{1 - C_0 D(z)} \tag{6.59}$$

where C_0 is the input chirp parameter.

The transformation of the inhomogeneous NLSE, given by Eq. (6.49), into the homogeneous one, given by Eq. (6.52), is only possible if the pulse and fiber parameters satisfy the following compatibility condition [51, 52]:

$$g'(z) = C(z)\frac{\beta_2(z)}{t_0^2} + \frac{d}{dz}\ln\left[\frac{\beta_2(z)}{\gamma(z)}\right]. \tag{6.60}$$

Equation (6.52) corresponds to the standard NLSE, which is well known to be integrable by the inverse scattering method [3]. As a consequence, any stable similariton solution of the inhomogeneous NLSE, given by Eq. (6.49), corresponds to a soliton solution of the homogeneous NLSE. The stability of the former follows from the stability of the latter.

In analogy with the standard solitons, a similariton is bright or dark depending on whether the dispersion is anomalous or normal. The fundamental bright similariton is given by:

$$U_B(\chi,\zeta) = B\text{sech}(\chi - v\zeta)\exp\left(iv\chi/2 + i\zeta - iv^2/4\right) \tag{6.61}$$

where v is the velocity of the corresponding bright soliton. The fundamental dark similariton is given by

$$U_D(\chi,\zeta) = u_0[\cos(\phi)\tanh(\Theta_D) + isen(\phi)]\exp\left(iu_0^2\zeta\right) \tag{6.62}$$

where

$$\Theta_D(\chi,\zeta) = u_0\cos(\phi)[\chi - u_0\zeta\sin(\phi)] \tag{6.63}$$

Here, u_0 is the background amplitude obeying the homogeneous NLS equation, and ϕ is the total phase shift across the dark soliton.

Considering a fiber with constant values of β_2 and γ, but a z-dependent gain coefficient, the compatibility condition becomes simply $g'(z) = \beta_2 C(z)/t_0^2$. For an unchirped initial pulse ($C_0 = 0$), this condition imposes that the amplifier net gain must be zero. In this case, all pulse parameters become constant, which corresponds to a standard soliton. On the other hand, for a chirped initial pulse ($C_0 \neq 0$), the compatibility condition can be written in the form [53]:

$$g'(z) = \frac{C_0\beta_2}{t_0^2 - C_0\beta_2 z} \tag{6.64}$$

Equation (6.64) shows that gain is required when $C_0\beta_2 > 0$. We conclude that a fiber amplifier whose gain increases with the propagation distance supports bright similaritons when $\beta_2 < 0$ and $C_0 < 0$, while it supports dark similaritons when $\beta_2 > 0$ and $C_0 > 0$. In both cases, the pulse width decreases along the fiber amplifier, as given by Eq. (6.56) with $D(z) = \left(\beta_2/t_0^2\right)z$.

6.8 Numerical Solution of the NLSE

It is generally difficult to obtain an analytical solution of the nonlinear Schrödinger equation in practical conditions. In such cases, the NLSE must be solved numerically, which is efficiently performed using the "split-step Fourier" analysis. In such approach, the waveguide is partitioned into n equally long segments, whose lengths h are adjusted according to the desired accuracy (Figure 6.5). For each segment, the nonlinear and dispersive effects are considered independently [54]. Since the dispersive effects are most naturally dealt in the frequency domain, whereas the nonlinear effects are easier to handle in the time domain, the propagation in each segment is treated in

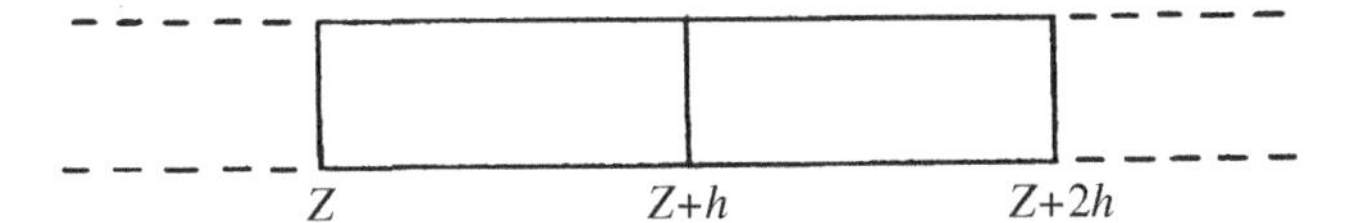

Figure 6.5 Illustration of the split-step Fourier method.

two consecutive steps. First, the pulse $q(Z, T)$ is transformed using the fast Fourier transform (FFT) into the frequency domain to $\widetilde{q}(Z,\omega)$. Then, to take into account the dispersive effects, an intermediate state is calculated

$$\widetilde{q}_{int}(Z+h,\omega) = \widetilde{q}(Z,\omega)e^{-i\frac{1}{2}\omega^2 h} \tag{6.65}$$

In the second step, the inverse FFT is used to transform $\widetilde{q}_{int}(Z+h,\omega)$ into an intermediate state in the time domain, $q_{int}(Z+h, T)$. Then, the nonlinear effects of the segment h are taken into account by computing

$$q(Z+h,T) = q_{int}(Z+h,T)e^{i|q|^2 h} \tag{6.66}$$

As long as the length of the segments is short enough, this method can provide very accurate results. Such accuracy can be improved using the symmetrized split-step Fourier method [55], in which the effect of nonlinearity is included in the middle of the segment rather than at the segment boundary.

Figure 6.6a shows the numerical solution of the NLSE obtained using the method described above for an initial condition $q(0, T) = \text{sech}(T)$, which corresponds to the fundamental soliton. This figure illustrates the most important property of the fundamental soliton: as long as the fiber loss is negligible, it propagates undistorted without any change in shape for arbitrarily long distances. It is this feature of fundamental solitons that makes them attractive for optical communication systems. Figure 6.6b illustrates the evolution of an initially rectangular pulse toward the fundamental soliton, demonstrating the truly stable nature of this solution. In the first stage of propagation, some energy of the initial pulse is radiated until the pulses achieve the fundamental soliton profile.

Figure 6.7 illustrates the evolution of an initially rectangular pulse with an amplitude 0.7 and width 10, which means that it has an energy greater than that in Figure 6.6b. In this case, the initial pulse evolves to a solution given by two coupled solitons, which exhibits a periodic behavior, with a period ~15.

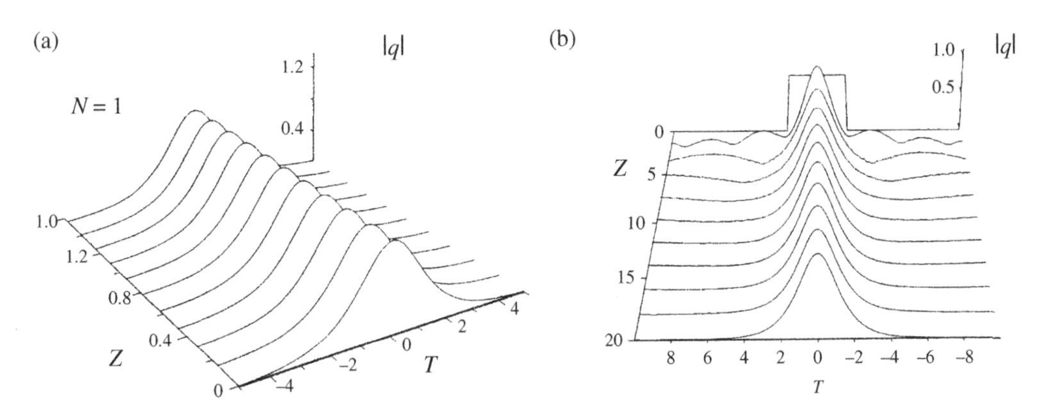

Figure 6.6 (a) Numerical solution of the NLSE for an initial condition $q(0, t) = \text{sech}(T)$. (b) Numerical solution of the NLSE for an initially rectangular pulse, with an amplitude 0.7626 and width 4.

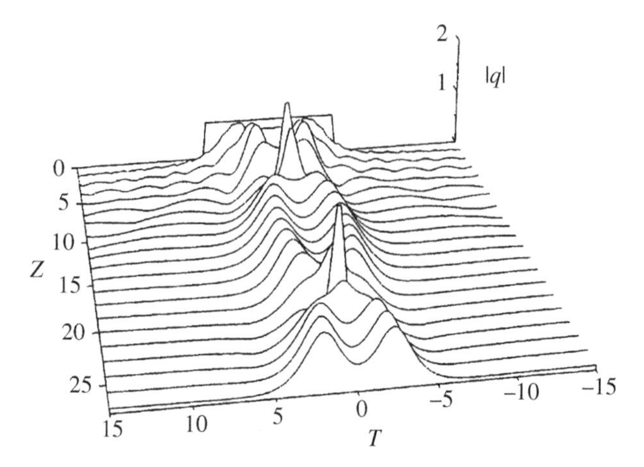

Figure 6.7 Numerical solution of the NLSE for an initially rectangular pulse, with an amplitude 0.7 and width 10.

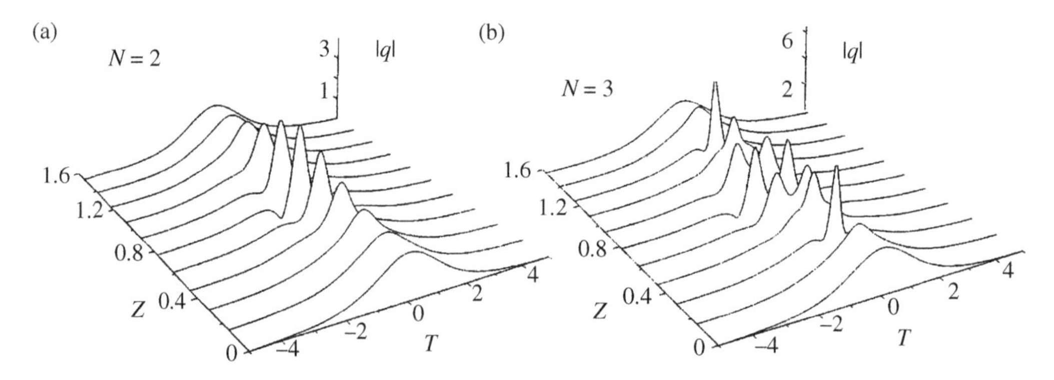

Figure 6.8 Numerical solution of the NLSE assuming an initial condition $q(0, T) = N\,\text{sech}(T)$, with (a) $N = 2$ and (b) $N = 3$.

Figure 6.8 shows the numerical solution of the NLSE assuming an initial condition $q(0, T) = N\,\text{sech}(T)$, with (a) $N = 2$ and (b) $N = 3$. As pointed out before, the pulse evolves during propagation, returning to its initial profile after one period. Meanwhile, it develops a multipeak structure with a maximum number of peaks equal to N-1

6.9 The Variational Approach

The variational technique is well known in several areas [56], and it is particularly useful in nonlinear optics for describing the propagation of chirped pulses in nonlinear media. The variational formulation was first used in 1977 to describe pulse propagation in nonlinear diffractive media [57]. Its use to describe pulse propagation in nonlinear dispersive media, such as optical fibers, was first realized by Anderson in 1983 [58]. This approach can be quite accurate even when the effect of nonlinearity is large [59, 60]. The accuracy depends essentially on the ansatz chosen to describe the evolution of the field experiencing nonlinearity. In this section, we apply this approach to discuss some combined effects of GVD, initial chirp, and SPM. The same approach is especially useful to describe the behavior of nonlinear pulses propagating in dispersion-managed links [61–66], as well as dissipative solitons [67–71],

The "action" functional is given by

$$S = \int L dz \tag{6.67}$$

where L the Lagrangian, which is related with the Lagrangian density L_d as

$$L = \int_{-\infty}^{\infty} L_d(v, v^*) d\tau \tag{6.68}$$

In Eq. (6.68), $v(z)$ and $v^*(z)$ represent the generalized coordinates. Minimization of the action S requires that L_d satisfy the Euler-Lagrange equation

$$\frac{\partial L_d}{\partial v} - \frac{\partial}{\partial \tau}\left(\frac{\partial L_d}{\partial v_\tau}\right) - \frac{\partial}{\partial z}\left(\frac{\partial L_d}{\partial v_z}\right) = 0 \tag{6.69}$$

where v_τ and v_z represent the derivative of v relative to τ and z, respectively. The Lagrangian density corresponding to Eq. (3.55) is

$$L_d = \frac{i}{2}\left(U\frac{\partial U^*}{\partial z} - U^*\frac{\partial U}{\partial z}\right) - \frac{\beta_2}{2}\left|\frac{\partial U}{\partial \tau}\right|^2 - \frac{\gamma}{2}|U|^4 \tag{6.70}$$

In fact, Eq. (3.55) can be obtained using Eq. (6.70) in Eq. (6.69) and considering U^* as the generalized coordinate v. Minimization of the action S with respect to the pulse parameters provides the reduced Euler-Lagrange equations

$$\frac{\partial L}{\partial v} - \frac{d}{dz}\left(\frac{\partial L}{\partial v_z}\right) = 0 \tag{6.71}$$

In the case of a chirped Gaussian pulse, we have

$$U(z,\tau) = U_p(z)\exp\left[-\frac{1}{2}(1 + iC_p(z))\left(\frac{\tau}{t_p(z)}\right)^2 - i\theta_p(z)\right] \tag{6.72}$$

where the parameters $U_p(z)$, $C_p(z)$, $t_p(z)$, and $\theta_p(z)$ represent the amplitude, the chirp parameter, the width, and the phase of the pulse, respectively. Substituting Eq. (6.72) into Eq. (6.70) and performing the time integration in Eq. (6.68) gives the following result for the Lagrangian:

$$L = -\frac{\beta_2 E_p}{4t_p^2}(1 + C^2) - \frac{\gamma E_p^2}{\sqrt{8\pi}t_p} - \frac{E_p}{4}\left(\frac{dC_p}{dz} - \frac{2C_p}{t_p}\frac{dt_p}{dz}\right) - E_p\frac{d\theta_p}{dz} \tag{6.73}$$

where $E_p = \sqrt{\pi}U_p^2 t_p$ is the pulse energy. Using $v = \theta_p$ in Eq. (6.71) gives $E_p = E_0 =$ constant. On the other hand, using $v = C_p$, t_p, and E_p provides the following equations for t_p, C_p, and θ_p, respectively:

$$\frac{dt_p}{dz} = \frac{\beta_2 C_p}{t_p} \tag{6.74}$$

$$\frac{dC_p}{dz} = \frac{\gamma E_0}{\sqrt{2\pi}t_p} + \left(1 + C_p^2\right)\frac{\beta_2}{t_p^2} \tag{6.75}$$

$$\frac{d\theta_p}{dz} = -\frac{5\gamma E_0}{4\sqrt{2\pi}t_p} - \frac{\beta_2}{2t_p^2} \tag{6.76}$$

Equations (6.74) and (6.75) show that the SPM affects the pulse width via the chirp parameter C_p. The two terms in Eq. (6.75) are due to the SPM and GVD effects, and they have the same or the opposite sign, depending if the dispersion is normal or anomalous, respectively. As a consequence, the combination of the SPM and GVD effects results in a reduced pulse broadening in the case of anomalous dispersion compared with the case of normal dispersion, because of smaller values of the chirp parameter in the first case.

The quantity $\left(1 + C_p^2\right)/t_p^2$ is related to the spectral width of the pulse and remains constant if we neglect the fiber nonlinearity. Replacing this quantity with its initial value $\left(1 + C_0^2\right)/t_0^2$ in Eq. (6.75) and integrating one obtains the solution

$$C_p = C_0 + \left(1 + C_0^2\right)\frac{\beta_2 z}{t_0^2} \tag{6.77}$$

which is similar to Eq. (3.18). Using this solution in Eq. (6.74), we find the following result for the pulse width:

$$t_p^2 = t_0^2\left[(1 + C_0\sigma)^2 + \sigma^2\right] \tag{6.78}$$

where

$$\sigma = \frac{\beta_2 z}{t_0^2} \tag{6.79}$$

Equations (6.78) and (6.79) are similar to Eqs. (3.19) and (3.17), respectively. They show that the input pulse broadens monotonously during propagation if C_0 and β_2 are of the same sign. However, if these parameters are of opposite sign, the pulse compresses until a given location, after which it starts to broaden, as illustrated in Figure 3.5.

Keeping the fiber nonlinearity but neglecting the GVD, Eq. (6.74) shows that the pulse width t_p remains constant, equal to its initial value t_0. Concerning the chirp parameter, it varies with distance as

$$C_p = C_0 + \frac{1}{\sqrt{2}}\gamma P_0 z \tag{6.80}$$

Equation (6.80) shows that the SPM-induced chirp is always positive. It adds with the initial pulse chirp when $C_0 > 0$, resulting in an enhanced oscillatory structure of the pulse spectrum. However, if $C_0 < 0$, the two contributions have opposite signs, which produces a spectral narrowing.

Let us consider, instead of the Gaussian ansatz given by Eq. (6.72), a sech ansatz of the form

$$U(z,\tau) = U_p(z)\mathrm{sech}\left(\frac{\tau}{t_p(z)}\right)\exp\left[-iC_p(z)\left(\frac{\tau}{t_p(z)}\right)^2 + i\theta_p(z)\right] \tag{6.81}$$

Substituting Eq. (6.81) into Eq. (6.70) and performing the time integration in Eq. (6.69) provides the following result for the Lagrangian:

$$L = -\frac{\beta_2 E_p}{6t_p^2}\left(1 + \frac{\pi^2}{4}C_p^2\right) - \frac{\gamma E_p^2}{6t_p} - \frac{\pi^2 E_p}{24}\left(\frac{dC_p}{dz} - \frac{2C_p}{t_p}\frac{dt_p}{dz}\right) - E_p\frac{d\theta_p}{dz} \tag{6.82}$$

where $E_p = 2U_p^2 t_p$. Using this expression in Eq. (6.71) and considering $v = C_p$, t_p, we re-obtain Eq. (6.74) for the pulse width, whereas the equation for the chirp parameter becomes

$$\frac{dC_p}{dz} = \left(C_p^2 + \frac{4}{\pi^2}\right)\frac{\beta_2}{t_p^2} + \gamma P_0 \frac{4}{\pi^2}\frac{t_0}{t_p} \tag{6.83}$$

where $P_0 = U_0^2$. Eq. (6.74) shows that the pulse width will not change if the chirp parameter C_p remains zero during propagation. On the other hand, it is clear from Eq. (6.83) that if $\beta_2 > 0$ both terms on the right side are positive and C_p cannot remain zero even if $C_p(0) = C_0 = 0$. However, in the case of anomalous dispersion ($\beta_2 < 0$), the two terms cancel each other when the initial pulse parameters satisfy the condition

$$\gamma P_0 t_0^2 = |\beta_2| \tag{6.84}$$

Equation (6.84) is in accordance with Eq. (6.28), giving the peak power for the fundamental soliton. In this case, if the pulse chirp is initially zero, its width will not change during propagation.

6.10 The Method of Moments

The NLS equation can also be solved approximately, using the method of moments developed by Maimisov [72]. This method has been applied to other types of pulses, namely, to dissipative solitons [73–77]. The main idea of this method is that if we know a finite number of the moments of a function, then we can recover this function to a certain accuracy, together with its evolution in Z. The complete evolution problem with an infinite number of degrees of freedom is thus reduced to the evolution of a finite set of pulse characteristics. If the pulse maintains its shape as it propagates along the fiber, these characteristics include its amplitude, width, and chirp.

Using the moment method, one can define the following power averaged quantities (moments):

$$E_p = \int_{-\infty}^{+\infty} |U|^2 d\tau \tag{6.85}$$

$$\sigma_p^2 = \frac{1}{E_p}\int_{-\infty}^{+\infty} \tau^2 |U|^2 d\tau \tag{6.86}$$

$$C_p = -\frac{2}{E_p}\int_{-\infty}^{+\infty} \mathrm{Im}\left(\tau U^* \frac{dU}{d\tau}\right) d\tau \tag{6.87}$$

where E_p is the pulse energy, σ_p is the rms pulse width, and C_p is the chirp parameter. By combining Eqs. (6.85)–(6.87) and using Eq. (3.55) we find that:

$$\frac{dE_p}{dz} = 0 \tag{6.88}$$

$$\frac{d\sigma_p^2}{dz} = \frac{\beta_2}{E_p}\int_{-\infty}^{\infty} \tau^2 \mathrm{Im}\left(U^* \frac{\partial^2 U}{\partial \tau^2}\right) d\tau \tag{6.89}$$

$$\frac{dC_p}{dz} = \frac{2\beta_2}{E_p} \int_{-\infty}^{\infty} \left|\frac{\partial U}{\partial \tau}\right|^2 d\tau + \frac{\gamma P_0}{E_p} \int_{-\infty}^{\infty} |U|^4 d\tau \tag{6.90}$$

Considering the case of a chirped Gaussian pulse, given by Eq. (6.72), the rms width is related to the width parameter t_p as $\sigma_p = t_p/\sqrt{2}$ and the pulse energy is given by $E_p = \sqrt{\pi}U_p^2 t_p$. According to Eq. (6.88), the pulse energy remains constant and equal to its initial value $E_0 = \sqrt{\pi}U_0^2 t_0$. Performing the integrals in Eqs. (6.89) and (6.90), one arrive again to Eqs. (6.74) and (6.75) for the evolution of the pulse width and chirp, respectively. So, we conclude that both the method of moments and the variational approach provide the possibility of reducing the NLSE to the same dynamical system for the pulse parameters.

References

1 Zabusky, N.J. and Kruskal, M.D. (1965). *Phys. Rev. Lett.* **15**: 240.
2 Hasegawa, A. and Tappert, F.D. (1973). *Appl. Phys. Lett.* **23**: 142.
3 Mollenauer, L.F., Stolen, R.H., and Gordon, J.P. (1980). *Phys. Rev. Lett.* **45**: 1095.
4 Gardner, C.S., Greene, J.M., Kruskal, M.D., and Miura, R.M. (1967). *Phys. Rev. Lett.* **19**: 1095.
5 Ablowitz, M.J. and Clarkson, P.A. (1991). *Solitons, Nonlinear Evolution Equations, and Inverse Scattering*. New York: Cambridge University Press.
6 Zakharov, V.E. and Shabat, A. (1972). *Sov. Phys. JETP* **34**: 62.
7 Taylor, J.T. (ed.) (1992). *Optical Solitons—Theory and Experiment*. New York: Cambridge University Press.
8 Drazin, P.G. (1993). *Solitons: An Introduction*. Cambridge, UK: Cambridge University Press.
9 Lamb, G.L. Jr. (1994). *Elements of Soliton Theory*. New York: Dover.
10 Gu, C.H. (1995). *Soliton Theory and Its Applications*. New York: Springer-Verlag.
11 Hasegawa, H. and Kodama, Y. (1995). *Solitons in Optical Communications*. New York: Oxford University Press.
12 Akhmediev, N. and Ankiewicz, A. (1997). *Solitons: Nonlinear Pulses and Beams*. New York: Chapman and Hall.
13 Satsuma, J. and Yajima, N. (1974). *Prog. Theor. Phys. Suppl.* **55**: 284.
14 Hasegawa, A. and Tappert, F. (1973). *Appl. Phy. Lett.* **23**: 171.
15 Emplit, P., Hamaide, J.P., Reynaud, F. et al. (1987). *Opt. Commun.* **62**: 374.
16 Kröbel, D., Halas, N.J., Giuliani, G., and Grischkowsky, D. (1988). *Phys. Rev. Lett.* **60**: 29.
17 Tomlinson, W.J., Hawkins, R.J., Weiner, A.M. et al. (1989). *J. Opt. Soc. Am. B* **6**: 329.
18 Giannini, J.A. and Joseph, R.I. (1990). *IEEE J. Quantum Electron.* **26**: 2109.
19 Kivshar, Y.S. and Afanasjev, V.V. (1991). *Phys. Rev. A* **44**: R1446.
20 Rothenberg, J.E. and Heinrich, H.K. (1992). *Opt. Lett.* **17**: 261.
21 Kivshar, Y.S. (1993). *IEEE J. Quantum Electron.* **29**: 250.
22 Williams, J.A.R., Allen, K.M., Doran, N.J., and Emplit, P. (1994). *Opt. Commun.* **112**: 333.
23 Richardson, D.J., Chamberlain, R.P., Dong, L., and Payne, D.N. (1994). *Electron. Lett.* **30**: 1326.
24 Kivshar, Y.S. and Yang, X. (1994). *Phys. Rev. E* **49**: 1657.
25 Nakazawa, M. and Suzuki, K. (1995). *Electron. Lett.* **31**: 1076.
26 Atieh, A.K., Myslinski, P., Chrostowski, J., and Galko, P. (1997). *Opt. Commun.* **133**: 541.
27 Emplit, P., Haelterman, M., Kashyap, R., and De Lathouwer, M. (1997). *IEEE Photon. Technol. Lett.* **9**: 1122.

28 Leners, R., Emplit, P., Foursa, D. et al. (1997). *J. Opt. Soc. Am. B* **14**: 2339.
29 Kivshar, Y.S. and Luther-Davies, B. (1998). *Phys. Rep.* **298**: 81.
30 Kivshar, Y.S. and Agrawal, G.P. (2003). *Optical Solitons: From Fibers to Photonic Crystals*. San Diego, CA: Academic Press.
31 Enns, R.H. and Rangnekar, S.S. (1987). *Opt. Lett.* **12**: 108.
32 Kaplan, A.E. (1985). *Phys. Rev. Lett.* **55**: 1291.
33 Enns, R.H., Rangnekar, S.S., and Kaplan, A.E. (1987). *Phys. Rev. A* **35**: 446.
34 Enns, R.H., Fung, R., and Rangnekar, S.S. (1990). *Opt. Lett.* **15**: 162.
35 Enns, R.H. and Rangnekar, S.S. (1991). *Phys. Rev. A* **43**: 4047.
36 Enns, R.H., Edmundson, D.E., Rangnekar, S.S., and Kaplan, A.E. (1992). *Opt. Quantum Electron.* **24**: S2195.
37 Eix, S.L., Enns, R.H., and Rangnekar, S.S. (1993). *Phys. Rev. A* **47**: 5009.
38 Deangelis, C. (1994). *IEEE J. Quantum Electron.* **30**: 818.
39 Aicklen, G.H. and Tamil, L.S. (1996). *J. Opt. Soc. Am. B* **13**: 1999.
40 Tanev, S. and Pushkarov, D.I. (1997). *Opt. Commun.* **141**: 322.
41 Kumar, A. (1998). *Phys. Rev. E* **58**: 5021; A. Kumar and T. Kurz, *Opt. Lett.* **24**, 373 (1999).
42 Trillo, S., Wabnitz, S., Wright, E.M., and Stegeman, G.I. (1988). *Opt. Lett.* **13**: 871.
43 Afanasjev, V.V., Kivshar, Y.S., Konotop, V.V., and Serkin, V.N. (1989). *Opt. Lett.* **14**: 805.
44 Lisak, M., Höök, A., and Anderson, D. (1990). *J. Opt. Soc. B* **7**: 810.
45 Afanasjev, V.V., Kivshar, Y.S., Konotop, V.V., and Serkin, V.N. (1989). *Opt. Lett.* **14**: 805.
46 Florjanczyk, M. and Tremblay, R. (1989). *Phys. Lett.* **141**: 34.
47 Hioe, F.T. (1997). *Phys. Rev.* **56**: 2373.
48 Kivshar, Y.S. and Agrawal, G.P. (2003). *Optical Solitons: From Fibers to Photonic Crystals*. San Diego, CA: Academic Press, Chap. 9.
49 Barenblatt, G.I. (1996). *Scaling, Self-Similarity, and Intermediate Asymptotics*. Cambridge, UK: Cambridge University Press.
50 Kruglov, V.I., Peacock, A.C., Dudley, J.M., and Harvey, J.D. (2000). *Opt. Lett.* **25**: 1753.
51 Serkin, V.N. and Hasegawa, A. (2000). *Phys. Rev. Lett.* **85**: 4502.
52 Kruglov, V.I., Peacock, A.C., and Harvey, J.D. (2003). *Phys. Rev. Lett.* **90**: 113902.
53 Moores, J.D. (1996). *Opt. Lett.* **21**: 555.
54 Fisher, R.A. and Bischel, W.K. (1975). *J. Appl. Phys.* **46**: 4921.
55 Fleck, J.A., Morris, J.R., and Feit, M.D. (1976). *Appl. Phys.* **10**: 129.
56 Nesbet, R.K. (2003). *Variational Principles and Methods in Theoretical Physics and Chemistry*. New York: CambridgeUniversity Press.
57 Firth, W.J. (1977). *Opt. Commun.* **22**: 226.
58 Anderson, D. (1983). *Phys. Rev. A* **27**: 3135.
59 Anderson, D., Lisak, M., and Reichel, T. (1988). *Phys. Rev. A* **38**: 1618.
60 Desaix, M., Anderson, D., and Lisak, M. (1989). *Phys. Rev. A* **40**: 2441.
61 Turitsyn, S.K. and Shapiro, E.G. (1998). *Opt. Fiber Tech.* **4**: 151.
62 Turitsyn, S.K., Gabitov, I., Laedke, E.W. et al. (1998). *Opt. Commun.* **151**: 117.
63 Sousa, M.H., Ferreira, M.F., and Panameño, E.M. (2004). *SPIE Proc.* **5622**: 944.
64 Jackson, R., Jones, C., and Zharnitsky, V. (2004). *Physica D* **190**: 63.
65 Ferreira, M.F., Facão, M.V., Latas, S.V., and Sousa, M.H. (2005). *Fiber Int. Opt.* **24**: 287.
66 Konar, S., Mishra, M., and Jana, S. (2006). *Chaos, Solitons Fractals* **29**: 823.
67 Manousakis, M., Papagiannis, P., Moshonas, N., and Hizanidis, K. (2001). *Opt. Commun.* **198**: 351.
68 Ankiewicz, A., Akhmediev, N., and Devine, N. (2007). *Opt. Fib. Technol.* **13**: 91.
69 Mancas, S.C. and Choudhury, S.R. (2007). *Theor. Math. Phys.* **152**: 1160.

70 Mancas, S.C. and Choudhury, S.R. (2009). *Chaos, Solitons Fractals* **40**: 91.

71 Ferreira, M.F. (2018). *IET Optoelectronics* **12**: 122.

72 Maimistov, A.I. (1993). *Zh. Eksp. Teor. Fiz.* **104**: 3620; JETP 77, 727 (1993).

73 Tsoy, E., Ankiewicz, A., and Akhmediev, N. (2006). *Phys. Rev. E* **73**: 036621.

74 Chang, W., Ankiewicz, A., Soto-Crespo, J.M., and Akhmediev, N. (2008). *Phys. Rev. A* **78**: 023830.

75 Akhmediev, N. and Ankiewicz, A. (2008). *Dissipative Solitons: 'From Optics to Biology and Medicine'*. Springer.

76 Uzunov, I.M., Arabadzhiev, T.N., and Georgiev, Z.D. (2015). *Opt. Fib. Tech.* **24**: 15.

77 Latas, S.C. and Ferreira, M.F. (2019). *J. Opt. Soc. Am. B* **36**: 3016.

7

Soliton Transmission Systems

High-speed fiber-optic communication systems are generally limited both by the fiber nonlinearity and by the group dispersion, which causes the pulse broadening. However, since fundamental solitons are obtained by the balance between the group dispersion and Kerr nonlinearity, their width can be maintained over long distances. In other words, fundamental solitons are free from either the dispersive distortion or from self-phase modulation. Thus, it is quite natural to use solitons as information carrier in fibers since all other format will face distortion either from dispersion or nonlinearity.

In a realistic communication system, the soliton propagation is affected by many perturbations related to the input pulse shape, chirp and power, amplifiers noise, optical filters, modulators, etc. However, because of its particle-like nature, it turns out that the soliton remains stable under most of these perturbations.

The possibility of using solitons in all-optical transmission systems was made by Hasegawa in 1973 [1]. Moreover, he suggested using the Raman effect of transmission fiber itself for optical amplification, instead of regenerative repeaters, which were considered standard until 1990. The idea was used in the first long-distance all-optical transmission experiment performed by Mollenauer and Smith [2] in 1988. Since then, a rapid progress during the 1990s has converted optical solitons into a practical candidate for modern lightwave systems [3–8].

7.1 Soliton Perturbation Theory

As seen in Chapter 6, the fundamental soliton solution can be written in the form:

$$q(Z,T) = \eta \operatorname{sech}[\eta(T + \kappa Z - T_0)] \exp\left(- i\kappa T + \frac{i}{2}(\eta^2 - \kappa^2)Z + i\sigma \right) \tag{7.1}$$

where η, κ, T_0, and σ represent the amplitude, the frequency, the time position, and the phase, respectively. Under the influence of arbitrary perturbations, the four soliton parameters used in Eq. (7.1) do not remain constant but evolve during the soliton propagation along the fiber. Soliton propagation in these circumstances can be described by a perturbed NLSE of the form

$$i\frac{\partial q}{\partial Z} + \frac{1}{2}\frac{\partial^2 q}{\partial T^2} + |q|^2 q = iP(q) \tag{7.2}$$

where $P(q)$ represents the various perturbations. When these perturbations are relatively small, the evolution of the four soliton parameters can be described using the soliton perturbation theory [9, 10].

Solitons in Optical Fiber Systems, First Edition. Mario F. S. Ferreira.

Considering the variational approach discussed in Section 6.9, we find that Eq. (7.2) can be obtained from the Euler-Lagrange equation using the Lagrangian density [11]:

$$L_d = \frac{i}{2}\left(q\frac{\partial q^*}{\partial z} - q^*\frac{\partial q}{\partial z}\right) + \frac{1}{2}\left|\frac{\partial q}{\partial T}\right|^2 - \frac{1}{2}|q|^4 + i\left(Pq^* - P^*q\right) \tag{7.3}$$

Minimization of the action S, given by Eq. (6.67), with respect to the pulse parameters provides the reduced Euler-Lagrange equations

$$\frac{\partial L}{\partial v} - \frac{d}{dz}\left(\frac{\partial L}{\partial v_z}\right) = \int_{-\infty}^{\infty}\left(P^*\frac{\partial q}{\partial v} + P\frac{\partial q^*}{\partial v}\right)dT \tag{7.4}$$

Using the sech ansatz given by Eq. (7.1), we obtain the following evolution equations for the soliton parameters:

$$\frac{d\eta}{dZ} = \mathrm{Re}\int_{-\infty}^{\infty} P(q)q^* dT \tag{7.5}$$

$$\frac{d\kappa}{dZ} = -\mathrm{Im}\int_{-\infty}^{\infty} P(q)\tanh\left(\eta(T-T_0)\right)q^* dT \tag{7.6}$$

$$\frac{dT_0}{dZ} = -\kappa + \frac{1}{\eta^2}\mathrm{Re}\int_{-\infty}^{\infty} P(q)(T-T_0)q^* dT \tag{7.7}$$

$$\frac{d\sigma}{dZ} = \frac{1}{2}\left(\eta^2 - \kappa^2\right) + T_0\frac{d\kappa}{dZ} + \frac{1}{\eta}\mathrm{Im}\int_{-\infty}^{\infty} P(q)(1-\eta(T-T_0)\tanh\left[\eta(T-T_0)\right]q^* dT \tag{7.8}$$

where Re and Im stand for the real and imaginary parts, respectively. Equations (7.5)–(7.8) are used extensively in the theory of soliton communication systems [3, 12, 13].

7.2 Effect of Fiber Losses

Fiber loss is the first perturbation affecting the soliton propagation and one of the main causes of signal degradation in long-distance fiber-optic communication systems. This limitation can be minimized by operating near $\lambda = 1.55\ \mu m$. However, even with fiber losses as low as 0.2 dB/km, the signal power is reduced by 20 dB after transmission over 100 km of fiber.

Fiber loss can be taken into account theoretically by adding a loss term to the NLSE, which becomes:

$$i\frac{\partial q}{\partial Z} + \frac{1}{2}\frac{\partial^2 q}{\partial T^2} + |q|^2 q = -\frac{i}{2}\Gamma q \tag{7.9}$$

where $\Gamma = \alpha L_D$ is the loss rate per dispersion distance, α being the fiber loss coefficient, and L_D the dispersion length. Assuming that $\Gamma << 1$ and using $P(q) = -\Gamma q/2$ in Eqs. (7.5)–(7.8), we find the following results for the soliton parameters:

$$\eta(Z) = \eta(0)\exp(-\Gamma Z) \equiv \eta_i \exp(-\Gamma Z) \tag{7.10a}$$

$$\kappa(Z) = \kappa(0) \equiv \kappa_i \tag{7.10b}$$

$$T_0(Z) = -\kappa_i Z + T_{0i} \tag{7.10c}$$

$$\sigma(Z) = -\frac{1}{2}\kappa_i^2 Z + \frac{\eta_i^2}{4\Gamma}\left(1 - \exp(-2\Gamma Z)\right) + \sigma_i \tag{7.10d}$$

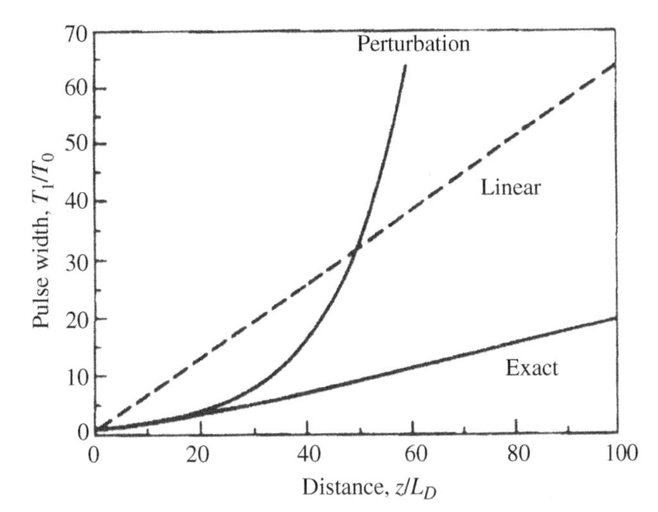

Figure 7.1 Broadening factor $t_p(Z)/t_0$ against propagation distance obtained numerically and using the perturbation theory. The dashed curve corresponds to the case of a linear pulse. *Source:* After Ref. [14].

where η_i, κ_i, T_{0i}, and σ_i are the initial values (at $Z = 0$) of the soliton amplitude, frequency, position, and phase, respectively. Eq. (7.10a) indicates that the soliton amplitude decreases along the fiber length at the same rate as the power amplitude. The decrease in the amplitude at a rate twice as fast as a linear pulse is a consequence of the nonlinear property of a soliton. Since the amplitude and width are inversely related, the soliton width will increase exponentially according to

$$t_p(Z) = t_0 \exp(\Gamma Z) \tag{7.11}$$

In fact, the exponential increase of the soliton width predicted by the perturbation theory occurs only for relatively short propagation distances. Figure 7.1 shows the variation of the broadening factor $t_p(Z)/t_0$ with distance, obtained numerically from Eq. (7.9), for the fundamental soliton when $\Gamma = 0.07$ [14]. Clearly, the result predicted by the perturbation theory is valid only for distances up to $Z \approx \Gamma^{-1}$. The numerical result displayed in Figure 7.1 shows that for large distances, the soliton width increases linearly with distance, but at a rate slower than the linear pulse.

7.3 Soliton Amplification

Several means of compensating for the fiber loss using optical amplification have been suggested since the 1970s. In general, solitons can be amplified periodically using either lumped or distributed amplification [15–18]. The use of distributed Raman gain provided by the fiber itself was proposed by Hasegawa [15] in 1983. However, the relatively high pump powers required to achieve the Raman amplification prevented its implementation at that time. As a more practical alternative, erbium-doped fiber amplifiers (EDFAs) have been used in several soliton transmission experiments since 1989, when they became available. Such experiments demonstrated that, despite the lumped nature of this amplification scheme, solitons could be maintained over very long distances.

The behavior of a soliton in the presence of fiber losses and optical amplification depends strongly on the relative magnitudes of dispersion length, L_D, and amplifier spacing, L_A, as well as on the loss over each dispersion length. If the conditions $L_D << L_A$ and $\Gamma << 1$ are satisfied, we have the so-called *quasi-adiabatic regime*. In such regime, the soliton can adapt adiabatically to losses by increasing the width and decreasing its peak power. Inversely, if the amplifier spacing is much smaller than the dispersion distance ($L_A << L_D$), the soliton remains almost undistorted between successive amplifications, even if large loss-induced energy variations occur. In this case, the soliton preserves its shape and width even if it is amplified hundreds of times. Such solitons are referred to as *path-averaged solitons*, since their properties are given by the soliton energy averaged over one amplifier spacing.

7.3.1 Lumped Amplification

Taking into account the fiber loss, periodic lumped-amplification, and possible variations in fiber dispersion, Eq. (7.9) is generalized and becomes:

$$i\frac{\partial q}{\partial Z} + \frac{d(Z)}{2}\frac{\partial^2 q}{\partial T^2} + |q|^2 q = -i\frac{\Gamma}{2}q + i\left(\sqrt{G}-1\right)\sum_{n=1}^{N}\delta(Z - nZ_A)q \tag{7.12}$$

where $d(Z)$ is the dispersion normalized so that its average becomes unity, N is the number of amplifiers, $G = \exp(\Gamma Z_A)$ is the amplifier gain, and $Z_A = L_A/L_D$ is the normalized amplifier spacing.

Equation (7.12) may be transformed to a Hamiltonian form by introducing a new amplitude u through

$$u(Z,T) = \frac{q(Z,T)}{a(Z)} \tag{7.13}$$

where $a(Z)$ contains rapid amplitude variations and $u(Z, T)$ is a slowly varying function of Z. Substituting (7.13) into (7.12), one obtains that $a(Z)$ satisfies

$$\frac{da}{dZ} = \left[-\frac{\Gamma}{2} + \left(\sqrt{G}-1\right)\sum_{n=1}^{N}\delta(Z - nZ_A)\right]a \tag{7.14}$$

while the new amplitude u satisfies

$$i\frac{\partial u}{\partial Z} + \frac{d(Z)}{2}\frac{\partial^2 u}{\partial T^2} + a^2(Z)|u|^2 u = 0 \tag{7.15}$$

In the following, we will consider the case with no variation in dispersion, $d(Z) = 1$. The concept of the average soliton makes use of the fact that $a^2(Z)$ in Eq. (7.15) varies rapidly with a period $Z_A << 1$. Since solitons evolve little over a short distance Z_A, one can replace $a^2(Z)$ by its average value $\langle a^2\rangle$. The resulting equation is then satisfied by the average soliton $\bar{u}$ such that $u = \bar{u} + \delta u$. The perturbation δu is relatively small, since the leading-order correction varies as Z_A^2 rather than Z_A [19]. The average-soliton description proves to be accurate even for $Z_A = 0.2$.

The normalization $\langle a^2\rangle = 1$ is achieved by choosing the integration constant a_0 in (7.14), leading to

$$a_0 = \left(\frac{\Gamma Z_A}{1 - \exp(-\Gamma Z_A)}\right)^{1/2} = \left(\frac{G\ln G}{G-1}\right)^{1/2} \tag{7.16}$$

Thus, soliton evolution in lossy fibers with periodic lumped amplification is identical to that in lossless fibers provided that the amplifier spacing L_A is much less than the dispersion distance L_D and the initial amplitude is enhanced by a factor a_0 given in Eq. (7.16).

Using $L_D = t_0^2/|\beta_2|$, the condition $L_A << L_D$ can be expressed in terms of the soliton width t_0 as

$$t_0 >> \sqrt{|\beta_2| L_A} \tag{7.17}$$

Considering typical values, $\beta_2 = -0.5\ \text{ps}^2/\text{km}$ and $L_A = 50\ \text{km}$, we obtain from Eq. (7.17) that $t_0 >> 5\ \text{ps}$. This condition imposes a limitation on the bit rate for soliton communication systems.

Higher bit rate soliton communication systems can be realized using pulses of sufficiently short width, which leads to the condition $L_D << L_A$. If the condition $\Gamma << 1$ is also satisfied, we have the quasi-adiabatic regime, in which the pulse amplitude decreases exponentially between two amplifiers, according to Eq. (7.10a). Simultaneously, the pulse width increases until reaching the next amplifier, where it becomes:

$$t(Z_A) = t(0)\exp(\Gamma Z_A) = t(0)\exp(\alpha L_A) \tag{7.18}$$

In order to minimize the emission of dispersive waves during this process, we must have $\alpha L_A \leq 1$, which limits the amplifier spacing to L_A~20 km.

7.3.2 Distributed Amplification

The use of distributed amplification scheme allows the alleviation of the constraints mentioned above imposed by the lumped-amplification scheme. As discussed in Section 5.5, the distributed-amplification configuration can provide a nearly lossless fiber by compensating losses locally at every point along the fiber link. Indeed, following an early suggestion of Hasegawa [15], the first soliton transmission experiment was realized in 1985 by Mollenauer and colleagues [17] making use of Raman amplification along a 10 km transmission fiber. Two color-center lasers were used in this experiment. One laser produced 10 ps-solitons with a peak power of 375 mW at a wavelength of 1.5 μm. The other laser was operated continuously as pump wave at a wavelength of 1.46 μm and injected from the output side of the fiber. Using a pump power of 125 mW, such that the Raman gain exactly compensates for the fiber loss of 1.8 dB, it was possible to avoid the soliton broadening and regain its original shape. The experimental result of a soliton reshaping using Raman amplification is shown in Figure 7.2.

The same scheme was extended in 1988 by Mollenauer and Smith [2], which successfully transmitted solitons over 4000 km in a fiber with loss periodically compensated by Raman gain (Figure 7.3). In this experiment, a 42 km length of standard single-mode fiber with a loss of 0.22 dB/km and dispersion of 17 ps/nm/km at 1600 nm was used. The loop was closed on itself with nearly 95% efficiency at the signal wavelength using an all-fiber Mach-Zhender interferometer.

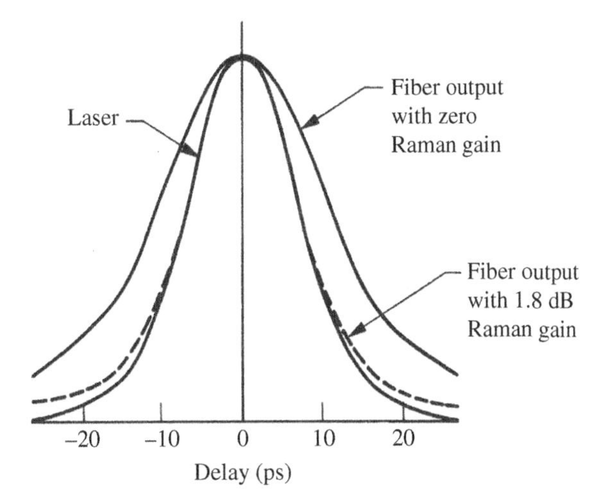

Figure 7.2 Experimental observation of soliton reshaping using Raman amplification. *Source:* After Ref. [17].

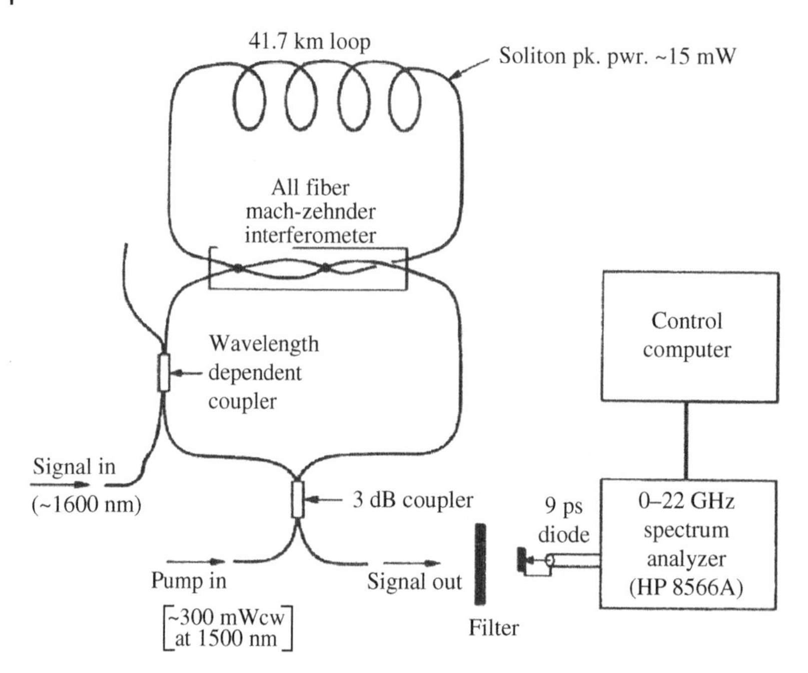

Figure 7.3 Schematic diagram of the experiment of Mollenauer and Smith of optical soliton transmission by repeated Raman amplifications. *Source:* After Ref. [2].

A color-center laser, operating at 1600 nm, provided 55 ps signal pulses at a 100-MHz repetition rate. The interferometer enabled nearly 100% efficient coupling into the loop of 300 mW of cw pump power at 1500 nm, provided by a second color-center laser. This experiment demonstrated that soliton communication systems over transoceanic distances were feasible.

In alternative to stimulated Raman scattering, the distributed amplification can be realized if the transmission fiber is doped lightly with erbium ions and pumped periodically to provide distributed gain. Several experiments have demonstrated that solitons can be propagated in such distributed EDFAs over relatively long distances [20–24].

In the presence of a distributed gain with $G << 1$, such that provided by the Raman amplification or a distributed EDFA, the soliton amplitude given by Eq. (7.10a) is simply modified to

$$\eta(Z) = \eta_0 \exp\left(\int_0^Z [2G(Z) - \Gamma]dZ\right) \tag{7.19}$$

By designing the gain so that the exponent of Eq. (7.19) vanishes – namely, making $G(Z)$ constant and equal to $\Gamma/2$ for all Z, one can achieve a system in which a soliton propagates without any distortion.

In practice, the effective gain cannot be made constant along the fiber, since the pump power also suffers from fiber loss. The evolution of the soliton energy is given in physical units as [18]:

$$\frac{dE_s}{dz} = [g(z) - \alpha]E_s \tag{7.20}$$

Assuming a bidirectional pumping scheme and neglecting the gain saturation, the gain coefficient $g(z)$ can be approximated by Eq. (5.49). Using such result, the integration of Eq. (7.20) provides the variation of the soliton energy according to Eq. (5.50). It is found that the range of energy variations increases with the pump station spacing, L_A, but it remains generally much smaller than that occurring in the lumped-amplification scheme. Pump-station spacings of 40–50 km can be used if energy deviations of 20% are acceptable [18].

Like in the case of lumped amplification, the effects of energy variations on solitons depend on the relative magnitudes of the dispersion length, L_D, and of the pump-station spacing, L_A. When $L_D << L_A$, we have the so-called *quasi-adiabatic regime*, in which solitons evolve adiabatically with a negligible emission of dispersive waves. On the other hand, for $L_A << L_D$ little soliton reshaping is observed. A more complicated behavior occurs when the values of L_A and L_D become of the same order. In particular, it has been verified that for $L_A \approx 4\pi L_D$ dispersive waves and solitons are resonantly amplified, leading to unstable soliton evolution [18].

7.4 Soliton Interaction

The soliton solution (7.1) exists in the range $-\infty < T < \infty$ and it remains valid for a train of solitons only if individual solitons are well separated. In fact, the combined optical field of neighboring solitons is not a solution of the NLSE. This phenomenon, referred to as *soliton interaction*, has been studied extensively [25–29]. In order to minimize such interaction, the soliton width must be a small fraction of the bit slot. The bit rate R is related to the soliton width t_0 as

$$R = \frac{1}{t_B} = \frac{1}{s_0 t_0} \tag{7.21}$$

where t_B is the duration of the bit slot and $s_0 = t_B/t_0$ is the separation between neighboring solitons in normalized units.

Assuming that the separation between the two pulses is sufficiently large and following the quasi-particle approach of Karpaman and Solovev [25], they can be expressed as a linear superposition of two fundamental solitons:

$$q(Z, T) = q_1(Z, T) + q_2(Z, T) \tag{7.22}$$

where

$$q_i(Z, T) = A_i \text{sech}[A_i(T - T_i)] \exp(-iB_i(T - T_i) + i\sigma_i), \quad i = 1, 2 \tag{7.23}$$

Considering the ansatz (7.22), the perturbed NLSE (7.2) corresponds to two equations of the form:

$$i\frac{\partial q_i}{\partial Z} + \frac{1}{2}\frac{\partial^2 q_i}{\partial T^2} + |q_i|^2 q_i = -\left(q_i^2 q_j^* + 2|q_i|^2 q_j\right) \tag{7.24}$$

with $i, j = 1, 2$ and $i \neq j$.

Using Eqs. (7.5)–(7.8) with $P(q)$ replaced by $i\left(q_i^2 q_j^* + 2|q_i|^2 q_j\right)$, we obtain the following evolution equations:

$$\frac{dA_i}{dZ} = (-1)^{i+1} 4A^3 e^{-A\Delta T} \sin(\Delta\phi) \tag{7.25}$$

$$\frac{dB_i}{dZ} = (-1)^{i+1} 4A^3 e^{-A\Delta T} \cos(\Delta\phi) \tag{7.26}$$

$$\frac{dT_i}{dZ} = -B_i - 2Ae^{-A\Delta T} \sin(\Delta\phi) \tag{7.27}$$

$$\frac{d\sigma_i}{dZ} = \frac{1}{2}\left(A_i^2 + B_i^2\right) - 2ABe^{-A\Delta T}\sin(\Delta\phi) + 6A^2e^{-A\Delta T}\cos(\Delta\phi) \tag{7.28}$$

where $A = (A_1 + A_2)/2$, $B = (B_1 + B_2)/2$, $\Delta\phi = B\Delta T + \Delta\sigma$, $\Delta T = T_1 - T_2 > 0$, and $\Delta\sigma = \sigma_1 - \sigma_2$. In obtaining the above equations, it has been assumed that $\Delta A = A_1 - A_2$, $\Delta B = B_1 - B_2$, and ΔT satisfy the conditions:

$$|\Delta A| << A,\ |\Delta B| << B,\ |\Delta A|\Delta T << 1,\ \text{and}\ A\Delta T >> 1 \tag{7.29}$$

From Eqs. (7.25)–(7.28), we can derive the following evolution equations:

$$\frac{dA}{dZ} = 0 \tag{7.30}$$

$$\frac{dB}{dZ} = 0 \tag{7.31}$$

$$\frac{d(\Delta A)}{dZ} = 8A^3e^{-A\Delta T}\sin(\Delta\phi) \tag{7.32}$$

$$\frac{d(\Delta B)}{dZ} = 8A^3e^{-A\Delta T}\cos(\Delta\phi) \tag{7.33}$$

$$\frac{d(\Delta T)}{dZ} = -\Delta B \tag{7.34}$$

$$\frac{d(\Delta\phi)}{dZ} = A\Delta A \tag{7.35}$$

Finally, from Eqs. (7.30)–(7.35), the following equations can be obtained for the soliton spacing, ΔT, and phase difference, $\Delta\phi$:

$$\frac{d^2(\Delta T)}{dZ^2} = -8A^3e^{-A\Delta T}\cos(\Delta\phi) \tag{7.36}$$

$$\frac{d^2(\Delta\phi)}{dZ^2} = 8A^4e^{-A\Delta T}\sin(\Delta\phi) \tag{7.37}$$

Equations (7.36) and (7.37) show that two adjacent solitons having the same phase attract each other initially, whereas their separation increases if they are opposite in phase. Figure 7.4a illustrates the evolution of the separation between two in-phase and equal amplitude solitons for different values of the initial separation $\Delta T_0 = s_0$, obtained from Eq. (7.36). The separation $s = \Delta T$ varies with the propagation distance according to [26]:

$$s = \ln\left[\frac{1}{2}\left(1 + \cos\left(4Ze^{-s_0/2}\right)\right)e^{s_0}\right] \tag{7.38}$$

From Eq. (7.38), we verify that collision occurs after a propagation distance Z_c given by:

$$Z_c \approx \frac{\pi}{4}e^{s_0/2} \tag{7.39}$$

It has been found that Eq. (7.39) is quite accurate for relatively large initial separations ($s_0 > 6$). A more accurate result, valid for arbitrary values of s_0, is given by [28]:

$$Z_c = \frac{\pi\sinh(s_0)\cosh(s_0/2)}{2[s_0 + \sinh(s_0)]} \tag{7.40}$$

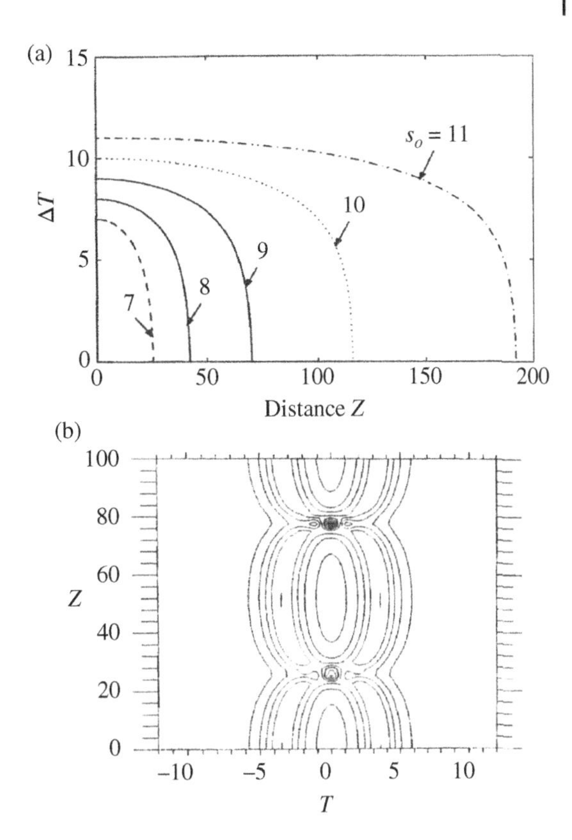

Figure 7.4 (a) Evolution of the separation between two in-phase and equal amplitude solitons for different values of the initial separation $\Delta T_0 = s_0$. (b) Contour plot for the case $s_0 = 7$.

Figure 7.4b shows the contour plot for the case $s_0 = 7$, obtained by solving numerically the NLSE, considering the initial condition given by Eqs. (7.22) and (7.23). We observe that the two solitons collide periodically, with a collision distance $Z_c \approx 26.3$, in good accordance with Eq. (7.39).

The effect of soliton interaction can be neglected if the total transmission distance L_T is much smaller than $Z_c L_D$. Using Eq. (7.39), together with the relations $L_D = t_0^2/|\beta_2|$ and $t_0 = 1/(s_0 R)$, such condition provides the following design criterion:

$$R^2 L_T << \frac{\pi \exp(s_0/2)}{2 s_0^2 |\beta_2|} \tag{7.41}$$

Considering $\beta_2 = -5\ \text{ps}^2/\text{km}$, $s_0 = 10$, and a system operating at 40 Gb/s, the total transmission distance is limited to $L_T << 3000\ \text{km}$.

Figure 7.5a illustrates the evolution of the separation between two equal amplitude solitons, with an initial separation $s_0 = 7$, for different values of their initial phase difference $\Delta\phi$. It can be seen that the existence of a small phase difference is sufficient to avoid the collision between the two pulses. For values of $\Delta\phi$ larger than $\sim\pi/8$ rad, the solitons begin separating after an initial attracting phase. The increase in separation becomes monotonous for higher values of $\Delta\phi$. This is illustrated in Figure 7.5b, which shows the contour plot for the case $\Delta\phi = \pi/2$ rad, obtained by solving numerically the NLSE.

Introducing a small difference in the amplitudes of two adjacent soliton becomes also an effective way of controlling the interaction between them. This can be observed from Figure 7.6a, which shows the evolution of the separation between the two initially in-phase solitons, for several values

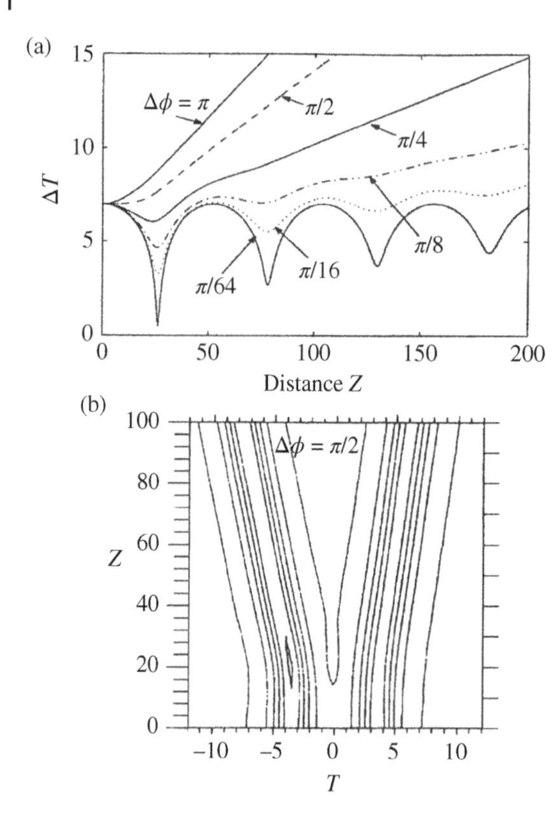

Figure 7.5 (a) Evolution of the separation between two equal amplitude solitons for $s_0 = 7$ and different values of their initial phase difference. (b) Contour plot for the case $\Delta\phi = \pi/2$ rad.

of their amplitude difference ΔA. The soliton separation varies periodically in this case, but such variation is reduced for larger values of the amplitude difference. This is due to a continuous variation of the phase difference between the two pulses, as illustrated in Figure 7.6b. Such phase variation determines a periodical variation of the force between the two solitons, which becomes alternately attractive and repulsive. Figure 7.7 shows the evolution of two solitons with different amplitudes obtained by numerically solving the NLSE for the cases $s_0 = 7$, $\Delta A = 0.01$ (a) and $\Delta A = 0.1$ (b). In the last case, the separation between the two pulses becomes practically constant.

7.5 Timing Jitter

When one soliton is used to represent one digit in a communication system, the major cause that leads to the loss of information is the timing jitter, which corresponds to the random variation of the soliton position T_0 [30–40]. This situation is quite different from a non-soliton pulse, where the major cause of information loss is due to the deformation of the waveform itself. In this section, we will discuss the main physical mechanisms inducing the soliton timing jitter and the limitations imposed by them on performance of soliton transmission systems.

7.5.1 Gordon-Haus Jitter

Among the physical mechanisms responsible for the soliton timing jitter, the amplified spontaneous emission (ASE) noise added by the optical amplifiers is often dominant in practice. The amplifier noise normally deteriorates the signal-to-noise ratio (SNR) for a linear signal due to the

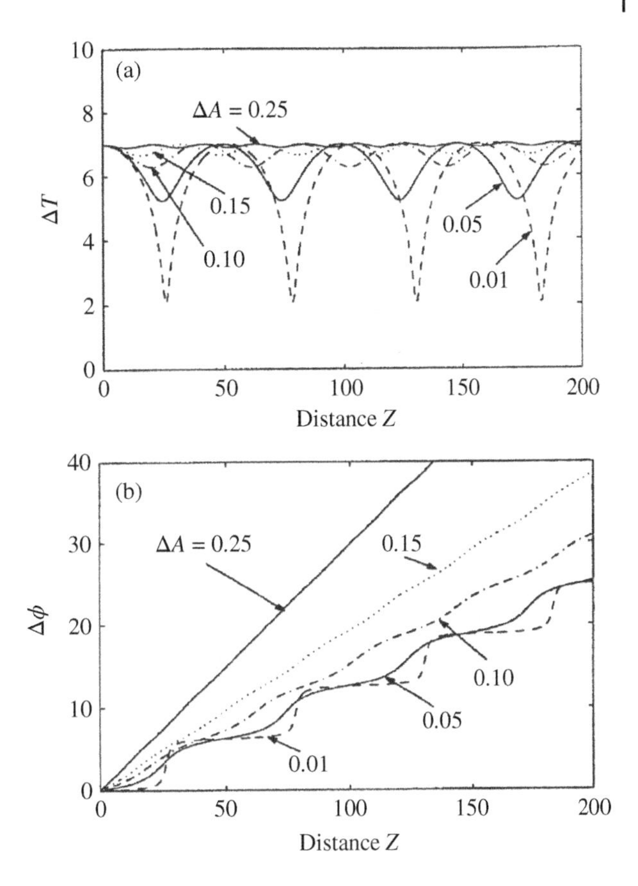

Figure 7.6 (a) Evolution of the separation and (b) of the phase difference between two initially in-phase solitons, for several values of their amplitude difference ΔA.

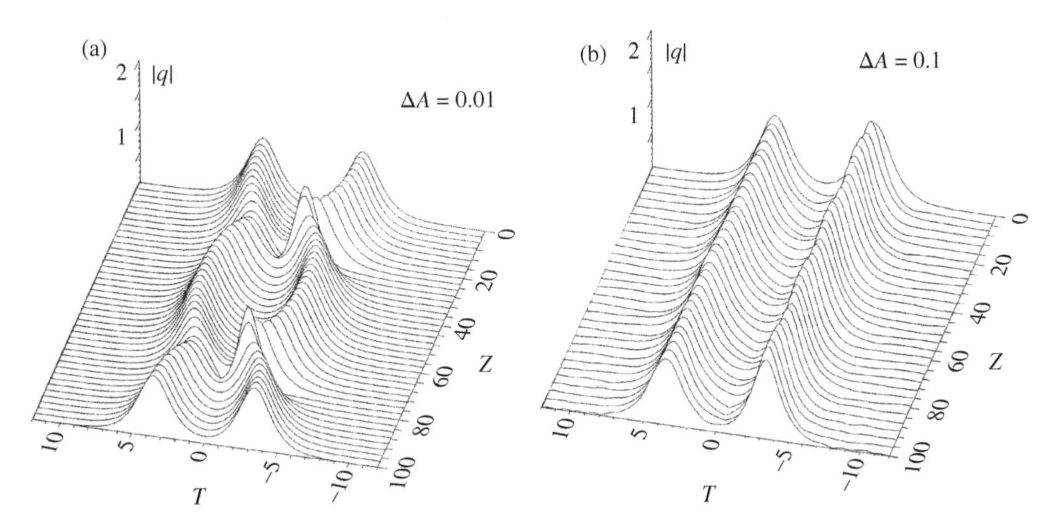

Figure 7.7 Interaction between two solitons with different amplitudes when $s_0 = 7$, $\Delta\phi = 0$, $\Delta A = 0.01$ (a), and $\Delta A = 0.1$ (b).

superposition of noise with the signal. For soliton transmission, this problem is generally not serious because the signal amplitude can be made large with proper choice of the fiber dispersion. The variation of the soliton position T_0 originates from the frequency shift κ due to the amplifier noise and the finite fiber dispersion. When the soliton frequency shifts, it modifies the group velocity through the group dispersion, resulting in jitter in arrival times of the soliton pulses. This is known as the *Gordon-Haus timing jitter* [30, 31, 41].

Under the effect of ASE, the values of the four soliton parameters in Eq. (7.1) change randomly at the output of each amplifier. The soliton perturbation theory can be used to find the variances of such fluctuations, assuming that the perturbation term due to ASE is given by

$$P(q) = n(Z, T) \exp(-i\kappa T + i\phi) \tag{7.42}$$

where $n(Z, T)$ is a Markovian stochastic process with Gaussian statistics with a zero average value and variance related with the spectral noise density S_{ASE}:

$$S_{ASE} = (G-1)n_{sp}h\nu_0 \tag{7.43}$$

In Eq. (7.43), n_{sp} is the spontaneous emission factor and G is the amplifier gain. Using Eqs. (7.5)–(7.8), we find the following results for the variances of fluctuations of the four soliton parameters [41, 42]:

$$\sigma_\eta^2 = 2\eta n_{sp}F(G)/N_s \tag{7.44}$$

$$\sigma_\kappa^2 = 2\eta n_{sp}F(G)/(3N_s) \tag{7.45}$$

$$\sigma_{T_0}^2 = \pi^2 n_{sp}F(G)/(6\eta N_s) \tag{7.46}$$

$$\sigma_\phi^2 = (\pi^2/6 + 2)\eta n_{sp}F(G)/(3\eta N_s) \tag{7.47}$$

where $F(G) = (G-1)^2/(G\ln G)$ and N_s is the number of photons in the soliton, given by

$$N_s = \frac{2P_0 t_0}{h\nu_0} \tag{7.48}$$

P_0 being the soliton peak power and t_0 the soliton pulse width.

Neglecting other perturbations, the soliton position variation at the end of one amplifier spacing due to the frequency shift κ is given from Eq. (7.7) as $T_0 = -\kappa Z_A$. Considering a link with N amplifiers, the total timing jitter is given by

$$T_0 = -Z_A \sum_{j=1}^{N} \sum_{i=1}^{j} \kappa_i \tag{7.49}$$

where κ_i is the frequency shift induced by the ith amplifier. When N is sufficiently large, the summations in Eq. (7.49) can be replaced by an integral and the variance of the timing jitter becomes [30]:

$$\sigma_{GH}^2 = \frac{\sigma_\kappa^2}{3} \frac{Z^3}{Z_A} \tag{7.50}$$

In Eq. (7.50), $Z = NZ_A$ is the total transmission distance and σ_κ^2 is the normalized variance of frequency fluctuations, given by Eq. (7.45).

As observed from Eq. (7.50), the Gordon-Haus timing jitter increases with the cube of distance and, in practice, it sets a limit on the bit-rate-distance product of a communication system.

Substituting Eq. (7.48), $P_0 = 1/(\gamma L_D)$, $z = NZ_A L_D$, and $R = 1/(s_0 t_0)$ in Eq. (7.50), we find that the total bit-rate-distance product Rz is limited by

$$(Rz)^3 < \frac{18\pi f_b^2 L_A}{n_{sp} F(G) s_0 \lambda h \gamma D} \tag{7.51}$$

where D is the dispersion parameter and f_b is the tolerable fraction of the bit slot corresponding to the arrival-time jitter. Considering a 10-Gb/s soliton communication system operating at 1.55 μm, with typical parameter values $L_A = 50$ km, $D = 0.5$ ps/(km−nm), $\alpha = 0.2$ dB/km, $\gamma = 3$ W^{-1}/km, $n_{sp} = 2$, $s_0 = 12$, and $f_b = 0.1$, the transmission distance is limited to $z = 6000$ km.

Besides the frequency, the ASE noise added by the amplifiers affects also the other three soliton parameters (amplitude, position, and phase), as well as its polarization. In the case of ultrashort solitons, ASE noise-induced amplitude fluctuations originate also a timing jitter, as will be seen in Chapter 9. Such timing jitter increases with the fifth power of distance and becomes more important than that due to the Gordon-Haus effect [43].

7.5.2 Polarization-Mode Dispersion Jitter

Polarization-mode dispersion (PMD) is an additional cause of timing jitter in soliton transmission systems [44]. As the solitons are periodically amplified, their state of polarization is scattered by the ASE noise added by optical amplifiers. Such polarization fluctuations lead to timing jitter in the arrival time of individual solitons through fiber birefringence because the two orthogonally polarized components travel with slightly different group velocities. The variance of the timing jitter introduced by the combination of ASE and PMD is given by [44]:

$$\sigma_P^2 = \frac{\pi}{16} \frac{F(G) n_{sp}}{N_s} \frac{D_p^2 z^2}{L_A} \tag{7.52}$$

where D_p is the so-called PMD parameter. It must be noted that σ_P increases linearly with both the transmission distance z and the PMD parameter D_p. Generally, the timing jitter due to PMD is much less than the Gordon-Haus jitter. However, the PMD-induced timing jitter becomes significant for fibers having large values of the PMD parameter and for soliton communication systems with bit rates of 40 Gbit/s or above.

7.5.3 Acoustic Jitter

Another contribution to the timing jitter, which was observed in the earliest long-distance soliton transmission experiments [45], arises from an acoustic interaction among the solitons. As a soliton propagates down the fiber, an acoustic wave is generated through electrostriction [46]. This acoustic wave induces a variation in the refractive index, which affects the speed of other pulses following in the wake of the soliton. As a consequence, they experience a steady acceleration proportional to the local slope of the induced index change. In practice, since a bit stream consists of a random string of 1 and 0 bits, changes in the speed of a given soliton depend on the presence or absence of solitons in the preceding bit slots. The timing jitter arises because, in these circumstances, different solitons experience different changes of speed.

Considering a fiber with $A_{eff} = 50$ μm^2, the standard deviation of the acoustic jitter can be approximated by [47]:

$$\sigma_{acou} \approx 4.3 \frac{D^2}{t_{FWHM}} z^2 \sqrt{R - 0.99} \tag{7.53}$$

where σ_{acou} is in ps, D is in ps/(km-nm), t_{FWHM} is in ps, R is in Gbit/s, and z is in thousands of km. For example, if $D = 0.5$ ps/(km−nm), $t_{FWHM} = 10$ ps, $z = 9000$ km (trans-Pacific distance), and $R = 10$ Gbit/s, we have $\sigma_{acou} = 26$ ps. Unlike the Gordon-Haus jitter, the acoustic jitter increases with bit rate and also increases as a higher power of the distance. As a result, the acoustic jitter becomes particularly important in high-bit rate long-distance communication systems. In practice, since the acoustic jitter has a deterministic nature, it is possible to reduce its impact by moving the detection window at the receiver through an automatic tracking circuit [48] or by using a coding scheme [49].

7.5.4 Soliton Interaction Jitter

As seen in Section 7.4, the interaction between two neighboring solitons depends strongly on the separation and relative phase between them. In principle, both aspects affect the soliton position in a deterministic way. However, in the presence of amplifier noise, the relative phase between the two solitons fluctuates randomly and the soliton timing jitter is generally enhanced [50].

It was found that, as a consequence of soliton interaction, the statistics of timing jitter deviates from the Gaussian statistics assumed in Section 7.5.1 to describe the Gordon-Hauss jitter [51]. In particular, Eq. (7.51) cannot be used to evaluate the timing jitter in such case. It has been found that the deviations from the Gaussian statistics become relevant even for a relatively weak soliton interaction ($s_0 > 10$). When the separation between adjacent solitons is so short that their interaction becomes quite important, the probability density function of the timing jitter develops a five-peak structure [52]. The use of numerical simulations becomes necessary to study the impact of timing jitter on a bit stream composed of interacting solitons.

7.6 WDM Soliton Systems

The capacity of a lightwave transmission system can be increased considerably by using the wavelength-division multiplexing (WDM) technique. A WDM soliton system transmits over the same fiber several soliton bit streams, distinguishable through their different carrier frequencies. The new feature that becomes important for WDM soliton systems is the possibility of collisions among solitons belonging to different channels because of their different group velocities.

7.6.1 Lossless Soliton Collisions

If two initially separated solitons have different carrier frequencies, they can collide during propagation and become again well separated. In the collision, the solitons experience a timing shift and a phase shift, but otherwise they recover fully. A collision between two solitons is illustrated in Figure 7.8. When the two pulse envelops overlap, beating between the two carrier frequencies is clearly discernible.

An important parameter in this context is the collision length, z_I, which is the distance the solitons must travel down the fiber together before the fast-moving soliton overtakes the slower one. Assuming that a collision begins and ends with overlap at the half-power points, z_I is given by

$$z_I = \frac{2t_{FWHM}}{D\Delta\lambda} \tag{7.54}$$

where $\Delta\lambda$ is the wavelength difference between the two pulses. For example, if $t_{FWHM} = 5$ ps, $D = 0.5$ ps/(km−nm), and $\Delta\lambda = 0.2$ nm, $z_I = 100$ km.

Figure 7.8 Collision between two solitons with different velocities.

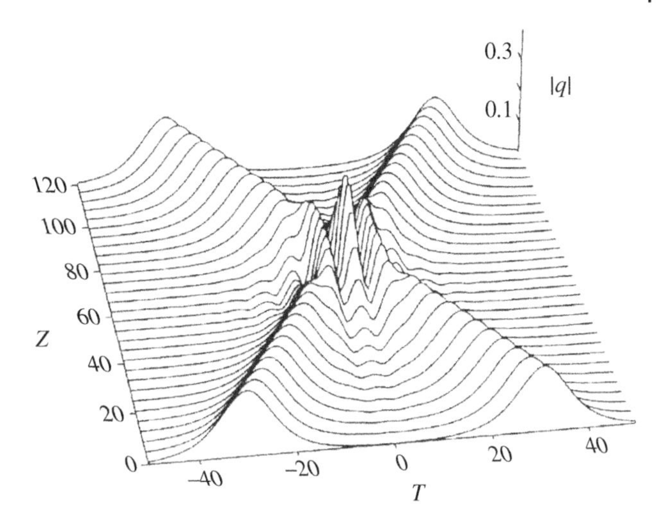

Considering the importance of dispersion management in the design of WDM systems, we use Eq. (7.15) to study the effects of collisions between solitons. Introducing a new propagation variable Z', related with Z in the form $dZ' = d(Z)dZ$, such equation becomes

$$i\frac{\partial u}{\partial Z'} + \frac{1}{2}\frac{\partial^2 u}{\partial T^2} + b|u|^2 u = 0 \tag{7.55}$$

where $b(Z) = a^2(z)/d(Z)$. To simplify the notation, the prime in Z' will be omitted in the following.

Consider the collision of two solitons of equal amplitude in different channels separated in frequency by $\omega_{ch} = 2\pi f_{ch} t_0$. The actual angular frequencies of the two channels are $\omega_0 \pm \omega_{ch}/2t_0$. Now insert $u = u_1 + u_2$ into Eq. (7.55), expand, and group terms according to their frequency dependencies. Neglecting the four-wave mixing (FWM) terms, the following two coupled equations for u_1 and u_2 are obtained [53]:

$$i\frac{\partial u_1}{\partial Z} + \frac{1}{2}\frac{\partial^2 u_1}{\partial T^2} + b\left(|u_1|^2 + 2|u_2|^2\right)u_1 = 0 \tag{7.56}$$

$$i\frac{\partial u_2}{\partial Z} + \frac{1}{2}\frac{\partial^2 u_2}{\partial T^2} + b\left(|u_2|^2 + 2|u_1|^2\right)u_2 = 0 \tag{7.57}$$

These equations are identical to the coupled NLSEs which describe the interaction of two pulses through cross-phase modulation. The second contribution in the nonlinear terms of Eqs. (7.56) and (7.57) corresponds to the cross-phase modulation and is zero except when the pulses overlap. It produces a frequency shift in the slower-moving soliton given by [53]:

$$\frac{d\kappa_1}{dZ} = \frac{b(Z)}{\omega_{ch}}\frac{d}{dZ}\left[\int_{-\infty}^{\infty} \operatorname{sech}^2(T - \omega_{ch}Z/2)\operatorname{sech}^2(T + \omega_{ch}Z/2)dT\right] \tag{7.58}$$

The change in κ_2 is symmetric of the change in κ_1. For a constant-dispersion and lossless fiber ($b = 1$), the integration of Eq. (7.58) can be performed analytically and the frequency shift is given by

$$\delta\kappa_1(Z) = -\delta\kappa_2(Z) = \frac{4}{\omega_{ch}}\frac{[\omega_{ch}Z\cosh(\omega_{ch}Z) - \sinh(\omega_{ch}Z)]}{[\sinh(\omega_{ch}Z)]^3} \tag{7.59}$$

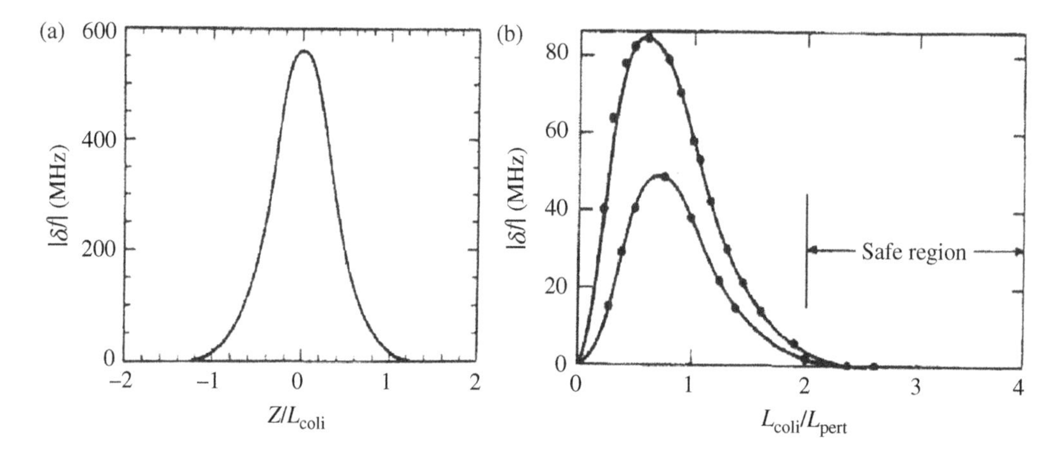

Figure 7.9 (a) Frequency shift during collision of two 50-ps solitons with 75-GHz channel spacing in a lossless fiber. (b) Residual frequency shift pertaining after collision because of lumped amplification. The amplifier spacing is L_A = 20 and 40 km for lower and upper curves, respectively. *Source:* After Ref. [53].

The peak frequency shift, at $Z = 0$, is

$$\delta\kappa_{\max} = 4/(3\omega_{ch}). \tag{7.60}$$

Equation (7.59) can be integrated to give the net temporal shift of the colliding soliton:

$$\delta T = 4/\omega_{ch}^2 \tag{7.61}$$

Figure 7.9 a shows the changes in the soliton frequency for the slower-moving soliton during collision. The maximum frequency shift occurs at the point of maximum overlap and has a value of about 0.6 GHz. The completed collision in a lossless fiber leave the soliton intact, with the same frequency and amplitude it had before the collision. The only changes concern their timings and phases.

Since these temporal shifts depend on the sequence of 1 and 0 bits, which occur randomly in real bit streams, different solitons of a channel shift by different amounts. As a consequence, soliton collisions cause some timing jitter even in lossless fibers.

7.6.2 Soliton Collisions in Perturbed Fiber Spans

Contrary to what happens in a lossless and constant-dispersion fiber system, soliton collisions are not symmetric in a system with real fibers and lumped amplifiers. As a result, solitons do not always recover their original frequency and velocity after the collision is over. In fact, the residual frequency shift increases rapidly as the collision length, z_I, approaches the amplifier spacing, L_A. However, the effects of gain-loss variations begin to average out and the residual frequency shift decreases when collisions occur over several amplifier spacings. Figure 7.9b shows the residual frequency shift remaining after a complete collision of initially well-separated solitons, given as a function of the ratio z_I/z_{pert}, where z_{pert} is equal to the amplifier spacing L_A. We observe that the residual frequency shift increases as z_I approaches L_A and can become ~0.1 GHz. However, it becomes essentially zero as long as the condition

$$z_I > 2L_A \tag{7.62}$$

is satisfied.

When combined with Eq. (7.54), the condition (7.62) puts an upper bound on the maximum separation between the two outermost channels of a WDM system:

$$\Delta\lambda_{max} = \frac{t_{FWHM}}{DL_A} \tag{7.63}$$

For example, if $t_{FWHM} = 10$ ps, $D = 0.5$ ps/(km−nm), and $L_A = 20$ km, Eq. (7.63) gives $\Delta\lambda_{max} = 1$nm. Assuming a channel wavelength spacing of 0.5 nm, the maximum allowable number of channels is just three.

It can be verified that Eq. (7.55) reduces to the standard NLSE when $b(Z) = 1$, that is, when the dispersion between lumped amplifiers varies at the same exponential rate as does the signal power. As a consequence, we conclude that the case of perfect transparency discussed above for a lossless and constant dispersion fiber also applies in this situation.

A technique to produce a dispersion decreasing fiber consists in reducing the core diameter along the fiber length during the fabrication process. In this way, the waveguide contribution to the fiber GVD is also progressively reduced. An accuracy better than 0.1 ps^2/km was estimated using this technique [54].

7.6.3 Timing Jitter

The timing jitter in WDM systems can be produced by different sources, besides those already discussed in Section 7.5 for the case of single-channel systems [55–60]. One of these sources is the collision-induced temporal shift given by Eq. (7.61), which occurs even in lossless fibers. Since such temporal shifts are bit-pattern dependent, different solitons of a channel shift by different amounts, thus inducing timing jitter. Another source is related with the relative time shift which occurs between two adjacent solitons in a given channel, since they experience a slightly different number of collisions. This happens because such adjacent solitons interact with different soliton sequences, shifted by one bit period, belonging to the copropagating channels [58]. Finally, multi-soliton collisions also contribute in a proper way to timing jitter. This contribution can be calculated by considering the summation of pairwise collisions when the channel spacing is sufficiently large.

Asymmetric collisions produce residual frequency shifts, which result also in temporal shifts due to corresponding change in the soliton velocity. Such asymmetric collisions occur as a consequence of energy variations in systems operating in the average-soliton regime. As seen above, this effect can be minimized if the collision distance satisfies the condition $z_I > 2L_A$, L_A being the amplifier spacing. Another important source of timing jitter is the residual frequency shift which results from an incomplete collision [55, 56]. This situation occurs when solitons from different channels are injected synchronously into the fiber link. In such case, the first half of the collision is absent and the frequency shift acquired during the second half is not cancelled.

References

1 Hasegawa, A. and Tappert, F. (1973). *Appl. Phys. Lett.* **23**: 142.

2 Mollenauer, L.F. and Smith, K. (1988). *Opt. Lett.* **13**: 675.

3 Hasegawa, A. and Kodama, Y. (1995). *Solitons in Optical Communications*. Oxford: Clarendon Press.

4 Mollenauer, L.F., Gordon, J.P., and Mamychev, P.V. (1997). Solitons in high bit-rate, long-distance transmission. In: *Optical Fiber Telecommunications*, vol. **IIIA** (ed. I.P. Kaminow and T.L. Koch). San Diego, CA: Academic Press, Chap. 12).

5 Essiambre, R.J. and Agrawal, G.P. (1997). Soliton communication systems. In: *Progress in Optics*, vol. **37** (ed. E. Wolf). Amsterdam: Elsevier, Chap. 4).
6 Iannone, E., Matera, F., Mecozzi, A., and Settembre, M. (1998). *Nonlinear Optical Communication Networks*. New York: Wiley, Chap. 5).
7 Agrawal, G.P. (2001). *Applications of Nonlinear Fiber Optics*. San Diego, CA: Academic Press, Chap. 8).
8 Ferreira, M.F. (2011). *Nonlinear Effects in Optical Fibers*. Hoboken, NJ: John Wiley & Sons, Chap. 8).
9 Karpman, V.I. and Maslov, E.M. (1977). *Zh. Eksp. Teor. Fiz.* **73**: 537. (*Sov. Phys. JETP* **46**, 281 (1977)).
10 Georges, T. (1995). *Opt. Fiber Technol.* **1**: 97.
11 Anderson, D. and Lisak, M. (1983). *Phys. Rev. A* **27**: 1393.
12 Hasegawa, A. (1995). *Pure Appl. Opt.* **4**: 265.
13 Ferreira, M.F., Facão, M.V., Latas, S.V., and Sousa, M.H. (2005). *Fiber Integr. Opt.* **24**: 287.
14 Blow, K.J. and Doran, N.J. (1985). *Opt. Commun.* **52**: 367.
15 Hasegawa, A. (1983). *Opt. Lett.* **8**: 650.
16 Kodama, Y. and Hasegawa, A. (1982). *Opt. Lett.* **7**: 339; 8, 342 (1983).
17 Mollenauer, L.F., Stolen, R.H., and Islam, M.N. (1985). *Opt. Lett.* **10**: 229.
18 Mollenauer, L.F., Gordon, J.P., and Islam, M.N. (1986). *IEEE J. Quantum Electron.* **22**: 157.
19 Hasegawa, A. and Kodama, Y. (1991). *Phys. Rev. Lett.* **66**: 161.
20 Spirit, D.M., Marshall, I.W., Constantine, P.D. et al. (1991). *Electron. Lett.* **27**: 222.
21 Nakazawa, M., Kubota, H., Kurakawa, K., and Yamada, E. (1991). *J. Opt. Soc. Am. B* **8**: 1811.
22 Kurokawa, K. and Nakazawa, M. (1992). *IEEE J. Quantum Electron.* **28**: 1922.
23 Rottwitt, K., Povlsen, J.H., Gundersen, S., and Bjarklev, A. (1993). *Opt. Lett.* **18**: 867.
24 Lester, C., Bertilsson, K., Rottwitt, K. et al. (1995). *Electron. Lett.* **31**: 219.
25 Karpman, V.I. and Solovev, V.V. (1981). *Physica* **3D**: 487.
26 Gordon, J.P. (1983). *Opt. Lett.* **8**: 596.
27 Mitschke, F.M. and Mollenauer, L.F. (1987). *Opt. Lett.* **12**: 355.
28 Desem, C. and Chu, P.L. (1987). *IEE Proc.* **134**: 145.
29 Kodama, Y. and Nozaki, K. (1987). *Opt. Lett.* **12**: 1038.
30 Gordon, J.P. and Haus, H.A. (1986). *Opt. Lett.* **11**: 665.
31 Marcuse, D. (1992). *J. Lightwave Technol.* **10**: 273.
32 Smith, N.J., Foryisak, W., and Doran, N.J. (1996). *Electron. Lett.* **32**: 2085.
33 Carter, G.M., Jacob, J.M., Menyuk, C.R. et al. (1997). *Opt. Lett.* **22**: 513.
34 Kumar, S. and Lederer, F. (1997). *Opt. Lett.* **22**: 1870.
35 Kutz, J.N. and Wai, P.K.A. (1998). *IEEE Photon. Technol. Lett.* **10**: 702.
36 Okamawari, T., Maruta, A., and Kodama, Y. (1998). *Opt. Lett.* **23**: 694; *Opt. Commun.* **149**, 261 (1998).
37 Grigoryan, V.S., Menyuk, C.R., and Mu, R.M. (1999). *J. Lightwave Technol.* **17**: 1347.
38 Ferreira, M.F. and Latas, S.V. (2001). *J. Lightwave Technol.* **19**: 332.
39 Santhanam, J., McKinstrie, C.J., Lakoba, T.I., and Agrawal, G.P. (2001). *Opt. Lett.* **26**: 1131.
40 Poutrina, E. and Agrawal, G.P. (2002). *IEEE Photon. Technol. Lett.* **14**: 39.
41 Georges, T. and Favre, F. (1993). *J. Opt. Soc. Am. B* **10**: 1880.
42 Haus, H.A. and Lai, Y. (1990). *J. Opt. Soc. Am. B* **7**: 386.
43 Facão, M.V. and Ferreira, M.F. (2001). *J. Nonlinear Math. Phys.* **8**: 112.
44 Mollenauer, L.F. and Gordon, J.P. (1994). *Opt. Lett.* **19**: 375.
45 Smith, K. and Mollenauer, L.F. (1989). *Opt. Lett.* **14**: 1284.
46 Dianov, E.M., Luchnikov, A.V., Pilipetskii, A.N., and Prokhorov, A.M. (1992). *Appl. Phys B* **54**: 175.
47 Mollenauer, L.F., Mamyshev, P.V., and Neubelt, M.J. (1994). *Opt. Lett.* **19**: 704.
48 Mollenauer, L.F. (1996). *Opt. Lett.* **21**: 384.

49 Adali, T., Wang, B., Pilipetskii, A.N., and Menyuk, C.R. (1998). *J. Lightwave Technol.* **16**: 986.

50 Georges, T. and Fabre, F. (1991). *Opt. Lett.* **16**: 1656.

51 Menyuk, C.R. (1995). *Opt. Lett.* **20**: 285.

52 Pinto, A.N., Agrawal, G.P., and da Rocha, J.F. (1998). *J. Lightwave Technol.* **16**: 515.

53 Mollenauer, L.F., Evangelides, S.G., and Gordon, J.P. (1991). *J. Lightwave Technol.* **9**: 362.

54 Richardson, D.J., Chaberlin, R.P., Dong, L., and Payne, D.N. (1995). *Electron. Lett.* **31**: 1681.

55 Kodama, Y. and Hasegawa, A. (1991). *Opt. Lett.* **16**: 208.

56 Aakjer, T., Povlsen, J.H., and Rottwitt, K. (1993). *Opt. Lett.* **18**: 1908.

57 Chakravarty, S., Ablowitz, M.J., Sauer, J.R., and Jenkins, R.B. (1995). *Opt. Lett.* **20**: 136.

58 Jenkins, R.B., Sauer, J.R., Chakravarty, S., and Ablowitz, M.J. (1995). *Opt. Lett.* **20**: 1964.

59 Tang, X.Y. and Chin, M.K. (1995). *Opt. Commun.* **119**: 41.

60 Wai, P.K.A., Menyuk, C.R., and Raghavan, B. (1996). *J. Lightwave Technol.* **14**: 1449.

8

Soliton Transmission Control

Although solitons are clearly a good choice as the information carrier in optical fibers because of their robust nature, their interaction and the timing jitter can lead to the loss of information in transmission systems. These problems can be suppressed using some techniques to control the soliton parameters, namely, its amplitude (or width), time position, velocity (or frequency), and phase. In fact, while the original wave equation has infinite-dimensional parameters, control of finite-dimensional parameters is sufficient to control the soliton transmission system. Several techniques were developed during the 1990s for controlling both the interaction and the timing jitter in solitons systems [1–15].

8.1 Fixed-Frequency Filters

The use of narrow-band filters was early suggested to control the timing jitter and other noise effects [1, 2, 4]. The basic idea is that any soliton whose central frequency has strayed from the filter peak will return to the peak by virtue of the differential loss induced by the filters across its spectrum. The timing jitter is suppressed as a consequence of the frequency jitter damping. Using etalon guiding filters, the multiple peaks present in their response provide the possibility to use them in WDM soliton transmission systems. Actually, the spectral filtering effect can also be provided by the optical amplifier itself, due to the limited bandwidth of its gain curve.

When expanded in a Taylor series, the logarithm of the filter transfer function H takes the form:

$$\ln\left[H\left(\omega-\omega_f\right)\right] = i\alpha_1\left(\omega-\omega_f\right) + \alpha_2\left(\omega-\omega_f\right)^2 + i\alpha_3\left(\omega-\omega_f\right)^3 + \cdots \tag{8.1}$$

where the constants α_j are all real and ω_f is the filter peak frequency. For example, in the case of a lumped Fabry-Perot filter, the complex transfer function is

$$H\left(\omega-\omega_f\right) = \frac{1}{1+2i\left(\omega-\omega_f\right)/B} \tag{8.2}$$

where B is the bandwidth of the filter. In this case, we have $\alpha_1 = -2/B$, $\alpha_2 = -2/B^2$, and $\alpha_3 = 8/(3B^3)$. In practice, a filter is often inserted together with each optical amplifier. Then, the effect of spectral filtering can be considered in a distributed form as $h(\omega-\omega_f) = \text{In}[H(\omega-\omega_f)]/L_A$, L_A being the amplifier spacing.

Solitons in Optical Fiber Systems, First Edition. Mario F. S. Ferreira.

The pulse propagation along a transmission line in the presence of optical amplification and frequency filtering can be described by the following perturbed nonlinear Schrödinger equation [4]:

$$i\frac{\partial U}{\partial z} - \frac{1}{2}\beta_2\frac{\partial^2 U}{\partial \tau^2} + \gamma|U|^2 U = ig_1 U + ig_2\frac{\partial^2 U}{\partial \tau^2} \tag{8.3}$$

where g_1 is the excess gain, given by the difference between the amplifier gain rate and the fiber loss rate, while g_2 is related to the filter bandwidth.

Equation (8.3) has an exact soliton solution given by [5, 16, 17]:

$$U(z, \tau) = u \exp(ikz)\left[\text{sech}\left(\frac{\tau}{w}\right)\right]^{1 + iC} \tag{8.4}$$

provided that the amplitude, u, wavenumber, k, width, w, chirp, C, and excess gain, g_1, are related by the equations:

$$0 = 3\beta_2 C + g_2(2 - C^2) \tag{8.5}$$

$$\gamma u^2 = -\beta_2(2 - C^2)/2w^2 + 3g_2 C/2w^2 \tag{8.6}$$

$$g_1 = -2\beta_2 C/w^2 - g_2(1 - C^2)/w^2 \tag{8.7}$$

$$k = -\beta_2(1 - C^2)/2w^2 + g_2 C/w^2 \tag{8.8}$$

By combining Eqs. (8.5) and (8.7), one finds that

$$g_1 = g_2(1 + C^2)/3w^2 \tag{8.9}$$

We verify from Eq. (8.9) that excess gain is required to compensate for filtering-induced loss.

Equation (8.4) is known as the solution of Pereira and Stenflo [16]. Although the amplitude profile of this solution is a hyperbolic secant as in the case of the fundamental NLSE soliton, two important differences exist between the two solutions. First, for the soliton given by Eq. (8.4), the amplitude and width are independently fixed by the parameters of Eq. (8.3), whereas for the fundamental NLSE soliton, we have $u = 1/w$. The second difference is that the soliton in Eq. (8.4) is chirped.

Equation (8.3) can be written in following normalized form:

$$i\frac{\partial q}{\partial Z} + \frac{1}{2}\frac{\partial^2 q}{\partial T^2} + |q|^2 q = i\delta q + i\beta\frac{\partial^2 q}{\partial T^2} \tag{8.10}$$

where δ is the normalized excess gain necessary to compensate for the filter-induced loss and β is the normalized filter strength.

8.1.1 Control of Timing Jitter

If both coefficients in the right-hand side of Eq. (8.10) are small, it can be considered as a perturbed version of the NLSE. In such case, the control of timing jitter and/or soliton interaction by means of frequency filtering can be demonstrated using the adiabatic soliton perturbation theory presented in Section 7.1. In this case, the perturbation term, $P(q)$, in Eqs. (7.5)–(7.8) reads:

$$P(q) = \delta q + \beta\frac{\partial^2 q}{\partial T^2} \tag{8.11}$$

In particular, we obtain the following evolution equations for the soliton amplitude, η, and the frequency, κ:

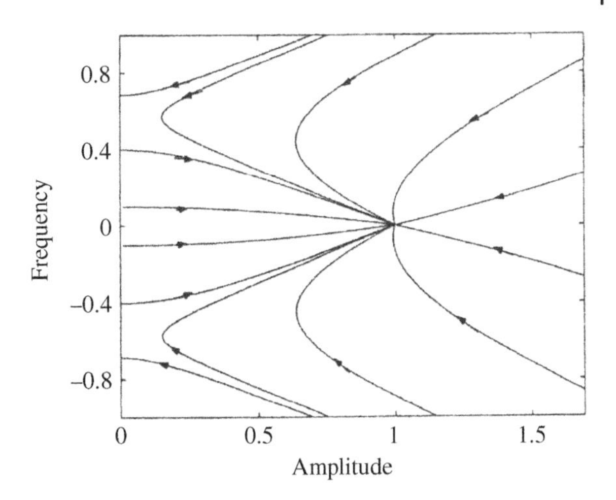

Figure 8.1 phase-plane of Eqs. (8.12) and (8.13) for the case $\beta = 3\delta$ and $\beta = 0.15$.

$$\frac{\partial \eta}{\partial Z} = 2\delta\eta - 2\beta\eta\left(\frac{1}{3}\eta^2 + \kappa^2\right) \tag{8.12}$$

$$\frac{\partial \kappa}{\partial Z} = -\frac{4}{3}\beta\eta^2\kappa \tag{8.13}$$

From Eq. (8.13), it can be seen that the soliton frequency approaches asymptotically to $\kappa = 0$ (stable fixed point) if $\eta \neq 0$. In such limit, Eq. (8.12) provides a stationary amplitude $\eta = 1$ if the condition $\beta = 3\delta$ is satisfied.

Figure 8.1 shows the phase-plane of Eqs. (8.12) and (8.13) for the case $\beta = 3\delta$ and $\beta = 0.15$. We observe that solitons having an initial range of amplitudes and frequencies emerge as solitons with an identical amplitude and frequency imposed by the attractor at $\eta = 1$ and $\kappa = 0$ after repeated amplifications. This process may be interpreted as soliton cooling and can be used to overcome the Gordon-Haus effect. In fact, the use of fixed-frequency guiding filters provides an effective suppression of the soliton timing jitter, whose variance is reduced relatively to the uncontrolled case, given by Eq. (7.51), by the factor [4, 18].

$$F_r(x) = \frac{3}{2}\frac{1}{x^3}[2x - 3 + 4\exp(-x) - \exp(-2x)] \tag{8.14}$$

where $x = 4\beta Z/3$. If $x >> 1$, we have $f(x) \sim 3x^{-2}$ and the variance of the timing jitter increases linearly with distance, instead of the cubic dependence shown in the uncontrolled case [2, 4, 18]. For a vanishing filter strength ($\beta \to 0$), we have $F_r = 1$ and the Gordon-Haus result for the timing jitter is recovered.

8.1.2 Control of Soliton Interaction

The interaction between adjacent solitons in the presence of spectral filtering and excess gain can be studied using the quasi-particle approach described in Section 7.4. The following set of equations is derived for the parameters A, B, ΔA, ΔB, ΔT, and $\Delta\phi$:

$$\frac{dA}{dz} = 2\delta A - \frac{2}{3}\beta A^3 - 2\beta AB^2 \tag{8.15}$$

$$\frac{dB}{dz} = -\frac{4}{3}\beta A^2 B \tag{8.16}$$

$$\frac{d(\Delta A)}{dz} = 8A^3 e^{-A\Delta T} \sin(\Delta\phi) + \left[2\delta - 2\beta\left(A^2 + B^2\right)\right]\Delta A - 4\beta AB\Delta B \tag{8.17}$$

$$\frac{d(\Delta B)}{dz} = 8A^3 e^{-A\Delta T} \cos(\Delta\phi) - \frac{4}{3}\beta A^2 \Delta B - \frac{8}{3}\beta AB\Delta A \tag{8.18}$$

$$\frac{d(\Delta T)}{dz} = -\Delta B \tag{8.19}$$

$$\frac{d(\Delta\phi)}{dz} = A\Delta A - \frac{4}{3}\beta A^2 B\Delta T \tag{8.20}$$

From Eqs. (8.15) and (8.16), we have a stationary point ($A = 1$, $B = 0$) if the condition $\beta = 3\delta$ is satisfied. In this case, considering two adjacent solitons with equal amplitudes but a nonzero initial phase difference, Eqs. (8.17)–(8.20) are reduced to the following equations:

$$\frac{d^2(\Delta T)}{dz^2} = -\frac{4}{3}\beta\frac{d(\Delta T)}{dz} - 8e^{-\Delta T}\cos(\Delta\phi) \tag{8.21}$$

$$\frac{d^2(\Delta\phi)}{dz^2} = -\frac{4}{3}\beta\frac{d(\Delta\phi)}{dz} + 8e^{\Delta T}\sin(\Delta\phi) \tag{8.22}$$

We observe from Eqs. (8.21) and (8.22) that the introduction of frequency filters determines a damping in the evolution of both the separation and the phase difference between the two pulses. In particular, if the pulses are initially in phase, $\Delta\phi$ remains zero and the filters determine a reduction of the attracting forces between them. This is illustrated in Figure 8.2a, which shows the evolution of the separation ΔT for several values of the filter strength, considering an initial separation $\Delta T_0 = 7$. Figure 8.2b shows the evolution of the separation for $\beta = 0.3$, considering several values of the initial phase difference $\Delta\phi$. We verify that the collision is avoided in all cases, even if a mutual attraction between the pulses occurs in the initial stage of propagation.

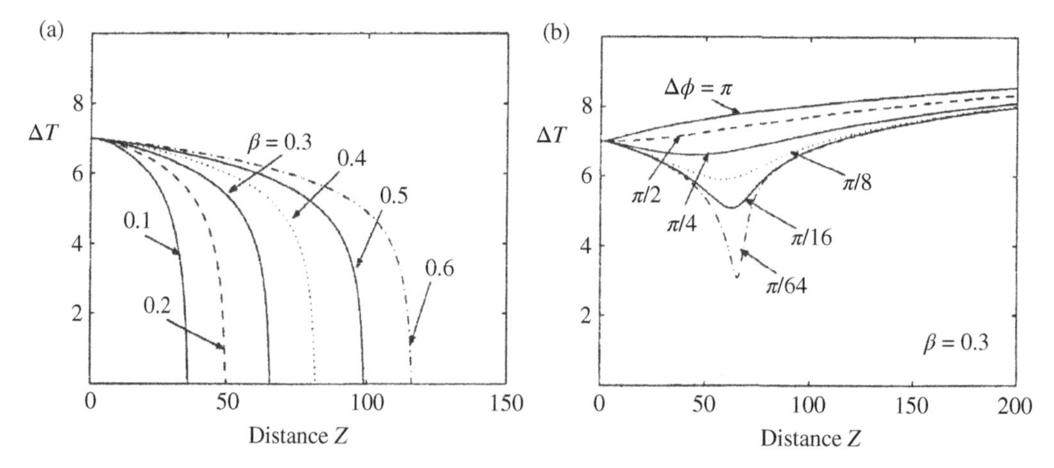

Figure 8.2 Evolution of the separation, ΔT, between two pulses initially separated by $\Delta T_0 = 7$ for (a) several values of the filter strength, β, considering $\Delta\phi_0 = 0$, and (b) different values of the initial phase difference, $\Delta\phi_0$, considering $\beta = 0.3$.

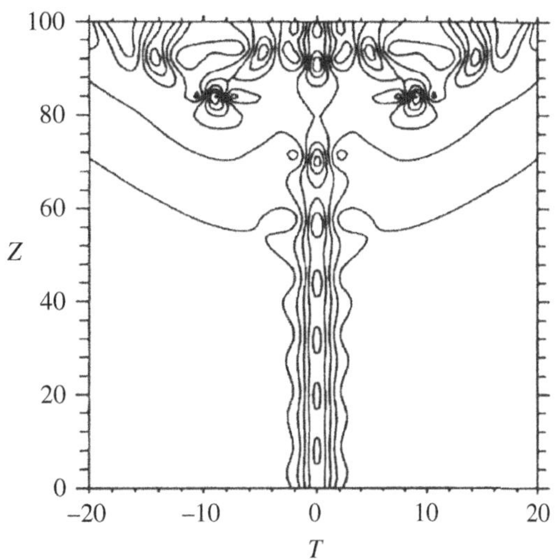

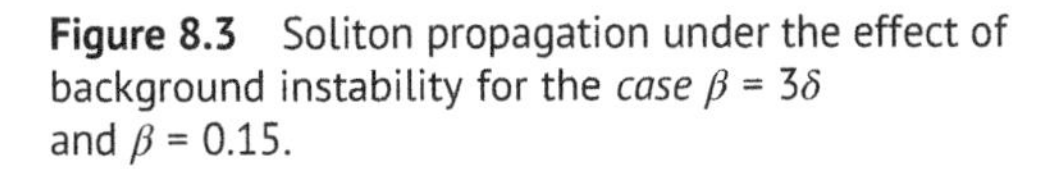
Figure 8.3 Soliton propagation under the effect of background instability for the *case* $\beta = 3\delta$ and $\beta = 0.15$.

8.1.3 Background Instability

As mentioned above, in the presence of spectral filtering, some excess gain must be provided around the filter center frequency to compensate for the loss that solitons suffer at wings of their spectrum. However, this excess gain amplifies also linear waves coexistent with the soliton trains, leading to instability of the background. The unstable linear waves degrade the SNR and if their power grows comparable to that of the soliton, they can affect significantly and even destroy the soliton itself. This process is illustrated in Figure 8.3 for $\beta = 3\delta$ and $\delta = 0.05$. In this case, we observe that the soliton is severely disturbed for transmission distances $Z > 80$.

Figure 8.4 shows the evolution of a pair of in-phase solitons with an initial separation $\Delta T_0 = 7$ considering the condition $\beta = 3\delta$, for the cases a) $\beta = 0.15$ and b) $\beta = 0.3$. In the case a), we observe the collision of the two solitons at $Z \sim 38$, after which they are destroyed due to the background instability. In the case b), such instability is stronger and its impact becomes significant even before the first collision. This is due to the higher value of the filter strength, and the corresponding higher value of the excess gain, which amplifies the linear waves.

8.2 Sliding-Frequency Filters

An approach to avoid the background instability was suggested in 1992 by Mollenauer et al. [5] and consists in using filters whose central frequency is gradually shifted along the transmission line. In this scheme, the linear waves, which initially grow near $\kappa = 0$, eventually fall into negative gain region of the filter and are dissipated while the soliton central frequency is shifted, following the central frequency of the filter. This example illustrates a remarkable property of solitons, since the signal carried by them can be effectively separated from noise which has the same frequency components as the signal. The benefits of using sliding filters have been experimentally demonstrated by Mollenauer et al. [19, 20]. Actually, sliding-frequency filters are useful also in WDM systems [21–25]. Filtering of all channels can be realized simultaneously using Fabry–Perot filters due to their periodic transmission windows.

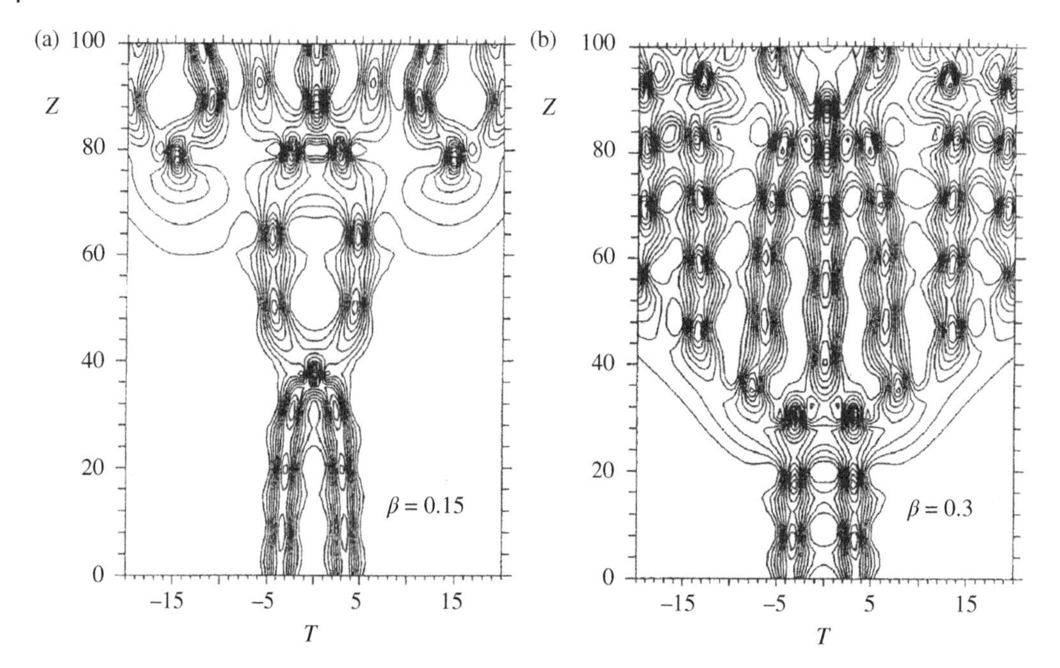

Figure 8.4 Evolution of a pair of in-phase solitons with an initial separation $\Delta T_0 = 7$, considering the condition $\beta = 3\delta$, for the cases (a) $\beta = 0.15$ and (b) $\beta = 0.3$.

8.2.1 Evolution of Soliton Parameters

Keeping terms through the third-order in Eq. (8.1), the soliton propagation in a system with sliding-frequency guiding filters can be described by the following averaged perturbed nonlinear Schrödinger equation [26]:

$$\frac{\partial \nu}{\partial Z} - \frac{i}{2}\frac{\partial^2 \nu}{\partial T^2} - i|\nu|^2\nu = \delta\nu + \beta\left\{\frac{\partial}{\partial T} + i\omega_f(Z)\right\}^2 \nu - \rho\left\{\frac{\partial}{\partial t} + i\omega_f(z)\right\}^3 \nu \tag{8.23}$$

swhere $\omega_f(Z)$ is the filter peak frequency, δ is the normalized excess gain, β is the normalized filter strength, and ρ is the third-order term, which for typical parameters has a value $\rho \approx \beta/2$ [26]. Assuming that the peak frequency $\omega_f(Z)$ is a linear function in Z, we can write:

$$\omega_f(Z) = -\alpha_s Z, \tag{8.24}$$

where α_s is the sliding rate.

Let us consider the following ansatz for the soliton pulse shape of Eq. (8.23):

$$\nu = \eta \mathrm{sech}(\eta T - pZ).\exp\left[i\frac{1}{2}\eta^2 Z + 3i\rho\eta\tanh(\eta T - pZ) - i\Omega T\right] \tag{8.25}$$

where $p = -\eta(\Omega + \rho\eta^2)$.

The mean frequency ω_0 of the soliton given by Eq. (8.25) is related to Ω by the relation:

$$\omega_0 = \frac{\mathrm{Im}\int dT \nu \partial\nu * /\partial T}{\int dT|\nu|^2} = \Omega - 2\rho\eta^2 \tag{8.26}$$

Substituting Eq. (8.25) into Eq. (8.23) we obtain through the first-order perturbation theory the following pair of coupled equations for the amplitude η and the mean frequency ω_0 of the soliton:

$$\frac{d\eta}{dZ} = 2\delta\eta - 2\beta\eta\left[\left(\omega_0 - \omega_f\right)^2 + \frac{1}{3}\eta^2\right] \tag{8.27}$$

$$\frac{d\omega_0}{dZ} = -\frac{4}{3}\beta\eta^2\left[\left(\omega_0 - \omega_f\right) - \frac{6}{5}\rho\eta^2\right] \tag{8.28}$$

Combining with Eqs. (8.24), (8.27), and (8.28), assume the following aspect:

$$\frac{\partial\eta}{\partial Z} = 2\delta\eta - 2\beta\eta\left(\frac{1}{3}\eta^2 + \kappa^2\right) \tag{8.29}$$

$$\frac{\partial\kappa}{\partial Z} = \alpha_s - \frac{4}{3}\beta\eta^2\left(\kappa - \frac{6}{5}\rho\eta^2\right) \tag{8.30}$$

where $\kappa = \omega_0 - \omega_f$. We note that Eq. (8.29) is equal to Eq. (8.12). At equilibrium and for $\eta = 1$, we have:

$$\kappa^2 = \frac{\delta}{\beta} - \frac{1}{3} \tag{8.31}$$

$$\kappa = \frac{3\alpha_s}{4\beta} + \frac{6}{5}\rho \tag{8.32}$$

If the third-order contribution is neglected, Eq. (8.32) shows a symmetric behavior of the mean frequency offset $\kappa = \omega_0 - \omega_f$ for up-sliding ($\alpha_s < 0$) and down-sliding ($\alpha_s > 0$). Physically, sliding leads to an offset because the soliton has inertia so that its central frequency lags behind the central frequency of the filter as it slides. However, a different behavior is observed when the third-order contribution is taken into account. In this case, we observe from Eq. (8.32) that the third-order term reduces the frequency offset for up-sliding, whereas the opposite occurs for down-sliding. As a consequence, Eq. (8.31) shows that the excess gain δ needed to compensate for the filter loss becomes smaller in the case of up-sliding and larger for down-sliding. In the first case, an optimum situation is achieved for a sliding rate given by:

$$\alpha_s = -\frac{8\beta\rho}{5} \tag{8.33}$$

when the mean frequency offset becomes zero and the excess gain has a minimum value $\delta_m = \beta/3$.

There exists a critical value for the sliding rate, α_c, such that the number of stationary points of Eqs. (8.29) and (8.30) is two for $|\alpha_s| < \alpha_c$, one at $|\alpha_s| = \alpha_c$, and none for $|\alpha_s| > \alpha_c$. This means that for strong sliding the soliton is pushed outside of the filter and is destroyed. In the case $|\alpha_s| < \alpha_c$, we have one stable point, (κ_s, η_s), and one unstable point, (κ_u, η_u), such that $|\kappa_s| \leq |\kappa_u|$ and $|\eta_s| \geq |\eta_u|$. This case is illustrated in Figure 8.5, which shows the phase-plane of Eqs. (8.29) and (8.30) considering the values $\alpha_s = 0.075$, $\beta = 0.3$, $\delta = 0.11$, and $\rho = 0$. These values provide a stable point ($\kappa_s = 0.1875,\ \eta_s = 1$).

Figure 8.5 corresponds to a case of down-sliding, as given by Eq. (8.24). By reversing he sign of sliding, we could obtain a symmetrical picture relatively to the line $\kappa = 0$. However, such symmetry doesn't occur if the third-order term in the filtering effect is included, as can be anticipated from Eq. (8.32).

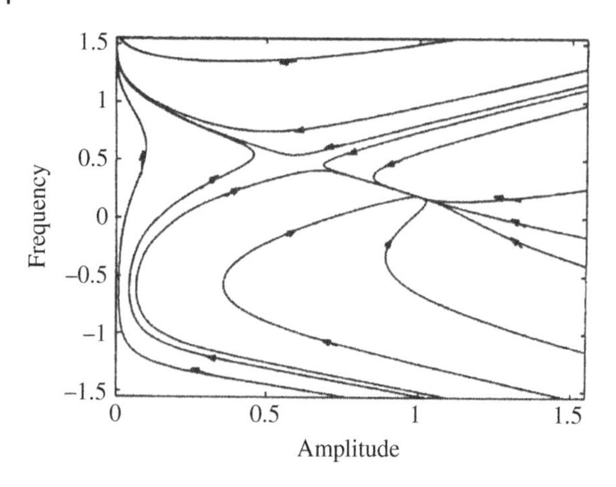

Figure 8.5 phase-plane of Eqs. (8.29) and (8.30) for the case $\alpha_s = 0.075, \beta = 0.3, \delta = 0.111$, and $\rho = 0$.

Performing in Eq. (8.25) a change of variables such as [27, 28],

$$q(Z', T) = v(Z.T)\exp\left[-i\alpha_s ZT + \frac{i}{6}\alpha_s^2 Z^3\right] \tag{8.34}$$

where $Z' = Z$ and $T' = T - \alpha_s Z^2/2$ we obtain the following perturbed NLSE [18]:

$$i\frac{\partial q}{\partial Z'} + \frac{1}{2}\frac{\partial^2 q}{\partial T'^2} + |q|^2 q = \alpha_s T' q + i\delta q + i\beta\frac{\partial^2 q}{\partial T'^2} + \rho\frac{\partial^3 q}{\partial T'^3} \tag{8.35}$$

Figure 8.6 illustrates the evolution of the soliton peak amplitude by numerically solving Eq. (8.35), assuming different values for the excess linear gain δ and considering the parameter values $\beta = 0.3$, $\rho = 0.15$, $\alpha_s = 0.075$ (a), and $\alpha_s = -0.075$ (b). In both cases, we observe that the pulse decays when the excess gain is below a given critical value δ_c, In such case, the gain is not sufficient to compensate for the filtering induced loss and the soliton is not able to follow the filter sliding. Figure 8.6 shows that the critical value δ_c is higher for down-sliding than for up-sliding.

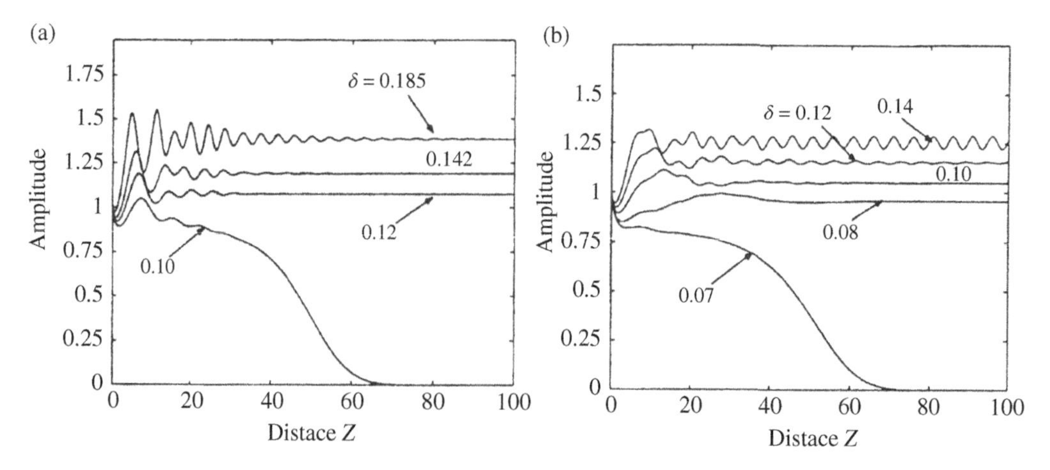

Figure 8.6 Evolution of the soliton peak amplitude for the case $\beta = 0.3, \rho = 0.15$, and (a) $\alpha_s = 0.075$ or (b) $\alpha_s = -0.075$. The values of the excess linear gain δ are given in the figures.

8.2.2 Control of Timing Jitter

Introducing the first-order quantities $\omega_0 = \omega_{0s} + \Delta\omega$ and $\eta = 1 + a$ and linearizing Eqs. (8.27) and (8.28) about the steady state, we obtain:

$$\frac{da}{dZ} = -\frac{4}{3}\beta a - 4\beta\Delta\omega_1\Delta\omega \tag{8.36}$$

$$\frac{d\Delta\omega}{dZ} = -\frac{8}{3}\beta\Delta\omega_2 a - \frac{4}{3}\beta\Delta\omega \tag{8.37}$$

where

$$\Delta\omega_1 = 3\alpha_s/4\beta + 6\rho/5, \tag{8.38}$$

$$\Delta\omega_2 = 3\alpha_s/4\beta - 6\rho/5 \tag{8.39}$$

The normal modes of Eqs. (8.36) and (8.37) have the following damping coefficients

$$\gamma_1 = \frac{4}{3}\beta(1 - \omega_r), \quad \gamma_2 = \frac{4}{3}\beta(1 + \omega_r) \tag{8.40}$$

where

$$\omega_r = \sqrt{6\Delta\omega_1\Delta\omega_2} \tag{8.41}$$

For $|\omega_r| > 1$one of the damping coefficients becomes negative and the system becomes unstable. This condition determines a maximum absolute value of the sliding rate, given by:

$$\alpha_{s,max} = \frac{4}{3}\sqrt{\frac{1}{6} + \frac{36}{25}\rho^2} \tag{8.42}$$

Adding to Eqs. (8.36) and (8.37) appropriate delta-correlated noise functions to take into account for the ASE noise of the in-line amplifiers and using standard techniques, it is found that the variance of the variance of the Gordon-Haus timing jitter is reduced relatively to the uncontrolled case, discussed in Section 7.5, by a factor [29]:

$$F_r(x_1, x_2) = \frac{3k_1}{4}[g(x_1, x_1) + g(x_2, x_2)] - \frac{k_2}{2}g(x_1, x_2) \tag{8.43}$$

where $x_i = \gamma_i Z$, $i = 1{,}2$ and

$$g(x_1, x_2) = \frac{3}{x_1 x_2}[1 - h(x_1) - h(x_2) + h(x_1 + x_2)] \tag{8.44}$$

$$h(x_i) = \frac{1 - \exp(-x_i)}{x}, \quad i = 1, 2 \tag{8.45}$$

$$k_1 = \frac{2\Delta\omega_2 + \Delta\omega_1}{3\Delta\omega_1} \tag{8.46}$$

$$k_2 = \frac{2\Delta\omega_2 - \Delta\omega_1}{\Delta\omega_1} \tag{8.47}$$

Figure 8.7 illustrates the variation with the propagation distance of the reduction factor F_r of the timing jitter for $\rho = 0$, $\beta = 0.1$, and several values of the sliding rate: $\alpha_s = 0$ (curve a), $\alpha_s = 0.03$ (curve b), $\alpha_s = 0.04$ (curve c), and $\alpha_s = 0.05$ (curve d). Besides the positive effect in controlling the background instability, Figure 8.7 shows that the use of sliding frequency filters increases the timing jitter relatively to the case of fixed-filters ($\alpha_s = 0$). The enhancement of timing jitter with sliding

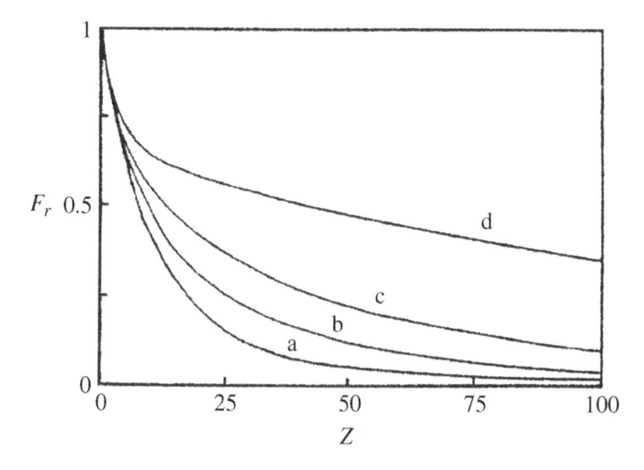

Figure 8.7 Reduction factor of the Gordon-Haus timing jitter, given by Eqs. (8.43)–(8.47) for $\rho = 0$, $\beta = 0.1$, and different values of the sliding rate: $\alpha_s = 0$ (curve a), $\alpha_s = 0.03$ (curve b), $\alpha_s = 0.04$ (curve c), and $\alpha_s = 0.05$ (curve d).

was verified experimentally by Mollenauer et al. [20], and it is due to the coupling between amplitude and frequency fluctuations.

It can be verified from Eqs. (8.43)–(8.47) that, for large values of Z, the reduction factor F_r is given by:

$$F_r = f\frac{27}{(4\beta_2 Z)^2} \tag{8.48}$$

where

$$f = \frac{1 + 12(\Delta\omega_2)^2}{(1 - 6\Delta\omega_1\Delta\omega_2)^2} \tag{8.49}$$

The factor f is an enhancement factor of the timing jitter that is due to sliding and the third-order filter term.

Figure 8.8 illustrates the effect of the third-order term on the reduction factor of the timing jitter, for the cases $\beta = 0.1$ (a) and $\beta = 0.3$ (b). In both cases the dashed curve corresponds to $\rho = 0$ and $\alpha_s = \pm\beta/2$, while the full curves were obtained considering $\rho = \beta/2$ and $\alpha_s = \pm\beta/2$. We observe that the timing jitter is larger in the case of up-sliding, assuming that the sliding rate is the same for both up- and down-sliding. This behavior is due to coupling strength between amplitude and frequency fluctuations, which is greater for up-sliding than for down-sliding, as it can be seen from Eq. (8.37). In fact, considering the numerator of Eq. (8.49), we see that for up-sliding the two terms in $\Delta\omega_2$ have the same sign (see the Eq. (8.39)), whereas they are of opposite signs for down-sliding, resulting a higher value of f in the first case. A particularly interesting case occurs for a down-sliding rate $\alpha_s = 8\beta\rho/5$, for which we have $\Delta\omega_2 = 0$ and the enhancement factor given by Eq. (8.49) is $f = 1$. On the other hand, considering the up-sliding case at the same sliding rate we have $\Delta\omega_1 = 0$ and the enhancement factor becomes $f = 1 + 12(\Delta\omega_2)^2$. For example, if we assume $\rho = \beta/2$ and $\beta = 0.3$ this gives $f = 2.55$.

Regarding the timing jitter enhancement due to the sliding, the reference is assumed to be the fixed filters with the same strength. However, such comparison becomes significant only if it is not possible to use fixed filters with a strength usually used with sliding, because the ASE noise background takes over the solitons. In fact, the suppression of the exponential growth of noise at the

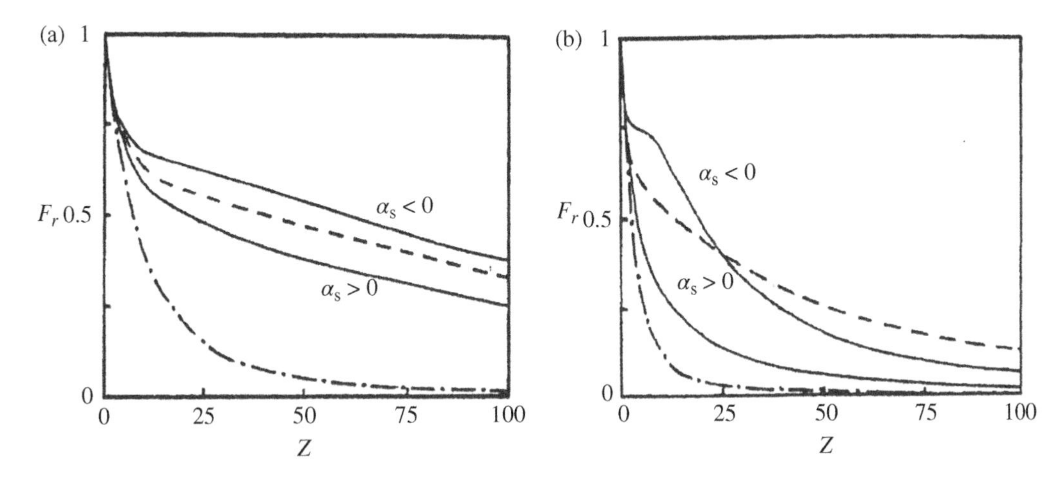

Figure 8.8 Reduction factor F_r of the variance of timing jitter against the propagation distance for the cases (a) $\beta = 0.1$ and (b) $\beta = 0.3$. The dashed curves corresponds to $\rho = 0$ and $\alpha_s = \pm\beta/2$, while the full curves were obtained considering $\rho = \beta/2$ and $\alpha_s = \pm\beta/2$. The dash-dotted curve corresponds to $\rho = 0$ and $\alpha_s = 0$.

signal frequency in presence of sliding makes it possible to use filters with a much lower bandwidth than with fixed filters. This lower bandwidth supported by sliding filters counter-balances to some extent the timing jitter enhancement reported above. As a consequence, the Gordon-Haus jitter can be in practice only slightly enhanced by sliding, particularly if the sliding rate is not too close to the critical value.

Sliding-frequency filters can be used also to reduce the timing jitter in WDM systems [21–25]. Actually, the periodic transmission windows of Fabry-Perot filters allow the simultaneous filtering of all channels. In particular, the timing jitter arising from the collision-induced frequency shifts is considerably reduced because the filter forces the soliton frequency to move toward its transmission peak [22]. Moreover, it is also found that the condition given by Eq. (7.61) can be relaxed, thus providing the possibility to increase the number of channels in a WDM system [24].

8.2.3 Control of Soliton Interaction

Considering the Eq. (8.28) for the evolution of the soliton central frequency and assuming that $\rho = 0$, the dynamical equations describing the interaction between two solitons are similar to Eqs. (8.15)–(8.20), except that Eq. (8.16) becomes:

$$\frac{dB}{dz} = \alpha_s - \frac{4}{3}\beta A^2 B \tag{8.50}$$

In order to have a stationary point with $A = 1$, the following conditions must be verified:

$$B^2 = \frac{\delta}{\beta} - \frac{1}{3} > 0 \tag{8.51}$$

$$\alpha_s = \frac{4}{3}\beta B \tag{8.52}$$

Figure 8.9a shows the evolution of the soliton pair separation ΔT for $\beta = 0.15$, $\alpha_s = -0.05$, $\delta = 0.059$, assuming several values for the initial separation ΔT_0. The value of δ was found in order

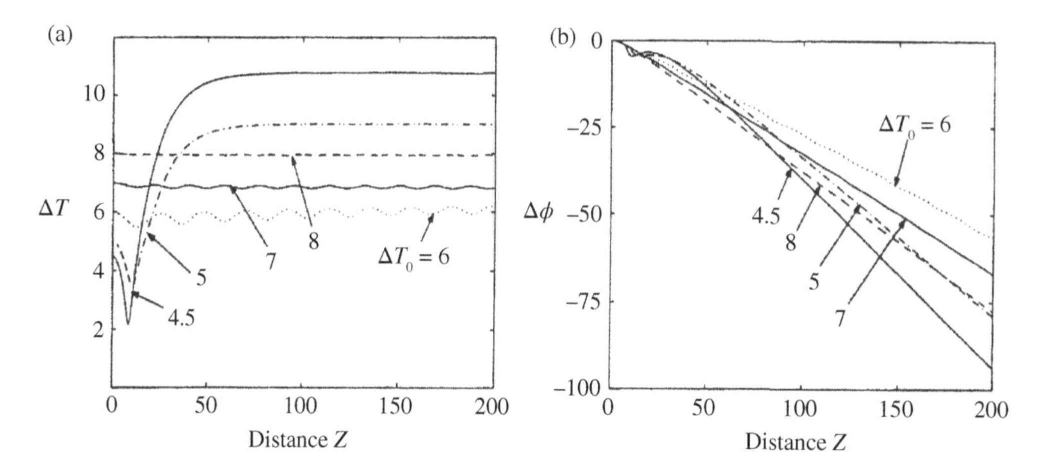

Figure 8.9 Evolution of (a) the pulse separation ΔT and (b) the phase difference $\Delta\phi$ for $\beta = 0.15$, $\alpha_s = -0.05$, $\delta = 0.059$ and different values of ΔT_0.

to provide an amplitude $A = 1$. In contrast with the case of fixed filters, we observe that the soliton collision is avoided using sliding-frequency filters. Moreover, the separation remains nearly constant for $\Delta T_0 \geq 6$. For lower values of ΔT_0, we observe an initial attractive phase, followed by a repulsion and a stabilization of the soliton separation. Such stabilization corresponds to a quasi-linear variation of the phase-difference between the two solitons, as illustrated in Figure 8.9b.

8.3 Synchronous Modulators

An alternative way to control the timing jitter due to the Gordon-Haus effect was suggested by Nakazawa et al. [3] and consisted in the use of a modulator, which is timed to pass solitons at the peak of its transmission. Synchronous modulators work by forcing the soliton to move toward their transmission peak, where loss is minimum, and such forcing reduces timing jitter considerably. This technique can be implemented using a $LiNbO_3$ modulator [3].

Synchronous modulation can be combined with optical filters to control solitons simultaneously in both the time and frequency domains. The evolution equations for the soliton amplitude η, frequency κ, and soliton position T_0 in the presence of amplitude modulators and fixed-frequency filters can be obtained from Eqs. (7.5) to (7.7) with:

$$P(q) = \delta q + \beta \frac{\partial^2 q}{\partial T^2} + \mu_R \left\{ \cos\left(\frac{2\pi}{T_M} T\right) - 1 \right\} q \tag{8.53}$$

where μ_R is the average extinction ratio and T_M is the modulator period. The resulting evolution equations are [18]:

$$\frac{\partial \eta}{\partial Z} = 2(\delta - \mu_R)\eta - 2\beta\eta\left(\frac{1}{3}\eta^2 + \kappa^2\right) + \frac{2\pi^2 \mu_R}{T_M} \cos\left(2\pi \frac{T_0}{T_M}\right) \operatorname{cosech}\left(\frac{\pi^2}{\eta T_M}\right) \tag{8.54}$$

$$\frac{\partial \kappa}{\partial Z} = -\frac{4}{3}\beta\eta^2\kappa \tag{8.55}$$

$$\frac{\partial T_0}{\partial Z} = -\kappa + \frac{\pi \mu_R}{\eta} \sin\left(2\pi \frac{T_0}{T_M}\right) \operatorname{cosech}\left(\frac{\pi^2}{\eta T_M}\right) \left\{ 1 - \frac{\pi^2}{\eta T_M} \coth\left(\frac{\pi^2}{\eta T_M}\right) \right\} \tag{8.56}$$

The suppression of the Gordon-Haus timing jitter results of the fact that both κ and T_0 tend exponentially to zero. Considering the case $T_M \gg \eta^{-1} = 1$, a stable fixed point ($\eta = 1$, $\kappa = 0$, $T_0 = 0$) is achieved if the following conditions are satisfied:

$$\delta - \frac{1}{3}\beta - \frac{\pi^4}{6T_M^2}\mu_R = 0 \tag{8.57}$$

$$0 < \frac{\pi^4}{2T_M^2}\mu_R < \beta \tag{8.58}$$

Equation (8.58) shows that in order to stabilize the amplitude fluctuations of the soliton, a filter must indeed be used together with the modulator. This is due to the fact that when the amplitude increases, its width decreases, which determines a reduced loss of the soliton in the modulator. Actually, the combined use of filters and modulators provides an effective control of the propagating solitons. This fact explains the experimental results obtained by Nakazawa et al. [7], which indicated the possibility of achieving arbitrarily large transmission distances using such combination. However, this scheme of soliton control requires active devices, having drawbacks of complexity, reduced reliability, and high cost. The same technique can be applied to WDM systems for controlling the timing jitter [30–32]. However, demultiplexing of individual channels and the use of a clock signal synchronized to the bit stream are required, which makes this technique less attractive in practice.

The synchronous modulation technique can also be implemented by using a phase modulator [12]. In this case, the soliton experiences a frequency shift only if it moves away from the center of the bit slot. The system performance can be further improved by combining intensity and phase synchronous modulators [13].

8.4 Amplifiers with Nonlinear Gain

An alternative approach to avoid the background instability consists in the use of an amplifier having a nonlinear property of gain, or gain and saturable absorption in combination [33, 34]. The key property of the nonlinearity in gain is to give an effective gain to the soliton and a suppression (or very small gain) to the noise. Nonlinear optical loop mirrors (NOLM's) [10, 35], nonlinear amplifying loop mirrors (NALM's) [36, 37], multi-quantum-well absorbers [38], and nonlinear birefringent fibers with polarizers [39] may be used as a saturable absorber for this purpose. Using the nonlinear gain approach may be particularly useful for transmission of solitons with subpicosecond or femtosecond durations, where the gain bandwidth of amplifiers will not be wide enough for the sliding of the filter frequency to be allowed [40].

The pulse propagation in optical fiber systems where narrow-band filters, linear, and nonlinear gain are used may be described by the following modified NLSE [34, 41–44]:

$$i\frac{\partial q}{\partial Z} + \frac{1}{2}\frac{\partial^2 q}{\partial T^2} + |q|^2 q = i\delta q + i\beta\frac{\partial^2 q}{\partial T^2} + i\varepsilon|q|^2 q + i\mu|q|^4 q \tag{8.59}$$

where β stands for the normalized filtering strength, δ is the linear excess gain, ε accounts for nonlinear gain-absorption processes, and μ represents a higher-order correction to the nonlinear gain-absorption.

Equation (8.59) is known as the complex Ginzburg-Landau equation (CGLE), so-called cubic for $\mu = 0$ and quintic for $\mu \neq 0$. It describes to a good approximation the soliton behavior both in optical transmission lines and in passively mode-locked fiber lasers [45–50].

8.4.1 Stationary Solutions

When all the coefficients on the right-hand side of Eq. (8.59) are small, the dynamical evolution of the soliton parameters can be obtained using Eqs. (7.5)–(7.8) with

$$P(q) = \delta q + \beta \frac{\partial^2 q}{\partial T^2} + \varepsilon |q|^2 q + \mu |q|^4 q \tag{8.60}$$

In particular, the soliton amplitude, η, and frequency, κ, satisfy the following evolution equations:

$$\frac{\partial \eta}{\partial Z} = 2\delta\eta - 2\beta\eta\left(\frac{1}{3}\eta^2 + \kappa^2\right) + \frac{4}{3}\varepsilon\eta^3 + \frac{16}{15}\mu\eta^5 \tag{8.61}$$

$$\frac{d\kappa}{dZ} = -\frac{4}{3}\beta\eta^2\kappa \tag{8.62}$$

As can be seen from Eq. (8.62), the soliton frequency approaches asymptotically to $\kappa = 0$ (stable fixed point) if $\eta \neq 0$. Considering the case of cubic CGLE ($\mu = 0$), a stationary soliton amplitude $\eta_s = 1$ is achieved when the following relation is verified:

$$3\delta + 2\varepsilon - \beta = 0 \tag{8.63}$$

In order to satisfy Eq. (8.63), the parameters δ and ε can be written as a function of the filter strength β, in the form

$$\delta = (1-K)\beta/3 \tag{8.64a}$$

$$\varepsilon = K\beta/2 \tag{8.64b}$$

The parameter K characterizes the relative contribution of the nonlinear gain to compensate for the filter-induced loss. The substitution of (8.63) into (8.61) leads to the following equation for the soliton amplitude in the vicinity of the stationary point ($\kappa = 0$):

$$\begin{aligned}\frac{d\eta}{dZ} &= 2\delta\eta(1-\eta^2)\\ &= \frac{2}{3}(1-K)\beta\eta(1-\eta^2)\end{aligned} \tag{8.65}$$

We can verify that the soliton amplitude is stable to small perturbations for $\delta > 0$, though the system returns to the steady-state more slowly when δ decreases. However, the amplitude $\eta = 1$ is no more a stable state when $\delta < 0$, since any perturbation would lead to the collapse or to the decayment of the soliton.

Figure 8.10 shows the evolution of the soliton amplitude for three values of K, and two values of the initial amplitude: $\eta = 0.7$ and $\eta = 1.3$. The dashed curves were obtained from Eq. (8.65), whereas the full curves were obtained by solving numerically Eq. (8.59). There is a good agreement between the two results, except for the case $K = 0.95$ and $\eta = 1.3$. We verify that as K decreases (δ increases), the amplitude tends more rapidly to the stationary value $\eta = 1$.

In general, the stable fixed points for the soliton amplitude can be found by looking for the minimums of the potential function ϕ defined by:

$$\frac{d\eta}{dZ} = -\frac{d\phi}{d\eta} \tag{8.66}$$

Considering the Eq. (8.61), we have the following expression for the potential function:

$$\phi(\eta) = -\delta\eta^2 + \frac{1}{6}(\beta - 2\varepsilon)\eta^4 - \frac{8}{45}\mu\eta^6 \tag{8.67}$$

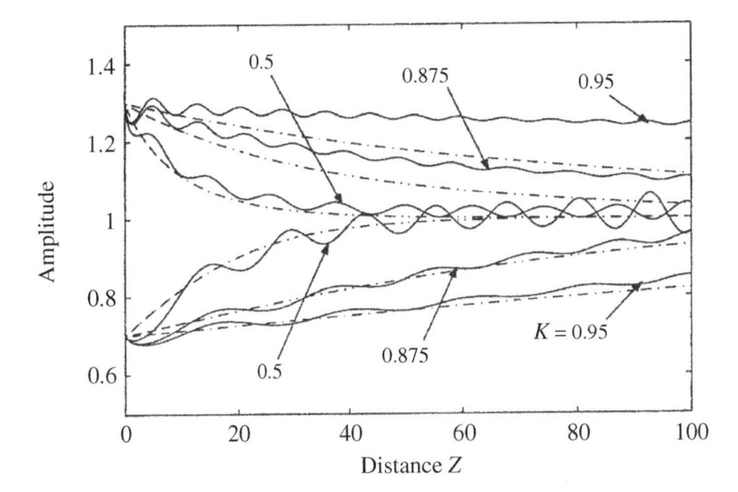

Figure 8.10 Evolution of the soliton amplitude for $\beta = 0.1$, $K = 0.5, 0.875$ and 0.95, and two values of the initial amplitude: $\eta = 0.7$ and $\eta = 1.3$. The dashed curves were obtained from Eq. (8.65), whereas the full curves were obtained by solving numerically Eq. (8.59).

For the zero-amplitude state to be stable, the potential function must have a minimum at $\eta = 0$, in addition to a minimum at $\eta = \eta_s \neq 0$. These objectives can be achieved if the following conditions are verified [51]:

$$\delta < 0, \ \mu < 0, \ \varepsilon > \beta/2, \ 15\delta > 8\mu\eta_s^4 \tag{8.68}$$

We can verify from the above conditions that the inclusion of the quintic term in Eq. (8.59) is necessary to have the double minimum potential.

The stationary value for the soliton amplitude can be obtained from Eq. (8.61) and is given by:

$$\eta_s^2 = \frac{-5(\varepsilon - \beta/2) \pm 5\sqrt{(\varepsilon - \beta/2)^2 - 24\delta\mu/5}}{8\mu} \tag{8.69}$$

From Eq. (8.69) it can be seen that, for $\varepsilon > \beta/2$, there are two real solutions for the stationary amplitude. Moreover, we observe that the higher amplitude solution grows to infinity when the nonlinear gain saturation μ tends to zero. Such singularity was predicted in Refs [52, 53] and provides the possibility of observing the very high amplitude pulses [54, 55] discussed in Chapter 12.

Equation (8.61) shows that a stationary amplitude $\eta_s = 1$ occurs when the coefficients satisfy the relation:

$$15\delta + 5(2\varepsilon - \beta) + 8\mu = 0 \tag{8.70}$$

The discriminant in Eq. (8.69) must be greater than or equal to zero for the solution to exist. For given values of β, μ, and ε, the allowed values of δ to guarantee a stable pulse propagation must satisfy the condition $0 > \delta > \delta_{min}$, where

$$\delta_{\min} = \frac{5(\varepsilon - \varepsilon_s)^2}{24\mu} \tag{8.71}$$

and $\varepsilon_s = \beta/2$. When $\delta = 0$, the peak amplitude is found to achieve a maximum value:

$$\eta_{\max} = \sqrt{-\frac{5}{4}\frac{(\varepsilon - \varepsilon_s)}{\mu}} \tag{8.72}$$

On the other hand, for given values of β, μ, and δ, the minimum value of allowed ε becomes

$$\varepsilon_{\min} = \varepsilon_s + \sqrt{24\delta\mu/5} \tag{8.73}$$

Considering the last condition in Eq. (8.68) or, alternatively, from Eqs. (8.69) and (8.73), we find that there is a minimum value for the peak amplitude given by:

$$\eta_{\min} = \sqrt[4]{\frac{15\delta}{8\mu}} \tag{8.74}$$

Figure 8.11 shows the potential function given by Eq. (8.67) when the relation (8.70) is satisfied for $\beta = 0.3$, $\varepsilon = 0.5$, $\mu = -0.25$ (curve a), $\mu = -0.34375$ (curve b), and $\mu = -0.5$ (curve c). Curves a and b present a minimum at $\eta = 1$ and $\eta = 0$ since they satisfy the conditions (8.68), corresponding to negative values of the linear gain ($\delta = -0.05$ and $\delta = -0.1$, respectively). However, curve c has no minimum at $\eta = 0$, since the corresponding value of δ is positive ($\delta = 0.033$).

Figure 8.12 illustrates the stability characteristics of the stationary solutions using the phase-plane formalism. Figure 8.12a corresponds to curve c in Figure 8.11, and we observe that, in this case, soliton propagation can be affected by background instability due to the amplification of small-amplitude waves. The steady-state solution shows a limited basin of attraction. For example, initial conditions with $\eta_i = 0.7$ and $\kappa_i = \pm 1$ evolve toward the trivial solution $\eta_s = 0$ of Eqs. (8.61) and (8.62). For these initial conditions, the nonlinearity is not sufficiently strong to balance dispersion, and the pulse disperses away. The dashed curves in Figure 8.12a give approximate limits between different basins of attraction. From a perturbation analysis of Eqs. (8.61) and (8.62) around $\eta = 0$, one can show that these curves cross the $\eta = 0$ axis at $\kappa_c = \pm 0.33$. Thus, waves with weak initial amplitudes grow up to $\eta_s = 1$ if $|\kappa_i| < 0.33$. In this case, soliton propagation can be severely affected by the background instability. Figure 8.12b corresponds to curve b in Figure 8.11, and we can see that, in this case, the background instability is avoided, since the small-amplitude waves are attenuated, irrespective of their frequency κ. Besides the stable stationary point at $\eta_s = 1$, we note, in this case, the existence of another stationary point at $\eta_s \approx 0.5$, which is unstable.

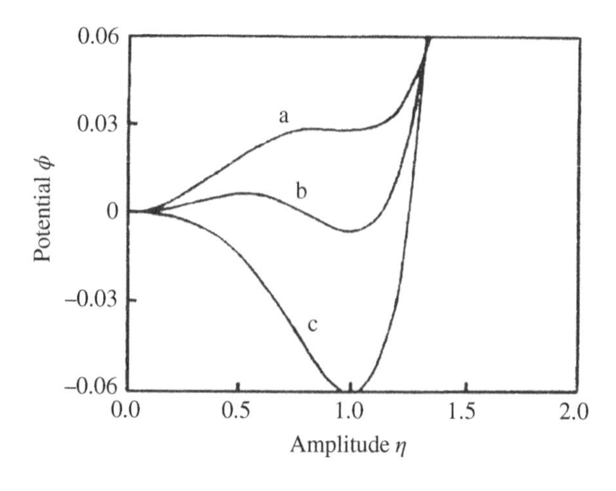

Figure 8.11 Potential ϕ versus soliton amplitude when the relation (8.70) is satisfied for $\beta = 0.3$, $\varepsilon = 0.5$, $v = 0$, $\mu = -0.5$ (curve a), $\mu = -0.343\ 75$ (curve b), and $\mu = -0.25$ (curve c).

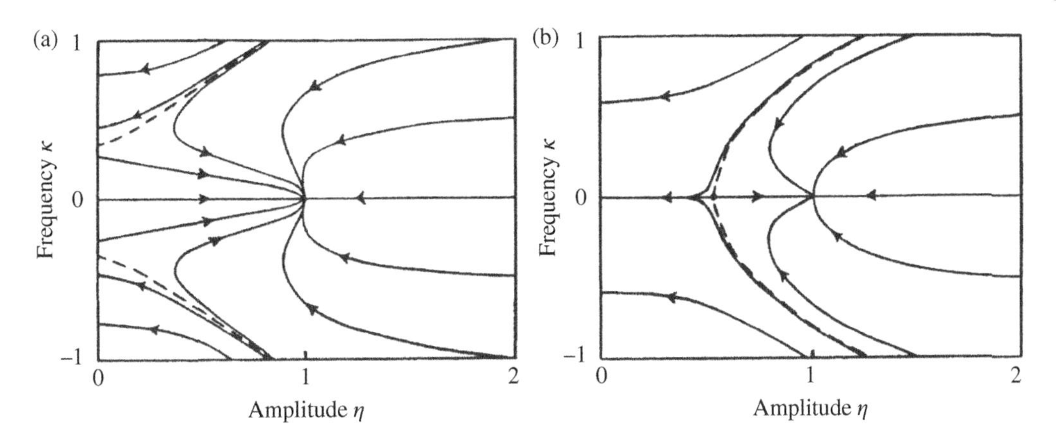

Figure 8.12 phase-plane of Eqs. (8.61) and (8.62) corresponding (a) to curve c and (b) to curve b of Figure 8.11.

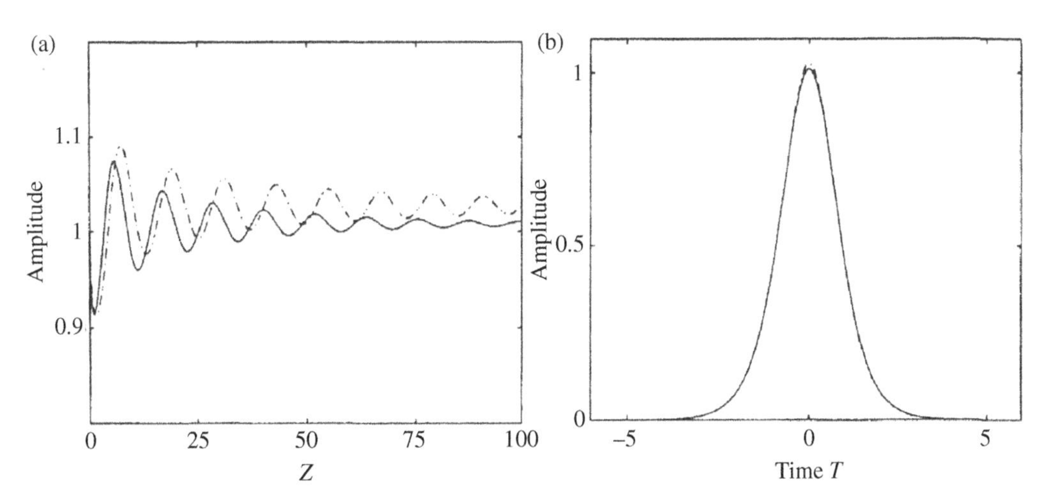

Figure 8.13 (a) Evolution of the peak amplitude and (b) the final pulse profile for $\delta = -0.01$, $\beta = 0.15$, $\varepsilon = 0.2$, and $\mu = -0.1375$ (dashed curves) or $\varepsilon = 0.4$ and $\mu = -0.3875$ (full curves), considering an input pulse $q(0, T) = \text{sech}(T)$.

Figure 8.13 shows (a) the evolution of the peak amplitude and (b) the final pulse profile obtained numerically from Eq. (8.59), assuming an input pulse with a sech profile and considering the following parameter values: $\delta = -0.01$, $\beta = 0.15$, $\varepsilon = 0.2$, and$\mu = -0.1375$ (dashed curves) or $\varepsilon = 0.4$ and $\mu = -0.3875$ (full curves). According to the perturbation theory, these values provide a stationary amplitude $\eta_s = 1$. This prediction and the stability of the stationary solution are confirmed by the numerical results of Figure 8.13.

8.4.2 Control of Soliton Interaction

The interaction between two well separated adjacent solitons in the presence of spectral filtering, linear, and nonlinear gains can be studied following the quasi-particle approach developed by

Karpman and V. V. Solovev [56], as described in Section 7.4. The following set of equations is derived for the parameters A, B, ΔA, ΔB, ΔT, and $\Delta\phi$:

$$\frac{dA}{dZ} = 2\delta A - 2\beta A\left(\frac{1}{3}A^2 + B^2\right) + \frac{4}{3}\varepsilon A^3 + \frac{16}{15}\mu A^5 \tag{8.75}$$

$$\frac{dB}{dZ} = -\frac{4}{3}\beta A^2 B \tag{8.76}$$

$$\frac{d(\Delta A)}{dZ} = 8A^3 e^{-A\Delta T}\sin(\Delta\phi) + 2\Delta A\left(\delta - \beta\left(A^2 + B^2\right) + 2\varepsilon A^2 + \frac{8}{3}\mu A^4\right) - 4\beta AB\Delta B \tag{8.77}$$

$$\frac{d(\Delta B)}{dZ} = 8A^3 e^{-A\Delta T}\cos(\Delta\phi) - \frac{4}{3}\beta A^2 \Delta B - \frac{8}{3}\beta AB\Delta A \tag{8.78}$$

$$\frac{d(\Delta T)}{dZ} = -\Delta B \tag{8.79}$$

$$\frac{d(\Delta\phi)}{dZ} = A\Delta A - \frac{4}{3}\beta A^2 B\Delta T \tag{8.80}$$

We verify that Eqs. (8.75) and (8.76) have a stable fixed point $A = A_s$ and $B = B_s = 0$. Assuming these stationary values and using the normalized propagation distance $\xi = A_s^2 Z$ and the normalized pulse separation $\Delta\tau = A_s\Delta T$, the system (8.77)–(8.80) can be reduced to:

$$\frac{d^2(\Delta\tau)}{d\xi^2} = -\frac{4}{3}\beta\frac{d(\Delta\tau)}{d\xi} - 8e^{-\Delta\tau}\cos(\Delta\phi) \tag{8.81}$$

$$\frac{d^2(\Delta\phi)}{d\xi^2} = -\frac{4}{3}\gamma\frac{d(\Delta\phi)}{d\xi} + 8e^{-\Delta\tau}\sin(\Delta\phi) \tag{8.82}$$

where

$$\gamma = \beta - 2\varepsilon - \frac{16}{5}\mu A_s^2 \tag{8.83}$$

Equations (8.81) and (8.82) correspond to the equations of motion for a mechanical system in the potential

$$U = -8e^{-\Delta\tau}\cos(\Delta\phi) \tag{8.84}$$

It was verified by Uzunov et al. [57] that the perturbation theory based on the NLSE soliton ansatz is not able to describe accurately all the features of soliton interaction in a system with nonlinear gain, due principally to the chirped nature of the CGLE soliton. This characteristic renders the tails of the pulses exponentially decaying with oscillations rather than simply decaying. As a consequence, the potential given by Eq. (8.84) must be modified, assuming the form [58]:

$$U' = -8e^{-\Delta\tau}\cos(\Delta\phi)\cos(k\Delta\tau) \tag{8.85}$$

where

$$k = \beta + \delta/A_s^2 \tag{8.86}$$

When the potential (8.84) is replaced by the potential (8.85), it is found that the system (8.81) and (8.82) has both saddle stationary points and unstable spiral stationary points. However, provided that the perturbation parameters in the right-hand side of Eq. (8.59) are all small, the instability of the spiral is very weak and the corresponding bound state will be practically stable. This fact is

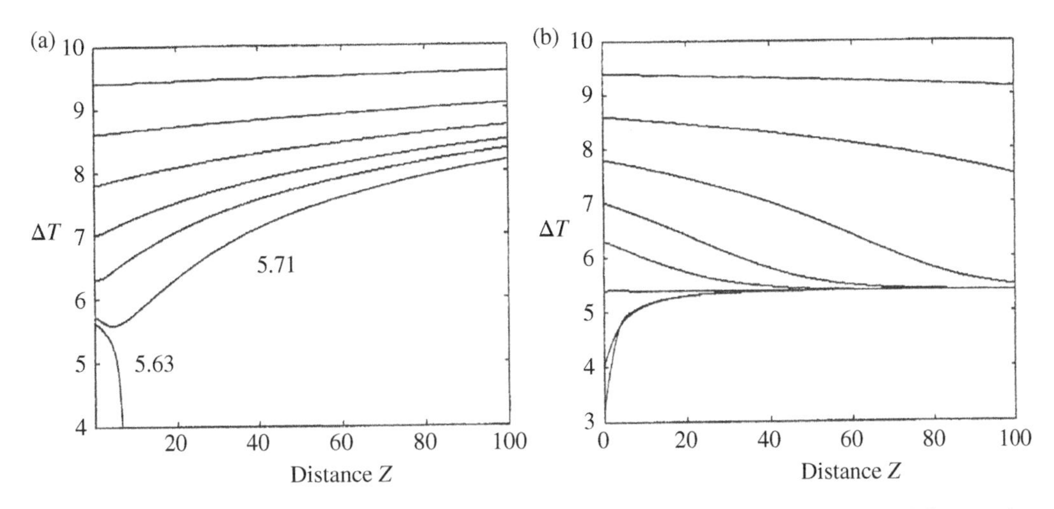

Figure 8.14 Evolution of the separation between two pulses: (a) initially in-phase and (b) initially opposite-phase. The following parameter values are assumed: $\beta = 0.1$, $\delta = -0.166$, $\varepsilon = 1$, and $\mu = -0.3125$.

illustrated in Figure 8.14, which shows the separation between two pulses (a) initially in-phase and (b) initially opposite-phase. We observe from Figure 8.14a that the two in-phase pulses attract or repel each other depending if the initial separation is below or above 5.7, respectively. In the case of two initially opposite phase pulses, Figure 8.14b shows that they evolve in order to form a bound state, with a separation $\Delta T = 5.4$. The formation of bound states of CGLE solitons will be further discussed in Section 12.5.

References

1. Mollenauer, L.F., Neubelt, M.J., Haner, M. et al. (1991). *Electron. Lett.* **27**: 2055.
2. Mecozzi, A., Moores, J.D., Haus, H.A., and Lai, Y. (1991). *Opt. Lett.* **16**: 1841.
3. Nakazawa, M., Yamada, E., Kubota, H., and Suzuki, K. (1991). *Electron. Lett.* **27**: 1270.
4. Kodama, Y. and Hasegawa, A. (1992). *Opt. Lett.* **17**: 31.
5. Mollenauer, L.F., Gordon, J.P., and Evangelides, S.G. (1992). *Opt. Lett.* **17**: 1575.
6. Afanasjev, V.V. (1993). *Opt. Lett.* **18**: 790.
7. Nakazawa, M., Suzuki, K., Yamada, E. et al. (1993). *Electron. Lett.* **29**: 729.
8. Forysiak, W., Blow, K.J., and Doran, N.J. (1993). *Electron. Lett.* **29**: 1225.
9. Romagnoli, M., Wabnitz, S., and Midrio, M. (1994). *Opt. Commun.* **104**: 293.
10. Matsumoto, M., Ikeda, H., and Hasegawa, A. (1994). *Opt. Lett.* **19**: 183.
11. Widdowson, T., Malyon, D.J., Ellis, A.D. et al. (1994). *Electron. Lett.* **30**: 990.
12. Smith, N.J., Firth, W.J., Blow, K.J., and Smith, K. (1994). *Opt. Lett.* **19**: 16.
13. Bigo, S., Audouin, O., and Desurvire, E. (1995). *Electron. Lett.* **31**: 2191.
14. Aubin, G., Jeanny, E., Montalant, T. et al. (1995). *Electron. Lett.* **31**: 1079.
15. Kumar, S. and Hasegawa, A. (1995). *Opt. Lett.* **20**: 1856.
16. Pereira, N.R. and Stenflo, L. (1977). *Phys. Fluids* **20**: 1733.
17. Nozaki, K. and Bekki, N. (1984). *Phys. Soc. Japan* **53**: 1581.
18. Hasegawa, A. and Kodama, Y. (1995). *Solitons in Optical Communications*. Oxford, UK: Oxford University Press.

19 Mollenauer, L.F., Lichtman, E., Neubelt, M.J., and Harvey, G.T. (1993). *Electron. Lett.* **29**: 910.
20 Mollenauer, L.F., Mamyshev, P.V., and Neubelt, M.J. (1994). *Opt. Lett.* **19**: 704.
21 Mollenauer, L.F., Lichtman, E., Harvey, G.T. et al. (1992). *Electron. Lett.* **28**: 792.
22 Mecozzi, A. and Haus, H.A. (1992). *Opt. Lett.* **17**: 988.
23 Ohhira, R., Matsumoto, M., and Hasegawa, A. (1994). *Opt. Commun.* **111**: 39.
24 Midrio, M., Franco, P., Matera, F. et al. (1994). *Opt. Commun.* **112**: 283.
25 Golovchenko, E.A., Pilipetskii, A.N., and Menyuk, C.R. (1996). *Opt. Lett.* **21**: 195.
26 Golovchenko, E.A., Pilipetskii, A.N., Menyuk, C.R. et al. (1995). *Opt. Lett.* **20**: 539.
27 Chen, H.H. and Liu, C.C. (1976). *Phys. Rev. Lett.* **37**: 693.
28 Kodama, Y. and Wabnitz, S. (1994). *Opt. Lett.* **19**: 162.
29 Ferreira, M.F. and Latas, S.V. (2001). *J. Lightwave Technol.* **19**: 332.
30 Desurvire, E., Leclerc, O., and Audouin, O. (1996). *Opt. Lett.* **21**: 1026.
31 Nakazawa, M., Suzuki, K., Kubota, H. et al. (1996). *Electron. Lett.* **32**: 828.
32 Nakazawa, M., Suzuki, K., Kubota, H., and Yamada, E. (1996). *Electron. Lett.* **32**: 1686.
33 Kodama, Y., Romagnoli, M., and Wabnitz, S. (1992). *Electron. Lett.* **28**: 1981.
34 Matsumoto, M., Ikeda, H., Uda, T., and Hasegawa, A. (1995). *J. Lightwave Technol.* **13**: 658.
35 Atkinson, D., Loh, W.H., Afanasjev, V.V. et al. (1994). *Opt. Lett.* **19**: 1514.
36 Hofer, M., Fermann, M.E., Haberl, F. et al. (1991). *Opt. Lett.* **16**: 502.
37 Mollenauer, L.F., Gordon, J.P., and Islam, M.N. (1986). *IEEE J. Quantum Electron.* **22**: 157.
38 Mollenauer, L.F., Evangelides, S.G., and Gordon, J.P. (1991). *J. Lightwave Technol.* **9**: 362.
39 Hasegawa, A. and Kodama, Y. (1991). *Phys. Rev. Lett.* **66**: 161.
40 Ferreira, M.F. (1997). Ultrashort soliton stability in distributed fiber amplifiers with different pumping configurations. In: *Applications of Photonic Technology*, vol. **2** (ed. G. Lampropoulos and R. Lessard), 249. New York: Plenum Press.
41 Akhmediev, N.N., Afanasjev, V.V., and Soto-Crespo, J.M. (1996). *Phys. Rev. E* **53**: 1190.
42 Soto-Crespo, J.M., Akhmediev, N.N., and Afanasjev, V.V. (1996). *J. Opt. Soc. Am. B* **13**: 1439.
43 Akhmediev, N.N., Ankiewicz, A., and Soto-Crespo, J.M. (1998). *J. Opt. Soc. Am. B* **15**: 515.
44 Ferreira, M.F. and Latas, S.V. (2002). *Optical Eng.* **41**: 1696.
45 Akhmediev, N. and Ankiewicz, A. (1997). *Solitons: Nonlinear Pulses and Beams*. London: Chapman and Hall.
46 Nelson, L.E., Jones, D.J., Tamura, K. et al. (1997). *Appl. Phys. B* **65**: 277.
47 Akhmediev, N. and Ankiewicz, A. (2005). Dissipative solitons in the complex Ginzburg-Landau and Swift-Hohenberg equations. In: *Dissipative Solitons*, Lecture Notes in Physics, vol. **661** (ed. N. Akhmediev and A. Ankiewicz). Berlin: Springer.
48 Wise, F.W., Chong, A., and Renninger, W.H. (2008). *Laser Phot. Rev.* **2**: 58.
49 Renninger, W.H., Chong, A., and Wise, F.W. (2011). *Opt. Express* **19**: 22496.
50 Ferreira, M.F. (2011). *Nonlinear Effects in Optical Fibers*. Hoboken, NJ: John Wiley & Sons.
51 Ferreira, M.F., Facão, M.V., and Latas, S.V. (2000). *Fiber Integrat. Opt.* **19**: 31.
52 Soto-Crespo, J.M., Akhmediev, N.N., Afanasjev, V.V., and Wabnitz, S. (1997). *Phys. Rev. E* **55**: 4783.
53 Afanasjev, V.V. (1995). *Opt. Lett.* **20**: 704.
54 Latas, S.C., Ferreira, M.F.S., and Facão, M. (2017). *J. Opt. Soc. Am. B* **34**: 1033.
55 Latas, S.C. and Ferreira, M.F. (2019). *J. Opt. Soc. of Am. B* **36**: 3016.
56 Karpman, V.I. and Solovev, V.V. (1981). *Physica* **3D**: 487.
57 Uzunov, I.M., Muschall, R., Golles, M. et al. (1995). *Opt. Commun.* **118**: 577.
58 Malomed, B.A. (1991). *Phys. Rev. A* **44**: 6954.

9

Propagation of Ultrashort Solitons

The propagation of picosecond optical solitons in fibers is governed by the nonlinear Schrödinger equation (NLSE), as described in Chapter 3. However, in the case of ultrashort solitons ($t_0 < 5$ ps), the spectral width of these pulses becomes large enough that it becomes necessary to take into account the Raman effect, the frequency dependence of the nonlinear parameter γ, and some higher-order terms in the expansion of the propagation constant, given by Eq. (3.4). The use of such ultrashort pulses becomes necessary in soliton transmission systems operating at high bit rates. This was the case, for example, of a soliton-based system implemented in 1993 using the optical time-division multiplexing (OTDM) technique and operating at 80 Gb/s [1]. In this chapter, we discuss the impact on pulse propagation of three main higher-order effects, namely, the third-order dispersion (TOD), the self-steepening (SST), and the intrapulse Raman scattering (IRS) effects.

9.1 Generalized NLSE

The inclusion of TOD, SST, and IRS effects determines a generalization of Eq. (3.57) in the form [2, 3]:

$$i\frac{\partial Q}{\partial z} - \frac{1}{2}\beta_2\frac{\partial^2 Q}{\partial \tau^2} + \gamma P_0|Q|^2 Q = i\frac{\beta_3}{6}\frac{\partial^3 Q}{\partial \tau^3} - \gamma P_0\left(\frac{i}{\omega_0}\frac{\partial}{\partial \tau}(|Q|^2 Q) - t_R Q\frac{\partial |Q|^2}{\partial \tau}\right) \tag{9.1}$$

where β_2 and β_3 are the second-order and the third-order dispersion coefficients in the expansion of Eq. (3.4a), ω_0 is the carrier frequency, t_0 is the initial pulse width, and t_R is a time constant, which depends on the Raman gain slope. Typically, we have $t_R \approx 5$ fs.

Assuming that the higher-order effects are relatively weak, we can use the method of moments presented in Section 6.10 to obtain some useful results. Let us consider the following sech ansatz:

$$Q(z,\tau) = Q_p \mathrm{sech}\left(\frac{\tau - q_p}{t_p}\right)\exp\left[-i\Omega_p\left(\tau - q_p\right) - i\frac{C_p}{2}\left(\frac{\tau - q_p}{t_p}\right)^2 + i\theta_p\right] \tag{9.2}$$

where the parameters Q_p, C_p, t_p, θ_p, Ω_p, and q_p may evolve with z and represent the amplitude, chirp, width, phase, frequency shift, and temporal shift of the pulse, respectively. The following evolution equations can be obtained for the pulse parameters [4, 5]:

$$\frac{dt_p}{dz} = \left(\beta_2 + \beta_3\Omega_p\right)\frac{C_p}{t_p} \tag{9.3}$$

Solitons in Optical Fiber Systems, First Edition. Mario F. S. Ferreira.

$$\frac{dC_p}{dz} = \left(C_p^2 + \frac{4}{\pi^2}\right)\frac{(\beta_2 + \beta_3\Omega_p)}{t_p^2} + \frac{4}{\pi^2}\frac{t_0}{t_p}(\gamma + \Omega_p/\omega_0)P_0 \tag{9.4}$$

$$\frac{dq_p}{dz} = \beta_2\Omega_p + \frac{\beta_3}{2}\Omega_p^2 + \frac{\beta_3}{6t_p^2}\left(1 + \frac{\pi^2}{4}C_p^2\right) + \frac{\gamma P_0}{\omega_0}\frac{t_0}{t_p} \tag{9.5}$$

$$\frac{d\Omega_p}{dz} = -\frac{8t_R\gamma P_0 t_0}{15t_p^3} + \frac{2\gamma P_0}{3\omega_0}\frac{t_0 C_p}{t_p^3} \tag{9.6}$$

The evolution equation for the phase θ_p has been ignored, since it does not affect the other pulse parameters. On the other hand, the pulse amplitude Q_p can be obtained from the relation $E_0 = 2P_0t_0 = 2Q_p^2(z)t_p(z)$.

Using the normalized variables Z, T, and q defined in Section 3.3, Eq. (9.1) can be rewritten as [2, 3]:

$$i\frac{\partial q}{\partial Z} + \frac{1}{2}\frac{\partial^2 q}{\partial T^2} + |q|^2 q = i\delta_3\frac{\partial^3 q}{\partial T^3} - is\frac{\partial}{\partial T}\left(|q|^2 q\right) + \tau_R q\frac{\partial |q|^2}{T} \tag{9.7}$$

where the pulse is assumed to propagate in the region of anomalous dispersion. The parameters δ_3, s, and τ_R, govern, respectively, the effects of third-order dispersion, self-steepening, and intrapulse Raman scattering, and are given by

$$\delta_3 = \frac{\beta_3}{6|\beta_2|t_0}, \quad s = \frac{1}{\omega_0 t_0} \quad \tau_R = \frac{t_R}{t_0} \tag{9.8}$$

The three parameters in Eq. (9.8) are inversely proportional to the initial pulse width t_0. Assuming a value $t_0 = 30$ fs for pulses propagating at 1550 nm in a standard silica fiber, we have $\delta_3 \approx 0.03$, $s \approx 0.03$, $\tau_R \approx 0.1$. In the case of dispersion-shifted fibers, the value of δ_3 becomes more significant, since $|\beta_2|$ is reduced. However, the IRS effect is often the most relevant among the higher-order effects.

9.1.1 Third-Order Dispersion

The derivation of the NLS equation in Chapter 3 assumed an expansion of the propagation constant β only up to the β_2 term in Eq. (3.4a). Although the contribution of this term dominates in most cases of practical interest, the inclusion of the term proportional to β_3 in this expansion becomes sometimes necessary. This is the case when the pulse propagates near the zero-dispersion wavelength, where β_2 nearly vanishes. Such inclusion is also necessary for sub-picosecond propagating pulses, since in such case the expansion parameter $\Delta\omega/\omega_0$ is no longer small enough to justify limiting the expansion in Eq. (3.4a) only up to the β_2 term.

As an asymmetric cubic function with respect to ω, TOD causes the pulse temporal profile to become also asymmetric, moving its peak from the original position. In order to isolate the TOD effects on pulse propagation, let us neglect both the IRS and SST effects, by considering $t_R = 0$ and $\omega_0 \to \infty$ in Eqs. (9.3)–(9.6). In this case and assuming the conditions for soliton formation, we have that $\Omega_p = C_p = 0$, $t_p = t_0$, and

$$q_p(z) = \frac{\beta_3}{6t_0^2}z \tag{9.9}$$

Equation (9.9) shows that, under the TOD effect, the soliton peak is shifted from its original position by an amount that increases linearly with the propagation distance. The soliton peak is advanced when $\beta_3 < 0$ and it is delayed when $\beta_3 > 0$. Such temporal shift is negligible for ps pulses, but becomes significant for fs pulses. For example, considering a typical value $\beta_3 = 0.1\ ps^3/km$, the peak shifts at a rate of 0.4 ps/km for $t_0 = 200$ fs.

In some cases, TOD can affect drastically the soliton dynamics. In particular, it can determine the breakup of second- and higher-order solitons into multiple fundamental solitons. This phenomenon is called soliton fission [2, 3, 6, 7] and will be discussed in detail in Section 16.4. During this process, TOD provides also the phase-matching between the solitons and the linear waves, enabling the resonant transfer of energy between them [8–12]. Such phase-matching occurs at particular frequencies satisfying the following condition:

$$\frac{1}{2}\beta_2(\omega-\omega_0)^2+\frac{1}{6}\beta_3(\omega-\omega_0)^3=\frac{1}{2}\gamma P_s \tag{9.10}$$

where P_s is the soliton power. If the propagation constant β is expanded only up to the β_2 term in Eq. (3.4), the phases of the soliton and of the linear waves will never be equal, as illustrated in Figure 9.1. However, by including the third-order dispersion, such phase-matching can be achieved at certain frequencies, since it is an odd function of frequency.

TOD can have also an important role on the dynamics of pulsating solitons, as it will be seen in Chapter 13. Depending on the signal of β_3, it can help in the transformation of the pulsating solitons into stationary solitons [13–15].

9.1.2 Self-Steepening

Self-steepening arises from the intensity dependence of the group velocity [16–18] and leads to an asymmetry in both temporal and spectral profiles of ultrashort pulses [19–21]. Neglecting the IRS effect, Eq. (9.6) can be integrated to obtain the following result for the frequency shift:

$$\Omega_p(z)=\frac{2\gamma P_0 t_0}{3\omega_0}\int_0^z\frac{C_p(z)}{t_p^3(z)}dz \tag{9.11}$$

We verify from Eq. (9.11) that a frequency shift occurs as soon as the propagating pulse becomes chirped.

The self-steepening effect produces also a temporal shift of the soliton peak, even in the absence of the frequency shift. Actually, neglecting the TOD effect and assuming that the soliton maintains

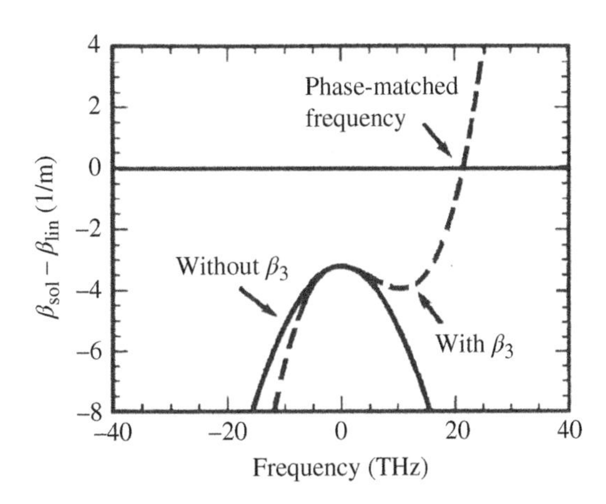

Figure 9.1 Phase mismatch $\Delta\beta = \beta_{sol} - \beta_{lin}$ between the soliton and the linear waves in the absence (full curve) and in the presence (dashed curve) of TOD.

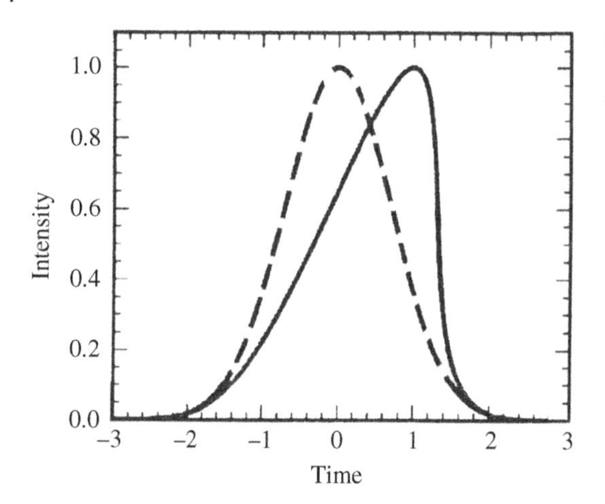

Figure 9.2 The effect of self-steepening on an initially Gaussian pulse propagating in the absence of dispersion.

its initial width, we obtain from Eq. (9.5) that such temporal shift increases linearly with the propagation distance, as:

$$q_p(z) = \frac{\gamma P_0}{\omega_0} z \tag{9.12}$$

As the pulse propagates inside the fiber, it becomes asymmetric, with its peak falling behind the pedestal, whereas the pulse's trailing edge becomes steepened (Figure 9.2). In the absence of dispersion, the pulse will continue to steepen until a shock front is formed. However, this process is counteracted in the presence of dispersion. Higher frequencies arise to support the steep trailing edge, which leads also to an asymmetric pulse spectrum [19–21]. The SST effect plays an important role in the generation of dispersive waves and supercontinuum, which will be discussed in later chapters. In particular, it can lead to instability of higher-order soliton propagation [22]. SST affects also the dynamics of pulsating solitons, as will be seen in Chapter 13 [13–15].

9.1.3 Intrapulse Raman Scattering

In order to study the IRS effect on pulse propagation, we consider $\beta_3 = 0$ and $\omega_0 \to \infty$ in Eqs. (9.3)–(9.6). In such case, the integration of Eq. (9.6) provides the following result for the frequency shift:

$$\Omega_p(z) = -\frac{8t_R\gamma P_0 t_0}{15} \int_0^z \frac{1}{t_p^3(z)} dz \tag{9.13}$$

In general, the pulse width $t_p(z)$ evolves along the fiber, according with Eq. (9.3). However, neglecting the TOD and SST effects and assuming that the coefficients in Eq. (9.4) satisfy the condition given by Eq. (6.84) together with $C_p(0) = 0$, we have $t_p(z) = t_0 = constant$ and

$$\Omega_p(z) = -\frac{8t_R\gamma P_0}{15t_0^2} z = -\frac{8t_R|\beta_2|}{15t_0^4} z \tag{9.14}$$

Equation (9.14) shows that the IRS effect leads to a continuous downshift of the carrier frequency, an effect known as the *soliton self-frequency shift* (SSFS) [23]. This effect was observed for the first time by Mitschke and Mollenauer in 1986 [24] and has been studied extensively since then [25–36]. The origin of SSFS can be understood by noting that for ultrashort solitons, the pulse spectrum becomes so broad that the high-frequency components of the pulse can transfer energy

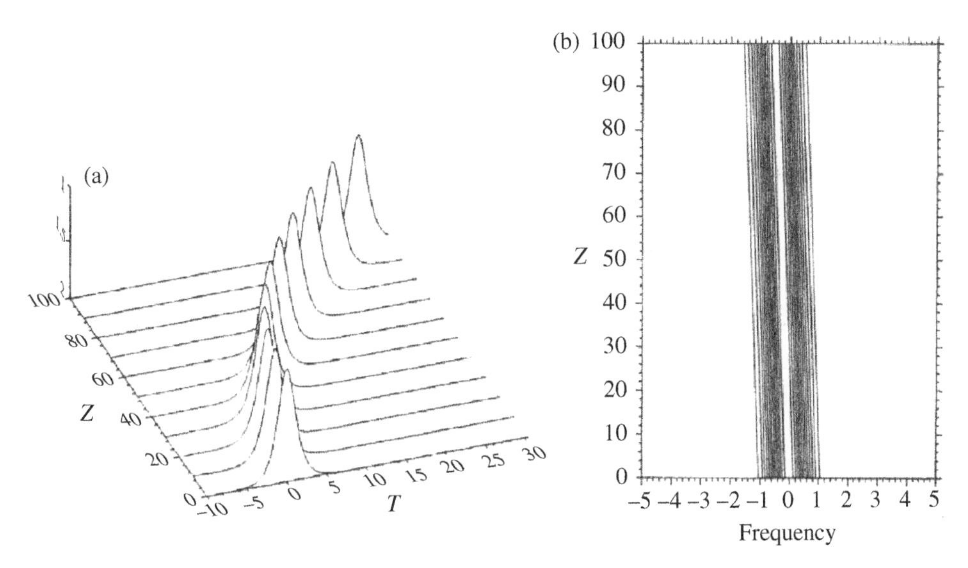

Figure 9.3 IRS effect on the propagation of the fundamental soliton for $\tau_R = 0.01$: (a) evolution of the amplitude and (b) contour plot of the soliton spectrum.

through Raman amplification to the low-frequency components of the same pulse. Such an energy transfer appears as a red shift of the soliton spectrum, with shift increasing with distance. Equation (9.14) shows that the frequency shift scales as t_0^{-4}, indicating that it can become quite large for short pulses. The SSFS effect can be significantly enhanced in some highly nonlinear fibers, where it has been used during the recent years for producing femtosecond pulses [37–40], as well as to realize several signal processing functions [7, 41–43].

Figure 9.3 illustrates the impact of the IRS effect on the propagation of the fundamental soliton, obtained by numerically solving Eq. (9.7) for the case $\delta_3 = s = 0$ and $\tau_R = 0.01$. Figure 9.3b shows a linear downshift of the soliton frequency with distance, which agrees with the prediction of Eq. (9.14). As a consequence, there is a continuous variation of the soliton velocity and temporal position, as observed in Figure 9.3a.

Similarly to the TOD and SST effects, IRS also leads to the breakup of the second and higher-order solitons into multiple fundamental solitons of different amplitudes. Figure 9.4 shows this behavior for the case of a second-order soliton ($N = 2$), by displaying the temporal and spectral evolutions for $\delta_3 = s = 0$ and $\tau_R = 0.01$. We observe that the high-amplitude soliton generated from the initial soliton fission is shifted much more than the low-amplitude soliton. Considering that the amplitude is equal to the inverse width of the fundamental soliton, such feature reflects the t_0^{-4} dependence of the SSFS on the soliton width, as seen from Eq. (9.14).

9.2 Timing Jitter of Ultrashort Solitons

Both the intrapulse Raman scattering and the third-order dispersion effects represent new sources of timing jitter, which increases rapidly with the bit rate [44–47]. In particular, the Raman-induced jitter becomes dominant in the case of ultrashort pulses. Actually, the amplifier noise produces fluctuations in the pulse energy, which are converted into frequency fluctuations through the Raman effect, resulting in position fluctuations by the fiber dispersion.

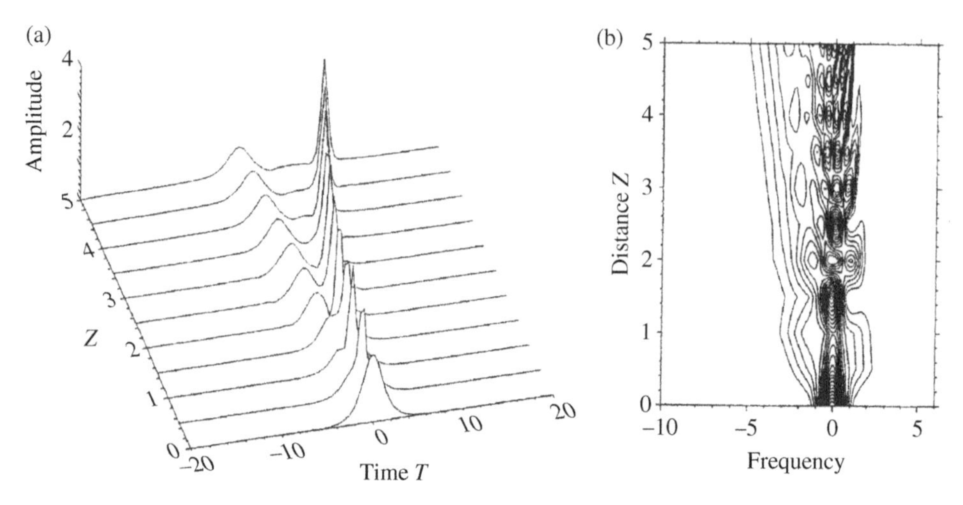

Figure 9.4 IRS effect on the propagation of the second-order soliton for $\tau_R = 0.01$ and $\delta_3 = s = 0$: (a) evolution of the amplitude and (b) contour plot of the second-order soliton spectrum.

The evolution of the soliton parameters in the presence of higher-order effects can be described using the perturbation theory presented in Chapter 7. Neglecting the self-steepening effect in Eq. (9.7), the perturbation term $P(q)$ in Eqs. (7.5)–(7.8) reads:

$$P(q) = \delta_3 \frac{\partial^3 q}{\partial T^3} - i\tau_R q \frac{\partial |q|^2}{T} \tag{9.15}$$

In particular, we obtain the following evolution equations for the soliton normalized amplitude, η, frequency, κ, and temporal position, T_0:

$$\frac{d\eta}{dZ} = 0 \tag{9.16}$$

$$\frac{d\kappa}{dZ} = -\frac{8}{15}\tau_R \eta^4 \tag{9.17}$$

$$\frac{dT_0}{dZ} = -\kappa + \delta_3\left(\eta^2 + 3\kappa^2\right) \tag{9.18}$$

The following solutions are obtained from Eqs. (9.16)–(9.18):

$$\eta(Z) = \eta_i \tag{9.19}$$

$$\kappa(Z) = -\frac{8}{15}\tau_R \eta_i^4 Z + \kappa_i \tag{9.20}$$

$$T_0(Z) = -\kappa_i Z + \frac{4}{15}\tau_R^2 \eta_i^4 Z^2 + \delta_3 \eta_i^2 Z + 3\delta_3 \kappa_i^2 Z + \frac{64}{225}\delta_3 \tau_R^2 \eta_i^8 Z^3 - \frac{8}{5}\delta_3 \tau_R \eta_i^4 \kappa_i Z^2 + T_{0i} \tag{9.21}$$

where η_i, κ_i, and T_{0i} are the initial values (at $Z = 0$) of the soliton amplitude, frequency, and position, respectively. The fifth and the sixth terms in Eq. (9.21) represent cross effects between TOD and the IRS effect. Neglecting the IRS effect, we verify from Eq. (9.21) that the main effect of TOD is to shift the soliton temporal position. Depending on the sign of δ_3, the soliton is delayed or advanced, as observed already from Eq. (9.9).

Let us consider a transmission link with normalized amplifier spacing Z_a. The time shift $\delta T_0(Z_a)$ induced by the amplifier noise at the end of a fiber section can be obtained by differentiating Eq. (9.21) and substituting $Z = Z_a$ in the resulting expression. If the fluctuations in the soliton amplitude, mean frequency, and position induced by the amplifier noise are represented by $\delta\eta_i$, $\delta\kappa_i$, and δT_{0i}, respectively, we have [46, 48]:

$$\delta T_0(Z_a) = \left(\frac{16}{15}\tau_R\eta_i^3 Z_a^2 + 2\delta_3\eta_i Z_a + \frac{2^9}{225}\delta_3\tau_R^2\eta_i^7 Z_a^3 - \frac{32}{5}\delta_3\tau_R\eta_i^3\kappa_i Z_a^2\right)\delta\eta_i + \left(-\frac{8}{5}\delta_3\tau_R Z_a^2\eta_i^4 + 6\delta_3\kappa_i Z_a - Z_a\right)\delta\kappa_i + \delta T_{0i} \tag{9.22}$$

The variance of arrival time jitter, $\sigma_{T_0}^2$, obtained by summing the individual displacements over a chain of $N = Z/Z_a$ amplifiers composing the link, is given by [46]:

$$\sigma_{T_0}^2 = \left(\frac{256}{1125}\tau_R^2\eta_i^6\frac{Z^5}{Z_a} + \frac{16}{15}\tau_R\delta_3\eta_i^4\frac{Z^4}{Z_a} + \frac{4}{3}\delta_3^2\eta_i^2\frac{Z^3}{Z_a}\right)\sigma_\eta^2 + \left(\frac{1}{3}\frac{Z^3}{Z_a} + \frac{2}{5}\delta_3\tau_R\eta_i^4\frac{Z^4}{Z_a}\right)\sigma_\kappa^2 + \frac{Z}{Z_a}\sigma_{T_{0i}}^2 \tag{9.23}$$

where σ_η^2, σ_κ^2, and $\sigma_{T_{0i}}^2$ are the variances of the soliton amplitude, frequency, and position fluctuations, respectively, at the output of an amplifier due to the noise added during the amplification, which are given by Eqs. (7.44)–(7.46).

Figure 9.5 illustrates the timing jitter as a function of transmission distance for three values of the pulses width: 20 ps (a), 3 ps (b), and 1 ps (c). The following values are assumed for the fiber parameters: $\beta_2 = -0.5\ \text{ps}^2/\text{km}$, $\beta_3 = 0.05\ \text{ps}^3/\text{km}$, $t_R = 6$ fs, and $\alpha = 0.2$ dB/km. In the case of 20 ps pulses, the timing jitter is determined mainly by the frequency fluctuations, corresponding to the Gordon-Haus effect discussed in Section 7.5. For 3 ps pulses, both the amplitude and frequency fluctuations are important, in spite of the fact that amplitude fluctuations become more relevant for large propagation distances. Finally, in the case of 1 ps pulses, the fluctuations of amplitude dominate clearly over the frequency fluctuations, even for relatively short distances.

The timing jitter arising from amplitude fluctuations are due mainly to the IRS effect and increases proportionally to Z^5, where $Z = NZ_a$, as can be seen from the first term in Eq. (9.23). If the tolerable timing jitter is less than 1 ps, we observe from Figure 9.5 that the total transmission distance must be limited to below 1500 km in the case of 3 ps pulses, or below 300 km in the case of 1 ps pulses. Longer transmission distances will be possible using some kind of jitter-control technique. These results show clearly that timing jitter becomes the limiting factor for soliton transmission systems operating at bit rates above 80 Gb/s, where solitons shorter than 5 ps are required.

9.3 Bandwidth-Limited Amplification of Ultrashort Solitons

Considering the limitations referred in Section 7.3 concerning the average soliton, the distributed amplification scheme appears as a better option compared with the lumped amplification scheme in the case of ultrashort pulses. In the following, we consider the case of distributed amplification provided by a periodically pumped fiber lightly doped with rare-earth ions. The evolution of ultrashort solitons in such active fiber, taking into account the IRS effect, can be described by the following generalized NLSE [49]:

$$i\frac{\partial q}{\partial Z} + \frac{1}{2}\frac{\partial^2 q}{\partial T^2} + |q|^2 q = i\delta q + \rho\frac{\partial q}{\partial T} + i\beta\frac{\partial^2 q}{\partial T^2} + \tau_R q\frac{\partial |q|^2}{\partial T} \tag{9.24}$$

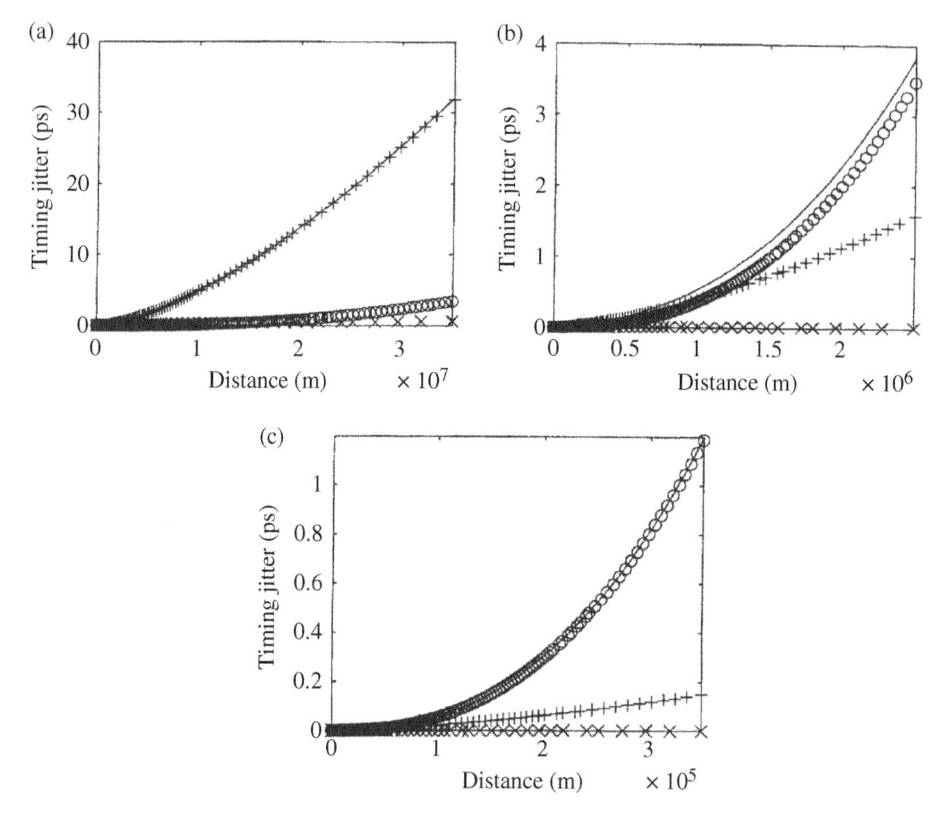

Figure 9.5 Timing jitter as a function of transmission distance for (a) $t_0 = 20$ ps, (b) $t_0 = 3$ ps, and (c) $t_0 = 1$ ps. Different kind of contributions (ooo) amplitude fluctuations; (+++) frequency fluctuations; (xxx) position fluctuations, whereas the full line represents the total timing jitter. The following parameter values are assumed: $\beta_2 = -0.5\ \text{ps}^2/\text{km}$, $\beta_3 = 0.05\ \text{ps}^3/\text{km}$, $t_R = 6$ fs, and $\alpha = 0.2$ dB/km.

where

$$Z = \beta_{2eff} z / t_0^2 \tag{9.25a}$$

$$\beta_{2eff} = \beta_2 - 6 g_0 T_2^3 \Delta \tag{9.25b}$$

$$T = \left[t - \beta_{1eff} z \right] / t_0 \tag{9.25c}$$

$$\beta_{1eff} = \beta_1 - g_0 T_2 \tag{9.25d}$$

$$\delta = (g_0 - \alpha/2) t_0^2 / \beta_{2eff} \tag{9.25e}$$

$$\rho = -2 g_0 T_2^2 \Delta / \beta_{2eff} \tag{9.25f}$$

$$\beta = g_0 T_2^2 / \beta_{2eff} \tag{9.25g}$$

g_0 being the gain coefficient at line center, α the fiber loss, t_0 is the initial pulse width, T_2 the relaxation time of the gain medium, which is inversely related with the gain bandwidth, $\Delta = \omega_a - \omega_0$ the detuning of the carrier frequency (ω_0) from the gain peak frequency (ω_a), $\beta_1 = \partial k/\partial \omega$, $\beta_2 = \partial^2 k/\partial \omega^2$, k being the propagation constant. The parameter ρ represents the effect of detuning of the carrier frequency from the gain peak frequency, whereas the parameter β is the normalized filtering strength due to the limited amplifier gain bandwidth.

The evolution of the soliton parameters along the fiber amplifier can be obtained from Eqs. (7.5)–(7.8) considering the perturbation term

$$P(q) = \delta q + \rho \frac{\partial q}{\partial T} + \beta \frac{\partial^2 q}{\partial T^2} - i\tau_R q \frac{\partial |q|^2}{T} \tag{9.26}$$

The following evolution equations are derived for the soliton amplitude, frequency, and position:

$$\frac{d\eta}{dZ} = 2\eta\left(\delta - \rho\kappa - \beta\kappa^2\right) - \frac{2}{3}\beta\eta^3 \tag{9.27}$$

$$\frac{d\kappa}{dZ} = -\frac{2}{3}\eta^2(\rho + 2\beta\kappa) - \frac{8}{15}\tau_R\eta^4 \tag{9.28}$$

$$\frac{dT_0}{dZ} = -\kappa \tag{9.29}$$

From Eqs. (9.27) and (9.28), it can be seen that both the amplitude and the frequency evolve along the distributed amplifier, approaching the stationary values $\eta = \eta_s$ and $\kappa = \kappa_s$. In the case $\rho = 0$, we have:

$$\eta_s^2 = \frac{25\beta^2}{24\tau_R^2}\left[-1 \pm \sqrt{1 + \frac{144\delta\tau_R^2}{25\beta^3}}\right] \tag{9.30}$$

$$\kappa_s = -\frac{2\tau_R\eta_s^2}{5\beta} \tag{9.31}$$

Figure 9.6 shows the evolution of the pulse width ($1/\eta$) and frequency κ given by Eqs. (9.27) and (9.28), when the soliton central frequency coincides with the peak of the gain curve ($\rho = 0$). In the case $\beta = 0$ (dashed curves), which corresponds to ignoring the amplifier filtering effect, the soliton width decreases continuously along the active fiber, in agreement with the result obtained from Eq. (9.27):

$$t_p(Z) = \frac{1}{\eta(Z)} = \frac{1}{\eta_i}\exp(-2\delta Z) \tag{9.32}$$

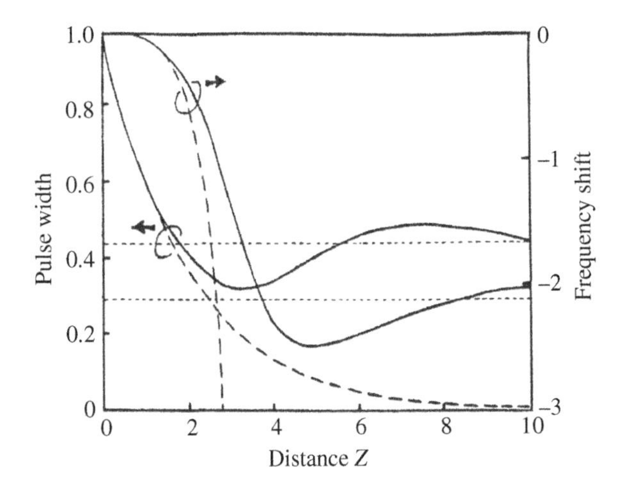

Figure 9.6 Evolution of the soliton pulse width (left scale) and frequency shift (right scale) for $\delta = 0.25$, $\tau_R = 0.04$, $\rho = 0$, and $\beta = 0$ (dashed curves) or $\beta = 0.04$ (full curves). The dotted lines represent the steady-state solutions obtained from Eqs. (9.30) and (9.31).

Simultaneously, the frequency decreases due to the IRS effect, according with the following result, obtained from Eqs. (9.28) and (9.32):

$$\kappa = \kappa_i - \frac{8\tau_R \eta_i^4 [\exp(8\delta Z) - 1]}{120\delta} \tag{9.33}$$

However, when $\beta \neq 0$, both the soliton pulse width reduction and the frequency shift are balanced by the limited gain bandwidth of the fiber amplifier, as can be observed from the full curves in Figure 9.6. The stationary values of the pulse width and frequency shift are given by $t_s = 1/\eta_s$ and κ_s as obtained from Eqs. (9.30) and (9.31), respectively. The trapping of ultrashort optical solitons in fiber amplifiers was analyzed theoretically and observed experimentally by several authors [49–53].

Figure 9.7 illustrates the impact of detuning the carrier frequency from the gain peak, for the same situation of Figure 9.6. The detuning is included in the parameter ρ, which is proportional to the first derivative of the gain relatively to the frequency. It can be observed that the soliton pulse width is practically unaffected by the carrier frequency detuning. However, the frequency shift is clearly enhanced for $\rho > 0$, whereas it can be compensated in the case $\rho < 0$, which corresponds to a positive gain slope. Moreover, Eqs. (9.26) and (9.27) show that a complete suppression of the frequency shift can be achieved with a detuning such that

$$\rho = -\frac{12\delta\tau_R}{5\beta} \tag{9.34}$$

This shows that the effective component of the gain spectrum for SSFS compensation is the linearly frequency-dependent gain.

Figure 9.8 illustrates the dynamics and stability properties of the steady-state solution of Eqs. (9.27) and (9.28) for the case of Figure 9.7 corresponding to $\rho = -0.04$. The sketch is not symmetric with respect to $\kappa = 0$, and the nonvanishing equilibrium solution has values $\eta = \eta_s = 2.32$ and $\kappa = \kappa_s = -1.67$, which is a stable sink of Eqs. (9.27) and (9.28). We observe that such steady-state solution has a limited basin of attraction. The dashed curve in Figure 9.8 represents an approximate limiting curve between different basins of attraction, From a perturbation analysis of Eqs. (9.27) and (9.28) around $\eta = \eta_s = 0$, one can verify that these curves cross the $\eta = 0$ at

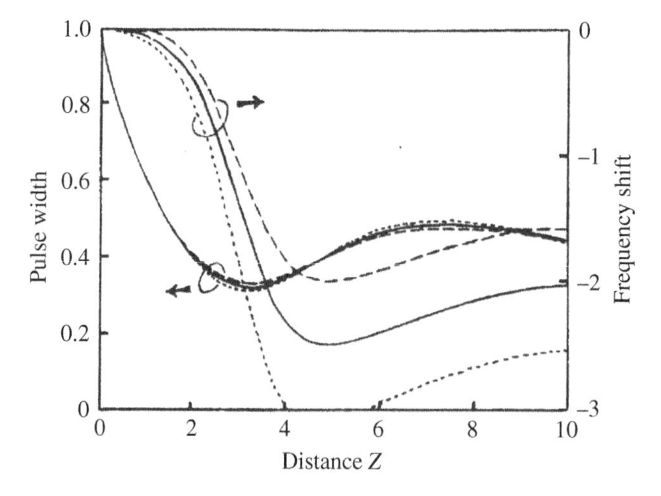

Figure 9.7 Evolution of the soliton pulse width (left scale) and frequency shift (right scale) for $\delta = 0.25$, $\tau_R = 0.04$, $\beta = 0.04$ e $\rho = 0$ (full curves), $\rho = 0.04$ (dotted curves), and $\rho = -0.04$ (dashed curves).

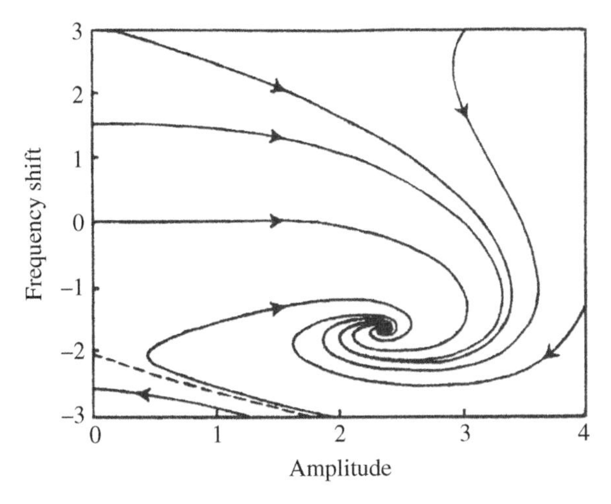

Figure 9.8 Phase-plane of Eqs. (9.27) and (9.28) for the situation of Figure 9.7 corresponding to $\rho = -0.04$.

$\kappa = 3.05$ and $\kappa = -2.05$. The asymmetry of these two values is due to the nonresonance of the carrier frequency.

9.4 Transmission Control Using Nonlinear Gain

As seen in previous sections, the IRS effect determines a frequency shift of the soliton during propagation, as well as a significant enhancement of the timing jitter of ultrashort pulses, which increases with the quintic power of distance. Both these effects can be controlled using bandwidth-limited amplification. However, in this case some excess gain must be provided, leading to instability of the background and limiting greatly the transmission distance. In these circumstances, the use of nonlinear gain appears particularly adequate to achieve a stable transmission of solitons with sub-picosecond or femtosecond durations. The key property of the nonlinear gain is to give an effective gain to the soliton and a suppression (or very small gain) to the noise [54–58].

The propagation of ultrashort solitons in the presence of intrapulse Raman scattering, spectral filtering, linear and nonlinear gain can be described by the following perturbed NLS equation [59, 60]:

$$i\frac{\partial q}{\partial Z} + \frac{1}{2}\frac{\partial^2 q}{\partial T^2} + |q|^2 q = i\delta q + i\beta\frac{\partial^2 q}{\partial T^2} + i\varepsilon|q|^2 q + i\mu|q|^4 q + \tau_R q\frac{\partial|q|^2}{\partial T} \tag{9.35}$$

Equation (9.35) is a generalized version of Eq. (8.59), the only difference being the inclusion of the last term in the right-hand side, which corresponds to the IRS effect. All the other coefficients have the same meaning as in Eq. (8.59).

9.4.1 Stationary Solutions

The evolution of the soliton parameters under the effect of the perturbations represented in the right-hand side of Eq. (9.35) can be obtained from Eqs. (7.5)–(7.8) considering that

$$P(q) = \delta q + \beta\frac{\partial^2 q}{\partial T^2} + \varepsilon|q|^2 q + \mu|q|^4 q - i\tau_R q\frac{\partial|q|^2}{T} \tag{9.36}$$

In particular, the following evolution equations for the soliton amplitude and frequency are given by:

$$\frac{d\eta}{dZ} = 2\delta\eta - 2\beta\eta\left(\frac{1}{3}\eta^2 + \kappa^2\right) + \frac{4}{3}\varepsilon\eta^3 + \frac{16}{15}\mu\eta^5 \tag{9.37}$$

$$\frac{d\kappa}{dZ} = -\frac{4}{3}\beta\eta^2\kappa - \frac{8}{15}\tau_R\eta^4 \tag{9.38}$$

The stationary solutions, η_s and κ_s, of Eqs. (9.37) and (9.38) are obtained considering $\frac{d\eta}{dZ} = \frac{d\kappa}{dZ} = 0$ and satisfy the conditions:

$$\delta - \beta\kappa_s^2 + \frac{1}{3}(2\varepsilon - \beta)\eta_s^2 + \frac{8}{15}\mu\eta_s^4 = 0 \tag{9.39}$$

$$\beta\kappa_s + \frac{2}{5}\tau_R\eta_s^2 = 0 \tag{9.40}$$

Substituting Eq. (9.40) in Eq. (9.39), we find that the stationary pulse amplitude is given by:

$$\eta_s = \left[A \pm \left[A^2 - B\right]^{1/2}\right]^{1/2} \tag{9.41}$$

where

$$A = \frac{\beta - 2\varepsilon}{2W} \tag{9.42}$$

$$B = \frac{3\delta}{W} \tag{9.43}$$

$$W = \frac{8}{5}\mu - \frac{12}{25}\frac{\tau_R^2}{\beta} \tag{9.44}$$

The stationary value of the amplitude must be real and positive. We find that Eq. (9.41) has two solutions if

i) $A > \sqrt{B} > 0$ (9.45)

On the other hand, there is only one stationary solution in the following cases:

ii) $B < 0$ (9.46)

iii) $A = \sqrt{B} > 0$ (9.47)

iv) $B = 0 \text{ and } A > 0$ (9.48)

Figure 9.9a illustrates the dependence of the stationary amplitude η_s, given by Eqs. (9.41)–(9.44), on the nonlinear gain parameter, ε, for $\beta = 0.1$, $\delta = -0.0025$, $\tau_R = 0.025$, and three values of the higher-order saturation parameter: $\mu = -0.006$, -0.01, and -0.015. Considering such values, there are two real solutions for the stationary amplitude when the nonlinear gain parameter is above ε~0.07. The high-amplitude solution increases with the nonlinear gain parameter and becomes higher when the magnitude of the nonlinear gain saturation parameter is reduced. Figure 9.9b shows that the magnitude of the stationary frequency shift also increases with the nonlinear gain parameter ε.

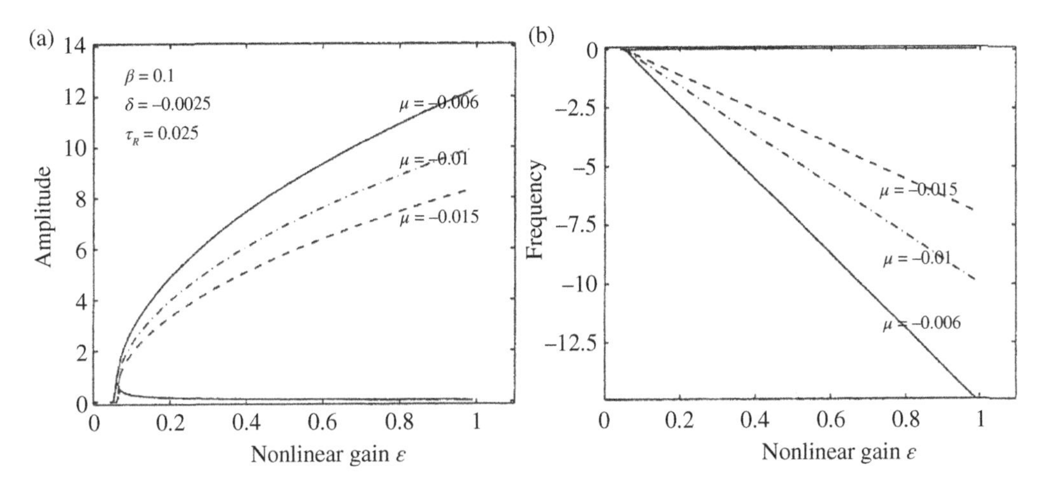

Figure 9.9 (a) Amplitude and (b) frequency shift of the stationary pulse as a function of the nonlinear gain parameter, ε, for $\beta = 0.1, \delta = -0.0025, \tau_R = 0.025$, and three values of the higher-order saturation parameter: $\mu = -0.006$, −0.01, and −0.015.

9.4.2 Linear Stability Analysis

Linearizing Eqs. (9.37) and (9.38) around the steady-state solution, we can write the system of equations describing the evolution of the small deviations $\Delta\eta$ and $\Delta\kappa$ in the following matrix form:

$$\begin{bmatrix} \dfrac{d\Delta\eta}{dZ} \\ \dfrac{d\Delta\kappa}{dZ} \end{bmatrix} = \begin{bmatrix} a_{11} & a_{12} \\ a_{21} & a_{22} \end{bmatrix} \begin{bmatrix} \Delta\eta \\ \Delta\kappa \end{bmatrix} \tag{9.49}$$

where

$$a_{11} = 2\delta - 2\beta\kappa_s^2 + (4\varepsilon - 2\beta)\eta_s^2 + \frac{16}{3}\mu\eta_s^4 \tag{9.50}$$

$$a_{12} = -4\beta\kappa_s\eta_s \tag{9.51}$$

$$a_{21} = -\frac{8}{3}\beta\kappa_s\eta_s - \frac{32}{15}\tau_R\eta_s^3 \tag{9.52}$$

$$a_{22} = -\frac{4}{3}\beta\eta_s^2 \tag{9.53}$$

The two eigenvalues (λ_1, λ_2) of the matrix $\begin{bmatrix} a_{11} & a_{12} \\ a_{21} & a_{22} \end{bmatrix}$ satisfy the characteristic equation:

$$\lambda^2 - \frac{8}{15}\eta_s^2[5(2\varepsilon - \beta) + 8\mu\eta_s^2]\lambda - \frac{16}{9}\beta\eta_s^4\left[2\varepsilon - \beta + \left(\frac{16}{5}\mu + \frac{24}{25}\frac{\tau_R^2}{\beta}\right)\eta_s^2\right] = 0 \tag{9.54}$$

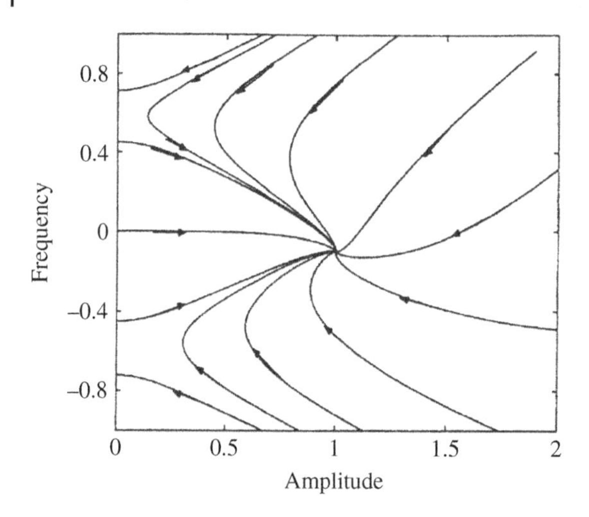

Figure 9.10 Phase-plain of Eqs. (9.37) and (9.38) for $\tau_R = 0.025$, $\beta = 0.1$, $\delta = 0.0343(3)$, and $\varepsilon = \mu = 0$.

The steady-state solution (η_s, κ_s) is linearly stable if the real parts of the two eigenvalues are negative. This is the case if the following conditions are verified:

$$8\mu\eta_s^2 + 5(\varepsilon - \beta) < 0 \tag{9.55}$$

$$2\varepsilon - \beta + \left(\frac{16}{5}\mu - \frac{24}{25}\frac{\tau_R^2}{\beta}\right)\eta_s^2 < 0 \tag{9.56}$$

In the absence of nonlinear gain ($\varepsilon = \mu = 0$), there is one steady-state solution for the pulse amplitude, corresponding to the case ii) given by Eq. (9.46). Assuming the parameter values $\tau_R = 0.025$, $\beta = 0.1$, and $\delta = 0.0343(3)$, we have the steady-state solution ($\eta_s = 1$, $\kappa_s = -0.1$). The two eigenvalues corresponding to this solution are $\lambda_1 = -0133 - i0.033$ and $\lambda_2 = -0133 + i0.033$. Since the real parts of both eigenvalues are negative, we can infer that the steady-state solution is linearly stable. Figure 9.10 shows that the equilibrium solution is indeed a stable sink of Eqs. (9.37) and (9.38). The sketch is not symmetric with respect to $\kappa = 0$, which is due to the SSFS effect. We observe also from Figure 9.10 that small amplitude waves within the basin of attraction of the nontrivial stationary solution are amplified, which will lead to instability of the background and eventually to the breakup of the soliton pulse. Figure 9.11 confirms that the pulse is destroyed after having propagated a distance of about 120 dispersion lengths.

In the presence of nonlinear gain proportional to the square of the amplitude ($\varepsilon \neq 0$, $\mu = 0$), the parameter B given by Eq. (9.43) becomes negative, and there is only one steady-state solution if $\delta > 0$, corresponding to case ii) given by Eq. (9.46). However, the background instability will arise in this-case due to the amplification of the linear waves within the basin of attraction of the stationary solution. This is illustrated in Figure 9.12, which shows the phase plain of Eqs. (9.37) and (9.38) assuming the following parameter values: $\beta = 0.025$, $\delta = 0.0025$, $\varepsilon = 0.01475$, and $\mu = 0$. Such background instability can be avoided only if the linear gain δ is negative. In this case, the parameter B is positive, and there are eventually two steady-state solutions corresponding to case i) given by Eq. (9.45).

Considering the parameter values $\tau_R = 0.025$, $\beta = 0.1$, $\delta = -0.0025$, $\varepsilon = 0.0553$, and $\mu = 0$, we have two steady-state solutions: ($\eta_{s1} = 1$, $\kappa_{s1} = -0.1$) and ($\eta_{s2} = 1.6$, $\kappa_{s2} = -0.256$). The two eigenvalues corresponding to the first solution are $\lambda_1 = 0.0653$ and $\lambda_2 = -0.1231$, so it is a saddle point. Concerning the second stationary solution, the two eigenvalues

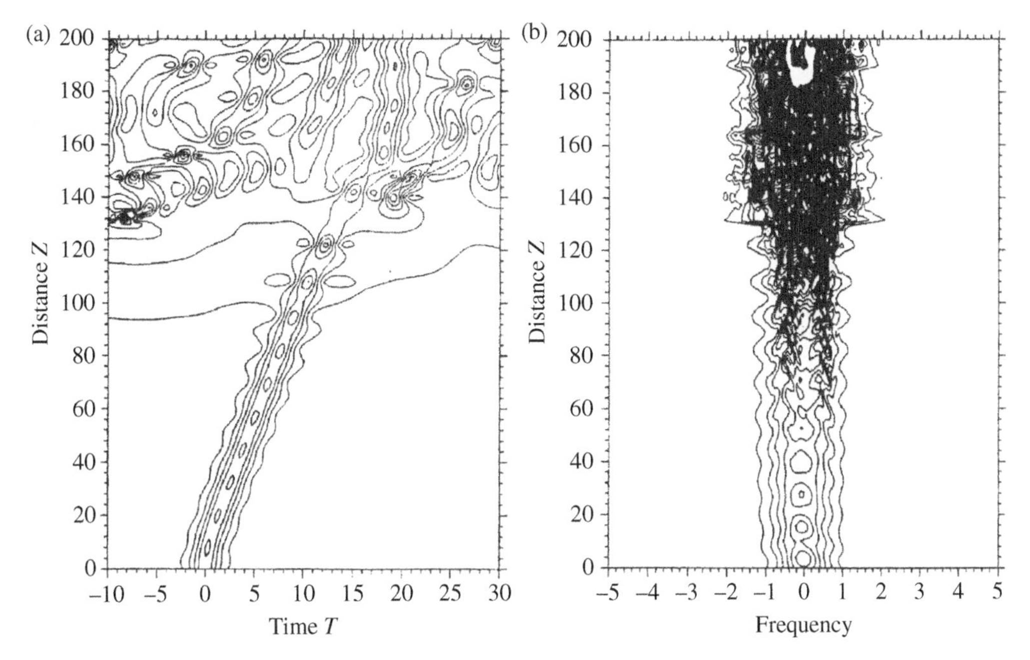

Figure 9.11 Contour maps for (a) the amplitude and (b) the spectrum for τ_R = 0.025, β = 0.1, δ = 0.0343(3), and ε = μ = 0.

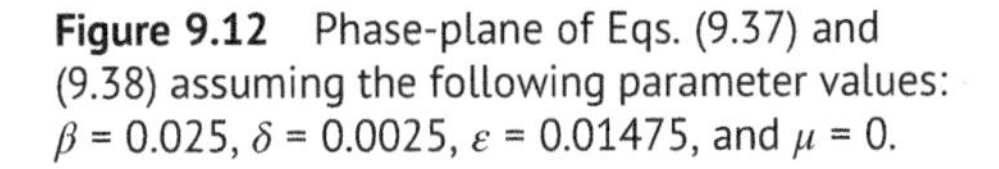
Figure 9.12 Phase-plane of Eqs. (9.37) and (9.38) assuming the following parameter values: β = 0.025, δ = 0.0025, ε = 0.01475, and μ = 0.

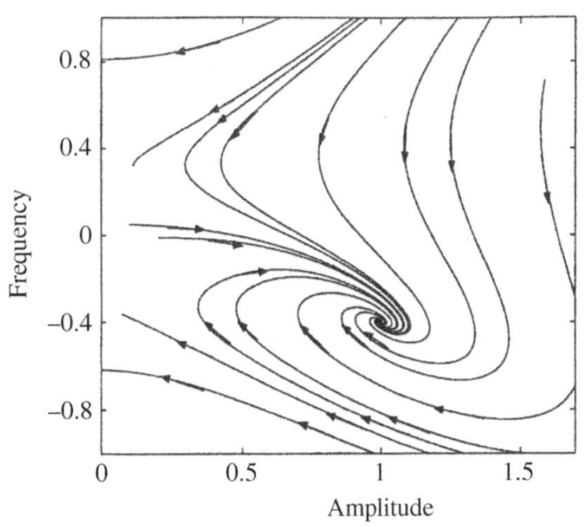

are $\lambda_1 = -0.0192$ and $\lambda_2 = -0.2853$, corresponding to a linearly stable point. The stability properties of these two steady-state solutions are illustrated in Figure 9.13, from which we confirm that in this case the background instability is effectively suppressed, since the small amplitude waves are attenuated irrespective of their frequency. Figure 9.14 shows the evolution of the soliton pulse toward the stable solution. The initially curved trajectory in Figure 9.14a corresponds to a decelerating pulse, which is determined by the increase of the pulse amplitude. Simultaneously, the pulse width is reduced and its spectrum broadens. The initial and final pulse profiles are shown in Figure 9.14c.

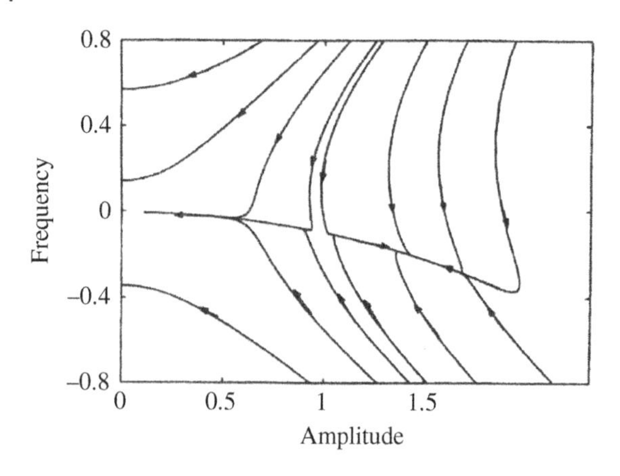

Figure 9.13 Phase-plane of Eqs. (9.37) and (9.38) for $\tau_R = 0.025$, $\beta = 0.1$, $\delta = -0.0025$, $\varepsilon = 0.0553$, and $\mu = 0$.

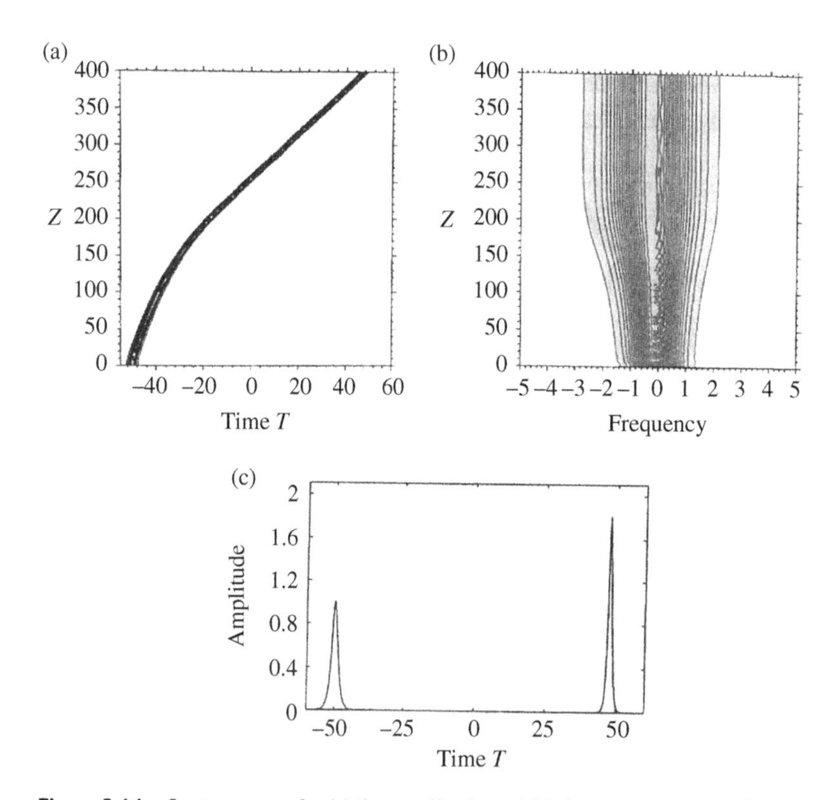

Figure 9.14 Contour maps for (a) the amplitude and (b) the spectrum, as well as (c) the initial and final pulse profiles for $\tau_R = 0.025$, $\beta = 0.1$, $\delta = -0.0025$, $\varepsilon = 0.0553$, and $\mu = 0$.

The existence of a stable pulse in a stable background observed above is possible only due to the presence of the IRS effect. In fact, as observed in Section 8.4, in the absence of IRS and considering only the nonlinear gain proportional to the second-order of the amplitude, when the pulse is stable, the background is unstable, and vice-versa, so that the pulse propagation is globally unstable [56, 61, 62].

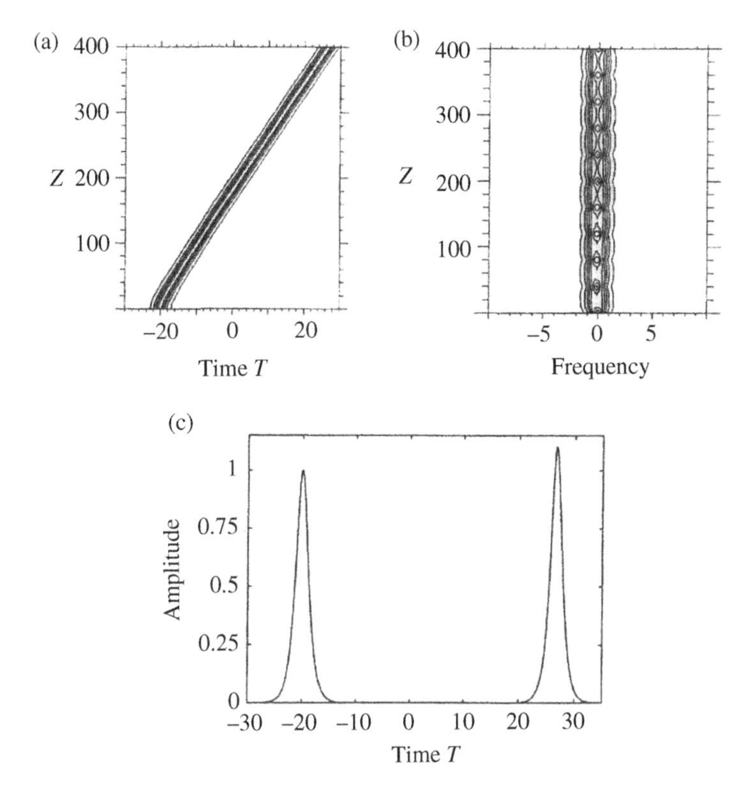

Figure 9.15 Contour maps for (a) the amplitude and (b) the spectrum, as well as (c) the initial and final pulse profiles for τ_R = 0.025, β = 0.1, δ = −0.0025, ε = 0.06, and μ = −0.006.

The stable propagation of ultrashort pulses can also occur when including the effect of saturation of nonlinear gain. For example, considering the parameter values $\tau_R = 0.025$, $\beta = 0.1$, $\delta = -0.0025$, $\varepsilon = 0.06$, and $\mu = -0.006$, we have two steady-state solutions in Eq. (9.41): $\eta_{s1} = 0.61$ and $\eta_{s2} = 1$. The two eigenvalues corresponding the first solution are $\lambda_1 = 0.0047$ and $\lambda_2 = -0.0780$, so it is a saddle point. On the other hand, the two eigenvalues corresponding to the second solution are $\lambda_1 = -0.0045$ and $\lambda_2 = -0.1227$, which indicates a linearly stable point. Figure 9.15 illustrates the stable propagation of the soliton pulse with amplitude $\eta_{s2} = 1$. Except in the initial stage, the trajectory is a straight line, which indicates a constant velocity of the pulse.

References

1 Iwatsuki, K., Suzuki, K., Nishi, S., and Saruwatari, M. (1993). *IEEE Photon. Technol. Lett.* **5**: 245.
2 Agrawal, G.P. (2001). *Nonlinear Fiber Optics*, 3e. San Diego: Academic Press.
3 Ferreira, M. (2011). *Nonlinear Effects in Optical Fibers*. Hoboken: John Wiley & Sons.
4 Santhanam, J. and Agrawal, G.P. (2003). *Opt. Commun.* **222**: 413.
5 Chen, Z., Taylor, A.J., and Efimov, A. (2010). *J. Opt. Soc. Am B* **27**: 1022.
6 Beaud, P., Hodel, W., Zysset, B., and Weber, H.P. (1987). *IEEE J. Quantum Electron.* **23**: 1938.
7 Ferreira, M.F. (2020). *Optical Signal Processing in Highly Nonlinear Fibers*. Oxon, UK: CRC Press.
8 Wai, P.K., Menyuk, C.R., Lee, Y.C., and Chen, H.H. (1986). *Opt. Lett.* **11**: 464.

9 Karpman, V.I. (1993). *Phys. Rev. E* **47**: 2073.
10 Akhmediev, N. and Karlsson, M. (1995). *Phys. Rev. A* **51**: 2602.
11 Husakou, A. and Herrmann, J. (2001). *Phys. Rev. Lett.* **87**: 203901.
12 Nishizawa, N. and Goto, T. (2001). *Opt. Express* **8**: 328.
13 Latas, S.C., Ferreira, M.F., and Facão, M. (2011). *Appl. Phys. B* **104**: 131.
14 Latas, S.C. and Ferreira, M.F. (2012). *Opt. Lett.* **37**: 3897.
15 Latas, S.V. and Ferreira, M.F. (2018). *J. Nonlinear Opt. Phys. Mat.* **27**: 1850008.
16 Jonek, R.J. and Landauer, R. (1967). *Phys. Lett.* **24A**: 228.
17 DeMartini, F., Townes, C.H., Gustafson, T.K., and Kelley, P.L. (1967). *Phys. Rev.* **164**: 312.
18 Grischkowsky, D., Courtens, E., and Armstrong, J.A. (1973). *Phys. Rev. Lett.* **31**: 422.
19 Yang, G. and Shen, Y.R. (1984). *Opt. Lett.* **9**: 510.
20 Manassah, J.T., Mustafa, M.A., Alfano, R.R., and Ho, P.P. (1985). *Phys. Lett.* **113A**: 242; *IEEE J. Quantum Electron.* **22**, 197 (1986).
21 Mestdagh, D. and Haelterman, M. (1987). *Opt. Commun.* **61**: 291.
22 Golovchenko, E.A., Dianov, E.M., Prokhorov, A.M., and Serkin, V.N. (1985). *Pis'ma Zh. Eksp. Teor. Fiz.* **42**: 74. [*JETP Lett.* **42**, 87 (1985)].
23 Gordon, J.P. (1986). *Opt. Lett.* **11**: 662.
24 Mitschke, F.M. and Mollenauer, L.F. (1986). *Opt. Lett.* **11**: 659.
25 Kodama, Y. and Hasegawa, A. (1987). *IEEE J. Quantum Electron.* **QE-23**: 510.
26 Zysset, B., Beaud, P., and Hodel, W. (1987). *Appl. Phys. Lett.* **50**: 1027.
27 Gouveia-Neto, A.S., Gomes, A.S.L., and Taylor, J.R. (1988). *IEEE J. Quantum Electron.* **24**: 332.
28 Tai, K., Hasegawa, A., and Bekki, N. (1988). *Opt. Lett.* **13**: 392.
29 Stolen, R.H., Gordon, J.P., Tomlinson, W.J., and Haus, H.A. (1989). *J. Opt. Soc. Am. B* **6**: 1159.
30 Blow, K.J. and Wood, D. (1989). *IEEE J. Quantum Electron.* **25**: 2665.
31 Afansasyev, V.V., Vysloukh, V.A., and Serkin, V.N. (1990). *Opt. Lett.* **15**: 489.
32 Mamyshev, P.V. and Chernikov, S.V. (1990). *Opt. Lett.* **15**: 1076.
33 Hong, B.J. and Yang, C.C. (1991). *J. Opt. Soc. Am. B* **8**: 1114.
34 Stolen, R.H. and Tomlinson, W.J. (1992). *J. Opt. Soc. Am. B* **9**: 565.
35 Kurokawa, K., Kubota, H., and Nakazawa, M. (1992). *Electron. Lett.* **28**: 2050.
36 Kivshar, Y.S. and Malomed, B.A. (1993). *Opt. Lett.* **18**: 485.
37 Nishizawa, N., Ito, Y., and Goto, T. (2002). *IEEE Photon. Technol. Lett.* **14**: 986.
38 Efimov, A., Taylor, A.J., Omenetto, F.G., and Vanin, E. (2004). *Opt. Lett.* **29**: 271.
39 Abedin, K.S. and Kubota, F. (2004). *IEEE J. Sel. Topics Quantum Electron.* **10**: 1203.
40 Lee, J.H., Howe, J., Xu, C., and Liu, X. (2008). *IEEE J. Sel. Quantum Electron.* **14**: 713.
41 Nishizawa, N. and Goto, T. (2003). *Opt. Express* **11**: 359.
42 Kato, M., Fujiura, K., and Kurihara, T. (2004). *Electron. Lett.* **40**: 381.
43 Oda, S. and Maruta, A. (2006). *Opt. Express* **14**: 7895.
44 Baboiu, D.M., Mihalache, D., and Panoiu, N.C. (1995). *Opt. Lett.* **20**: 1865.
45 Essiambre, R.-J. and Agrawal, G.P. (1997). *J. Opt. Soc. Am. B* **14**: 314.
46 Facão, M. and Ferreira, M. (2001). *J. Nonlinear Math. Phys.* **8**: 112.
47 Corney, J.F. and Drummond, P.D. (2001). *J. Opt. Soc. Am. B* **18**: 153.
48 Essiambre, R.-J. and Agrawal, G.P. (1996). *Opt. Commun.* **131**: 274.
49 Ferreira, M.F. (1994). *Optics Commun.* **107**: 365.
50 Nakazawa, M., Kurokawa, K., Kubota, H., and Yamada, E. (1986). *Phys. Rev. Lett.* **65**: 662.
51 Blow, K.J., Doran, N.J., and Wood, D. (1988). *J. Opt. Soc. Am. B* **5**: 1301.
52 Nakazawa, M., Kubota, H., Kurokawa, K., and Yamada, E. (1991). *J. Opt. Soc. Am.* **8**: 1811.
53 Ding, M. and Kikuchi, K. (1992). *IEEE Photonics Technol. Lett.* **4**: 497.

54 Matsumoto, M., Ikeda, H., Uda, T., and Hasegawa, A. (1995). *J. Lightwave Technol.* **13**: 658.

55 Ferreira, M.F. (1997). Ultrashort soliton stability in distributed fiber amplifiers with different pumping configurations. In: *Applications of Photonic Technology*, vol. **2** (ed. G. Lampropoulos and R. Lessard), 249. New York: Plenum Press.

56 Ferreira, M.F., Facão, M.V., and Latas, S.C. (2000). *Fiber Int Opt.* **19**: 31.

57 Ferreira, M.F. and Latas, S.C. (2002). *Opt. Eng.* **41**: 1696.

58 Tian, H., Li, Z., Xu, Z. et al. (2003). *J. Opt. Soc. Am. B* **20**: 59.

59 Ferreira, M.F. and Facão, M.V. (1997). *SPIE Proc.* **2841**: 154.

60 Latas, S.V. and Ferreira, M.F. (2005). *Opt. Commun.* **251**: 415.

61 Akhmediev, N.N., Afanasjev, V.V., and Soto-Crespo, J.M. (1996). *Phys. Rev. E* **53**: 1190.

62 Soto-Crespo, J.M., Akhmediev, N.N., and Afanasjev, V.V. (1996). *J. Opt. Soc. Am. B* **13**: 1439.

10

Dispersion-Managed Solitons

To overcome the limitations arising from timing jitter, much effort has been placed in the development of soliton control techniques. Some of these techniques were reviewed in Chapter 8 and rely on the placement of additional components within the transmission path, namely spectral filters, synchronous modulators, or different nonlinear amplification schemes.

An alternative and simple technique to realize soliton transmission control consists in using dispersion management (DM). This technique can be realized using different approaches. In particular, Smith et al. [1] proposed in 1996 the use of a periodic map using both anomalous dispersion and normal dispersion fibers alternately. The main characteristic of this proposal is to combine a high local group velocity dispersion with low path-average dispersion. The former feature results in the reduction of some nonlinear effects, namely the four-wave mixing (FWM) effect, while the latter one reduces the Gordon-Haus (GH) timing jitter. An intense theoretical effort has been devoted to understanding the characteristics of the nonlinear stationary pulse that propagates in a transmission line with a periodic dispersion map [1–10].

10.1 Dispersion Management

As referred to in Chapter 3, standard single-mode fibers have a zero-dispersion wavelength of 1.31 μm. On the other hand, most of the installed communication systems using these fibers operate in the 1.55 μm, where the fiber loss is minimum but the typical dispersion is +16 ps/km-nm. In the presence of this dispersion, the linear optical pulses will suffer from broadening and tend to overlap. One approach to avoid such pulse broadening is to use dispersion-shift fibers, where the fiber group velocity dispersion (GVD) is almost constantly equal to zero along the system. However, in this case, the fiber nonlinearity severely degrades the performance of wavelength-division multiplexing (WDM) systems, mainly because of the FWM-induced cross talk between channels [11, 12].

In order to reduce the nonlinearity-induced degradation of system performance, nonzero dispersion fibers should be used as a transmission fiber, which requires some form of dispersion compensation [13]. Actually, Eq. (3.67) shows that, in a linear transmission system, the dispersion-induced degradation of optical signals can be suppressed using multiple fiber segments with different dispersion characteristics. All dispersion-management schemes attempt to achieve such an objective. The basic idea can be understood by considering just two fiber segments: a transmission fiber with anomalous dispersion and a *dispersion compensating fiber* (DCF) with normal dispersion.

Solitons in Optical Fiber Systems, First Edition. Mario F. S. Ferreira.

If the transmission and compensating fibers have dispersions D_1 and D_2 and lengths L_1 and L_2, respectively, the requirement for a complete dispersion compensation at a given wavelength λ is

$$D_2L_2 = -D_1L_1 \tag{10.1}$$

There are two main options to implement the dispersion compensation. In one scheme, dispersion accumulates along most of the link length and is compensated using dispersion compensation fibers only at the transmitter (pre-compensation) and receiver (post-compensation) ends. In the other option, dispersion is compensated periodically along the link, either completely or partially. In this case, the pulse broadens and recompresses periodically, according to the dispersion characteristics of the fiber segments.

The first experimental demonstration of nonlinear pulse transmission in dispersion-compensated systems was reported in 1995 by Suzuki et al. [14]. Soon after, Smith et al. [1] performed a numerical study demonstrating that a periodically stationary soliton-like pulse exists in such systems. Since then, numerous studies have recognized that these periodically stationary soliton-like pulses have several advantages over conventional solitons in uniform dispersion fibers.

Figure 10.1 shows an example of a dispersion-managed (DM) system. The GVD is locally large and alternates signs in consecutive fiber segments, which determines a rapid oscillation of the pulse temporal width and chirp. The averaged pulse width is broadened if there exists a nonzero residual averaged dispersion unless such broadening is compensated by the fiber nonlinearity. By choosing appropriately the initial pulse amplitude, a balance between the nonlinearity and dispersion can be achieved and the pulse evolution becomes periodically stationary. We call such a periodically stationary pulse a *dispersion-managed (DM) soliton*. It was found that the power required to maintain a DM soliton increases with the GVD difference between the two fiber segments.

Lowering of average GVD through dispersion compensation not only reduces pulse distortion but also lowers the timing jitter [12, 15]. Actually, the timing jitter in the dispersion-compensated system is reduced to a value even smaller than the value calculated by using the average value of dispersion [16]. This behavior is a consequence of the fact that the average power launched into the DM system is higher than that in a system having a uniform dispersion of the same average GVD.

One disadvantage of using DCFs in dispersion-compensated systems is the added loss associated with the increased fiber span. Concerning this aspect, a useful figure of merit is the ratio of the DCF dispersion magnitude to its loss in dB/km, expressed in units of ps/nm/dB. Clearly, a high value of this parameter is desired. Another disadvantage is related to nonlinear effects, which may degrade the signal over the long length of the fiber if its intensity is relatively high. This problem

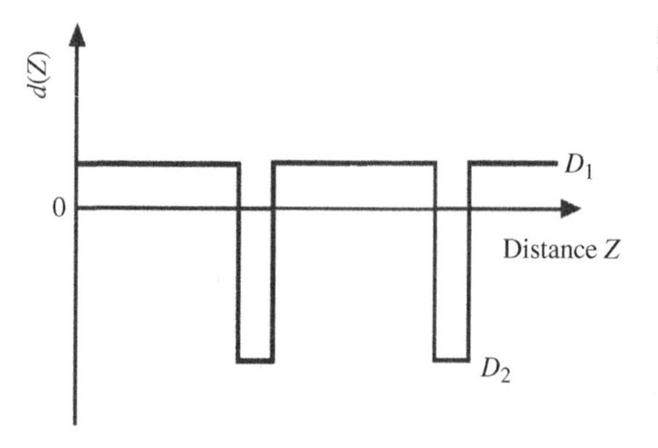

Figure 10.1 A two-step dispersion-managed (DM) transmission line.

arises because the effective core area of a DCF is usually much smaller than that of a conventional single-mode fiber. However, this fact can be used with advantage to achieve Raman amplification along the DCF. In such a case, by pumping a DCF with a high-power Raman pump, simultaneous dispersion compensation and amplification are possible [17].

10.2 Characteristics of the Dispersion-Managed Soliton

In order to illustrate the effects of programmed dispersion, let us consider the canonical dispersion map for DM soliton systems represented in Figure 10.2. This is a two-step symmetric dispersion map, constituted by an anomalous dispersion fiber ($D_1 > 0$) of length L_1 and a normal dispersion fiber ($D_2 < 0$) of equal length L_2, while $L_1 + L_2 = L$ is the map length.

The model equation describing the pulse propagation in a DM transmission line is the normalized nonlinear Schrödinger equation (NLSE), modified to include a spatially varying dispersion $d(z)$, fiber loss, and amplifier gain:

$$i\frac{\partial q}{\partial Z} + \frac{1}{2}d(Z)\frac{\partial^2 q}{\partial T^2} + |q|^2 q = -i\frac{\Gamma}{2}q + iG(Z)q \tag{10.2}$$

where Γ describes fiber losses, $G(Z)$ is the amplifier gain compensating for the fiber loss, and $d(Z)$ is the variable normalized dispersion, assuming the value D_1 in the anomalous dispersion fiber and the value D_2 in the normal dispersion fiber.

Equation (10.2) may be transformed to a Hamiltonian form by introducing a new amplitude u through

$$u(Z,T) = \frac{q(Z,T)}{a(Z)} \tag{10.3}$$

where $a(Z)$ contains rapid amplitude variations due to the loss and gain and $u(Z, T)$ is a slowly varying function of Z. Substituting (10.3) into (10.2), one obtains that $a(Z)$ satisfies the equation:

$$\frac{da}{dZ} = \left[-\frac{\Gamma}{2} + G(Z)\right]a \tag{10.4}$$

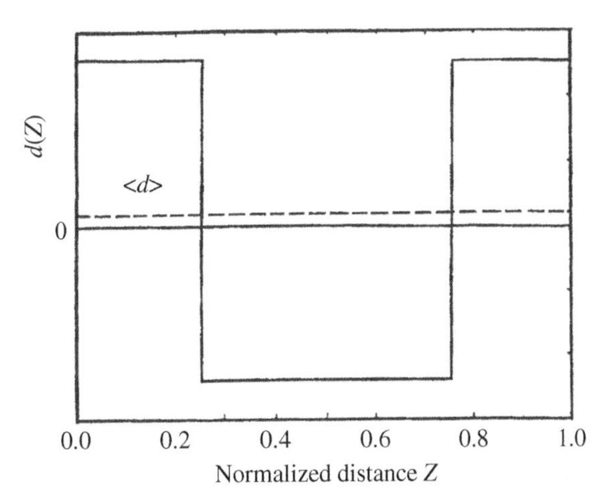

Figure 10.2 Dispersion map with anomalous- and normal-dispersion fibers. The distance Z is normalized by the map length L.

while the new amplitude u satisfies

$$i\frac{\partial u}{\partial Z} + \frac{1}{2}d(Z)\frac{\partial^2 u}{\partial T^2} + a^2(Z)|u|^2 u = 0 \tag{10.5}$$

Assuming a lumped amplification scheme, we have $a(Z) = \exp(-\Gamma(Z - nZ_a)/2)$ in each amplifier span $nZ_a < Z < (n+1)Z_a$.

Equation (10.5) describes the propagation of a DM soliton through a polarization-preserved optical fiber with distributed loss and periodic amplification. If $d(z) = a^2(z) = 1$, it becomes the standard NLSE, which is integrable by the method of inverse scattering transform and has an infinite number of conserved quantities. However, Eq. (10.5), as it appears, is no longer integrable because of inhomogeneous coefficients $d(z)$ and $a^2(z)$. Moreover, it contains only two integrals of motion, corresponding to the energy, E, and to the momentum, M:

$$E = \int_{-\infty}^{\infty} |u|^2 dT \tag{10.6}$$

$$M = \frac{i}{2}d(Z)\int_{-\infty}^{\infty}\left(u^*\frac{\partial u}{\partial T} - u\frac{\partial u^*}{\partial T}\right)dT \tag{10.7}$$

However, in general, the Hamiltonian, H, given by

$$H = \frac{1}{2}\int_{-\infty}^{\infty}\left(d(Z)\left|\frac{\partial u}{\partial T}\right|^2 - a^2(Z)|u|^4\right)dT, \tag{10.8}$$

is not a constant of motion.

Equation (10.5) can be solved numerically using the split-step Fourier method described in Section 6.8. In early numerical simulations, the exact shape of the periodic pulse solutions was unknown and an iterative method has been used to obtain the exact soliton solution. Given a Gaussian initial guess, of approximately the correct power, it is found that it converges upon successive iterations to a DM soliton solution. In general, this solution is almost periodic, but with some remaining oscillations. A numerical averaging algorithm is used to memorize the pulse on one oscillation period (typically a few dispersion maps) and to detect the extrema in pulse width. Both pulse widths, maximum and minimum, are used to define the pulses $u_{\max}$ and $u_{\min}$, respectively, which are then stored. The phase difference between $u_{\max}$ and $u_{\min}$, as well as the pulse energy, are then adjusted, so that the energy is conserved. The iterations are stopped when the variation of the extremum pulse width is smaller than the precision wanted.

Figure 10.3a shows the evolution of the pulse width in the middle of the anomalous dispersion fiber when performing the averaging procedure. A dispersion map with $L_1 = L_2$, $D_1 = 5.02$, $D_2 = -4.98$, and a chirp-free Gaussian initial pulse were assumed. We observe from Figure 10.3a that the pulse width converges to a fixed value after a few iterations. Figure 10.3b shows the final pulse shape obtained in the middle of the anomalous dispersion fiber. The pulse is exactly periodic down to a low power level of 10^{-22}.

After the averaging method is applied, a long-term stable pulse can be obtained [18–20]. Figure 10.4 shows such a long-term stable solution represented in a logarithmic scale. This result was obtained considering the lossless case ($a^2(Z) = 1$), which occurs in practice when distributed amplification is used so that fiber losses are nearly compensated by the local gain all along the fiber. Figure 10.4 shows that the exact DM soliton develops oscillatory tails in the wings.

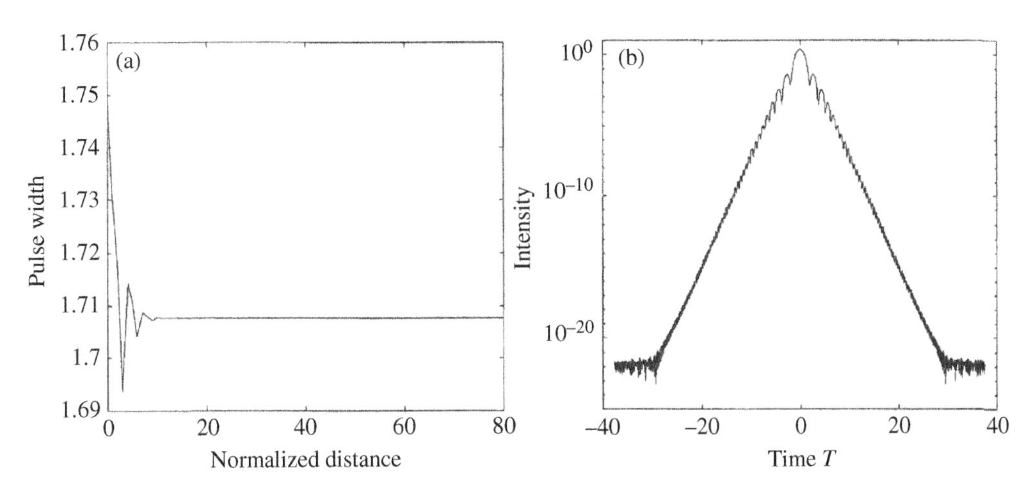

Figure 10.3 (a) Evolution of the pulse width with the propagation distance given in map periods, and (b) The pulse power of the periodic solution found after the averaging procedure, in a logarithmic scale.

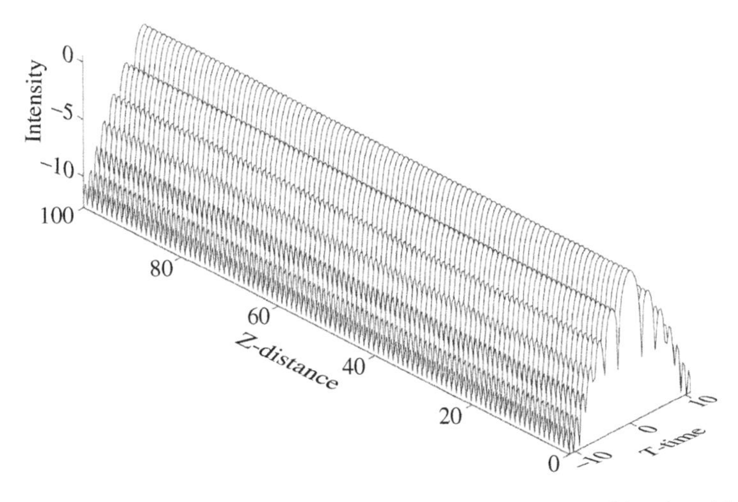

Figure 10.4 Long-term stable stationary solution, represented in a logarithmic scale.

DM solitons are entities whose shape varies substantially over one period of the dispersion map but which return to the same shape at the end of each period. Figure 10.5a and b shows the evolution of the DM soliton during one period of the dispersion map, in linear and logarithmic scales, respectively. The DM pulse alternately spreads and compresses as the sign of the dispersion is switched. The pulse peak power varies rapidly and the pulse width becomes minimum at the center of each fiber where frequency chirp vanishes.

When the pulse in Figure 10.5 is broadened, its intensity is reduced. As a consequence, the nonlinearity experienced by the pulse is also reduced and higher energy compared with that of the standard soliton is required to achieve the necessary balance between nonlinearity and average dispersion. The pulse-enhanced energy leads not only to an improved signal-to-noise ratio, but also contributes to reducing the GH timing jitter.

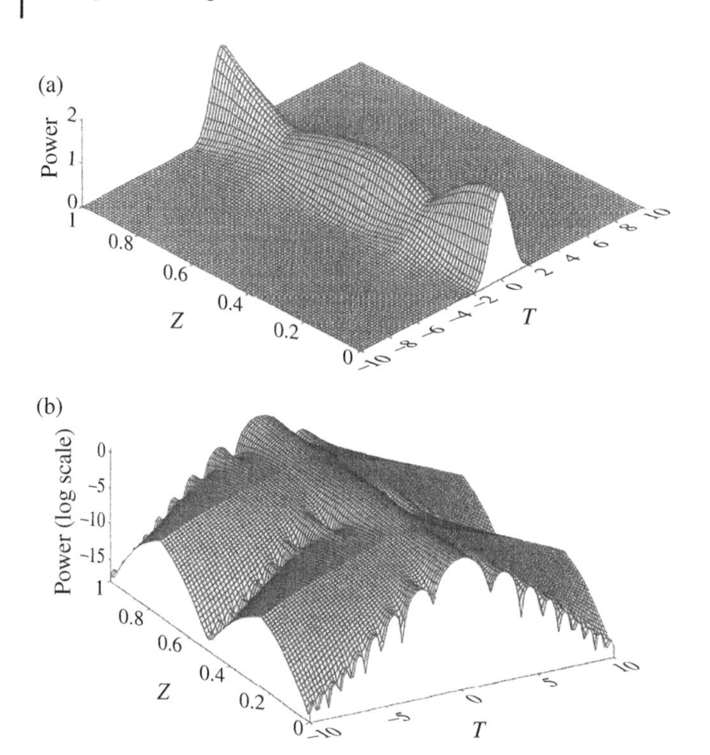

Figure 10.5 Evolution of the DM soliton along one dispersion map shown in (a) linear and (b) logarithmic scales.

An empirical expression for the energy enhancement factor, defined as the ratio between the energy of the DM soliton to that of a standard soliton having the same pulse width considering the same average dispersion, was first obtained by Smith et al. [21] as:

$$\Delta E = 1 + 0.7S^2 \tag{10.9}$$

where

$$S = \frac{(\beta_{21} - \overline{\beta}_2)L_1 - (\beta_{22} - \overline{\beta}_2)L_2}{t_0^2} \tag{10.10}$$

is the map strength. In (10.10), β_{2i} and L_i ($i = 1,2$) are the GVD parameter (in ps^2/km) and the length (in km) of each fiber segment constituting the periodic dispersion map, t_0 (in ps) is the minimum pulse width in the dispersion map, and

$$\overline{\beta}_2 = \frac{\beta_{21}L_1 + \beta_{22}L_2}{L_1 + L_2} \tag{10.11}$$

is the average dispersion.

Equation (10.9) is valid for values of S smaller than about three. For higher values of the map strength, the energy enhancement factor also depends on $|\beta_{2i}/\overline{\beta}_2|$ ($i = 1,2$) and saturates when S increases [22].

10.3 The Variational Approach to DM Solitons

The pulse dynamics in DM systems can be accurately described using a numerical approach to solve the governing propagation equation. However, design optimization using this approach requires intensive computation, which makes the process generally much too time-consuming. Thus, an analytic framework becomes necessary to understand the pulse dynamics, as well as to achieve an optimum design of DM fiber systems.

The variational approach is one of the most powerful techniques capable of providing an insight into the main characteristics of the DM soliton [5, 23–28]. As discussed in Section 6.9, this approximation involves choosing an initial ansatz for the shape of solution sought, but leaves in the ansatz a set of free parameters, which may evolve with Z.

10.3.1 Generic Ansatz

Let us consider the following generic ansatz for the stationary solution of Eq. (10.5):

$$u(Z,T) = Af[(T-T_0)/w]\exp\left[-\frac{i}{2}C(T-T_0)^2 - i\kappa(T-T_0) + i\theta\right] \tag{10.12}$$

where $A(Z)$ is the amplitude, $f(s)$ is an appropriate function describing the pulse profile, $T_0(Z)$ is the soliton position, $w(Z) = t_p(Z)/t_0$ is the normalized pulse width, $\kappa(Z)$ is the frequency shift, $C(Z)$ is the chirp parameter, and $\theta(Z)$ is the phase. The form $f(s)$ of the pulse is still arbitrary. Actually, an important feature of DM solitons is that their profile, in general, is not given by a sech, as it happens for the NLSE fundamental soliton, but varies from sech to a Gaussian function, and even more flat distributions [2, 3].

Equation (10.5) can be constructed by the variation of the Lagrangian density, L_d, given by:

$$L_d(Z,T) = \frac{i}{2}\left(\frac{\partial u^*}{\partial Z}u - \frac{\partial u}{\partial Z}u^*\right) - \frac{1}{2}a^2(Z)|u|^4 + \frac{1}{2}d(Z)\left|\frac{\partial u}{\partial T}\right|^2. \tag{10.13}$$

It is straightforward to see that Eq. (10.5) can be obtained by taking the functional derivative of L_d with respect to u^*:

$$\frac{\delta L_d}{\delta u^*} = 0 \tag{10.14}$$

The merit of using the Lagrangian method is that once the Lagrangian density is known, one can construct evolution equations for parameters that characterize the solution (10.12) from the variational principle. If we substitute the ansatz (10.12) into (10.13) and integrate the result over T, we can construct the Lagrangian L for parameters T_0, w, κ, C, and θ:

$$\begin{aligned} L &\equiv \int_{-\infty}^{\infty} L_d(Z,T)dT \\ &= \frac{A^4a^4w}{2}F_4 - \frac{dA^2}{2w}F_3 - \frac{1}{2}dA^2C^2w^3F_2 - \frac{dwA^2\kappa^2}{2}F_1 + \frac{1}{2}A^2w^3F_2\frac{dC}{dZ} + A^2wF_1\left(T_0\frac{d\kappa}{dZ} - \frac{d\theta}{dZ}\right) \end{aligned} \tag{10.15}$$

where

$$F_1 = \int_{-\infty}^{\infty} f^2(s)\,ds, \quad F_2 = \int_{-\infty}^{\infty} s^2f^2(s)\,ds, \quad F_3 = \int_{-\infty}^{\infty}\left(\frac{df(s)}{ds}\right)^2 ds, \quad F_4 = \int_{-\infty}^{\infty} f^4(s)\,ds \tag{10.16}$$

Now, variation of the Lagrangian naturally gives the Euler–Lagrange equations of motion:

$$\frac{\partial}{\partial Z}\left(\frac{\partial L}{\partial v_Z}\right) - \frac{\partial L}{\partial v} = 0 \tag{10.17}$$

where

$$v = A, w, C, \kappa, T_0, \theta \tag{10.18}$$

and v_Z indicates the derivative of v with respect to Z. Actually, we can obtain the following evolution equations for the pulse parameters:

$$\frac{dA}{dZ} = \frac{1}{2}ACd(Z) \tag{10.19}$$

$$\frac{dw}{dZ} = -d(Z)wC \tag{10.20}$$

$$\frac{dC}{dZ} = -\frac{d(Z)K_1}{w^4} + 2\frac{a^2(Z)A^2K_2}{w^2} + d(Z)C^2 \tag{10.21}$$

$$\frac{d\kappa}{dZ} = 0 \tag{10.22}$$

$$\frac{dT_0}{dZ} = -\kappa d(Z) \tag{10.23}$$

$$\frac{d\theta}{dZ} = \frac{d(Z)\kappa^2}{2} - \frac{d(Z)}{w^2}K_3 + \frac{5a^2(Z)A^2}{4}K_4 \tag{10.24}$$

where $K_1 = F_3/F_2$, $K_2 = F_4/(4F_2)$, $K_3 = F_3/F_1$, $K_4 = F_4/F_1$. From (10.19) and (10.20), we conclude that $A^2w = k =$ constant; so, the number of independent parameters is reduced by one. The constant k is proportional to the pulse energy, given by $E = kF_1$.

10.3.2 Gaussian Pulses

In the case of a Gaussian pulse, we have $f(s) = \exp(-s^2/2)$ and the integrals of motion reduce to

$$E = \int_{-\infty}^{\infty} |u|^2 dT = A^2 w\sqrt{\pi} = k\sqrt{\pi} \tag{10.25}$$

$$M = \frac{i}{2}d(Z)\int_{-\infty}^{\infty}\left(u^*\frac{\partial u}{\partial T} - u\frac{\partial u^*}{\partial T}\right)dT = -\kappa d(Z)A^2w\sqrt{\pi} = -\kappa d(Z)k\sqrt{\pi} \tag{10.26}$$

On the other hand, the evolution equations for the pulse parameters become

$$\frac{dw}{dZ} = -d(Z)wC \tag{10.27}$$

$$\frac{dC}{dZ} = -\frac{d(Z)}{w^4} + \frac{a^2(z)E}{\sqrt{2\pi}w^3} + d(Z)C^2 \tag{10.28}$$

$$\frac{d\kappa}{dZ} = 0 \tag{10.29}$$

$$\frac{dT_0}{d} = -\kappa d(Z) \tag{10.30}$$

$$\frac{d\theta}{dZ} = \frac{d(Z)\kappa^2}{2} - \frac{d(Z)}{2w^2} + \frac{5a^2(z)E}{4\sqrt{2\pi}w} \tag{10.31}$$

10.3.3 Stationary Solutions

Introducing $I = wC$, we can obtain from Eqs. (10.20) and (10.21) the following evolution equation:

$$\frac{dI}{dZ} = -\frac{d(Z)K_1}{w^3} + \frac{2a^2(Z)K_2'}{w^2} \tag{10.32}$$

where $K_2' = kK_2$. One can obtain a periodic solution of Eqs. (10.20) and (10.32) by a proper choice of initial conditions $w(0)$ and $I(0)$ so that $w(1) = w(0)$ and $I(1) = I(0)$. Applying these conditions, we have:

$$\langle d(Z)I \rangle = 0 \tag{10.33}$$

and

$$\left\langle \frac{d(Z)K_1}{w^3} \right\rangle = \left\langle \frac{2a^2(Z)K_2'}{w^2} \right\rangle \tag{10.34}$$

where the angular brackets mean averaging over the dispersion map. We observe from Eq. (10.34) that, for a linear pulse ($a^2(Z) = 0$), we must have $\langle d(Z) \rangle \equiv \bar{d} = 0$. In this case, the periodic solution for f is a Gaussian with a periodically varying normalized chirp $I(Z)$. Moreover, the trajectories in w–I plane at the normal- and anomalous-dispersion fibers completely overlap themselves. However, in the presence of nonlinearity, a gap is produced between these trajectories because of the frequency chirp produced by the self-induced phase shift.

Using Eqs. (10.20) and (10.32), it can be verified also that, in order to have a periodic solution, the following condition must be satisfied:

$$\left\langle \frac{d(Z)}{w^2} \right\rangle = \left\langle \frac{2a^2(Z)K_2'}{w(K_1 + I^2w^2)} \right\rangle > 0 \tag{10.35}$$

The condition given by Eq. (10.35) can be satisfied even if the average dispersion is normal [5, 25–27, 29]. For example, Figure 10.6 shows the evolution of the pulse width (a) and chirp (b) in a case of normal average dispersion ($\bar{d} = -0.02$), for pulse energy $K_2' = 1.485$. The pulse width achieves a minimum value in the middle of each fiber segment, where the chirp becomes zero. However, the shortest pulse occurs in the middle of the anomalous dispersion fiber.

Figure 10.7 shows the phase-space diagram for the variables w and I, corresponding to the same situation of Figure 10.6. The gap between the two C-shaped curves is due to the frequency chirp produced by the self-induced phase shift. In fact, the pulse width in the middle of the normal dispersion fiber is higher than in the middle of the anomalous dispersion fiber (see Figure 10.6a) because of this phase shift. This makes $\langle d(Z)/w^2 \rangle > 0$ even if the average dispersion is zero or normal. Actually, this surprising feature of DM solitons is verified only if the map strength exceeds a critical value, Scr [6–8, 29, 30]. Stable transmission of DM solitons in a system with average zero or normal dispersion has been experimentally demonstrated [31, 32].

In contrast with the DM soliton, the linear stationary pulse is possible only if the average dispersion $\bar{d}$ is zero. As a consequence, DM solitons have a much larger tolerance relative to the fiber dispersion in comparison with the linear pulses. Several experiments have shown the benefits of DM solitons for lightwave systems [31–40].

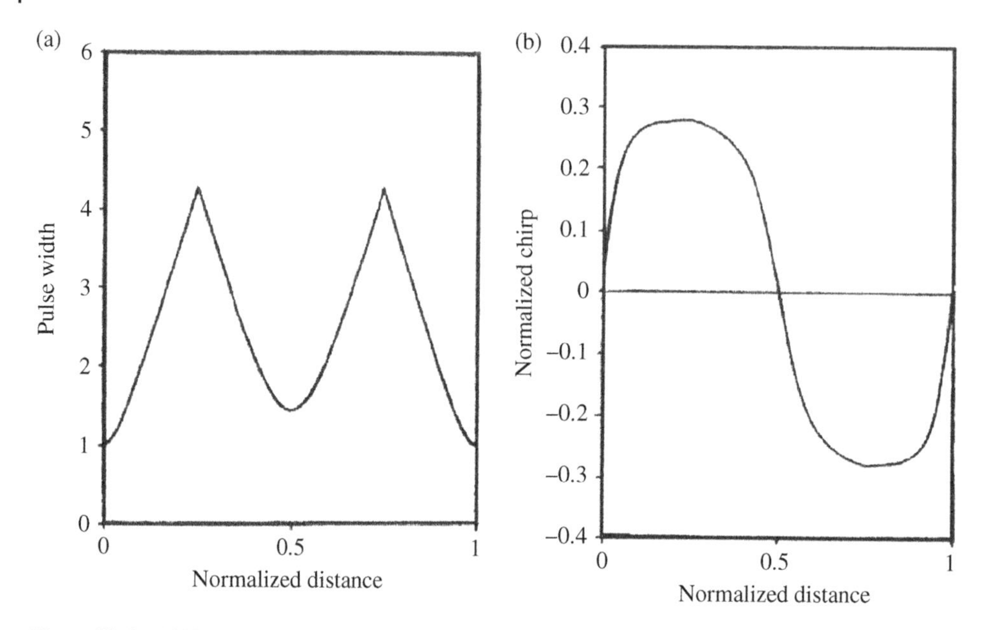

Figure 10.6 (a) Variation of the pulse width $w(Z)$ and (b) the normalized chirp $l(Z)$ along one dispersion map for an average dispersion $\overline{d} = -0.02$ and $K_2' = 1.485$.

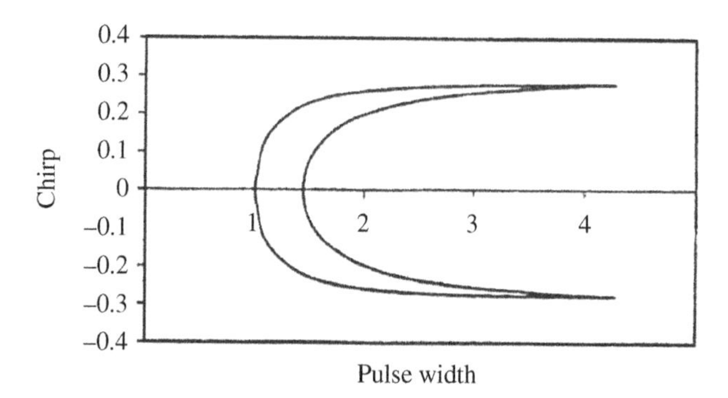

Figure 10.7 Trajectory of parameters w and l in the $w-l$ plane for the case represented in Figure 10.6.

10.4 Interaction Between DM Solitons

The interaction between adjacent DM solitons has been studied numerically or by using a variational technique [41–46]. Such studies show that the qualitative features are similar to those discussed in Chapter 7 for the standard solitons. However, it is found that the collision length of DM solitons depends on the particular characteristics of the dispersion map.

In order to study the interaction between DM solitons, we replace $u(Z,T)$ in (10.5) by a sum of two closely spaced pulses, $u(Z, T) = u_1(Z, T) + u_2(Z, T)$, and choose only terms of self-phase modulation and cross-phase modulation as nonlinear effects. Then the pulse propagation equation for $u_l(l = 1, 2)$ becomes:

$$i\frac{\partial u_l}{\partial Z} + \frac{1}{2}d(Z)\frac{\partial^2 u_l}{\partial T^2} + a^2(Z)|u_l|^2 u_l = -2a^2(Z)\left|u_{(3-l)}\right|^2 u_l \tag{10.36}$$

The Lagrangian density for the system of two pulses is given by:

$$L_d = \sum_{l=1}^{2}\left[\frac{i}{2}\left(u_l\frac{\partial u_l^*}{\partial Z} - u_l^*\frac{\partial u_l}{\partial Z}\right) + \frac{1}{2}d(Z)\left|\frac{\partial u_l}{\partial T}\right|^2 - \frac{1}{2}a^2|u_l|^4\right] - 2a^2|u_1|^2|u_2|^2 \tag{10.37}$$

Assuming a Gaussian ansatz:

$$u_l(Z,T) = A_l \exp\left[-\frac{1}{2}\left(\frac{1}{w_l^2} + iC_l\right)(T - T_{0l})^2 - i\kappa_l(T - T_{0l}) + i\theta_l\right], \quad (l = 1, 2) \tag{10.38}$$

provides the following evolution equations for the pulse parameters:

$$\frac{dw_l}{dZ} = -2d(Z)w_lC_l \tag{10.39}$$

$$\frac{dC_l}{dZ} = -\frac{d(Z)}{2w_l^4} + \frac{a^2(z)k_l}{2\sqrt{2}w_l^3} + 2d(Z)C_l^2 + 2a^2(Z)k_{3-l}\left(\frac{1}{W^3} - \frac{2(\Delta T)^2}{W^5}\right)\exp\left[-\frac{(\Delta T)^2}{W^2}\right] \tag{10.40}$$

$$\frac{d\kappa_l}{dZ} = (-1)^{3-l}4a^2(Z)k_{3-l}\frac{\Delta T}{W^3}\exp\left[-\frac{(\Delta T)^2}{W^2}\right] \tag{10.41}$$

$$\frac{dT_{0l}}{d} = -\kappa_l d(Z) \tag{10.42}$$

where

$$k_j = A_j^2 w_j \tag{10.43}$$

$$\Delta T = T_{01} - T_{02} \tag{10.44}$$

$$W = \sqrt{w_1^2 + w_2^2} \tag{10.45}$$

Equations (10.39)–(10.45) show that the main effect of the interaction between two adjacent DM solitons is to change the pulse frequency, which determines a time shift via the fiber dispersion. This behavior does not depend on the initial phase difference, as long as the cross-phase modulation is the main contribution to the interaction between the solitons.

It is found that the collision length of DM solitons decreases significantly with the dispersion difference of the two fibers constituting a two-step map. This is due to a large amount of overlap between adjacent pulses when they are stretched and because of their enhanced pulse energy. Actually, the interaction between adjacent pulses is one of the main factors limiting the performance of DM soliton transmission systems, especially in the case of strong DM.

In order to reduce the interaction between adjacent DM solitons, a new multiplexing technique, called intra-channel polarization multiplexing, has been implemented. In this technique, the solitons of a single-wavelength channel are interleaved in such a way that any two adjacent solitons are orthogonally polarized [47–52]. Actually, the interaction among orthogonally polarized solitons is much weaker compared with the case of co-polarized solitons. Moreover, it is found that the orthogonal nature of any two adjacent solitons is maintained, even if the polarization states of the soliton train change during propagation. In practice, polarization multiplexing can be implemented using the optical time-division multiplexing (OTDM) technique. To realize this, two-bit streams with orthogonally polarized optical carriers are generated and interleaved using an appropriate optical delay line.

10.5 The Gordon–Haus Effect for DM Solitons

One of the main limitations for the high-bit-rate soliton communication systems is imposed by the random jitter in the pulse arrival time due to the amplifier noise. As discussed in Chapter 7, such noise causes a random frequency shift of the soliton, which translates into a random group-velocity shift owing to dispersion. In the case of DM solitons, it is expected that their enhanced pulse energy and the near-zero path-average dispersion of DM systems will result in both higher signal-to-noise ratios and strongly reduced timing jitters when compared with standard solitons [1, 53, 54].

Let us consider a periodic DM map with period L and average dispersion $\overline{d}$ as described by Eq. (10.5). A lumped amplifier is inserted in every dispersion-map period, compensating for the fiber loss. The shape of the pulse propagating in such a system is assumed to be approximately Gaussian and can be described as:

$$u(Z,T) = A\exp\left\{-\frac{1}{2}(p^2 + iC)(T - T_0)^2 - i\kappa(T - T_0) + i\theta\right\} \tag{10.46}$$

where $A(Z)$, $p(Z)$, $C(Z)$, $T_0(Z)$, $\kappa(Z)$, and $\theta(Z)$ are the amplitude, reciprocal of pulse width, chirp, temporal position, frequency and phase of the pulse, respectively. These parameters fluctuate randomly as a result of the amplifier-added noise. The variances and covariances of the fluctuations of $\kappa(Z)$ and $T_0(Z)$ induced by the amplified spontaneous emission (ASE) noise at each amplification are given by [54, 55]:

$$\sigma_\kappa^2 = n_a\frac{p_a^4 + C_a^2}{\sqrt{\pi}A_a^2 p_a} \tag{10.47}$$

$$\sigma_{T_{0i}}^2 = n_a\frac{1}{\sqrt{\pi}A_a^2 p_a} \tag{10.48}$$

$$\sigma_{\kappa,T_{0i}}^2 = n_a\frac{C_a}{\sqrt{\pi}A_a^2 p_a} \tag{10.49}$$

where A_a, p_a, and C_a are the amplitude, inverse width, and chirp at the amplifier and n_a is the ASE noise power spectrum density, given in physical unities by:

$$n_a = \frac{4\pi^2 c^2 h n_{sp} n_2}{d\overline{D}\lambda^4 A_{eff} p_a}(G-1) \tag{10.50}$$

In Eq. (10.50) n_{sp} is the spontaneous emission coefficient of the amplifier, n_2 is the nonlinear coefficient of the fiber, λ is the wavelength, A_{eff} is the effective fiber core area, h is the Planck constant, c is the velocity of light, and G is the power gain of the amplifier.

After passing N amplifiers, the variance of the timing jitter becomes:

$$\sigma_{T_0}^2 = \frac{1}{6}(2N-1)(N-1)N\left(\overline{d}L\right)^2\sigma_\kappa^2 + N\sigma_{T_{0i}}^2 - (N-1)N\overline{d}L\sigma_{\kappa,T_{0i}}^2 \tag{10.51}$$

For large values of N, we have

$$\sigma_{T_0}^2 \approx \frac{1}{3}N^3\left(\overline{d}L\right)^2\sigma_\kappa^2 \tag{10.52}$$

Comparing with the case of uniform dispersion, it can be seen that the timing jitter of DM solitons can be suppressed for two reasons. First, it occurs due to the enhanced pulse energy, which reduces the variance of the soliton frequency fluctuations, as given in Eq. (10.47). Second, the conversion of the frequency fluctuations into the timing jitter depends on the average dispersion $\bar{d}$, which can be conveniently reduced. Anyway, Eq. (10.52) shows that DM by itself cannot transform the usual cubic growth with distance of the GH timing jitter, $\sigma_{T_0}^2 \sim N^3$.

In a 1996 experiment, solitons could be transmitted at 29 Gb/s over 5520 km in a fiber link with a periodic dispersion map containing optical amplifiers spaced of 40 km [33]. In another 20-Gb/s experiment, the periodic use of DCFs enabled the transmission over 900 km with a considerably reduced timing jitter [34].

10.6 Effects of a Spectral Filter

In order to control the timing jitter growth, it is essential to implement some technique of in-line control, namely the use of bandpass filters [56–58]. As seen in Chapter 8, the periodic insertion of fixed-frequency filters is effective in avoiding the accumulation of random frequency shifts of solitons in constant-dispersion systems. Actually, transmission controls that make use of the nonlinear nature of the pulses are still an effective and natural means to improve the performance of DM soliton systems [9, 29, 51, 59–66].

A unique feature of soliton pulse propagation in DM systems in contrast to constant-dispersion systems is that the spectral as well as temporal widths of the pulse are not constant along the fiber. The dynamics of the pulse cause the stability of DM soliton propagation to be dependent on the location of the control devices. Actually, the use of guiding filters can cause instability if their position in the dispersion map is inappropriate [51, 56, 62].

Let us assume that a control filter is inserted in every dispersion-map period, at a distance z_f from the beginning of the stage. The transfer function of the filter is given by a Gaussian function of the form $H(\Omega) = \exp(-\beta\Omega^2)$, where β (>0) is inversely proportional to the square root of the filter bandwidth and represents the strength of the filter. The effect of the filter on the pulse frequency and temporal position can be described by the following relations [51, 56, 62]:

$$\kappa_{\text{out}} = a\kappa_{\text{in}} \tag{10.53}$$

$$T_{\text{out}} = T_{\text{in}} - \frac{2a\beta C_{\text{in}}}{p_{\text{in}}^2}\kappa_{\text{in}} \tag{10.54}$$

where

$$a = \frac{p_{\text{in}}^2}{p_{\text{in}}^2 + 2\beta\left(p_{\text{in}}^4 + C_{\text{in}}^2\right)} \tag{10.55}$$

and the subscripts *in* and *out* indicate quantities at the entrance and the exit of the filter, respectively. It must be noted from Eq. (10.55) that $0 < a < 1$ for $\beta > 0$. On the other hand, Eq. (10.29) shows that the frequency does not change in the fiber. As a consequence, we conclude from Eq. (10.53) that the frequency is always stabilized at $\kappa = 0$, the center frequency of the filter. Moreover, we verify from Eqs. (10.53) and (10.55) that the stabilization of frequency is more efficient when the spectral width of the pulse, given by $W = (p^2 + (C/p)^2)^{1/2}$, is larger at the filter location.

We also assume that an optical amplifier is inserted in each dispersion-map period, providing an amplitude gain exp(g), which compensates for the filter-induced loss. A pulse that has a spectral width W_{in} at the entrance of the filter suffers an energy loss given by $\left(1 + 2\beta W_{\text{in}}^2\right)^{-1/2}$. The energy loss must be compensated for by the amplifier, giving the relation $\left(1 + 2\beta W_{\text{in}}^2\right)^{1/2} = \exp(2g)$. This relation can be written in the following form:

$$\frac{e^{4g} - 1}{2\beta} = p_{\text{in}}^2 + \left(\frac{C_{\text{in}}}{p_{\text{in}}}\right)^2 \tag{10.56}$$

As seen in Figure 10.6b, in the absence of filtering the pulse becomes transform-limited with no frequency chirp ($C = 0$) at midpoints of fiber segments. It can be shown that the spectral width at the midpoint of the anomalous-dispersion fiber segment increases monotonically with the strength of the DM, a behavior similar to that of pulse energy [51]. As a consequence, the spectral width at this position is an increasing function of the pulse energy. In the presence of filtering, this provides negative feedback that acts on the change in pulse energy, which is self-stabilized. On the contrary, the spectral width at the midpoint of the normal-dispersion fiber segment becomes a decreasing function of pulse energy when the dispersion-map strength is above a given level. In these circumstances, the bandpass filter acts on the pulse to destabilize its energy [51].

10.6.1 Timing Jitter Control

Equations (10.29), (10.30), (10.53), and (10.55) can be linearized around the stationary solution. The linearized analysis provides the following results for the evolution of the frequency and temporal position fluctuations in one period of the dispersion map

$$\Delta\kappa_n = a\Delta\kappa_{n-1} \tag{10.57}$$

$$\Delta T_n = b\Delta\kappa_{n-1} + \Delta T_{n-1} \tag{10.58}$$

where

$$b = -\left(2a\beta C_{\text{in}}/p_{\text{in}}^2 + 2d_{\mathit{eff}}\right) \tag{10.59}$$

and

$$d_{\mathit{eff}} = \bar{d}L + (a-1)\int_{Z_f}^{L} d(Z)dZ \tag{10.60}$$

is the effective cumulative dispersion and L is the dispersion-map period. Equation (10.60) shows that the effective cumulative dispersion depends on the filter position. In particular, if the filter is located at the beginning of the stage ($Z_f = 0$), we have $d_{\mathit{eff}} = a\bar{d}L$, whereas if it is located at the end of the stage ($Z_f = L$), we have $d_{\mathit{eff}} = \bar{d}L$.

After n stages, the timing fluctuation generated at an amplifier becomes:

$$\Delta T_n = b\frac{1-a^n}{1-a}\Delta\kappa_0 + \Delta T_0 \tag{10.61}$$

Assuming that the fluctuations induced by each amplification are independent, we obtain the following result for the variance of the timing jitter at the end of a transmission system corresponding to N dispersion maps:

$$\sigma_{T_0}^2 = \left(\frac{b}{1-a}\right)^2 \frac{N(1-a^2)+1-a^{2N}-2(1+a)(1-a^N)}{1-a^2}\sigma_\kappa^2 + N\sigma_{T_{0i}}^2 + \frac{2b[N(1-a)-(1-a^N)]}{(1-a)^2}\sigma_{\kappa,T_{0i}}^2 \tag{10.62}$$

The variances and covariances of the fluctuations induced by the ASE noise at each amplification are given by Eqs. (10.47)–(10.49).

Equation (10.62) describes the variance of timing jitter of a DM soliton in the presence of narrowband fixed-frequency filters for arbitrary values of the transmission distance (proportional to N) and of the filter strength. When the filter strength β tends to zero, we have from Eqs. (10.55) and (10.59) that $a \rightarrow 1$ and $b \rightarrow -\bar{d}L$, respectively. In this case, the variance of the timing jitter reduces to Eq. (10.51).

In the presence of spectral filtering and for large transmission distances, corresponding to $N \rightarrow \infty$, we obtain from Eq. (10.62) that

$$\sigma_{T_0}^2 = \left[\frac{b^2}{(1-a)^2}\sigma_\kappa^2 + \sigma_{T_{0i}}^2 + \frac{2b}{1-a}\sigma_{\kappa,T_{0i}}^2\right]N \tag{10.63}$$

Thus, contrary to the GH growth law, much slower growth of the DM soliton timing jitter takes place in the presence of the filters, $\sigma_{T_0}^2 \sim N$. We observe from Eq. (10.63) that the asymptotic timing jitter depends generally on the pulse parameters both at the amplifier and at the filter.

10.7 Effects of an Amplitude Modulator

It has been shown that the use of synchronous amplitude modulators, which always require the simultaneous use of narrowband filters to stabilize the energy of standard solitons, can provide the stabilization of DM solitons without the use of filters [61]. In this section, we analyze the combined effects of synchronous amplitude modulators and DM on the soliton timing jitter.

Let us assume that an amplitude modulator is inserted in every dispersion-map period and that its transfer function is given by a Gaussian function of the form $H(t) = \exp(-\sigma t^2)$, where σ represents the strength of the modulator, which is taken to be small. The effect of the modulator on the pulse inverse width, chirp, frequency, and temporal position can be described by the following relations [58, 65]:

$$p_{\text{out}}^2 = p_{\text{in}}^2 + 2\sigma \tag{10.64}$$

$$C_{\text{out}} = C_{\text{in}} \tag{10.65}$$

$$\kappa_{\text{out}} = \kappa_{\text{in}} - \frac{2\sigma C_{\text{in}}}{p_{\text{in}}^2 + 2\sigma}T_{\text{in}} \tag{10.66}$$

$$T_{\text{out}} = \frac{p_{\text{in}}^2}{p_{\text{in}}^2 + 2\sigma}T_{\text{in}} \tag{10.67}$$

where the subscripts *in* and *out* indicate quantities at the entrance and the exit of the modulator, respectively. Eqs. (10.64)–(10.67) can be linearized around the stationary solution. The linearized analysis provides the following results for the evolution of the frequency and temporal position fluctuations in one period of the dispersion map

$$\begin{bmatrix} \Delta\kappa_n \\ \Delta T_n \end{bmatrix} = \begin{bmatrix} a & b \\ c & d \end{bmatrix} \begin{bmatrix} \Delta\kappa_{n-1} \\ \Delta T_{n-1} \end{bmatrix} \tag{10.68}$$

with

$$a = 1 - bd_- \tag{10.69}$$

$$b = -2\sigma C_0/p_0^2 \tag{10.70}$$

$$c = (bd_- - 1)d_+ - Bd_- \tag{10.71}$$

$$d = B - bd_+ \tag{10.72}$$

where

$$B = 1 - 2\sigma/p_0^2 \tag{10.73}$$

and

$$d_- = \int_0^{Z_m} d(Z)dZ, \quad d_+ = \int_{Z_m}^{L} d(Z)dZ \tag{10.74}$$

are the cumulative dispersions between the beginning of the stage and the modulator and between the modulator and the end of the stage, respectively, while Z_m indicates the position of the modulator. The parameters p_0 and C_0 represent the stationary values of the inverse pulse width and chirp, respectively, at the exit of the nth modulator.

After n stages, the timing fluctuation generated at a given amplifier becomes:

$$\Delta T_n = c_n \Delta\kappa_0 + d_n \Delta T_0 \tag{10.75}$$

where c_n and d_n are elements of the matrix

$$\begin{bmatrix} a_n & b_n \\ c_n & d_n \end{bmatrix} = \begin{bmatrix} a & b \\ c & d \end{bmatrix}^n \tag{10.76}$$

Assuming that the fluctuations induced by each amplifier are independent, we obtain the following result for the variance of the timing jitter at the end of a transmission system corresponding to N dispersion maps:

$$\sigma_{T_0}^2 = \sigma_\kappa^2 \sum_{n=0}^{N-1} c_n^2 + \sigma_{T_{0i}}^2 \sum_{n=0}^{N-1} d_n^2 + 2\sigma_{\kappa,T_{0i}}^2 \sum_{n=0}^{N-1} c_n d_n \tag{10.77}$$

where the variances of the fluctuations induced by the ASE noise are given by Eqs. (10.47)–(10.49).

Let us consider the case in which modulator n is at the end of stage n, which follows amplifier $n-1$. In this case, we have from Eq. (10.74) that $d_+ = 0$ and $d_- = \overline{d}L$. Using these results, it can be verified that both c_n and d_n are proportional to d. As a consequence, the variance of the timing jitter given by Eq. (10.77) becomes proportional to d^2. Considering Eqs. (10.72) and (10.73), we observe that $d \to 0$ and the suppression of the timing jitter is achieved when

$$\sigma \to p_0^2/2 \tag{10.78}$$

Another situation of special interest occurs when the modulator is located at a chirp-free point ($C_0 = 0$). In the absence of any control device, this situation occurs in the middle of both the anomalous- and the normal-dispersion fibers in each dispersion map period, as observed in Figure 10.6b. In this case, we have:

$$a_n = 1, \;\; b_n = 0, \;\; c_n = c\frac{1-d^n}{1-d}, \;\; d_n = d^n \tag{10.79}$$

and the variance of the timing jitter is given by:

$$\begin{aligned}\sigma_{T_0}^2 = \sigma_\kappa^2\left(\frac{c}{1-d}\right)^2\left\{\frac{N(1-d^2)-d^{2N}+2d^{N+1}+2d^N-2d-1}{1-d^2}\right\}\\ +\sigma_{T_{0i}}^2\left\{\frac{1-d^{2N}}{1-d^2}\right\}+\sigma_{\kappa,T_{0i}}^2\left(\frac{c}{1-d}\right)\left\{\frac{d^{2N}-d^{N+1}-d^N+d}{1-d^2}\right\}\end{aligned} \tag{10.80}$$

When the modulator strength σ tends to zero, we have that $d \to 1$ and $c \to -\overline{d}L$, respectively, whereas the variance of the timing jitter reduces to Eq. (10.51).

In the presence of synchronous modulation such that $0 < \sigma < p_0^2$ and for large transmission distances, corresponding to $N \to \infty$, we obtain from Eqs. (10.71)–(10.73) and (10.80) that

$$\sigma_{T_0}^2 = \left(\frac{c}{1-d}\right)^2 N\sigma_\kappa^2 \tag{10.81}$$

Thus, contrary to the GH growth law, much slower growth of the DM soliton timing jitter takes place in the presence of the modulators, $\sigma_{T_0}^2 \sim N$.

Figure 10.8 illustrates the variation of the reduction factor of the timing jitter, $F(x, N)$, given by the ratio between Eqs. (10.80) and (10.51), against the number of amplifiers, N, for several values of the parameter $x = \sigma/p_0^2$. It is clearly observed that the suppression of the timing jitter becomes more significant both for higher values of N (longer transmission distances) and for higher values of x (higher modulator strengths).

10.8 WDM with DM Solitons

FWM and cross-phase modulation (XPM) among different channels impose the main nonlinear limitations to the transmission capacity of WDM transmission systems. Such limitations can be greatly attenuated through the use of DM and eventually some of the control techniques discussed in Chapter 8. Actually, the FWM effects are negligible in DM systems due to the large values of local dispersion. In these circumstances, the cross-phase modulation becomes the only relevant inter-channel nonlinear effect in DM WDM soliton systems. It has been verified that the XPM-induced interactions among solitons can be reduced using distributed Raman amplification [67].

In a WDM system, there are a number of wavelengths propagating simultaneously through the fiber. In such a case, if Eq. (10.1) is satisfied for a channel of wavelength λ, the residual dispersion at a nearby wavelength $\lambda + \Delta\lambda$ will be

$$\text{Dispersion} = D_1(\lambda+\Delta\lambda)L_1 + D_2(\lambda+\Delta\lambda)L_2 \approx D_1' L_1 \Delta\lambda\left(1-\frac{D_2'/D_2}{D_1'/D_1}\right) \tag{10.82}$$

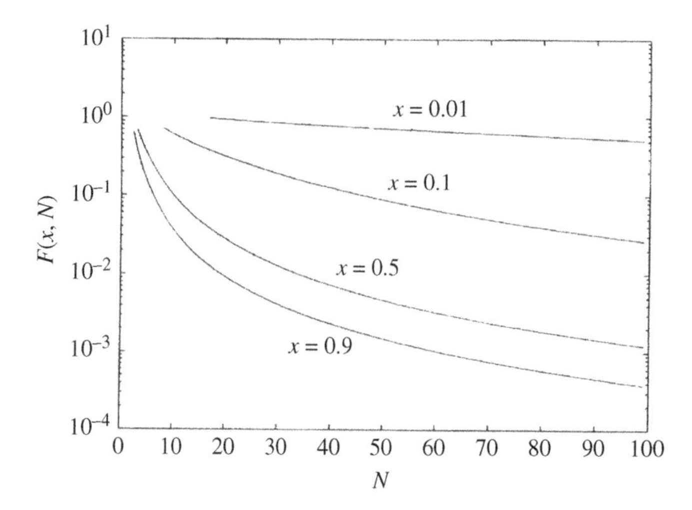

Figure 10.8 Variation of the reduction factor of the timing jitter (logarithmic scale) with N, for several values of the parameter $x = \sigma/p_0^2$.

where prime denotes differentiation with respect to wavelength. Thus, for the dispersion compensation over a band of wavelengths, the relative dispersion slopes D'/D of the two fibers must be equal:

$$\frac{D_2'}{D_2} = \frac{D_1'}{D_1} \tag{10.83}$$

If the condition (10.83) is not satisfied, the dispersion may accumulate and reach unacceptable levels in multichannel systems.

The main aspects affecting the performance of WDM soliton systems using periodic dispersion maps, namely the interchannel collisions and the timing jitter, have been studied extensively [68–80]. Considering two solitons belonging to different channels, it is found that the shorter-wavelength soliton travels faster in the anomalous-GVD section but slower in the normal-GVD section. Moreover, as seen in Section 10.3, the pulse width changes during propagation and can become quite large in some regions of the dispersion map. As a consequence, the two colliding solitons move in a zigzag fashion and pass through each other many times before they separate. The residual frequency shift resulting from the soliton collision depends on a large number of parameters, namely the amplifier spacing, the map period, the map strength, and the proximity to the junction of opposite-GVD fibers [73–75]. Moreover, since the effective collision length is much larger than the map period, the condition $z_I > 2L_A$, discussed in Section 7.6, is satisfied even when soliton wavelengths differ by 20 nm or more, which allows a large number of channels.

The design optimization of periodic dispersion maps like those considered in previous sections has resulted in WDM soliton systems capable of operating at bit rates close to 1 Tb/s [81–91]. In one experiment, Fukuchi et al. [86] have achieved error-free transmission of 1.1 Tb/s (55 × 20 Gbit/s) over 3000 km by using both L and C bands of the erbium-doped fiber amplifier. A system capacity of 1 Tb/s was reported by Nakazawa in a 2000 experiment by transmitting 25 channels at 40 Gb/s over 1500 km with 100-GHz channel spacing [89]. By 2001, a system capacity of 2.56 Tb/s was realized by transmitting 32 channels at 80 Gb/s over 120 km by interleaving two orthogonally polarized 40-Gb/s WDM pulse trains [91]. In another experiment, eight 20-Gb/s channels, corresponding to a 160-Gb/s capacity, were transmitted over 10 000 km using optical filters and synchronous modulators inside a 250-km recirculating fiber loop [87]. These results show that the use of DM solitons has the potential of realizing transoceanic lightwave systems capable of operating with a capacity of 1 Tb/s or more. Several other experiments have confirmed that DM solitons are, in fact, suitable for high-bit-rate and long-haul WDM transmission [92–94].

References

1 Smith, N.J., Knox, F.M., Doran, N.J. et al. (1996). *Electron. Lett.* **32**: 54.
2 Nakazawa, M., Kubota, H., and Tamura, K. (1996). *IEEE Photon. Technol. Lett.* **8**: 452.
3 Grudinin, A.B. and Goncharenko, I.A. (1996). *Electron. Lett.* **32**: 1602.
4 Kutz, J.N., Holmes, P., Evangelides, S.G., and Gordon, J.P. (1998). *J. Opt. Soc. Am. B* **15**: 87.
5 Turitsyn, S.K., Gabitov, I., Laedke, E.W. et al. (1998). *Opt. Commun.* **151**: 117.
6 Grigoryan, V.S. and Menyuk, C.R. (1998). *Opt. Lett.* **23**: 609.
7 Kutz, J.N. and Evangelides, S.G. Jr. (1998). *Opt. Lett.* **23**: 685.
8 Nijhof, J.H.B., Forysiak, W., and Doran, N.J. (1998). *Opt. Lett.* **23**: 1674.
9 Turitsyn, S.K. and Shapiro, E.G. (1999). *J. Opt. Soc. Am. B* **16**: 1321.
10 Turitsyn, S.K., Nijhof, J.H.B., Mezentsev, V.K., and Doran, N.J. (1999). *Opt. Lett.* **24**: 1871.
11 Forghieri, F., Tkach, R.W., and Chraplyvy, A.R. (1994). *IEEE Photon. Technol. Lett.* **6**: 754.
12 Ferreira, M.F. (2011). *Nonlinear Effects in Optical Fibers*. Hoboken, NJ: John Wiley & Sons.
13 Gnauck, A.H. and Jopson, R.M. (1977). Dispersion compensation for optical fiber systems. In: *Optical Fiber Telecommunications IIIA* (ed. I.P. Kaminow and T.L. Koch). Academic Press.
14 Suzuki, M., Morita, I., Edagawa, N. et al. (1995). *Electron. Lett.* **31**: 2027.
15 Kubota, H. and Nakazawa, M. (1992). *Opt. Commun.* **87**: 17.
16 Smith, N.J., Forysiak, W., and Doran, N.J. (1996). *Electron. Lett.* **32**: 2085.
17 Islam, M.N. (2002). *IEEE J. Sel. Top. Quantum Electron.* **8**: 548.
18 Nijhof, J.H., Forysiak, W., and Doran, N.J. (2000). *IEEE J. Select. Topics Quantum Electron.* **6**: 330.
19 Cautaerts, V., Maruta, A., and Kodama, Y. (2000). *Chaos* **10**: 515.
20 Sousa, M.H., Ferreira, M.F., and Panameño, E.M. (2004). *SPIE Proc.* **5622**: 1002.
21 Smith, N.J., Doran, N.J., Knox, F.M., and Forysiak, W. (1996). *Opt. Lett.* **21**: 1981.
22 Nijhof, J.H.B., Doran, N.J., Forysiak, W., and Berntson, A. (1998). *Electron. Lett.* **34**: 481.
23 Gabitov, I.R. and Turitsyn, S.K. (1996). *Opt. Lett.* **21**: 327.
24 Berntson, A., Doran, N.J., Forysiak, W., and Nijhof, J.H.B. (1998). *Opt. Lett.* **23**: 900.
25 Turitsyn, S.K. and Shapiro, E.G. (1998). *Opt. Fiber Technol.* **4**: 151.
26 Sousa, M.H., Ferreira, M.F., and Panameño, E.M. (2004). *SPIE Proc.* **5622**: 944.
27 Jackson, R., Jones, C., and Zharnitsky, V. (2004). *Physica D* **190**: 63.
28 Konar, S., Mishra, M., and Jana, S. (2006). *Chaos, Solitons Fractals* **29**: 823.
29 Nijhof, J.H.B., Doran, N.J., Forysiak, W., and Berntson, A. (1998). *Electron. Lett.* **33**: 1726.
30 Chen, Y. and Haus, H.A. (1998). *Opt. Lett.* **23**: 1013.
31 Jacob, J.M., Golovchenko, E.A., Pilipetskii, A.N. et al. (1997). *Technol. Lett.* **9**: 130.
32 Grigoryan, V.S., Mu, R.M., Carter, G.M., and Menyuk, C.R. (2000). *IEEE Photon. Technol. Lett.* **10**: 45.
33 Naka, A., Matsuda, T., and Saito, S. (1996). *Electron. Lett.* **32**: 1694.
34 Morita, I., Suzuki, M., Edagawa, N. et al. (1996). *IEEE Photon. Technol. Lett.* **8**: 1573.
35 Carter, G.M. and Jacob, J.M. (1998). *IEEE Photon. Technol. Lett.* **10**: 546.
36 Mu, R.M., Menyuk, C.R., Carter, G.M., and Jacob, J.M. (2000). *IEEE J. Sel. Topics Quantum Electron.* **6**: 248.

37 Grudinin, A.B., Durkin, M., Isben, M. et al. (1997). *Electron. Lett.* **33**: 1572.
38 Favre, F., Le Guen, D., and Georges, T. (1999). *J. Lightwave Technol.* **17**: 1032.
39 Penketh, I.S., Harper, P., Aleston, S.B. et al. (1999). *Opt. Lett.* **24**: 803.
40 Zitelli, M., Favre, F., Le Guen, D., and Del Burgo, S. (1999). *IEEE Photon. Technol. Lett.* **9**: 904.
41 Yu, T., Golovchenko, E.A., Pilipetskii, A.N., and Menyuk, C.R. (1997). *Opt. Lett.* **22**: 793.
42 Georges, T. (1998). *J. Opt. Soc. Am. B* **15**: 1553.
43 Kumar, S., Wald, M., Lederer, F., and Hasegawa, A. (1998). *Opt. Lett.* **23**: 1019.
44 Romagnoli, M., Socci, L., Midrio, M. et al. (1998). *Opt. Lett.* **23**: 1182.
45 Inoue, T., Sugahara, H., Maruta, A., and Kodama, Y. (2000). *IEEE Photon. Technol. Lett.* **12**: 299.
46 Takushima, Y., Douke, T., and Kikuchi, K. (2001). *Electron. Lett.* **37**: 849.
47 Wabnitz, S. (1995). *Opt. Lett.* **20**: 261.
48 De Angelis, C. and Wabnitz, S. (1996). *Opt. Commun.* **125**: 186.
49 Midrio, M., Franco, P., Crivellari, M. et al. (1996). *J. Opt. Soc. Am. B* **13**: 1526.
50 Chen, C.F. and Chi, S. (1998). *J. Opt.* **4**: 278.
51 Eleftherianos, C.A., Syvridis, D., Sphicopoulos, T., and Caroubalos, C. (1998). *Opt. Commun.* **154**: 14.
52 Silmon-Clyde, J.P. and Elgin, J.N. (1999). *J. Opt. Soc. Am. B* **16**: 1348.
53 Carter, G.M., Jacob, J.M., Menyuk, C.R. et al. (1997). *Opt. Lett.* **22**: 513.
54 Okamawari, T., Maruta, A., and Kodama, Y. (1998). *Opt. Commun.* **149**: 261.
55 Georges, T., Favre, F., and Le Guen, D. (1998). *IEICE Trans. Electron.* **E81-C**: 226.
56 Kumasako, J., Matsumoto, M., and Waiyapot, S. (2000). *J. Lightwave Technol.* **18**: 1064.
57 Dany, B., Brindel, P., Leclerc, O., and Desurvire, E. (2000). *Electron. Lett.* **35**: 418.
58 Leclerc, O., Brindel, P., Rouvillain, D. et al. (2000). *Electron. Lett.* **36**: 58.
59 Matsumoto, M. (1998). *Electron. Lett.* **22**: 2155.
60 Matsumoto, M. (1998). *J. Opt. Soc. Am. B* **15**: 2831.
61 Waiyapot, S. and Matsumoto, M. (1999). *IEEE Photon. Technol. Lett.* **11**: 1408.
62 Ferreira, M.F.S. and Sousa, S.H. (2001). *Electron. Lett.* **37**: 1184.
63 Matsumoto, M. (2001). *Opt. Lett.* **23**: 1901.
64 Ferreira, M.F. and Sousa, M.H. (2004). *Laser Phys. Lett.* **1**: 491.
65 Ferreira, M.F. and Sousa, M.H. (2005). *Nonl. Opt. Quantum Opt.* **33**: 51.
66 Ferreira, M.F. and Sousa, M.H. (2004). *Laser Phys. Lett.* **1**: 602.
67 Wabnitz, S. and Le Meur, G. (2001). *Opt. Lett.* **26**: 777.
68 Wabnitz, S. (1996). *Opt. Lett.* **21**: 638.
69 Kodama, Y., Mikhailov, A.V., and Wabnitz, S. (1997). *Opt. Commun.* **143**: 53.
70 Devaney, J.F.L., Forysiak, W., Niculae, A.M., and Doran, N.J. (1997). *Opt. Lett.* **22**: 1695.
71 Yang, T.S., Kath, W.L., and Turitsyn, S.K. (1998). *Opt. Lett.* **23**: 597.
72 Mecozzi, A. (1998). *J. Opt. Soc. Am. B* **15**: 152.
73 Niculae, A.M., Forysiak, W., Golag, A.J. et al. (1998). *Opt. Lett.* **23**: 1354.
74 Mamyshev, P.V. and Mollenauer, L.F. (1999). *Opt. Lett.* **24**: 1.
75 Sugahara, H., Maruta, A., and Kodama, Y. (1999). *Opt. Lett.* **24**: 145.
76 Sugahara, H., Kato, H., Inoue, T. et al. (1999). *J. Lightwave Technol.* **17**: 1547.
77 Hasegawa, A. and Hirooka, T. (2000). *Electron. Lett.* **36**: 68.
78 Grigoryan, V.S. and Richter, A. (2000). *J. Lightwave Technol.* **18**: 1148.
79 Wald, M., Malomed, B.A., and Lederer, F. (2001). *Opt. Lett.* **26**: 965.
80 Sugahara, H. (2001). *IEEE Photon. Technol. Lett.* **13**: 963.
81 Favre, F., Le Guen, D., Moulinard, M.L. et al. (1997). *Electron. Lett.* **33**: 2135.
82 Tanaka, K., Morita, I., Suzuki, M. et al. (1998). *Electron. Lett.* **34**: 2257.
83 Mollenauer, L.F., Bonney, R., Gordon, J.P., and Mamyshev, P.V. (1999). *Opt. Lett.* **24**: 285.

84 Dennis, M.L., Kaechele, W.I., Goldberg, L. et al. (1999). *IEEE Photon. Technol. Lett.* **11**: 1680.
85 Morita, I., Suzuki, M., Edagawa, N. et al. (1999). *J. Lightwave Technol.* **17**: 80.
86 Fukuchi, K., Kakui, M., Sasaki, A., et al. (1999). ECOC'99. Postdeadline Paper PD2-10, Nice, France.
87 Nakazawa, M., Kubota, H., Suzuki, K. et al. (2000). *IEEE J. Sel. Topics Quantum Electron.* **6**: 363.
88 Suzuki, K., Kubota, H., Sahara, A., and Nakazawa, M. (2000). *Electron. Lett.* **36**: 443.
89 Nakazawa, M. (2000). *IEEE J. Sel. Topics Quantum Electron.* **6**: 1332.
90 Mollenauer, L.F., Mamyshev, P.V., Gripp, J. et al. (2000). *Opt. Lett.* **25**: 704.
91 Lee, W.S., Zhu, Y., Shaw, B. et al. (2001). *Electron. Lett.* **37**: 964.
92 Suzuki, M. and Edagawa, N. (2003). *J. Lightwave Technol.* **21**: 916.
93 Mollenauer, L.F., Grant, A., Liu, X. et al. (2003). *Opt. Lett.* **28**: 2043.
94 Mollenauer, L.F. and Gordon, J.P. (2006). *Solitons in Optical Fibers – Fundamentals and Applications*. San Diego, CA: Elsevier Academic Press.

11

Polarization Effects

Polarization mode dispersion (PMD) has been studied extensively because it is one of the major factors that limit the performance of modem lightwave systems [1–17]. The origin of PMD is the small random birefringence in the optical fiber caused by imperfections in the manufacturing process and/or mechanical stress on the fiber after manufacture. This birefringence means that there are two orthogonal polarization modes that obey different dispersion relations, and hence, have different group velocities. The random change of this birefringence along the fiber results in random coupling between the modes.

In a linear system, if the input pulse excites both polarization components, it becomes broader at the fiber output because the two components disperse along the fiber as a result of their different group velocities. On the other hand, in soliton-based systems, polarization of the pulses propagating along a fiber also changes continually. PMD tends to separate the pulse components in the two principal polarizations; however, the soliton counteracts the separation through the effective potential well produced by the local index change that is due to the Kerr effect. It is to be expected, therefore, that solitons could withstand PMD if it is not excessive.

11.1 Fiber Birefringence and Polarization Mode Dispersion

The mechanisms inducing birefringence in optical fibers can be divided into two main categories: intrinsic and extrinsic. Intrinsic mechanisms are related to the manufacturing process and constitute a permanent feature of the fiber. The noncircular geometry of the core and a nonsymmetrical stress field in the glass around the core are two examples, as represented in Figure 11.1a. Birefringence can also be created in a fiber affected by external forces in handling or cabling. Figure 11.1b shows three examples of the so-called extrinsic mechanisms: lateral stress, bending, and twisting. Due to the extrinsic mechanisms, fiber birefringence will change randomly over time, reflecting the environmental conditions. This effect is responsible for the stochastic behavior of PMD.

In a birefringent fiber, the effective mode index varies continuously with the field orientation angle in the transverse plane. The directions that correspond to the maximum and minimum mode indices are orthogonal and define the *principal axes* of the fiber. Let us assume that these principal axes coincide with the x and y axes. A linearly polarized field along x or y has a propagation constant $\beta_x = n_x\omega/c$ or $\beta_y = n_y\omega/c$, where n_x and n_y are the effective mode indices associated with x and y polarizations, respectively. The fiber *birefringence* is given by

$$\Delta\beta = \beta_x - \beta_y = \frac{\omega}{c}\Delta n_{eff} \tag{11.1}$$

Solitons in Optical Fiber Systems, First Edition. Mario F. S. Ferreira.

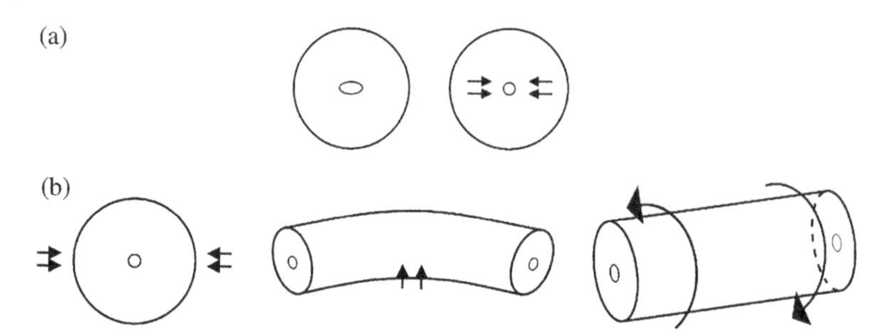

Figure 11.1 Different kinds of birefringence mechanisms that can be found in single-mode fibers: (a) intrinsic mechanisms; and (b) extrinsic mechanisms.

where $\Delta n_{eff} = n_x - n_y$. Typically, Δn_{eff} range between 10^{-7} and 10^{-5}, which is much smaller than the index difference between core and cladding (~3×10^{-3}).

Let us consider an input field given by

$$\mathbf{E}(z) = \exp(i\beta_x z)\left[E_{0x}\hat{\boldsymbol{i}} + E_{0y}\exp(-i\Delta\beta z)\hat{\boldsymbol{j}}\right] \tag{11.2}$$

where E_{0x} and E_{0y} are the input field amplitudes along the principal axes. Equation (11.2) describes a wave whose polarization state transforms from linear to elliptical and then returns to the original linear state when $\Delta\beta z = 2\pi$. The spatial period of this evolution process, known as the *beat length*, is given by:

$$L_b = \frac{2\pi}{\Delta\beta} = \frac{\lambda}{\Delta n_{eff}} \tag{11.3}$$

Equations (11.1) and (11.3) show that both the fiber birefringence and the beat length depend on the frequency. Therefore, the output polarization state also varies in a cyclic manner with frequency for a fixed fiber length. Changes in frequency or distance, therefore, produce equivalent effects on the polarization state.

A different behavior occurs when the input field is polarized along one of the two principal axes. In this case, the polarization remains along the given principal axis for the entire span, even if the frequency changes. This behavior provides the basis for the definition of what is called the *input principal polarization states* of the fiber. Such states correspond to the input polarizations that lead to output polarization states that are *invariant with frequency.*

The difference in the local propagation constants indicated by Eq. (11.1) is usually accompanied by a difference in the local group velocities for the two polarization modes. This differential group velocity is described by a *differential group delay* (DGD), $\Delta\tau$, per unit length between the two modes, often referred to as *intrinsic polarization mode dispersion* (PMD*i*), given by [1, 18–21]:

$$\mathrm{PMD}_i = \frac{\Delta\tau}{L} = \frac{d}{d\omega}(\Delta\beta) = \frac{1}{c}\left(\Delta n_{eff} + \omega\frac{\Delta n_{eff}}{d\omega}\right) \tag{11.4}$$

The quantity $\mathrm{PMD}_i i$ is often expressed in units of picoseconds per kilometer of fiber length. The linear length dependence of the DGD applies only when the birefringence can be considered uniform, as in a short fiber. Figure 11.2 illustrates the effect of PMD when a signal is introduced in such a short fiber with equal components along the two principal axes. The DGD can be interpreted as the effective pulse spread corresponding to the group delay difference between the two orthogonally polarized components.

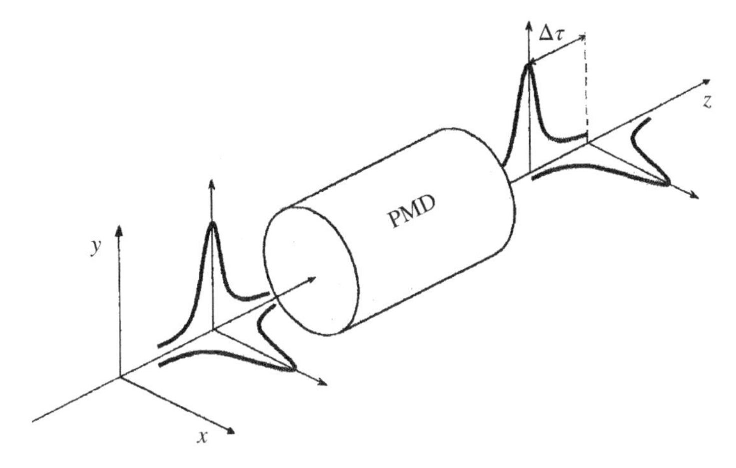

Figure 11.2 Effect of PMD when a signal is introduced in a short fiber with equal components along the two principal axes.

From Eq. (11.4) and ignoring the dispersion of Δn_{eff}, we can see that the DGD for a single beat length is given by

$$\Delta\tau_b = L_b \frac{\Delta n_{eff}}{c} = \frac{\lambda}{c} = \frac{1}{f} \tag{11.5}$$

which corresponds to an optical cycle. At $\lambda = 1.55$ μm, we have $\Delta\tau_b = 5.2$ fs.

11.1.1 PMD in Long Fiber Spans

The assumption of constant birefringence and polarization axis orientation is no longer valid in the case of long fiber spans. In fact, a long span is likely to possess a very complicated progression of features along its length that will modify the polarization of light. As a consequence, even if the light is linearly polarized along a principal axis of the fiber at the input, it couples into the orthogonal polarization during its progress along the fiber. Long fibers are often modelled as a concatenation of a large number of segments, as shown schematically in Figure 11.3. Both the degree of birefringence and the orientation of the principal axes remain constant in each segment, but change randomly from segment to segment. Actually, each fiber segment is treated as a phase plate with different birefringence characteristics.

Due to mode coupling, the birefringence of each section may either add to or subtract from the total birefringence. As a result, PMD in long fiber spans does not accumulate linearly with fiber length, but accumulates in a random-walk-like process that leads to a square root of length dependence.

Since most of the perturbations that act on a fiber depend on the temperature, the transmission properties also vary with ambient temperature. This manifests, for example, as a random, time-dependent

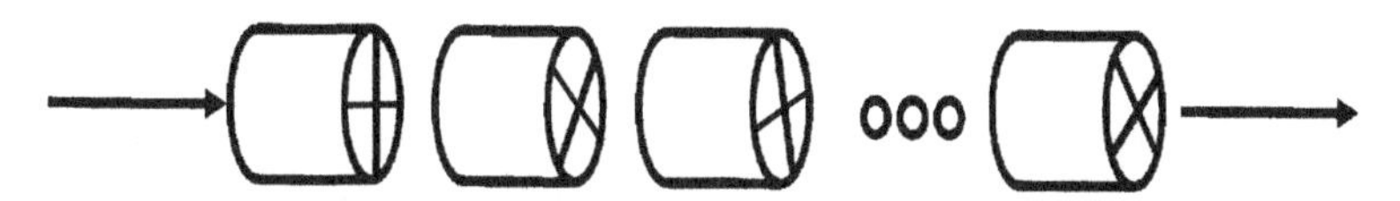

Figure 11.3 Schematic of the technique used for calculating the PMD effects in long fibers.

drifting of the state of polarization at the output of a fiber. Therefore, a statistical approach for PMD must be adopted.

To distinguish between the short- and the long-length regimes, it is important to consider a parameter called the *correlation length*, L_c. One considers the evolution of the polarizations as a function of length in an ensemble of fibers with statistically equivalent perturbations. While the input polarization is fixed (e.g. along the x axis), it is equally probable to observe any polarization state at large lengths. The correlation length is defined to be the length at which the average power in the orthogonal polarization mode, $\langle P_y \rangle$, is within $1/e^2$ of the power in the starting mode $\langle P_x \rangle$, i.e.

$$\frac{\langle P_y \rangle - \langle P_x \rangle}{P_{tot}} = \frac{1}{e^2} \tag{11.6}$$

where $P_{tot} = \langle P_x \rangle + \langle P_y \rangle$. The value of L_c can vary over a wide range from 1 m to 1 km for different fibers, typical values being ~ 10 m. When the fiber transmission distance L satisfies $L << L_c$, the fiber is in the short-length regime, whereas the case $L >> L_c$ corresponds to the long-length regime.

The mean square DGD of the fiber for arbitrary values of the fiber length is given by [3, 21]:

$$\langle \Delta\tau^2 \rangle = 2\left(\Delta\tau_b \frac{L_c}{L_b}\right)^2 \left(\frac{L}{L_c} + e^{-L/L_c} - 1\right) \tag{11.7}$$

For $L << L_c$, the root mean square of the DGD simplifies to

$$\sqrt{\langle \Delta\tau^2 \rangle} = \Delta\tau_{rms} = \Delta\tau_b L/L_b \tag{11.8}$$

On the other hand, for $L >> L_c$, Eq. (11.7) gives

$$\Delta\tau_{rms} = (\Delta\tau_b/L_b)\sqrt{2LL_c} \tag{11.9}$$

reflecting the length dependence discussed earlier.

Equations (11.7)–(11.9) give the mean (and most likely) DGD values that would be observed. However, in practice, significant variations of the DGD are observed about the mean, and so the form of this distribution is important. When $L >> L_c$, it was found that the PMD has a Maxwellian probability distribution given by [3, 22]:

$$p(\Delta\tau) = \frac{8}{\pi^2 \langle \Delta\tau \rangle}\left(\frac{2\Delta\tau}{\langle \Delta\tau \rangle}\right)^2 e^{-\left(\frac{2\Delta\tau}{\langle \Delta\tau \rangle}\right)^2 \frac{1}{\pi}} \quad (\Delta\tau > 0) \tag{11.10}$$

where $\langle \Delta\tau \rangle$ is the mean DGD, which is related to $\Delta\tau_{rms}$ by the expression $\langle \Delta\tau \rangle = \sqrt{\frac{8}{3\pi}}\Delta\tau_{rms}$.

The propagation of a pulse through a long span is very complicated due to the random sequence of perturbations along the length that produce local changes in the polarization. However, a surprising feature of PMD is that even for long fibers, there are two *input principal states of polarization* (PSPs) [23]. These two principal states are orthogonal and are defined as those input polarizations for which the output states of polarization (SOPs) are independent of frequency to first order, i.e. over a small frequency range. The later states, known as *output* PSPs, are also orthogonal and are not necessarily the same as the input states. Therefore, if a pulse having a sufficiently narrow frequency spectrum is coupled into the span in either one of the input PSPs, the output pulse (appearing in the associated output PSP) experiences no broadening or change of shape and presents a group delay τ_i (i = 1,2). The DGD between the PSPs is $\Delta\tau = \tau_1 - \tau_2$.

Using the principal states model, PMD can be characterized by the PMD vector, $\mathbf{\Omega}$, defined in the three-dimensional Stokes space, around which the SOP, $\mathbf{s}$, rotates when the carrier frequency is changed, such that $d\mathbf{s}/d\omega = \mathbf{\Omega} \times \mathbf{s}$. The direction of the vector $\mathbf{\Omega}$ defines an axis whose two intercepts with the surface of the Poincaré sphere correspond to the two output PSPs. The magnitude of the PMD vector is the DGD, $\Delta\tau$.

11.1.2 PMD-Induced Pulse Broadening in Linear Systems

A well-known manifestation of PMD is that optical pulses broaden by an amount that is dependent on the launched SOP. The rms pulse width broadened due to the PMD is given by [4]:

$$\sigma^2 = \sigma_o^2 + \frac{1}{4}\left[\langle \mathbf{\Omega}^2 \rangle - (\mathbf{s}\cdot\langle \mathbf{\Omega} \rangle)^2\right] \tag{11.11}$$

where σ_o is the initial rms pulse width defined as $\sigma_o^2 = \int_{-\infty}^{\infty} \tau^2 |f(\tau)|^2 d\tau$, $\mathbf{s}$ is the input SOP of the signal, and the bracket denotes frequency average such that $\langle a \rangle = (1/2\pi)\int_{-\infty}^{\infty} a\left|\tilde{f}(\omega)\right|^2 d\omega$. $f(\tau)$ and $\tilde{f}(\omega)$ are the initial pulse amplitude and its Fourier transform, respectively. Equation (11.11) includes all higher-order PMD effects.

The expected rms-broadening factor b is defined as:

$$b^2 = E\left\{\frac{\sigma^2}{\sigma_o^2}\right\} = 1 + \frac{1}{4\sigma_o^2}\left[E\{\langle \mathbf{\Omega}^2 \rangle\} - E\{(\mathbf{s}\cdot\langle \mathbf{\Omega} \rangle)^2\}\right] \tag{11.12}$$

where $E\{\cdot\}$ denotes the expectation value. On the other hand, the frequency correlation of the PMD vector is given by [24, 25]:

$$E\{\Omega_i(\omega_1)\Omega_j(\omega_2)\} = \frac{1}{3}\delta_{ij} g(\omega_1 - \omega_2) = \delta_{ij}\frac{1 - \exp\left[-\frac{E\{\Delta\tau^2\}}{3}(\omega_1 - \omega_2)^2\right]}{(\omega_1 - \omega_2)^2} \tag{11.13}$$

From Eq. (11.12) and using Eq. (11.13), the broadening factor due to the PMD becomes [26]:

$$b^2 = 1 + \frac{1}{4\sigma_o^2}\left[E\{\Delta\tau^2\} - \frac{1}{3}\int_{-\infty}^{\infty}\int_{-\infty}^{\infty} g(\omega_1 - \omega_2)\frac{\left|\tilde{f}(\omega_1)\right|^2\left|\tilde{f}(\omega_2)\right|^2}{4\pi^2} d\omega_1 d\omega_2\right] \tag{11.14}$$

In deriving Eq. (11.13), the order of the frequency integration and averaging was changed.

For an unchirped Gaussian initial pulse with amplitude $f(\tau) = A\exp(-\tau^2/2t_o^2)$, where $t_o = \sqrt{2}\sigma_o$ is the initial pulse width and A is a constant, the broadening factor is calculated analytically and given by

$$b^2 = 1 + x - \frac{1}{2}\left[\left(1 + \frac{4x}{3}\right)^{1/2} - 1\right] \tag{11.15}$$

where $x = E\{\Delta\tau^2\}/4\sigma_o^2$ is a measure of the amount of average PMD relative to the pulse width. In the long-pulse regime ($x << 1$), the normalized broadening approaches $b^2 \approx 1 + 2x/3$, while for $x >> 1$, the PMD dominates the broadening and $b^2 \approx x$.

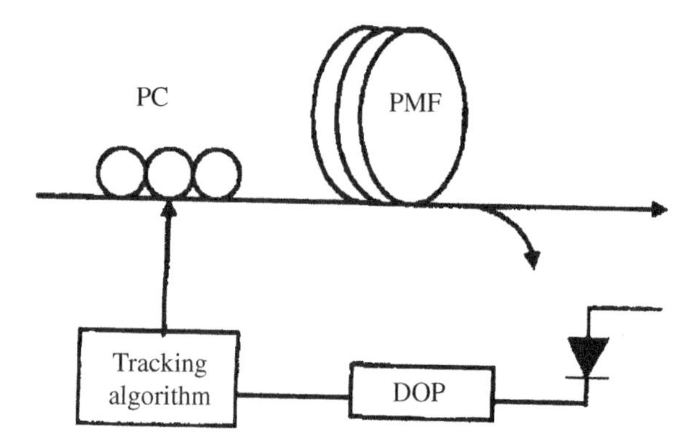

Figure 11.4 Schematic of an optical PMD compensator. PC, PMF, and DOP stand for polarization controller, polarization-maintaining fiber, and degree of polarization, respectively. *Source:* After Ref. [42].

11.1.3 PMD Compensation

Broadening due to PMD occurs in addition to the group velocity dispersion (GVD)-induced pulse broadening. However, whereas the GVD-induced broadening can be eliminated through the use of dispersion management, this technique is not effective to compensate for the PMD-induced degradation. Actually, the control of PMD has become a major issue for modern lightwave systems [1, 10, 13, 27–36]. Several optical and electrical techniques have been proposed for compensating PMD [37–47].

Figure 11.4 shows a basic optical PMD compensator design, consisting of a polarization controller (PC) and a birefringent element, such as a polarization-maintaining fiber (PMF). A feedback loop that measures the degree of polarization uses this information to adjust the PC. Instead of the fixed DGD provided by the birefringent element, other compensators employ a variable DGD using a tunable delay line [42].

The first approach considered for PMD compensation was based on launching the signal into a PSP, which results in less output signal distortion. In practice, a control signal must be fed back to the input from the receiver. The analysis of such a system is performed considering $\mathbf{s} = \langle\mathbf{\Omega}\rangle/|\langle\mathbf{\Omega}\rangle|$ and the resulting average broadening becomes [26]:

$$b_{PSP}^2 = 1 + x - \frac{3}{2}\left[\left(1 + \frac{4x}{3}\right)^{1/2} - 1\right] \tag{11.16}$$

which is always smaller than the uncompensated result. In particular, for small PMD values ($x << 1$), we have $b_{PSP}^2 \approx 1 + x^2/3$. On the other hand, for large PMD values, the squared broadening factor increases as x.

A simple compensation approach in the receiver is to cancel out the PMD vector at the carrier frequency $\mathbf{\Omega}(0)$ with a variable first-order PMD compensator. The input PMD vector of such a system is the vector sum of the PMD vector of the fiber and that of the compensator, i.e. $\mathbf{\Omega}_{tot,\,1} = \mathbf{\Omega} - \mathbf{\Omega}(0)$. The pulse broadening, in this case, can be calculated performing a similar analysis as earlier but now by using the vector $\mathbf{\Omega}_{tot,\,1}$ instead of $\mathbf{\Omega}$. The result is [26]:

$$b_{1st}^2 = 1 + \frac{5x}{3} - \frac{1}{2}\left[\left(1 + \frac{4x}{3}\right)^{1/2} - 1\right] - 4\left[\left(1 + \frac{2x}{3}\right)^{1/2} - 1\right] \tag{11.17}$$

b_{1st}^2 increases as $1 + x^2/3$ for small x, which is the same as for the PSP method. However, it increases as $5x/3$ for large x, which is actually faster than the uncompensated fiber. In fact, the first-order compensation deteriorates the system performance relative to the uncompensated case for $x > 12$. For such short pulses, it is not sufficient to compensate for only first-order PMD, but it is necessary to compensate also for higher orders. This is the case especially for lightwave systems operating at bit rates of 40 Gb/s or more.

In the second-order compensation at the carrier frequency, the total PMD vector is given by $\boldsymbol{\Omega}_{\text{tot},2} = \boldsymbol{\Omega} - \boldsymbol{\Omega}(0) - \boldsymbol{\Omega}^{(1)}(0)\omega$, where $\boldsymbol{\Omega}^{(1)}(0)$ denotes the first frequency derivative of the PMD vector at the carrier frequency. The broadening factor is calculated as in the previous cases and is given by:

$$b_{2nd}^2 = b_{1st}^2 + \frac{x^2}{3} - 12\left[\left(1 + \frac{2x}{3}\right)^{1/2} - \frac{x}{3}\left(1 + \frac{2x}{3}\right)^{-1/2} - 1\right] \tag{11.18}$$

From Eq. (11.18), it can be seen that b_{2nd}^2 increases as $1 + (8x^3/27)$ for small x so that the PMD impairment is reduced further than the first-order compensation case. However, for large x, b_{2nd}^2 increases as $x^2/3$, which is actually much faster than the first-order compensation and the uncompensated cases.

Clearly, the PMD compensation approach based on Taylor's expansion of the fiber link's PMD vector up to a certain order at a specific frequency does not work for large PMD-to-pulse width ratios. In this case, the valid frequency range of the expansion is much narrower than the signal bandwidth, and thus, the higher-order terms in the expansion rather increase the error between the fiber link's PMD vector and the compensation vector over the signal bandwidth. Since the higher-order term grows faster, a larger error will be incurred as the order of the expansion increases.

An alternative strategy makes use of a compensating vector equal to the frequency average of the higher-order PMD vector of the fiber in emulating the PMD vector. The compensation PMD vector will be then in the form of $\Omega_c = \langle\Omega\rangle + \langle\Omega^{(1)}\rangle\omega + \cdots$

In the first-order compensation, the total PMD vector is $\boldsymbol{\Omega}_{1st} = \boldsymbol{\Omega} - \langle\boldsymbol{\Omega}\rangle$, and the broadening factor turns out to be [48]:

$$b_{1st}^2 = 1 + x - \frac{3}{2}\left[\left(1 + \frac{4x}{3}\right)^{1/2} - 1\right] \tag{11.19}$$

which is always smaller than the uncompensated result. In this case, the squared broadening factor increases as $1 + x^2/3$ for small PMD values, while for large PMD values, it increases proportionally to x.

In the case of the second-order compensation, the total PMD vector becomes $\boldsymbol{\Omega}_{2nd} = \boldsymbol{\Omega} - \langle\boldsymbol{\Omega}\rangle - \langle\boldsymbol{\Omega}^{(1)}\rangle\omega$ and the broadening factor is given by [48]:

$$b_{2nd}^2 = b_{1st}^2 + \frac{3}{2}\left(1 + \frac{4x}{3}\right)^{3/2} - 3(x+1)\left(1 + \frac{4x}{3}\right)^{1/2} + \left(\frac{4x^2}{3} + 2x\right)\left(1 + \frac{4x}{3}\right)^{-1/2} + \frac{3}{2} \tag{11.20}$$

From Eq. (11.20), it can be seen that b_{2nd}^2 increases as $1 + (2x^3/9)$ for small x so that the PMD impairment is reduced further than the first-order compensation case. On the other hand, for large x, b_{2nd}^2 increases as x. Therefore, the second-order compensation has a smaller pulse broadening than the first-order compensation and the uncompensated cases for all PMD values.

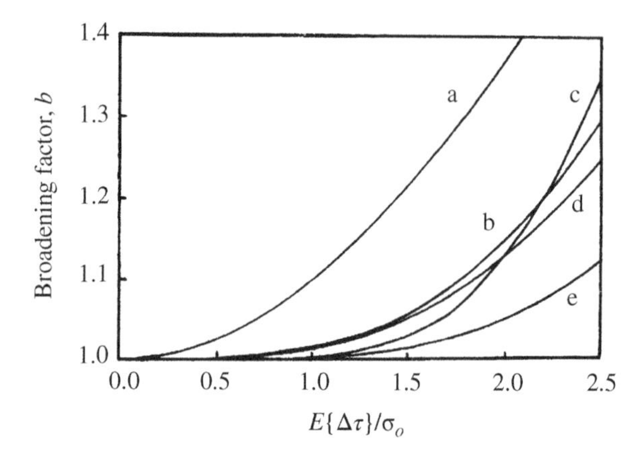

Figure 11.5 Average pulse-broadening factor as a function of the normalized average DGD for the uncompensated case (a), first-order PMD compensation at the carrier frequency (b), second-order PMD compensation at the carrier frequency (c), first-order (d) and second-order (e) compensation of the average PMD vector.

Figure 11.5 illustrates the calculated broadening factors obtained from Eqs. (11.15)–(11.20) as a function of the normalized average DGD, $E\{\Delta\tau\}/\sigma_o$. It can be observed that in the case of second-order PMD compensation at the carrier frequency (curve c), the broadening factor increases rapidly and the system performance is deteriorated for large values of the PMD-to-pulse width ratio. However, if the compensation is realized with the frequency averaged second-order PMD vector (curve e), the system performance is effectively improved for all values of the PMD-to-pulse width ratio.

11.2 Coupled Nonlinear Schrödinger Equations

As seen in Chapter 3, the lowest-order nonlinear effects in optical fibers originate from the third-order susceptibility $\chi^{(3)}$, which is also responsible for the formation of optical solitons. When the nonlinear response of the medium is assumed to be instantaneous, the third-order nonlinear polarization can be written as

$$\mathbf{P}_{NL}(\mathbf{r},t) = \varepsilon_0 \chi^{(3)} \vdots \mathbf{E}(\mathbf{r},t)\mathbf{E}(\mathbf{r},t)\mathbf{E}(\mathbf{r},t) \tag{11.21}$$

This expression neglects the delayed nonlinear response that accounts for the Raman effect, corresponding to the molecular vibrations. For silica fibers, the vibrational response occurs over a time scale of 60–70 fs, which makes Eq. (11.21) approximately valid for pulse widths greater than 1 ps. Let us consider the case where the electric field vector has both x and y components:

$$\mathbf{E}(\mathbf{r},t) = \frac{1}{2}\left(\hat{\mathbf{i}}E_x + \hat{\mathbf{j}}E_y\right)\exp(-i\omega_0 t) + c.c. \tag{11.22}$$

where $c.c.$ means the complex conjugate of the previous expression. By substituting (11.22) in (11.21), we obtain the result:

$$\mathbf{P}_{NL}(\mathbf{r},t) = \frac{1}{2}\left(\hat{\mathbf{i}}P_{NLx} + \hat{\mathbf{j}}P_{NLy}\right)\exp(-i\omega_0 t) + c.c. \tag{11.23}$$

where

$$\overline{P}_{NLx} = \frac{3}{4}\varepsilon_0\chi^{(3)}_{xxxx}\left[\left(|\overline{E}_x|^2 + \frac{2}{3}|\overline{E}_y|^2\right)\overline{E}_x + \frac{1}{3}\overline{E}_x^*\overline{E}_y^2\right] \tag{11.24}$$

$$\overline{P}_{NLy} = \frac{3}{4}\varepsilon_0\chi^{(3)}_{xxxx}\left[\left(|\overline{E}_y|^2 + \frac{2}{3}|\overline{E}_x|^2\right)\overline{E}_y + \frac{1}{3}\overline{E}_y^*\overline{E}_x^2\right] \tag{11.25}$$

In obtaining (11.24) and (11.25), we have used the relation

$$\chi^{(3)}_{xxxx} = \chi^{(3)}_{xxyy} + \chi^{(3)}_{xyxy} + \chi^{(3)}_{xyyx} \tag{11.26}$$

which is valid for an isotropic medium with rotational symmetry [49, 50]. Moreover, the three components in Eq. (11.26) were considered to have the same magnitude, as happens in the case of silica fibers.

Using the earlier expressions for the nonlinear polarization and adopting the procedure of Section 3.3, the slowly varying amplitudes U_x and U_y of the polarization components are found to satisfy the following coupled differential equations [51–54]:

$$\frac{\partial U_x}{\partial z} + \beta_{1x}\frac{\partial U_x}{\partial t} + \frac{i}{2}\beta_{2x}\frac{\partial^2 U_x}{\partial t^2} = i\gamma\left(|U_x|^2 + \frac{2}{3}|U_y|^2\right)U_x + \frac{i\gamma}{3}U_x^*U_y^2\exp\left[-2i(\beta_{0x} - \beta_{0y})z\right] \tag{11.27}$$

$$\frac{\partial U_y}{\partial z} + \beta_{1y}\frac{\partial U_y}{\partial t} + \frac{i}{2}\beta_{2y}\frac{\partial^2 U_y}{\partial t^2} = i\gamma\left(|U_y|^2 + \frac{2}{3}|U_x|^2\right)U_y + \frac{i\gamma}{3}U_y^*U_x^2\exp\left[2i(\beta_{0x} - \beta_{0y})z\right] \tag{11.28}$$

where the amplitudes U_j $(j = x, y)$ are such that their absolute square represents the optical power, γ is the nonlinear coefficient, β_{0j} $(j = x, y)$ represent the propagation constants of the two orthogonal linearly polarized waves, $\beta_{1j} = d\beta_j/d\omega|_{\omega=\omega_0}$, and $\beta_{2j} = d^2\beta_j/d\omega^2|_{\omega=\omega_0}$. Equations (11.27) and (11.28) govern the pulse propagation in linearly birefringent fibers. The first term on the right-hand side of these equations is responsible for SPM, whereas the second term is responsible for XPM. This term corresponds to nonlinear coupling between the two polarization components, through which the nonlinear phase shift acquired by one component depends on the intensity of the other component. Finally, the last term in Eqs. (11.27) and (11.28) results from the phenomenon of four-wave mixing. As seen in Section 4.3, the efficiency of this process depends on the matching of the phases of the involved fields. If the fiber length is much longer than the beat length $L_b = 2\pi/(\beta_{0x} - \beta_{0y})$, these terms change signs often and they can be neglected. This is particularly the case in highly birefringent fibers, for which L_b~1 cm typically.

11.3 Solitons in Fibers with Constant Birefringence

In the following, we will neglect the last term in the right-hand side of Eqs. (11.27) and (11.28) and consider that $\beta_{2x} = \beta_{2y} = \beta_2$, since the two orthogonally polarized waves are assumed to have the same frequency. Moreover, using the average group velocity $v_g = 2/(\beta_{1x} + \beta_{1y})$, the normalized time, T, distance, Z, and amplitude, q_j $(j = x, y)$, as defined in Section 3.3, we obtain the following normalized coupled nonlinear Schrödinger equations:

$$\frac{\partial q_x}{\partial Z} + \delta_g\frac{\partial q_x}{\partial T} - \frac{i}{2}d(Z)\frac{\partial^2 q_x}{\partial T^2} = i\left(|q_x|^2 + \hat{\varepsilon}|q_y|^2\right)q_x \tag{11.29}$$

$$\frac{\partial q_y}{\partial Z} - \delta_g \frac{\partial q_y}{\partial T} - \frac{i}{2} d(Z) \frac{\partial^2 q_y}{\partial T^2} = i\left(\left|q_y\right|^2 + \hat{\varepsilon}\left|q_x\right|^2\right) q_y \tag{11.30}$$

where $d(Z)$ is the normalized dispersion,

$$\Delta\beta = L_D\left(\beta_{0x} - \beta_{0y}\right) \tag{11.31}$$

is the differential wave number of the orthogonal components, whereas

$$\delta_g = \frac{L_D}{2t_0}\left(\beta_{1x} - \beta_{1y}\right) \tag{11.32}$$

represents the differential group velocity. Considering the general case of an elliptically birefringent fiber, the XPM coupling parameter $\hat{\varepsilon}$ depends on the ellipticity angle and can vary from $\hat{\varepsilon} = 2/3$, for a linearly birefringent fiber, to $\hat{\varepsilon} = 2$, in the case of a circularly birefringent fiber [55].

The coupled nonlinear Schrödinger Eqs. (11.29) and (11.30) can be restated as a variational problem [56] in terms of a Lagrangian density, L_d, given by

$$L_d = L_x + L_y - L_{xy} \tag{11.33}$$

where

$$L_x = \frac{i}{2}\left(q_x \frac{\partial q_x^*}{\partial Z} - q_x^* \frac{\partial q_x}{\partial Z}\right) + \frac{i}{2}\delta_g\left(q_x \frac{\partial q_x^*}{\partial T} - q_x^* \frac{\partial q_x}{\partial T}\right) - \frac{1}{2}\left|q_x\right|^4 + \frac{1}{2}\left|\frac{\partial q_x}{\partial T}\right|^2 \tag{11.34}$$

$$L_y = L_x \quad (x \to y) \tag{11.35}$$

$$L_{xy} = \hat{\varepsilon}\left|q_x\right|^2\left|q_y\right|^2 \tag{11.36}$$

When the initial conditions for q_x and q_y are given by

$$q_x(0, T) = q_y(0, T) = \frac{A}{\sqrt{2}} \operatorname{sech}(T) \tag{11.37}$$

the asymptotic soliton solution of Eqs. (11.29) and (11.30) with $d(Z) = 1$, $\delta_g = 0$ becomes

$$q_x = q_y = \frac{1}{\sqrt{1 + \hat{\varepsilon}}} \eta \operatorname{sech}(\eta T) \exp\left(\frac{i}{2}\eta^2 Z\right) \tag{11.38}$$

with

$$\eta = A\sqrt{2(1 + \hat{\varepsilon})} - 1 \tag{11.39}$$

Considering the birefringence as a perturbation, we can find approximate solutions of Eqs. (11.29) and (11.30) in the form:

$$\begin{bmatrix} q_x \\ q_y \end{bmatrix} = \frac{\eta_j}{\sqrt{1 + \hat{\varepsilon}}} \operatorname{sech}\left[\eta_j\left(T - T_{0j}\right)\right] \exp\left[-i\kappa_j\left(T - T_{0j}\right) + i\theta_j\right] \tag{11.40}$$

where $j = x, y$. Evolution of the parameters η_j, κ_j, T_{0j}, and θ_j may be found from the Euler equation

$$\frac{\partial L}{\partial \nu} = \frac{d}{dZ}\left[\frac{\partial L}{\partial \nu_Z}\right] \tag{11.41}$$

where ν stands for the aforementioned soliton parameters, $\nu_Z = \partial\nu/\partial Z$, and L is a Lagrangian, obtained from the Lagrangian density L_d

$$L = \int_{-\infty}^{+\infty} L_d dT \tag{11.42}$$

Substituting Eq. (11.40) into Eqs. (11.33)–(11.36) and using Eqs. (11.41) and (11.42), the following equations for the soliton parameters are obtained [56]:

$$\frac{d\eta_j}{dZ} = 0 \tag{11.43}$$

$$\frac{dT_{0j}}{dZ} = (-1)^{j+1}\delta_g - \kappa_j \tag{11.44}$$

$$\frac{d\kappa_j}{dZ} = -\frac{1}{2\eta_j}\frac{\partial\langle L_{xy}\rangle}{\partial T_{0j}} \tag{11.45}$$

$$\frac{d\theta_j}{dZ} = \frac{1}{2}\kappa_j^2 + \frac{1}{2}\left(\frac{1-\hat{\varepsilon}}{1+\hat{\varepsilon}}\right)\eta_j^2 + \frac{1}{2}(1+\hat{\varepsilon})\frac{\partial\langle L_{xy}\rangle}{\partial\eta_j} \tag{11.46}$$

where

$$\langle L_{xy}\rangle = \frac{\hat{\varepsilon}\eta_x^2\eta_y^2}{(1+\hat{\varepsilon})}\int_{-\infty}^{+\infty} \mathrm{sech}^2[\eta_x(T-T_{0x})]\,\mathrm{sech}^2\left[\eta_y\left(T-T_{0y}\right)\right]dT \tag{11.47}$$

Since the birefringence parameter appears in Eq. (10.44) in a symmetric way, we may consider the case corresponding to the perturbation-induced dynamics of a symmetric solution, assuming

$$\eta_x = \eta_y = \eta, \quad \kappa_x = -\kappa_y = -\kappa, \quad T_{0x} = -T_{0y} = \Delta T/2 \tag{11.48}$$

Then, Eqs. (11.44) and (11.45) are reduced to

$$\frac{d\Delta T}{dZ} = 2\delta_g + 2\kappa \tag{11.49}$$

$$\frac{d\kappa}{dZ} = -\frac{1}{2}\frac{d}{d\Delta T}U(\Delta T) \tag{11.50}$$

where

$$U(\Delta T) = \frac{8\hat{\varepsilon}\eta^2}{1+\hat{\varepsilon}}\frac{\sinh(\eta\Delta T) - \eta\Delta T\cosh(\eta\Delta T)}{\sinh^3(\eta\Delta T)} \tag{11.51}$$

From Eqs. (11.49) and (11.50), it follows that

$$\frac{1}{2}\left(\frac{d\Delta T}{dZ}\right)^2 + U(\Delta T) = \text{const.} \tag{11.52}$$

This result indicates that ΔT behaves as the position of a particle moving in the potential $U(\Delta T)$. The particle will be trapped in the potential well if the initial kinetic energy is smaller than the potential barrier:

$$\frac{1}{2}\left(\left.\frac{d\Delta T}{dZ}\right|_{Z=0}\right)^2 < U(\Delta T \to \infty) - U(\Delta T = 0) \tag{11.53}$$

This condition determines a threshold value for the soliton amplitude for capture of separate polarizations into a bound intermode state. Assuming that the initial frequency difference between these polarizations is zero, Eq. (11.53) becomes:

$$\eta^2 > \frac{15}{8}\delta_g^2 \tag{11.54}$$

Taking into account the relation (11.39), we obtain from Eq. (11.54) the threshold amplitude of the input pulse, A_{thr}, as a function of the birefringence, δ_g:

$$A_{thr} = \frac{1}{\sqrt{2(1+\hat{\varepsilon})}} + \frac{1}{2}\sqrt{\frac{3}{2\hat{\varepsilon}}}\delta_g \tag{11.55}$$

When $\hat{\varepsilon} = 2/3$, Eq. (11.55) takes the approximate form

$$A_{thr} \approx 0.55 + 0.75\delta_g \tag{11.56}$$

Figure 11.6 illustrates the temporal separation between the two polarization components, ΔT, against the propagation distance for the initial condition given by Eq. (11.37), considering two different values of A and $\delta_g = 0.5$. The full curves correspond to the numerical solution of (11.49) and (11.50), whereas the dashed curves were obtained from Eqs. (11.29) and (11.30) with an initial soliton amplitude given by Eq. (11.39) with $\hat{\varepsilon} = 2/3$. A qualitative agreement is observed between the full and the dashed curves in Figure 11.6. In the case $A = 0.8$, which is smaller than the threshold value $A_{th} = 0.925$ given by Eq. (11.55), the separation between the two polarization components increases monotonously. However, for $A = 1.1$, this separation becomes limited and oscillates around zero during propagation, which confirms the mutual trapping of the two polarization components by the nonlinear birefringent fiber. In this case, the two polarization components propagate at a common group velocity due to the XPM-induced nonlinear coupling between them. In order to realize such temporal synchronization, the two components shift their carrier frequencies in the opposite directions, such that the one along the fast axis slows down while the other along the slow axis speeds up. Soliton trapping in optical fibers was first theoretically predicted by Curtis R. Menyuk [57, 58] and later experimentally demonstrated [59, 60].

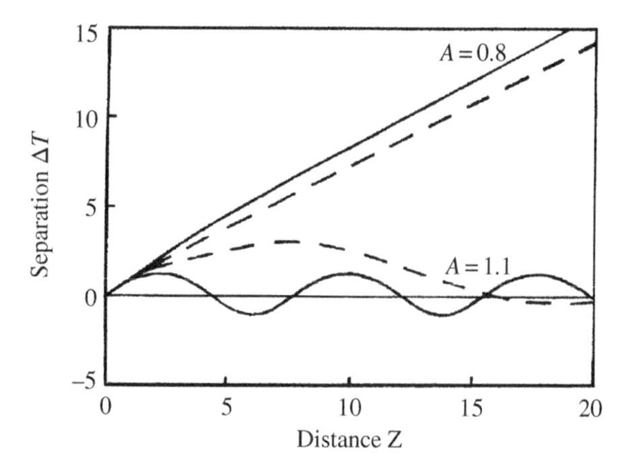

Figure 11.6 Pulse separation between the two polarization components, ΔT, against the propagation distance for the initial condition (11.37), considering two different values of A and $\delta = 0.5$. The full curves are the numerical solution of (11.49) and (11.50), while the dashed curves were obtained from Eqs. (11.29) and (11.30) with $\hat{\varepsilon} = 2/3$.

11.4 Vector Solitons

Vector solitons correspond to exact solitary wave solutions of the coupled NLS Eqs. (11.29) and (11.30) such that the polarization state and the shape of their orthogonally polarized components are preserved during propagation. Different types of vector solitons can be observed in birefringent single-mode fibers, including the polarization-rotating vector solitons, group velocity-locked vector solitons, and polarization-locked vector solitons [61–63]. Vector solitons were also theoretically predicted and experimentally confirmed in mode-locked fiber lasers [64–70].

Let us assume that $d(Z) = 1$ and use the transformations

$$q_x = \widetilde{q}_x \exp\left(i\delta_g^2 Z/2 - i\delta_g T\right) \tag{11.57}$$

$$q_y = \widetilde{q}_y \exp\left(i\delta_g^2 Z/2 + i\delta_g T\right) \tag{11.58}$$

in Eqs. (11.29) and (11.30). The following simplified equations are then obtained:

$$\frac{\partial \widetilde{q}_x}{\partial Z} - \frac{i}{2}\frac{\partial^2 \widetilde{q}_x}{\partial T^2} = i\left(|\widetilde{q}_x|^2 + \hat{\varepsilon}|\widetilde{q}_y|^2\right)\widetilde{q}_x \tag{11.59}$$

$$\frac{\partial \widetilde{q}_y}{\partial Z} - \frac{i}{2}\frac{\partial^2 \widetilde{q}_y}{\partial T^2} = i\left(|\widetilde{q}_y|^2 + \hat{\varepsilon}|\widetilde{q}_x|^2\right)\widetilde{q}_y \tag{11.60}$$

When $\hat{\varepsilon} = 0$, these two equations become decoupled and admit distinct soliton solutions, as discussed in Chapter 6. The case $\hat{\varepsilon} = 1$ is of special interest because Eqs. (11.59) and (11.60) can then be solved with the inverse scattering method. The corresponding solution was obtained by Manakov and can be written in the form [71]:

$$\widetilde{q}_x(Z,T) = \cos(\theta)\mathrm{sech}(T)\exp(iZ/2) \tag{11.61}$$

$$\widetilde{q}_y(Z,T) = \sin(\theta)\mathrm{sech}(T)\exp(iZ/2) \tag{11.62}$$

where θ is an arbitrary angle. The solution given by Eqs. (11.61) and (11.62) corresponds to a vector soliton that is similar to the fundamental soliton solution described in Chapter 6, the angle θ corresponding to the polarization angle.

Actually, the case $\hat{\varepsilon} = 1$ is very restrictive. Several solitary wave solutions of Eqs. (11.59) and (11.60) have been studied in the general case $\hat{\varepsilon} \neq 1$ [72–81]. Among them is the following solution [72]:

$$q_x(Z,T) = \eta\,\mathrm{sech}\left[(1+\hat{\varepsilon})^{1/2}\eta T\right]\exp\left[i\left(1+\hat{\varepsilon}+\delta_g^2\right)\eta^2 Z/2 - i\delta_g T\right] \tag{11.63}$$

$$q_y(Z,T) = \eta\,\mathrm{sech}\left[(1+\hat{\varepsilon})^{1/2}\eta T\right]\exp\left[i\left(1+\hat{\varepsilon}+\delta_g^2\right)\eta^2 Z/2 + i\delta_g T\right] \tag{11.64}$$

where η represents the amplitude. For $\hat{\varepsilon} = 0$, Eqs. (11.63) and (11.64) represent the scalar fundamental soliton solution discussed in Chapter 6. For $\hat{\varepsilon} \neq 1$, they correspond to a vector soliton polarized at 45° relatively to the principal axes of the fiber, presenting equal amplitude components. The only difference between the two polarization components is the change of sign in the last phase term, corresponding to a shift of the carrier frequency in opposite directions. This difference in the carrier frequencies compensates for the PMD, whereas the combination of SPM and XPM compensates for the GVD.

11.5 Solitons in Fibers with Randomly Varying Birefringence

The propagation of solitons in fibers with randomly varying birefringence can be described assuming that the fiber has linear birefringence of fixed strength with the randomly varying direction of axes, as proposed by Wai and Menyuk [3]. In this model, the coupled nonlinear Schrödinger equations for the lossless fiber whose birefringence direction is rotated by α from the fixed coordinates x and y can be written in the form [3]:

$$i\frac{\partial \mathbf{Q}}{\partial Z} + \frac{\Delta\beta}{2}\Sigma\mathbf{Q} + i\delta_g\Sigma\frac{\partial \mathbf{Q}}{\partial T} + \frac{1}{2}\frac{\partial^2 \mathbf{Q}}{\partial T^2} + \frac{5}{6}|\mathbf{Q}|^2\mathbf{Q} + \frac{1}{6}(\mathbf{Q}^t\sigma_3\mathbf{Q})\sigma_3\mathbf{Q} + \frac{1}{3}\mathbf{R} = 0 \tag{11.65}$$

where $\mathbf{Q} = \begin{bmatrix} q_x & q_y \end{bmatrix}^t, \mathbf{Q}^\dagger = \begin{bmatrix} q_x^* & q_y^* \end{bmatrix}, \mathbf{R} = \begin{bmatrix} q_x^* q_y^2 & q_x^2 q_y^* \end{bmatrix}$, and

$$\Sigma = \sigma_3\cos(2\alpha) + \sigma_1\sin(2\alpha) \tag{11.66}$$

is defined in terms of the Pauli's matrices

$$I = \begin{bmatrix} 1 & 0 \\ 0 & 1 \end{bmatrix}, \quad \sigma_1 = \begin{bmatrix} 0 & 1 \\ 1 & 0 \end{bmatrix}, \quad \sigma_2 = \begin{bmatrix} 0 & -i \\ i & 0 \end{bmatrix}, \quad \sigma_3 = \begin{bmatrix} 1 & 0 \\ 0 & -1 \end{bmatrix} \tag{11.67}$$

The field $\mathbf{Q}$ is then transformed to the field $\mathbf{\Psi} = \begin{bmatrix} U & V \end{bmatrix}^t$, obtained when one uses the polarization states of the linear and continuous waves propagating in the fiber as a base, according to the relations

$$Q = \begin{bmatrix} q_x \\ q_y \end{bmatrix} = \begin{bmatrix} \cos\alpha & -\sin\alpha \\ \sin\alpha & \cos\alpha \end{bmatrix}\begin{bmatrix} U' \\ V' \end{bmatrix} \text{ and } \begin{bmatrix} U' \\ V' \end{bmatrix} = \begin{bmatrix} u_1 & u_2 \\ -u_2^* & u_1^* \end{bmatrix}\begin{bmatrix} U \\ V \end{bmatrix} \tag{11.68}$$

with $|u_1|^2 + |u_2|^2 = 1$, where $\begin{bmatrix} U' & V' \end{bmatrix}^t$ is the electric field expressed in local axes of birefringence. The electric field $\mathbf{\Psi}$ satisfies the equation:

$$i\frac{\partial \mathbf{\Psi}}{\partial Z} + i\delta_g\bar{\sigma}\frac{\partial \mathbf{\Psi}}{\partial T} + \frac{1}{2}\frac{\partial^2 \mathbf{\Psi}}{\partial T^2} + \frac{5}{6}|\mathbf{\Psi}|^2\mathbf{\Psi} + \frac{1}{6}(\mathbf{\Psi}^t\sigma_3\mathbf{\Psi})\sigma_3\mathbf{\Psi} + \frac{1}{3}\mathbf{N} = 0 \tag{11.69}$$

where

$$\bar{\sigma} = \begin{bmatrix} a_1 & a_4^* \\ a_4 & -a_1 \end{bmatrix} \tag{11.70}$$

and $\mathbf{N} = \begin{bmatrix} \mathbf{N}_1 & \mathbf{N}_2 \end{bmatrix}^t$, with

$$\mathbf{N}_1 = a_3^2(2|V|^2 - |U|^2)U - a_3a_6^*(2|U|^2 - |V|^2)V - a_3a_6U^2V^* - a_6^{*2}V^2U^* \tag{11.71a}$$

$$\mathbf{N}_2 = a_3^2(2|U|^2 - |V|^2)V + a_3a_6(2|V|^2 - |U|^2)U + a_3a_6^*V^2U^* - a_6{}^2U^2V^* \tag{11.71b}$$

The coefficients a_i, $i = 1,\ldots,6$ are defined in terms of u_1 and u_2 as

$$a_1 = |u_1|^2 - |u_2|^2,\ a_2 = -\left(u_1u_2 + u_1^*u_2^*\right),\ a_3 = i\left(u_1u_2 - u_1^*u_2^*\right),\ a_4 = 2u_1u_2^*a_5 = u_1^2 - u_2^{*2},$$
$$a_6 = -i\left(u_1^2 + u_2^{*2}\right) \tag{11.72}$$

Since $\begin{bmatrix} u_1 & -u_2^* \end{bmatrix}^t$ is rapidly varying as it represents the evolution of the polarization state of a continuous wave in the fiber in response to birefringence, $\bar{\sigma}$ and $\mathbf{N}$ are also rapidly varying. The long-term average of these quantities is given by

$$\langle\bar{\sigma}\rangle = 0 \tag{11.73}$$

$$\langle\mathbf{N}\rangle = \begin{bmatrix} -\frac{1}{3}\left(|U|^2 - 2|V|^2\right)U \\ \frac{1}{3}\left(2|U|^2 - |V|^2\right)V \end{bmatrix} \tag{11.74}$$

Moving the rapid variation about the long-term average to the right-hand side of Eq. (11.69), we obtain the so-called Manakov-PMD equation [3, 82]:

$$i\frac{\partial\mathbf{\Psi}}{\partial Z} + \frac{1}{2}\frac{\partial^2\mathbf{\Psi}}{\partial T^2} + \frac{8}{9}|\mathbf{\Psi}|^2\mathbf{\Psi} = -i\delta_g\bar{\sigma}\frac{\partial\mathbf{\Psi}}{\partial T} - \frac{1}{3}(\mathbf{N} - \langle\mathbf{N}\rangle) \tag{11.75}$$

Regarding the left-hand side of (11.75), the second term includes the effect of chromatic dispersion, while the third term includes the effect of Kerr nonlinearity averaged over the Poincaré sphere with the well-known 8/9 factor [53, 83, 84]. Concerning the right-hand side of (11.75), the first term leads to the linear PMD, while the second term corresponds to the nonlinear PMD, which is usually negligible [82]. When the evolution of the polarization state is sufficiently rapid, the terms on the right-hand side of Eq. (11.75) can be neglected and the Manakov equation is obtained [72]. Since this equation has soliton solutions, it is expected that solitons may survive even in the presence of PMD.

11.6 PMD-Induced Soliton Pulse Broadening

In the following, the fiber will be modeled as a cascade of many short segments having randomly varying birefringence. Such random birefringence scatters a part of the energy of the solitons to dispersive radiation in both polarization states parallel and orthogonal to that of the soliton. Because of the generation of these dispersive radiations, the degree of polarization of the soliton is degraded and its energy decays during propagation over long distances.

Let us consider one of the fiber segments constituting the transmission link. We can denote the orthogonal PSPs of this piece of fiber by the unit vectors $\hat{\boldsymbol{i}}$ and $\hat{\boldsymbol{j}}$. At the entrance of the segment, the electric field of a Manakov soliton can be expressed as

$$\mathbf{q}(Z,T) = \left(r\hat{\boldsymbol{i}} + s\hat{\boldsymbol{j}}\right)\sqrt{\frac{9}{8}}\,\eta\,\mathrm{sech}(\eta T) \tag{11.76}$$

where $|r|^2 + |s|^2 = 1$ and the factor $\sqrt{9/8}$ accounts for the energy enhancement required for a soliton be formed in the randomly birefringent fiber. At the output of the fiber segment of normalized length $Z_f = z_f/L_D$, the soliton has the form

$$\mathbf{q}(Z + Z_f, T) = \sqrt{\frac{9}{8}}e^{i\theta}\eta\left\{re^{i\phi}\hat{\mathbf{i}}\,\mathrm{sech}[\eta(T - \Delta T/2)] + se^{-i\phi}\hat{\mathbf{j}}\,\mathrm{sech}[\eta(T + \Delta T/2)]\right\} \tag{11.77}$$

where ϕ arises from the wave-number birefringence and ΔT is the normalized DGD between the two polarization components of the soliton after traversing L. Assuming that ΔT is much smaller than the pulse width, we can expand (11.77) and rewrite $\mathbf{q}(Z + L, T)$ in the form [85]:

$$\mathbf{q}(Z + Z_f, T) = \sqrt{\frac{9}{8}}e^{i\theta}\eta\left\{\left(re^{i\phi}\hat{\boldsymbol{i}} + se^{-i\phi}\hat{\mathbf{j}}\right)\mathrm{sech}(\eta T) - \frac{1}{2}\eta\Delta T\left(re^{i\phi}\hat{\boldsymbol{i}} - se^{-i\phi}\hat{\mathbf{j}}\right)\mathrm{sech}(\eta T)\tanh(\eta T)\right\} \tag{11.78}$$

The first term in Eq. (11.78) corresponds to the majority of the soliton, whose polarization state is $\mathbf{u} = re^{i\phi}\hat{\boldsymbol{i}} + se^{-i\phi}\hat{\boldsymbol{j}}$. Projecting the second term on the polarization state $\mathbf{u}$ and its orthogonal polarization state $\boldsymbol{v} = se^{i\phi}\hat{\boldsymbol{i}} - re^{-i\phi}\hat{\boldsymbol{j}}$, we can write the field (10.78) as

$$\mathbf{q}(Z + Z_f, T) = A(T)\mathbf{u} + B(T)\mathbf{v} \tag{11.79}$$

where

$$A(T) = \sqrt{\frac{9}{8}}e^{i\theta}\eta\text{sech}\left[\eta T - \frac{1}{2}\eta\Delta T\left(|r|^2 - |s|^2\right)\right] \tag{11.80}$$

$$B(T) = \sqrt{\frac{9}{8}}e^{i\theta}\eta^2\Delta Trs\text{sech}(\eta T)\tanh(\eta T) \tag{11.81}$$

Equation (11.80) shows that $A(T)$ is the soliton displaced by the effect of birefringence. On the other hand, $B(T)$ can be interpreted as the radiation shed by the soliton over the length Z_f. The energy of this radiation is given by

$$\delta E = \int_{-\infty}^{\infty} |B(T)|^2 dT = \frac{3}{4}\eta^3|rs|^2\Delta T^2 \tag{11.82}$$

When the energy δE is averaged over all possible soliton polarization states (which gives $\langle |rs|^2\rangle = 1/6$), we obtain

$$\langle \delta E\rangle = \frac{1}{8}\eta^3\langle \Delta T^2\rangle = \frac{8}{729}E^3\langle \Delta T^2\rangle \tag{11.83}$$

where the relation between the soliton energy and amplitude $E = 9\eta/4$ has been used in the last equation.

The average of the square of the normalized DGD for the fiber of normalized length Z_f is given by:

$$\langle \Delta T^2\rangle = \frac{3\pi}{8}\langle |\Delta T|\rangle^2 = \frac{3\pi}{8t_0^2}D_p^2 z_f \tag{11.84}$$

where D_p is the so-called PMD parameter and $z_f = Z_f L_D$ is the real fiber length. Using (11.84) in (11.83), we can write the following equation for the average energy decay:

$$\frac{dE}{dz} = -\frac{\pi L_D D_p^2}{243t_0^2}E^3 \tag{11.85}$$

which has a solution

$$E(z) = \frac{E(0)}{\sqrt{1 + 2\pi L_D D_p^2 E^2(0)z/(243t_0^2)}} \tag{11.86}$$

This solution indicates that the soliton pulse is gradually broadened by a factor

$$b_{sol} = \sqrt{1 + \frac{\pi D_p^2 z}{24t_0^2}} \tag{11.87}$$

where we considered that $E(0) = 9/4$. Using the relation $\langle \Delta\tau^2\rangle = (3\pi/8)\langle \Delta\tau\rangle^2 = (3\pi/8)D_p^2 z$ and the rms initial pulse width σ_0, this broadening factor can be written in the form

$$b_{sol} = \sqrt{1 + \frac{\pi^2}{108}\frac{\langle \Delta\tau^2\rangle}{\sigma_0^2}} \tag{11.88}$$

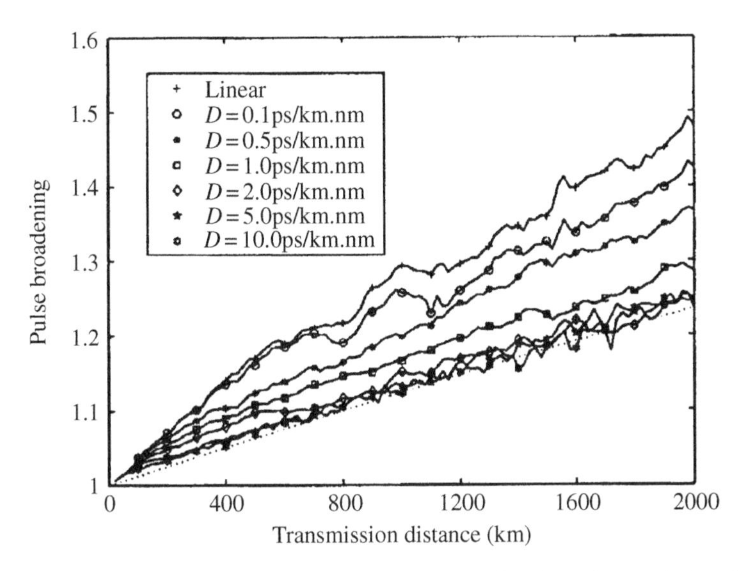

Figure 11.7 PMD-induced pulse broadening as a function of transmission distance for different GVD values. *Source:* After Ref. [86].

The broadening factor for the soliton pulse given by Eq. (11.87) or (11.88) has the same form as the broadening factor for a linear pulse but has a slower broadening rate, showing better robustness of solitons to PMD.

Figure 11.7 shows the simulation results for the PMD-induced pulse broadening as a function of transmission distance for different GVD values [86]. In the simulations, an initial soliton $A(0,t) = \sqrt{9/8}\,\text{sech}(t/t_0)$ was used. The initial pulse width $t_{FWHM0} = 20$ ps, coupling length $L_c = 500$ m, and the PMD coefficient $D_p = 0.5\text{ps}/\sqrt{\text{km}}$. The dotted line corresponds to the analytical result given by Eq. (11.87). The figure shows that the soliton pulse broadening decreases with the increase of GVD, and when GVD is large enough, the pulse broadening approaches the analytical result. The reason is that the analytical result was obtained assuming that the Manakov soliton is weakly perturbed. This is only the case if the birefringence terms are much smaller than the GVD terms in the coupled nonlinear Schrödinger equations.

Figure 11.8 illustrates the broadening of soliton pulses (curve b) and linear Gaussian pulses with (curve c) and without (curve a) first-order PMD compensation at the carrier frequency [4, 87, 88]. Comparing the curves a and b in Figure 11.8, we verify that the solitons broaden at a significantly lower rate than linear pulses without PMD compensation. Moreover, comparing curves b and c, we observe that the solitons broaden at a lower rate than linear pulses with first-order PMD compensation for $\langle\Delta\tau^2\rangle^{1/2}/\sigma_0 > 2.5$, demonstrating that soliton robustness to PMD would become definitively better at high values of the DGD.

The first experimental observation of soliton robustness to PMD in a real transmission system was reported in 1999 [89]. Such robustness was confirmed later both experimentally and numerically [90, 91]. Because of the self-trapping effects of the two polarization components, the soliton can adjust itself and maintain its shape after radiating some dispersive waves. As a result, the net broadening will be at a lower rate than linear pulses. However, significant degradations in soliton systems are caused by such dispersive waves. Besides the pulse broadening determined by the loss of energy, the dispersive waves interact with soliton pulses and cause timing jitter and pulse distortion [90]. It has been shown that some of the soliton control methods discussed in

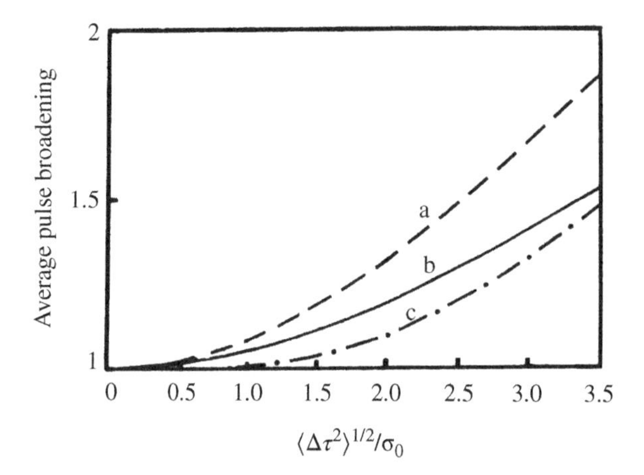

Figure 11.8 Broadening of soliton pulses (b) and linear Gaussian pulses with (c) and without (a) first-order PMD compensation at the carrier frequency.

Chapter 8 can significantly improve the soliton's robustness to PMD, particularly for long-distance systems [84, 92].

11.7 Dispersion-Managed Solitons and PMD

As discussed in Chapter 10, dispersion-managed (DM) solitons offer significant advantages with respect to conventional solitons in high-capacity transmission systems [93–95]. The main idea is to combine a high local group-velocity dispersion with low path-average dispersion. The former feature results in the reduction of the four-wave mixing while the latter one reduces the Gordon–Haus timing jitter effect [96, 97]. Moreover, the DM solitons show enhanced pulse energy [93], which leads not only to improved signal-to-noise ratio (SNR), but also contributes to reducing the Gordon–Haus timing jitter and to enhancing the nonlinear trapping effect. As a consequence, a DM soliton is expected to be more effective in reducing PMD-induced pulse broadening than a conventional soliton. The enhancement factor of the pulse energy is proportional to the map strength S, given by Eq. (10.9).

As seen in Chapter 10, we can derive the ordinary differential equations which describe the evolution of the DM pulse parameters by using the variational approach with a proper ansatz of the propagating pulse. We consider that such ansatz corresponds to a Gaussian-pulse shape, given by:

$$q_j(Z,T) = Q_j \exp\left\{-\left(\frac{1}{2w_j^2} + \frac{iC_j}{2}\right)\left(T - T_{0j}\right)^2 - i\kappa_j\left(T - T_{0j}\right) + i\theta_j\right\}, \quad (j = x, y) \tag{11.89}$$

where Q_j, w_j, C_j, κ_j, T_{0j}, and θ_j represent the amplitude, pulse width, frequency chirp, frequency, time position, and phase, respectively. Using the variational approach to Eqs. (11.29) and (11.30), in which the last terms on the left-hand sides are neglected, we obtain the following ordinary differential equations for the difference of frequency $\Delta\kappa = \kappa_x - \kappa_y$ and time position $\Delta T = T_{0x} - T_{0y}$:

$$\frac{d\Delta\kappa}{dZ} = \frac{2\hat{\varepsilon}E\Delta T}{\sqrt{\pi}\left(w_x^2 + w_y^2\right)^{3/2}} \exp\left\{-\frac{\Delta T^2}{w_x^2 + w_y^2}\right\} \tag{11.90}$$

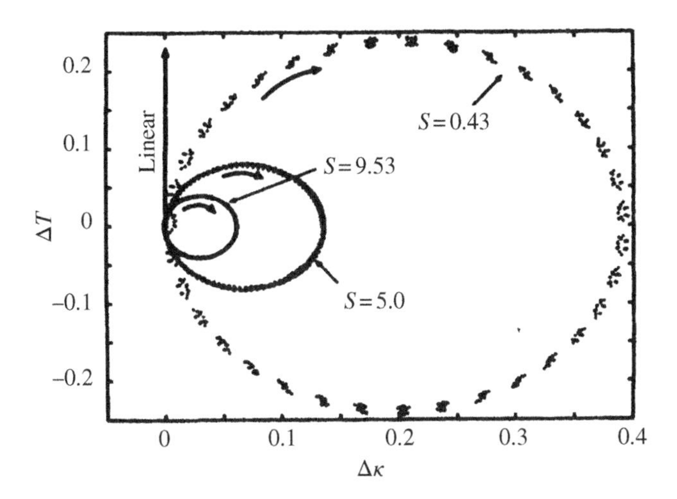

Figure 11.9 Trajectories in the ($\Delta\kappa$, ΔT) plane described in Eqs. (11.90) and (11.91) for $S = 0.43, 5.0, 9.53$ with $\delta_g = 0.1$. *Source:* After Ref [98].

$$\frac{d\Delta T}{dZ} = 2\delta_g - d(Z)\Delta\kappa \tag{11.91}$$

where $E = \sqrt{\pi}\left(Q_x^2 w_x + Q_y^2 w_y\right)$ is the pulse energy. From Eqs. (11.90) and (11.91), we can see that, in the presence of nonlinearity, $\Delta\kappa$ and ΔT behave periodically along Z, whereas in the linear case, $\Delta\kappa = 0$ and ΔT increases linearly. Figure 11.9 illustrates the trajectories in the ($\Delta\kappa$, ΔT) plane at every period of the dispersion map for various values of the map strength S [98]. The period of these trajectories becomes shorter for strong dispersion management due to large cross-phase modulation induced by higher nonlinearity. Hence, the nonlinear trap becomes effective in controlling the PMD in fibers with constant birefringence.

In order to analyze the effect of PMD in randomly varying birefringent fibers, let us expand Eqs. (11.90) and (11.91) around $\Delta T = 0$. Taking the first order in this expansion and assuming that δ_g is a white Gaussian process, such that $\langle\delta_g\rangle = 0$ and $\langle\delta_g(Z)\delta_g(Z')\rangle = \sigma^2\delta(Z - Z')$, we obtain [98]:

$$\sqrt{\langle\Delta T^2\rangle} = \left[2\sigma^2 Z\left\{1 + \frac{\sin\left(2b\left|\bar{d}\right|Z\right)}{2b\left|\bar{d}\right|Z}\right\}\right]^{1/2} \tag{11.92}$$

where $b = 2\hat{\varepsilon}E/\sqrt{\pi\left(w_x^2 + w_y^2\right)^3}$. In the linear case ($b = 0$), we have

$$\sqrt{\langle\Delta T^2\rangle} = 2\sqrt{\sigma^2 Z} \tag{11.93}$$

In obtaining Eq. (11.93), the local dispersion d was replaced with the average dispersion $\bar{d}$, which is reasonable for weak dispersion management. Equation (11.93) is consistent with the result obtained earlier for the case of an ideal soliton and it shows that the effect of PMD in DM soliton transmission systems is reduced with the help of nonlinear trap by $1/\sqrt{2}$ compared with linear systems. However, for strong dispersion management, this analysis does not apply. In such a case,

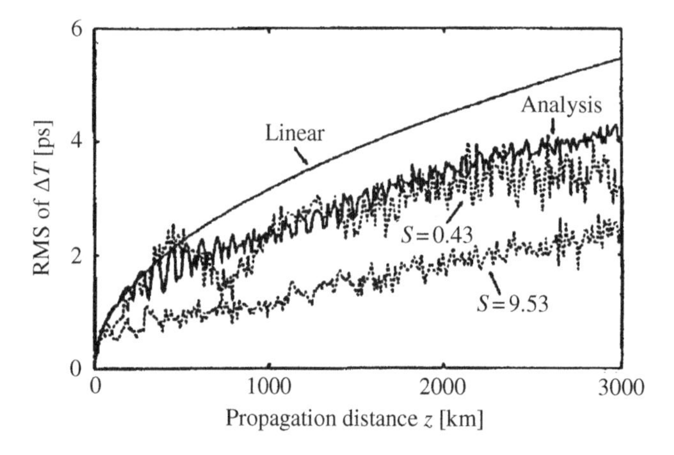

Figure 11.10 Comparison of the evolution of $\sqrt{\langle \Delta T^2 \rangle}$ in linear and nonlinear systems with the initial pulse width 2 ps. Solid lines represent the results of linear and nonlinear systems given by Eqs. (11.92) and (11.93), respectively, and dashed lines are the results of numerical simulations for $S = 0.43$ and 9.53, obtained by averaging over 50 trials. Dispersion management period $z_a = 10$ km, PMD coefficient $D_p = 0.1\text{ps}/\sqrt{\text{km}}$. *Source:* After Ref. [98].

the orthogonally polarized modes are completely trapped and the pulse broadening increases almost as $\ln(Z)$ [98, 99].

The robustness of DM solitons to PMD has been confirmed both experimentally [100] and numerically [98, 101, 102]. In general, DM solitons are more robust to PMD than conventional solitons. Such robustness increases with the map strength and average GVD since this determines an increased power of DM solitons.

Figure 11.10 shows a comparison of the evolution of $\sqrt{\langle \Delta T^2 \rangle}$ in linear and DM nonlinear systems obtained by direct numerical simulations of Eqs. (11.29) and (11.30) and the approximate results provided by Eqs. (11.92) and (11.93) [98]. The PMD-induced pulse broadening is suppressed considerably by increasing the map strength with the help of an enhanced nonlinear trap. The analytical result of the Langevin equation is in good agreement with the numerical results when $S = 0.43$. Indeed, Eq. (11.92) is valid especially for small values of the map strength S, when the pulse dynamics is dominated by the average dispersion.

References

1 Poole, C.D. and Nagel, J. (1997). Polarization effects in lightwave systems. In: *Optical Fiber Telecommunications*, vol. **IIIA** (ed. I.P. Kaminow and T.L. Koch). San Diego, CA: Academic Press.
2 Bruyère, F. (1996). *Opt. Fiber Technol.* **2**: 269.
3 Wai, P.K.A. and Menyuk, C.R. (1996). *J. Lightwave Technol.* **14**: 148.
4 Karlsson, M. (1998). *Opt. Lett.* **23**: 688.
5 Matera, F., Settembre, M., Tamburrini, M. et al. (1999). *J. Lightwave Technol.* **17**: 2225.
6 Kolltveit, E., Andrekson, P.A., Brentel, J. et al. (1999). *Electron. Lett.* **35**: 75.
7 Foschini, G.J., Jopson, R.M., Nelson, L.E., and Kogelnik, H. (1999). *J. Lightwave Technol.* **17**: 1560.
8 Huttner, B., Geiser, C., and Gisin, N. (2000). *IEEE J. Sei. Topics Quantum Electron.* **6**: 317.

9 Shtaif, M. and Mecozzi, A. (2000). *Opt. Lett.* **25**: 707.
10 Karlsson, M., Brentel, J., and Andrekson, P.A. (2000). *J. Lightwave Technol.* **18**: 941.
11 Li, Y. and Yariv, A. (2000). *Opt. Soc. Am. B* **17**: 1821.
12 Fini, J.M. and Haus, H.A. (2001). *IEEE Photon. Technol. Lett.* **13**: 124.
13 Khosravani, R. and Willner, A.E. (2001). *IEEE Photon. Technol. Lett.* **13**: 296.
14 Sunnerud, H., Karlsson, M., Xie, C., and Andrekson, P.A. (2002). *J. Lightwave Technol.* **20**: 2204.
15 Galtarrossa, A., Griggio, P., Palmieri, L., and Pizzinat, A. (2004). *J. Lightwave Technol.* **22**: 1127.
16 Chernyak, V., Chertkov, M., Gabitov, I. et al. (2004). *J. Lightwave Technol.* **22**: 1155.
17 Ning, G., Aditya, S., Shum, P. et al. (2006). *Opt. Commun.* **260**: 560.
18 Kaminow, I.P. (1981). *IEEE J. Quantum Electron.* **17**: 15.
19 Kogelnik, H., Jopson, R.M., and Nelson, L.E. (2002). Polarization-mode dispersion. In: *Optical Fiber Telecommunications*, vol. **IVA** (ed. I.P. Kaminow and T. Li). San Diego, CA: Academic Press.
20 Damask, J.N. (2005). *Polarization Optics in Telecommunications*. New York, NY: Springer.
21 Galtarrossa, A. and Menyuk, C.R. (ed.) (2005). *Polarization Mode Dispersion*. New York, NY: Springer.
22 Foschini, G.J. and Poole, C.D. (1991). *J. Lightwave Technol.* **9**: 1439.
23 Poole, C.D. and Wagner, R.E. (1986). *Electron. Lett.* **22**: 1029.
24 Karlson, M. and Brenel, J. (1999). *Opt. Lett.* **24**: 939.
25 Lin, Q. and Agrawal, G.P. (2003). *J. Opt. Soc. Am B* **20**: 292.
26 Sunnerud, H., Karlsson, M., and Andrekson, P.A. (2000). *IEEE Photon. Technol. Lett.* **12**: 50.
27 Willner, A.E., Reza, S.M., Nezam, M. et al. (2004). *J. Lightwave Technol.* **22**: 106.
28 Biilow, H. (1998). *IEEE Photon. Technol. Lett.* **10**: 696.
29 Lu, P., Chen, L., and Bao, X. (2002). *J. Lightwave Technol.* **20**: 1805.
30 Damask, J.N., Gray, G., Leo, P. et al. (2003). *IEEE Photon. Technol. Lett.* **15**: 48.
31 Kissing, J., Gravemann, T., and Voges, E. (2003). *IEEE Photon. Technol. Lett.* **15**: 611.
32 Winzer, P.J., Kogelnik, H., and Ramanan, K. (2004). *IEEE Photon. Technol. Lett.* **16**: 449.
33 Forestieri, E. and Prati, G. (2004). *J. Lightwave Technol.* **22**: 988.
34 Biondini, G., Kath, W.L., and Menyuk, C.R. (2004). *J. Lightwave Technol.* **22**: 1201.
35 Boroditsky, M., Brodsky, M., Frigo, N.J. et al. (2005). *IEEE Photon. Technol. Lett.* **17**: 345.
36 Cvijetic, N., Wilson, S.G., and Qian, D.Y. (2008). *J. Lightwave Technol.* **26**: 2118.
37 Francia, C., Bruyère, F., Thiéry, J.P., and Penninckx, D. (1999). *Electron. Lett.* **35**: 414.
38 Merker, T., Hahnenkamp, N., and Meissner, P. (2000). *Opt. Commun.* **182**: 135.
39 Pua, H.Y., Peddanarappagari, K., Zhu, B. et al. (2000). *J. Lightwave Technol.* **18**: 832.
40 Sunnerud, H., Xie, C., Karlsson, M. et al. (2002). *J. Lightwave Technol.* **20**: 368.
41 Noè, R., Sandel, D., and Mirvoda, V. (2004). *IEEE J. Sel. Topics Quantum Electron.* **10**: 341.
42 Lanne, S. and Corbel, E. (2004). *J. Lightwave Technol.* **22**: 1033.
43 Phua, P.B., Haus, H.A., and Ippen, E.P. (2004). *J. Lightwave Technol.* **22**: 1280.
44 Yan, L., Yao, X.S., Hauer, M.C., and Willner, A.E. (2006). *J. Lightwave Technol.* **24**: 3992.
45 Miao, H., Weiner, A.M., Mirkin, L., and Miller, P.J. (2008). *IEEE Photon. Technol. Lett.* **20**: 545.
46 Dogariu, A., Ji, P.N., Cimponeriu, L., and Wanga, T. (2009). *Opt. Commun.* **282**: 3706.
47 Daikoku, M., Miyazaki, T., Morita, I. et al. (2009). *J. Lightwave Technol.* **27**: 451.
48 Kim, S. (2002). *J. Lightwave Technol.* **20**: 1118.
49 Kleinman, D.A. (1962). *Phys. Rev.* **126**: 1977.
50 Boyd, R.W. (2003). *Nonlinear Optics*, 2e. San Diego: Academic Press.
51 Menyuk, C.R. (1987). *IEEE J. Quantum Electron.* **23**: 176.
52 Agrawal, G.P. (2001). *Nonlinear Fiber Optics*, 3e. San Diego: Academic Press.
53 Hasegawa, A. (2004). *Physica D* **188**: 241.

54 Ferreira, M.F. (2008). *Fiber Int. Opt.* **27**: 113.
55 Menyuk, C.R. (1989). *IEEE J. Quantum Electron.* **25**: 2674.
56 Kivshar, Y.S. (1990). *J. Opt. Soc. Am. B* **7**: 2204.
57 Menyuk, C.R. (1987). *Opt. Lett.* **12**: 614.
58 Menyuk, C.R. (1988). *J. Opt. Soc. Am. B* **5**: 392.
59 Islam, M.N., Poole, C.D., and Gordon, J.P. (1989). *Opt. Lett.* **14**: 1011.
60 Korolev, E., Nazarov, V.N., Nolan, D.A., and Truesdale, C.M. (2005). *Opt. Lett.* **30**: 132.
61 Cundiff, S.T., Collings, B.C., Akhmediev, N.N. et al. (1999). *Phys. Rev. Lett.* **82**: 3988.
62 Afanasjev, V.V. (1995). *Opt. Lett.* **20**: 270.
63 Akhmediev, N.N., Buryak, A.V., Soto-Crespo, J.M., and Andersen, D.R. (1995). *J. Opt. Soc. Am. B* **12**: 434.
64 Zhang, H., Tang, D.Y., Zhao, L.M., and Xiang, N. (2008). *Opt. Express* **16**: 12618.
65 Akhmediev, N.N., Soto-Crespo, J.M., Cundiff, S.T. et al. (1998). *Opt. Lett.* **23**: 852.
66 Song, Y.F., Li, L., Zhang, H. et al. (2013). *Opt. Express* **21**: 10010.
67 Wu, X., Tang, D.Y., Zhao, L.M., and Zhang, H. (2009). *Phys. Rev. A* **80**: 013804.
68 Zhang, H., Tang, D.Y., Zhao, L.M. et al. (2009). *Opt. Express* **17**: 455.
69 Zhang, H., Tang, D.Y., Zhao, L.M., and Tam, H.Y. (2008). *Opt. Lett.* **33**: 2317.
70 Jin, X.X., Wu, Z.C., Li, L. et al. (2016). *IEEE Photonics J.* **8**: 1.
71 Manakov, S.V. (1974). *Sov. Phys. JETP* **38**: 248.
72 Ueda, T. and Kath, W.L. (1990). *Phys. Rev. A* **42**: 563.
73 Trillo, S. and Wabnitz, S. (1991). *Phys. Lett.* **159**: 252.
74 Malomed, B.A. (1992). *Phys. Rev. A* **45**: R8821.
75 Tratnik, M.V. (1992). *Opt. Lett.* **17**: 917.
76 Potasek, M.J. (1993). *J. Opt. Soc. Am. B* **10**: 941.
77 Kivshar, Y.S. and Turitsyn, S.K. (1993). *Opt. Lett.* **18**: 337.
78 Bhakta, J.C. (1994). *Phys. Rev. E* **49**: 5731.
79 Haelterman, M. and Sheppard, A.P. (1994). *Phys. Rev. E* **49**: 3376.
80 Silberberg, Y. and Barad, Y. (1995). *Opt. Lett.* **20**: 246.
81 Chen, Y. and Atai, J. (1997). *Phys. Rev. E* **55**: 3652.
82 Marcuse, D., Menyuk, C.R., and Wai, P.K. (1997). *J. Lightwave Technol.* **15**: 1735.
83 Wai, P.K., Menyuk, C.R., and Chen, H.H. (1991). *Opt. Lett.* **16**: 1231.
84 Matsumoto, M., Akagi, Y., and Hasegawa, A. (1997). *J. Lightwave Technol.* **15**: 584.
85 Mollenauer, L.F., Gordon, J.P., and Mamyshev, P.V. (1997). Solitons in high bit-rate, long-distance transmission. In: *Optical Fiber Telecommunications*, vol. **III A** (ed. I.P. Kaminow and T.L. Koch). San Diego, CA: Academic Press, Chap. 12.
86 Xie, C., Karlsson, M., and Andrekson, P.A. (2000). *IEEE Photon. Technol. Lett.* **12**: 801.
87 Heismann, F., Fishman, D., and Wilson, D.L. (1998). *Proceedings of the European Conference on Optical Communications (ECOC '98)*, Madrid, Spain (20–24 September 1998), pp. 529–530.
88 Ferreira, M.F., Latas, S.V., Sousa, M.H. et al. (2005). *Fiber and Int. Opt.* **24**: 261.
89 Bakhshi, B., Hansryd, J., Andrekson, P.A. et al. (1999). *Electron. Lett.* **35**: 65.
90 Xie, C., Karlsson, M., Andrekson, P., and Sunnerud, H. (2002). *IEEE J. Selected Topics Quantum Electron.* **8**: 575.
91 Sunnerud, H., Lie, J., Xie, C., and Andrekson, P. (2001). *J. Lightwave Technol.* **19**: 1453.
92 Xie, C., Karlsson, M., and Sunnerud, H. (2001). *Opt. Lett.* **26**: 672.
93 Smith, N.J., Knox, F.M., Doran, N.J. et al. (1996). *Electron. Lett.* **32**: 54.
94 Hasegawa, A. (ed.) (1998). *New Trends in Optical Soliton Transmission Systems*. Dordrecht, Holland: Kluwer.

95 Ferreira, M.F., Facão, M.V., Latas, S.V., and Sousa, M.H. (2005). *Fiber Integrated Opt.* **24**: 287.

96 Suzuki, M., Morita, I., Edagawa, N. et al. (1995). *Electron. Lett.* **31**: 2027.

97 Carter, G.M., Jacob, J.M., Menyuk, C.R. et al. (1997). *Opt. Lett.* **22**: 513.

98 Nishioka, I., Hirooka, T., and Hasegawa, A. (2000). *IEEE Photon. Technol. Lett.* **12**: 1480.

99 Chen, Y. and Haus, H. (2000). *Opt. Lett.* **25**: 29.

100 Sunnerud, H., Li, J., Andrekson, P.A., and Xie, C. (2000). *IEEE Photon. Technol. Lett.* **13**: 118.

101 Zhang, X., Karlsson, M., Andrekson, P., and Brtilsson, K. (1998). *Electron. Lett.* **34**: 1122.

102 Xie, C., Karlsson, M., Andrekson, P., and Sunnerud, H. (2000). *IEEE Photon. Technol. Lett.* **13**: 221.

12

Stationary Dissipative Solitons

Dissipative optical solitons are confined wave packets of light whose existence and stability require a continuous energy supply from an external source, which is dissipated in the medium around the solitons. This corresponds to an extension of the conventional soliton concept, implying that the single balance between nonlinearity and dispersion is replaced by a double balance: between nonlinearity and dispersion and also between gain and loss. Although the dissipative soliton concept evolved initially from phenomena observed or predicted in nonlinear optics, it has received wide recognition in other disciplines, such as biology and medicine [1, 2].

A vast variety of dissipative systems and nonlinear phenomena in physics, chemistry, and biology can be described by the complex Ginzburg–Landau equation (CGLE) [1–3]. In the field of nonlinear optics, the CGLE has been used to describe optical parametric oscillators [4], free-electron laser oscillators [5], spatial and temporal soliton lasers [6–10], and all-optical long-haul soliton transmission lines [11–13]. As seen in Section 8.4, the active nonlinearity in the equation must be at least quintic in order to model the stable propagation of pulses. Thus, the simplest model that we can use is the complex cubic-quintic CGLE. Depending upon the particular system or the phenomena that we want to describe, the model can be complemented with additional terms in the equation.

A large variety of soliton solutions can be obtained by numerically solving the cubic-quintic CGLE. In general, such soliton solutions belong to one of the two classes: localized stationary solutions or localized pulsating solutions [1, 13–15]. Actually, the number of solutions to this equation is so large that the sphere of this knowledge by itself has been called "the world of the Ginzburg–Landau equation" [3, 14]. Several examples of localized stationary solutions are presented in this chapter, whereas the localized pulsating solutions will be discussed in the next chapter.

12.1 Balance Equations for the CGL Equation

The cubic-quintic CGLE presented in Section 8.4 can be further generalized taking into account the higher-order reactive nonlinearity and the possibility of anomalous or normal dispersion regimes of propagation, assuming the form:

$$i\frac{\partial q}{\partial Z} + \frac{D}{2}\frac{\partial^2 q}{\partial T^2} + |q|^2 q = i\delta q + i\beta\frac{\partial^2 q}{\partial T^2} + i\varepsilon|q|^2 q + i\mu|q|^4 q - \nu|q|^4 q \tag{12.1}$$

where ν is a higher-order correction term to the nonlinear refractive index and D is the dispersion parameter, with $D > 0$ in the anomalous regime and $D < 0$ in the normal regime. The other parameters have the same meaning as in Chapter 8.

Solitons in Optical Fiber Systems, First Edition. Mario F. S. Ferreira.

Equation (12.1) becomes the standard nonlinear Schrödinger equation (NLSE) when the right-hand side is set to zero. When this does not happen, Eq. (12.1) is non-integrable, and only particular solutions can be obtained. The soliton perturbation theory can be used in the anomalous dispersion regime ($D > 0$) for small values of the parameters [13]. In the case of the cubic CGLE ($\mu = \nu = 0$), exact solutions can be obtained using a special ansatz [16], Hirota bilinear method [17], or reduction to systems of linear partial differential equations (PDEs) [18]. Concerning the quintic CGLE, the existence of soliton-like solutions in the case $\varepsilon > 0$, as well as sources, sinks, and fronts with fixed velocity have been demonstrated both analytically and numerically [19–21]. Analytical solutions can be presented explicitly only for certain relations between the parameters of the equation. Furthermore, so far, only stationary solutions of the CGLE are known in analytical form. Actually, several types of localized pulsating solutions of the CGLE have also been found numerically [22–27]. Approximate expressions for some of these localized solutions can be derived for arbitrary values of the CGLE parameters by reducing this equation to finite-dimensional dynamical models. The reduced models can be obtained by applying the method of moments [28], or Lagrangian techniques [15, 29–32].

In spite of having no known conserved quantities, there are two rate equations that can be written for the CGLE: one for the energy,

$$E = \int_{-\infty}^{\infty} |q|^2 dT \tag{12.2}$$

and the other for the momentum,

$$M = \frac{1}{2}\int_{-\infty}^{\infty} \left(q\frac{\partial q^*}{\partial T} - q^*\frac{\partial q}{\partial T} \right) dT \tag{12.3}$$

For a dissipative system, E is a function of Z rather than a conserved quantity. The rate of change of the energy is [1]

$$\frac{d}{dZ}E = F[q] \tag{12.4}$$

where the real functional $F[q]$ is given by

$$F[q] = 2\int_{-\infty}^{\infty} \left[\delta|q|^2 + \varepsilon|q|^4 + \mu|q|^6 - \beta\left|\frac{\partial q}{\partial T}\right|^2 \right] dT \tag{12.5}$$

For localized solutions, E is finite and it takes a constant value when the solution is stationary. The rate of change of the momentum is given by [1]:

$$\frac{d}{dZ}M = J[q], \tag{12.6}$$

where the real functional $J[q]$ is given by

$$J[q] = 2\mathrm{Im}\int_{-\infty}^{\infty} \left[(\delta + \varepsilon|q|^2 + \mu|q|^4)q + \beta\frac{\partial^2 q}{\partial T^2} \right] \frac{\partial q*}{\partial T} dT \tag{12.7}$$

The rate Eqs. (12.4) and (12.6) can be used for studying different problems related to CGLE solitons. For example, assuming that the coefficients on the right-hand side of Eq. (12.1) are all small, they can be used to derive the evolution equations for the amplitude and the frequency of the perturbed soliton solution of the NLSE:

$$q(T, Z) = \eta \mathrm{sech}[\eta(T + \kappa Z)] \exp\left[-i\kappa T + i(\eta^2 - \kappa^2)Z/2\right] \tag{12.8}$$

The equation for the evolution of $\eta(Z)$ can be derived from the equation for the energy (12.4), whereas the equation for $\kappa(Z)$ can be obtained from the equation for the momentum (12.6). It turns out that both results are similar to Eqs. (8.61) and (8.62), respectively.

In a dissipative soliton, the energy flows from the parts where the energy is generated to the parts where it is dissipated. To illustrate this, we consider the continuity relation for the CGLE equation, which is given by [1]:

$$\frac{\partial \rho}{\partial Z} + \frac{\partial j}{\partial T} = P \tag{12.9}$$

where $\rho = |q|^2$ is the energy density,

$$j = \frac{i}{2}\left(q q_T^* - q_T q^*\right) \tag{12.10}$$

is the energy flux, and

$$P = 2\delta|q|^2 + 2\varepsilon|q|^4 + 2\mu|q|^6 - 2\beta\left|\frac{\partial q}{\partial T}\right|^2 + \beta\frac{\partial^2\left(|q|^2\right)}{\partial T^2} \tag{12.11}$$

is the density of energy generation.

For a stationary dissipative soliton, the first term in (12.9) is zero since the energy density remains constant along the propagation. Such stationary solution results from the continuous exchange of energy with the environment along with the concomitant redistribution of energy between different parts of the pulse. Let us consider that the stationary solution corresponds to a higher-order Gaussian-type pulse, given by:

$$q(Z,T) = A(m)\sqrt{\frac{E}{w}}\exp\left(-\frac{T^2}{w^2} - \frac{T^4}{mw^4} + iCT^2 - i\kappa Z\right) \tag{12.12}$$

where E, w, and C represent the pulse energy, width, and chirp factor, respectively, m is an adjustment parameter, and A is a constant, such that the total pulse energy is equal to E. Using Eq. (12.12), the flux and the density of energy generation become:

$$j(T) = P_0 T \exp\left(-2\frac{T^2}{w^2} - 2\frac{T^4}{mw^4}\right) \tag{12.13}$$

$$P(T) = P_0\left(1 - 4\frac{T^2}{w^2} - \frac{8}{m}\frac{T^4}{w^4}\right)\exp\left(-2\frac{T^2}{w^2} - 2\frac{T^4}{mw^4}\right) \tag{12.14}$$

where

$$P_0 = \frac{2A^2(m)CE}{w} \tag{12.15}$$

Figure 12.1 illustrates the normalized density of energy generation, $\hat{P} = P/P_0$, and the normalized energy flux, $\hat{j} = j/(wP_0)$, for $m = 1$ and $C > 0$. In this case, $\hat{P} < 0$ in the wings and $\hat{P} > 0$ around the pulse center. This means that energy is generated in the middle, and flows toward the wings, where it is dissipated. However, if $C < 0$, energy is generated in the wings and flows toward the middle where it is lost. This flux of energy leads to a dynamic equilibrium which allows the existence of the dissipative soliton.

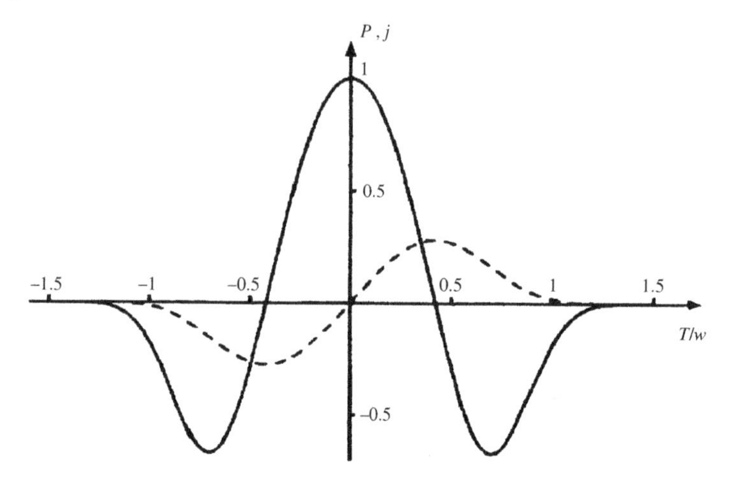

Figure 12.1 Normalized density of energy generation $\hat{P} = P/P_0$ (full line) and normalized energy flux $\hat{j} = j/(wP_0)$ (dashed line) for the case of a higher-order Gaussian pulse (m = 1).

12.2 Exact Analytical Solutions

Exact solutions of the quintic CGLE, including solitons, sinks, fronts and sources, were obtained in [33], using Painlevé analysis and symbolic computations. Other exact analytical solutions can be obtained considering a particular ansatz [19, 20, 34]. In the following, we look for stationary solutions of Eq. (12.1) in the form:

$$q(T,Z) = a(T)\exp\{ib\ln[a(T)] - i\kappa Z\} \tag{12.16}$$

where $a(T)$ is a real function, and b and ω are real constants.

12.2.1 Solutions of the Cubic CGLE

We will first consider the cubic CGLE, which is given by Eq. (12.1) with $\mu = \nu = 0$. Inserting Eq. (12.16) in this equation, the following solution for $a(T)$ is obtained:

$$a(T) = A\mathrm{sech}(BT) \tag{12.17}$$

where

$$A = \sqrt{\frac{B^2(2-b^2)}{2} + 3b\beta B^2}, \; B = \sqrt{\frac{\delta}{\beta b^2 + b - \beta}} \tag{12.18}$$

and b is given by

$$b = \frac{3(1+2\varepsilon\beta) \pm \sqrt{9(1+2\varepsilon\beta)^2 + 8(\varepsilon - 2\beta)^2}}{2(\varepsilon - 2\beta)} \tag{12.19}$$

On the other hand, we have

$$\kappa = -\frac{\delta(1-b^2+4\beta b)}{2(b-\beta+\beta b^2)} \tag{12.20}$$

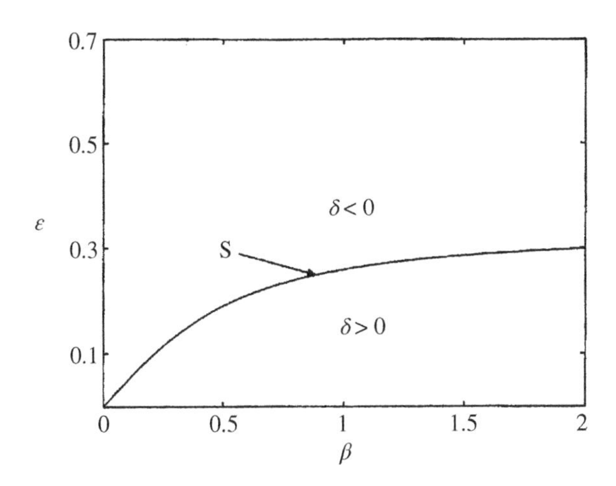

Figure 12.2 Illustration of the curve *S*, given by Eq. (12.21).

The solution (12.16)–(12.20) is known as the solution of Pereira and Stenflo [16], already discussed in Section 8.1.

The solution given by Eqs. (12.17)–(12.19) has a singularity at $b - \beta + \beta b^2 = 0$, which takes place on the line $\varepsilon_s(\beta)$ in the plane (β, ε) given by

$$\varepsilon_S = \frac{\beta}{2}\frac{3\sqrt{1+4\beta^2}-1}{2+9\beta^2} \tag{12.21}$$

Figure 12.2 shows $\varepsilon_s(\beta)$, as represented by line *S*. For $\beta << 1$, we have $\varepsilon_s(\beta) \sim \beta/2$, whereas, for $\beta >> 1$, it is $\varepsilon_s(\beta) \sim 1/3$. For a given value of β, the denominator in the expression for *B* in Eq. (12.18) is positive below the line *S* ($\varepsilon < \varepsilon_S$) and negative above it ($\varepsilon > \varepsilon_S$). Hence, for solution (12.16)–(12.20) to exist, the excess linear gain δ must be positive for $\varepsilon < \varepsilon_S$ and negative for $\varepsilon > \varepsilon_S$. In the last case, both numerical simulations and the soliton perturbation theory show that the soliton is unstable relatively to any small-amplitude fluctuations [19, 20, 34]. On the other hand, for $\delta > 0$ and $\varepsilon < \varepsilon_S$, the soliton is stable since after any small perturbation, it approaches the stationary state. However, the background state is unstable in this case since the positive excess gain also amplifies the linear waves coexistent with the soliton trains. The general conclusion is that either the soliton itself or the background state is unstable at any point in the plane (β, ε), which means that the total solution is always unstable.

If β and ε satisfy Eq. (12.21) and $\delta = 0$, a solution of the cubic CGLE with arbitrary amplitude exists, given by [19]:

$$a(T) = C\,\mathrm{sech}(DT) \tag{12.22}$$

where *C* is an arbitrary positive parameter and *C*/*D* is given by:

$$\frac{C}{D} = \sqrt{\frac{(2+9\beta^2)\sqrt{1+4\beta^2}\left(\sqrt{1+4\beta^2}-1\right)}{2\beta^2\left(3\sqrt{1+4\beta^2}-1\right)}} \tag{12.23}$$

We also have that the parameters *b* and κ in Eq. (12.16) are given by:

$$b = \frac{\sqrt{1+4\beta^2}-1}{2\beta} \tag{12.24}$$

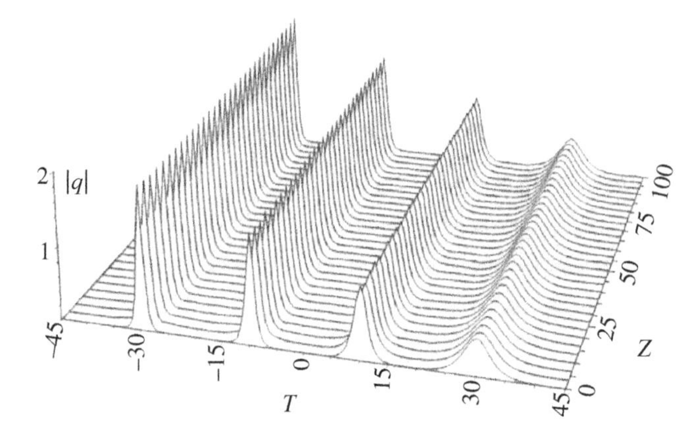

Figure 12.3 Simultaneous propagation of four solitons with amplitudes 2, 1.5, 1, and 0.5, for $\delta = \mu = 0, \beta = 0.2$, and $\varepsilon = \varepsilon_S$.

$$\kappa = -b\frac{1 + 4\beta^2}{2\beta}D^2 \tag{12.25}$$

Arbitrary amplitude solitons are stable pulses, which propagate in a stable background because $\delta = 0$. This feature is illustrated in Figure 12.3, which shows the simultaneous propagation of four stable solitons with amplitudes 2, 1.5, 1, and 0.5 for $\delta = 0$, $\beta = 0.2$, and $\varepsilon = \varepsilon_S$.

12.2.2 Solutions of the Quintic CGLE

Considering the quintic CGLE and inserting Eq. (12.16) in Eq. (12.1), the following general solution can be obtained for $f = a^2$ [19]:

$$f(T) = \frac{2f_1 f_2}{(f_1 + f_2) - (f_1 - f_2)\cosh\left(2\alpha\sqrt{f_1|f_2|T}\right)} \tag{12.26}$$

where

$$\alpha = \sqrt{\left|\frac{\mu}{3\beta - 2b - \beta b^2}\right|} \tag{12.27}$$

and b is given by Eq. (12.19). The parameters f_1 and f_1 are the roots of the equation:

$$\frac{2\nu}{8\beta b - b^2 + 3}f^2 + \frac{2(2\beta - \varepsilon)}{3b\left(1 + 4\beta^2\right)}f - \frac{\delta}{b - \beta + \beta b^2} = 0 \tag{12.28}$$

and the coefficients are connected by the relation:

$$\nu\left[\frac{12\varepsilon\beta^2 + 4\varepsilon - 2\beta}{\varepsilon - 2\beta}b - 2\beta\right] + \mu\left[\frac{2\varepsilon\beta - 16\beta^2 - 3}{\varepsilon - 2\beta}b + 1\right] = 0 \tag{12.29}$$

One of the roots of Eq. (12.28) must be positive for the solution (12.26) to exist, while the other can have either sign.

When the two roots are both positive, the general solution given by Eq. (12.26) becomes wider and flatter as they approach each other, as can be seen in Figure 12.4. These flat-top solitons correspond to stable pulses, whereas solution Eq. (12.26) is generally unstable for arbitrary choice

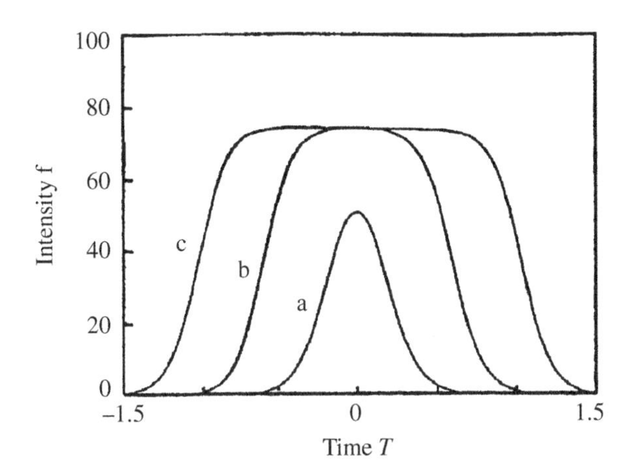

Figure 12.4 The pulse shapes given by Eq. (13.26) when the two roots f_1 and f_2 are close to each other. It is assumed that $(f_1 - f_2)/(f_1 + f_2) = 10^{(-n)}$ with n = 1 (curve a), n = 5 (curve b) and n = 9 (curve c).

of parameters. If $f_1 = f_2$, the width of the flat-top soliton tends to infinity and the soliton splits into two fronts. The formation and stable propagation of a flat-top soliton will be demonstrated numerically in Section 12.3.

If β and ε satisfy Eq. (12.21) and $\delta = 0$, a solution of the quintic CGLE with arbitrary amplitude exists, given by [19]:

$$f(T) = [a(T)]^2 = \frac{3b(1 + 4\beta^2)P}{(2\beta - \varepsilon) + F\cosh(2\sqrt{P}T)} \tag{12.30}$$

where P is an arbitrary positive parameter and

$$F = \sqrt{(2\beta - \varepsilon)^2 + \frac{9b^2\mu(1 + 4\beta^2)^2}{3\beta - 2b - \beta b^2}P} \tag{12.31}$$

The parameters b and κ are given by Eqs. (12.24) and (12.25), respectively. When $\mu \to 0$, the solution (12.30) transforms to the arbitrary-amplitude solution of the cubic CGLE, given by Eqs. (12.22) and (12.23).

12.3 Numerical Stationary Soliton Solutions

Stable solitons can be found numerically from the propagation Eq. (12.1) taking as the initial condition a pulse of somewhat arbitrary profile. In fact, such a profile appears to be of little importance. For example, Figure 12.5 illustrates the formation of a fixed amplitude soliton of the cubic CGLE starting from an initial pulse with a rectangular profile. In this case, the linear gain is positive but relatively small ($\delta = 0.003$) and the soliton propagation remains stable within the displayed distance.

In general, if the result of the numerical calculation converges to a stationary solution, it can be considered as a stable one, and the chosen set of parameters can be deemed to belong to the class of those which permit the existence of solitons. For small values of parameters in the right-hand side of Eq. (12.1), the stable soliton solutions of the CGLE have a sech profile similar to the soliton solutions of the NLSE, and correspond to the so-called plain pulses (PPs). However, rather different

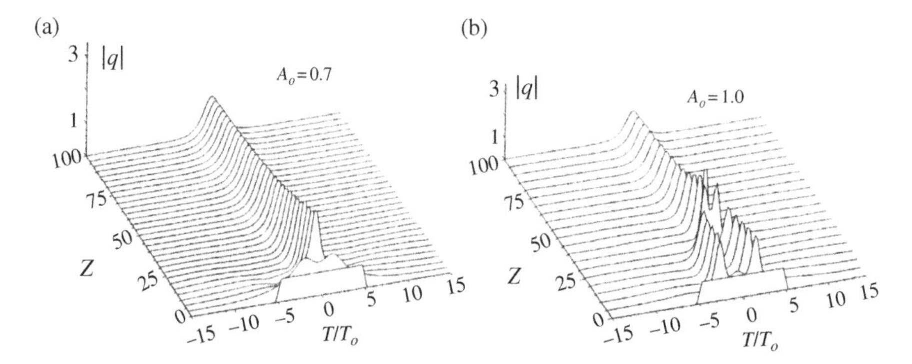

Figure 12.5 Formation of a fixed amplitude soliton of the cubic CGLE starting from an initial pulse with a rectangular profile of amplitude $A_0 = 0.7$ (a) and $A_0 = 1.0$ (b) for: $\delta = -0.003$, $\beta = 0.2$, and $\varepsilon = 0.09$.

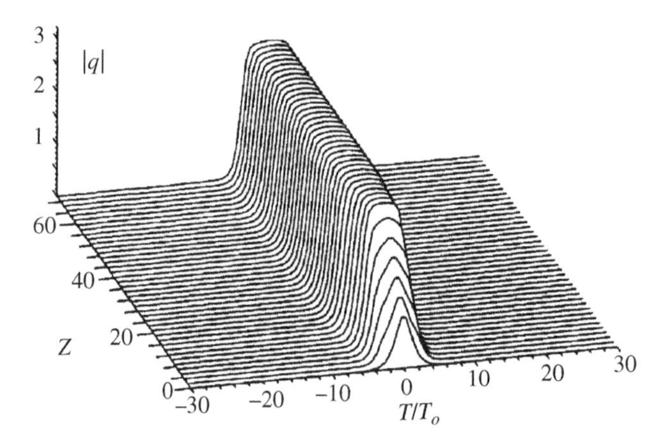

Figure 12.6 Formation and stable propagation of a flat-top soliton, starting from an initial pulse with a sech profile for: $\delta = -0.1$, $\beta = 0.5$, $\varepsilon = 0.66$, and $\mu = \nu = -0.01$.

pulse profiles can be obtained for non-small values of those parameters. For example, Figure 12.6 illustrates the formation and stable propagation of a flat-top soliton, starting from an initial pulse with a sech profile and assuming the following parameter values: $\delta = -0.1$, $\beta = 0.5$, $\varepsilon = 0.66$, and $\mu = \nu = -0.01$.

Figure 12.7 shows the amplitude and the spectrum of a plain pulse, considering the parameter values: $\delta = -0.01$, $\beta = 0.5$, $\mu = -0.03$, $\nu = 0$, and $\varepsilon = 1.5$. The plain pulse resembles a conventional soliton, exhibiting a bell-shaped amplitude profile and a single peak spectrum.

Another example is given in Figure 12.8, which shows (a) the amplitude profiles; and (b) the spectra of a PP, as well as of two composite pulses (CPs) [35]. The following parameter values were considered in this case: $\delta = -0.01$, $\beta = 0.5$, $\mu = -0.03$, $\nu = 0$, $\varepsilon = 1.5$ (PP), $\varepsilon = 2.0$ (narrow CP), and $\varepsilon = 2.15$ (wide CP). Figure 12.8c illustrates the formation and propagation of the wide CP starting from the PP solution represented in (a) and (b). A CP exhibits a dual-peak but symmetric spectrum (Figure 12.8b) and can be considered as a bound state of a PP and two fronts attached to it from both sides [20]. The "hill" between the two fronts should be counted as a source because it follows from the phase profile that energy flows from the center to the CP wings.

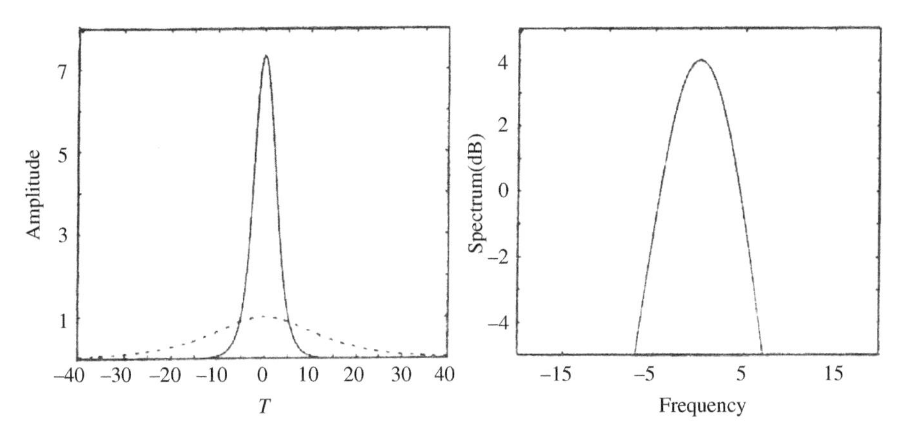

Figure 12.7 Amplitude profile and spectrum of a plain pulse (PP) for $\delta = -0.01$, $\beta = 0.5$, $\mu = -0.03$, $\nu = 0$, and $\varepsilon = 1.5$.

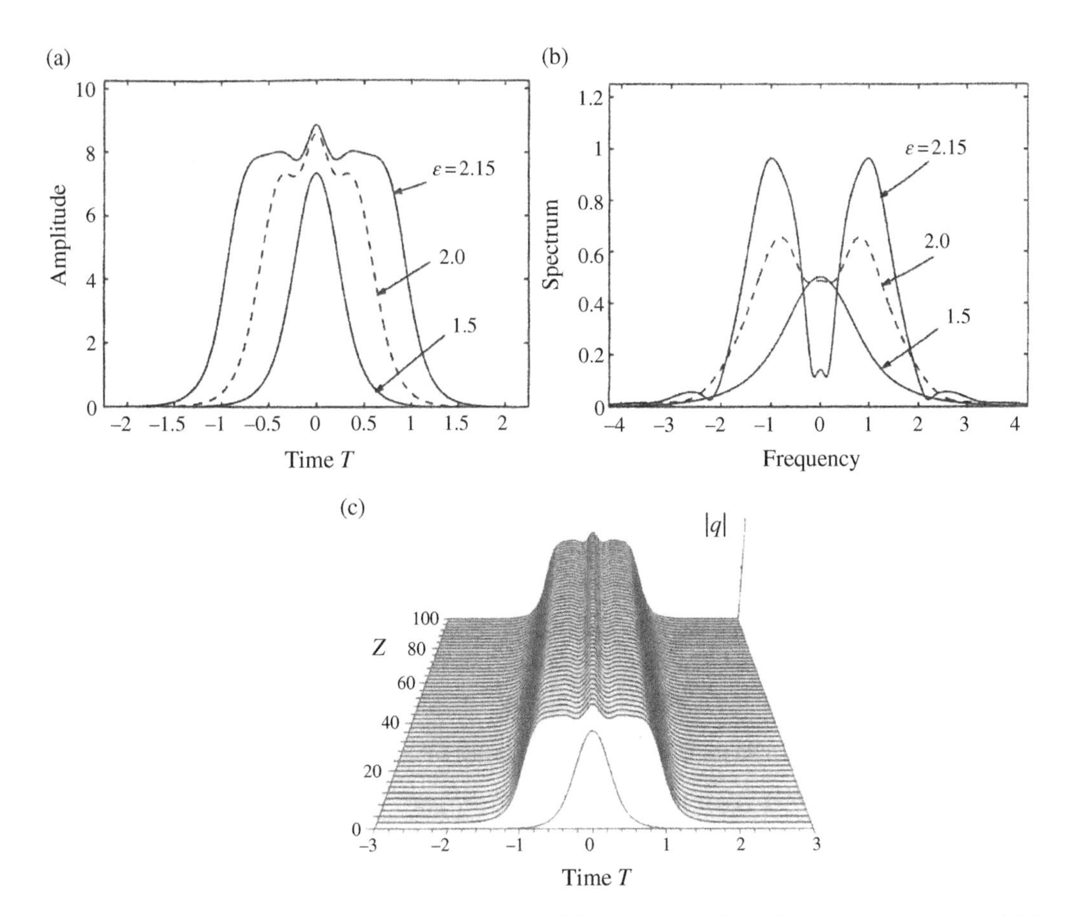

Figure 12.8 Amplitude profiles (a) and spectra of a plain pulse, as well as of two composite pulses (b) for $\delta = -0.01$, $\beta = 0.5$, $\mu = -0.03$, $\nu = 0$, $\varepsilon = 1.5$ (plain pulse), $\varepsilon = 2.0$ (narrow composite pulse), and $\varepsilon = 2.15$ (wide composite pulse). Figure 12.8c illustrates the formation and propagation of the wide composite pulse starting from the plain pulse solution represented in (a) and (b).

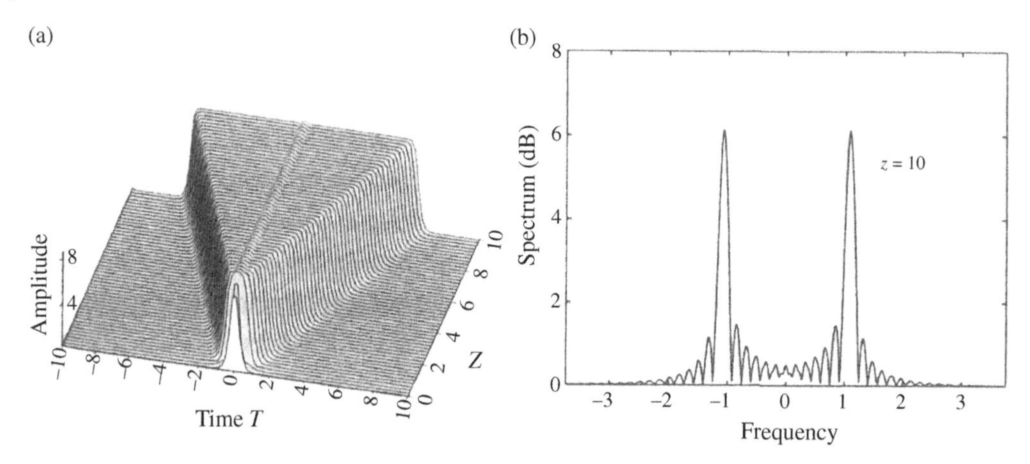

Figure 12.9 Nonstationary expanding structure obtained from an initial plain pulse when $\delta = -0.01$, $\beta = 0.5$, $\mu = -0.03$, $\nu = 0$, and $\varepsilon = 2.183$.

If one of the fronts of a CP is missing, one has a moving soliton (MS) [20]. The MS always moves with a velocity smaller than the velocity of the front for the same set of parameters. In fact, the front tends to move with its own velocity but the soliton tends to be stationary due to the spectral filtering. The resulting velocity of the MS is determined by competition between these two processes.

Increasing slightly the nonlinear gain coefficient and keeping the values of the other parameters equal to those used in Figure 12.8, the stationary wide CP shown in Figure 12.8c is lost and a nonstationary expanding structure appear, as illustrated in Figure 12.9.

12.4 High-Energy Dissipative Solitons

Ultrashort optical pulses with extremely high energy are of great importance for a variety of applications. They can be used in fields like medicine, biology, ultrasensitive laser spectroscopy, micromachining, ultrahigh bit rate communication systems, supercontinuum generation, metrology, and others [36–50]. Thus, producing shorter and more powerful pulses is vital for further progress in science and technology.

It has been found numerically that the energy of a dissipative soliton solution of the CGLE increases indefinitely when the equation parameters converge to a given region of the parameter space [51]. Such a set of parameters was called a dissipative soliton resonance (DSR) [52]. Found in the normal dispersion regime, it initially required a positive quintic reactive nonlinearity to appear [51, 52], but then DSR was also revealed in the anomalous dispersion and dispersion-free ($D = 0$) regimes, along with negative quintic reactive nonlinearity [53]. The concept of DSR has soon been proposed to achieve high-energy wave-breaking-free pulses [51–56].

An illustration of the DSR phenomenon is presented in Figure 12.10a, where the pulse energy increases dramatically by orders of magnitude as we approach the resonance [54]. Figure 12.10b shows that the energy of a DSR pulse increases mainly due to the increase of the pulse width while keeping the amplitude at a constant level [51–54]. Nevertheless, as the DSR pulse is highly chirped, this can be used to compress the pulse [57] and, hence, obtain high-energy ultrashort pulses. Actually, the DSR phenomenon has been frequently observed and reported in the area of fiber lasers [58–77]

(a)

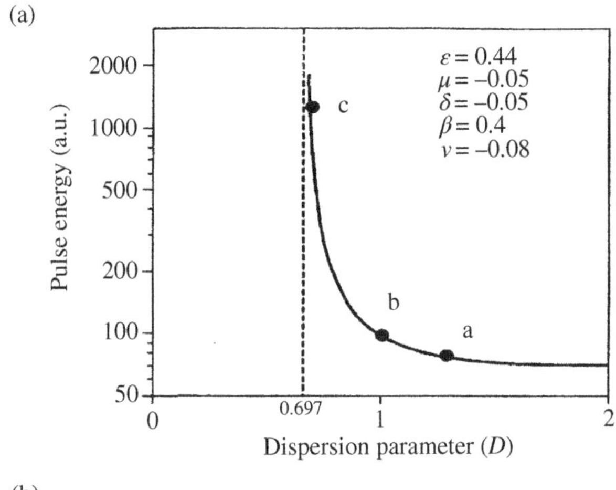

(b)

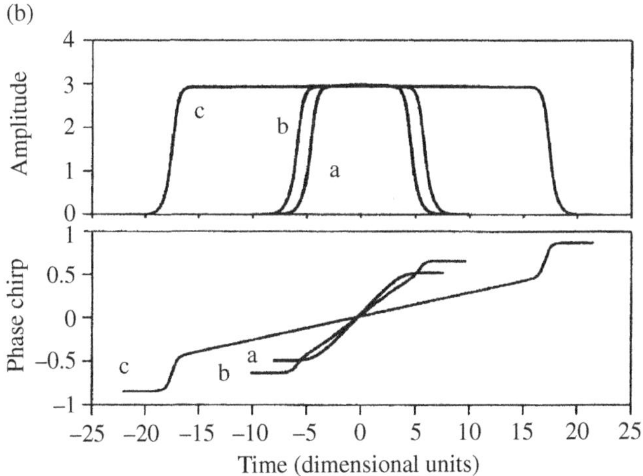

Figure 12.10 (a) Pulse energy against dispersion near the resonance; (b) Pulse profiles and phase chirp corresponding to points a, b, and c in (a). *Source:* After Ref. [54].

A different kind of high-energy ultrashort pulses, corresponding to the very high-amplitude (VHA) soliton solutions of the CGLE, have been reported recently [78, 79]. Such VHA solutions occur due to a singularity first predicted in Refs. [80, 81], namely as the nonlinear gain saturation effect tends to vanish. The increase in energy of these pulses is mainly due to the increase of the pulse amplitude, whereas the pulse width becomes narrower. The region of existence of VHA pulses has been found, considering both the normal and the anomalous dispersion regimes [79]. VHA pulses with high energy were found mainly in the normal dispersion region, which agrees with the large majority of experimental results, reporting the observation of high-energy pulses [82–84].

The stationary amplitude, η_S, of the soliton solution of the cubic-quintic CGLE is given by Eq. (8.69), obtained using the soliton perturbation theory. As discussed in Section 8.4, the stationary amplitude increases to infinity when the gain saturation parameter μ tends to zero. This can be observed from Figure 12.11, which represents the stationary amplitude against the nonlinear gain saturation parameter, μ, for $\delta = -0.1$, $\varepsilon = 0.35$, and four values of the filter strength: $\beta = 0.05$, 0.1, 0.2, and 0.5. For each set of parameter values, two different solutions are obtained: a small amplitude (SA) solution, i.e. $\eta_S \lesssim 1$, and a VHA solution, i.e. $\eta_S \gg 1$. As $\mu \to 0^-$, we observe that the SA

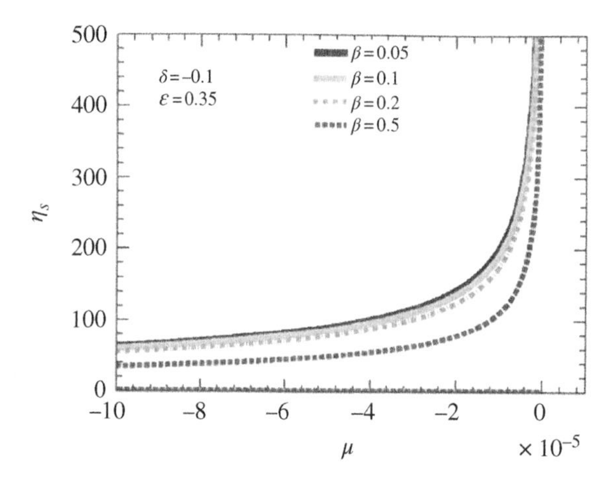

Figure 12.11 Stationary amplitude, η_S, versus nonlinear gain saturation parameter, μ, assuming four different values of the spectral filtering strength: $\beta = 0.05$ (thin solid curve), $\beta = 0.1$ (thick solid curve), $\beta = 0.2$ (dashed-dotted curve), and $\beta = 0.5$ (dashed curve). The other parameter values are: $\delta = -0.1$ and $\varepsilon = 0.35$.

solution remains practically constant, whereas the amplitude for the VHA solution tends to infinity. The VHA solution exhibits a higher amplitude for smaller values of filter strength.

Figure 12.12 shows the stationary amplitude, η_s, against the nonlinear gain, ε, for two different values of nonlinear gain saturation, μ, and four different values of the spectral filtering parameter, β. It can be seen that, for each specific value of the spectral filtering parameter, there is a minimum value of the nonlinear gain, $\varepsilon_0 \approx \beta/2$, below which there are no solutions. For values of ε above this threshold, a bifurcation occurs and both SA and VHA solutions exist. The amplitude of the VHA solutions increases with ε and becomes higher for lower values of μ and/or β.

Figure 12.13 illustrates some of the VHA stationary solutions obtained numerically from Eq. (12.1) for different values of the dispersion parameter D. The other parameter values are $\delta = -0.1$, $\beta = 0.2$, and $\varepsilon = 0.35$, and $\mu = -0.0001$.

Figure 12.13a illustrates the pulse profiles in the normal dispersion regime ($D < 0$), assuming the values $D = -0.5$, -1, and -1.5. Figure 12.13c shows the pulse profiles in the anomalous dispersion regime ($D > 0$), considering that the dispersion parameter, D, assumes symmetric values of the previous ones: 0.5, 1, and 1.5. In both regimes, the peak pulse amplitudes increase with an increase in the dispersion magnitude. The pulse width increases significantly with the magnitude of D in the normal dispersion regime, whereas it remains almost unchanged in the anomalous dispersion regime. From Figure 12.13b and d, we observe that the spectral range of the pulses power spectral density is almost the same for all cases in both regimes. On the other hand, the spectral peak power becomes higher when increasing the magnitude of D in the normal dispersion regime, as observed particularly for $D = -1.5$. Actually, this corresponds to the case with the highest pulse energy.

Figure 12.14 shows (a) the amplitude profiles; and (b) the power spectral density for $D = -1.5$ and several values of the nonlinear gain saturation parameter: $\mu = -0.0001$ (solid curves), $\mu = -0.000\,05$ (dashed-dot curves), and $\mu = -0.000\,01$ (dashed curves). As the magnitude of μ decreases, the pulses become narrower, whereas their amplitudes increase, assuming the values 57.5, 81.5, and 181.9, respectively. On the other hand, Figure 12.14b shows that both the power spectra peak values and the spectral range increase as $\mu \to 0^-$. The pulse energy, given by Eq. (12.2), increases and achieves the values $E = 2487$, 3469, and 7870 for $\mu = -0.0001$, $-0.000\,05$, and $-0.000\,01$, respectively.

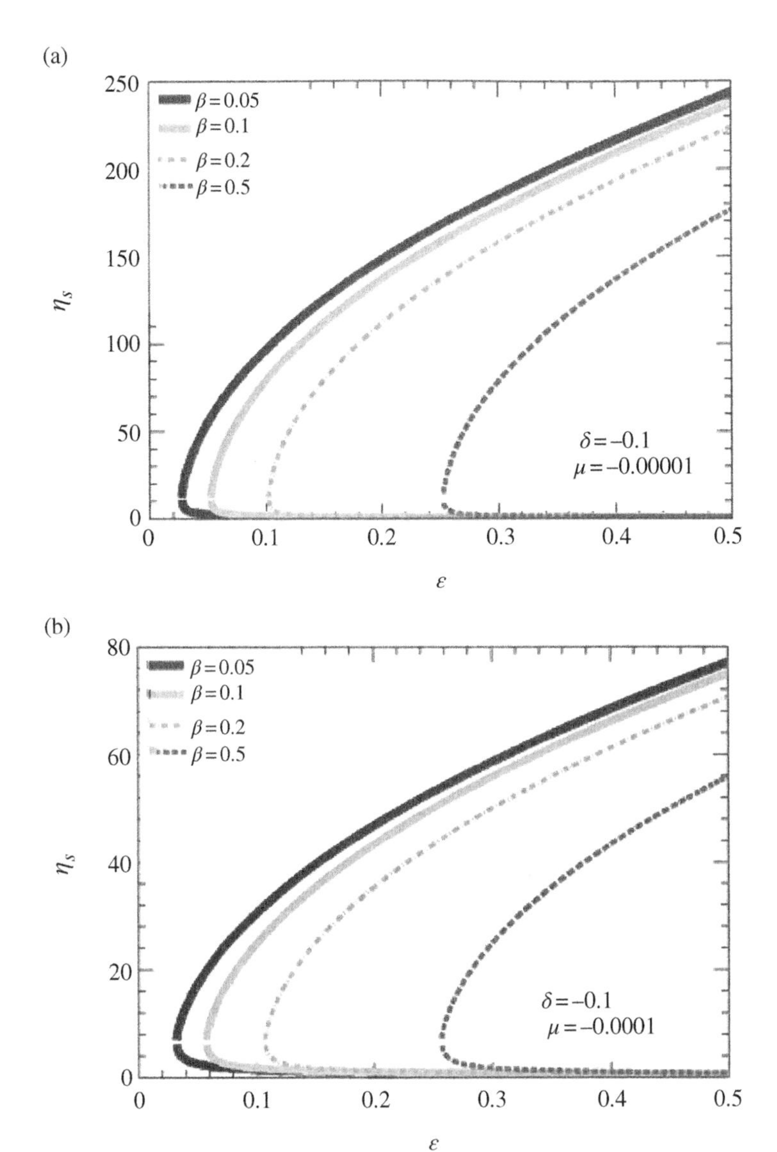

Figure 12.12 Stationary amplitude, η_S, versus nonlinear gain parameter, ε, for a linear gain δ = −0.1 and (a) μ = −0.00001 or (b) μ = −0.0001. Four different values of the spectral filtering parameter are considered: β = 0.05 (thin solid curve), β = 0.1 (thick solid curve), β = 0.2 (dashed-dotted curve), and β = 0.5 (dashed curve), respectively.

It is important to recognize a significant difference between the VHA pulses and the DSR pulses discussed above. In fact, the increase in energy of the DSR pulses is mainly due to the increase of the pulse width. As a consequence, in order to obtain high-energy ultrashort pulses, a linear pulse compression technique has to be used outside the laser cavity. However, the increase of the energy of VHA pulses occurring when the absolute value of μ decreases is mainly due to the increase of the pulse amplitude, whereas the pulse width becomes narrower, as illustrated by Figure 12.14a. Clearly, high-energy ultrashort pulses can be obtained in this case without using any additional

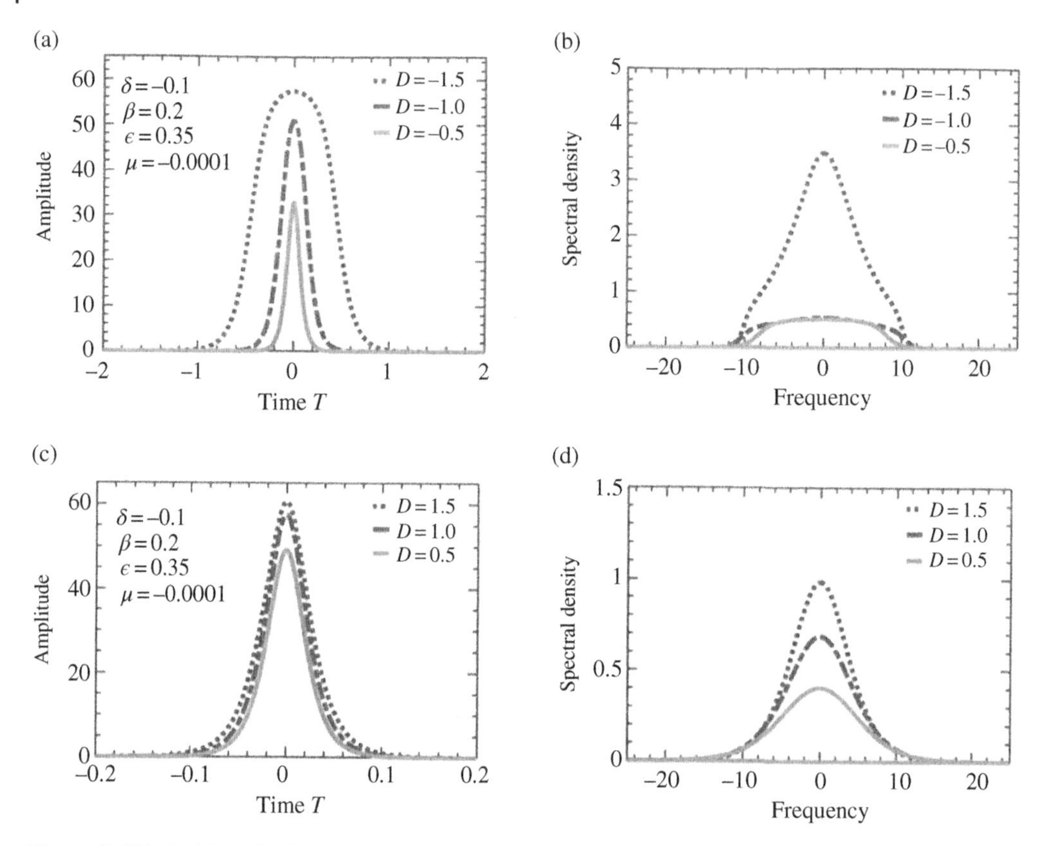

Figure 12.13 (a,c) Amplitude profiles and (b,d) power spectral density in the normal ($D < 0$) and anomalous ($D > 0$) dispersion regimes. The dispersion parameter assumes the following values: $D = \pm 1.5$ (dashed curves), $D = \pm 1$ (dashed-dotted curves), and $D = \pm 0.5$ (solid curves). The other parameter values are: $\delta = -0.1, \beta = 0.2, \varepsilon = 0.35$, $\mu = -0.0001$, and $\nu = 0$.

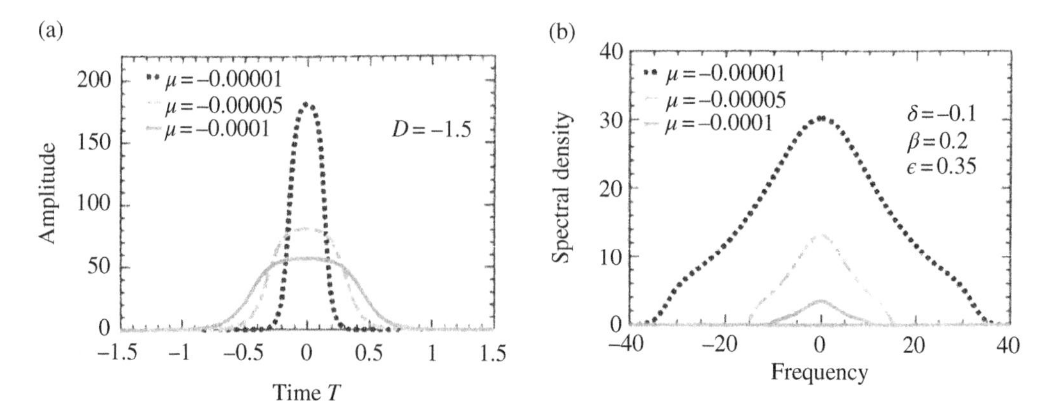

Figure 12.14 (a) Amplitude profiles and (b) power spectral density for $D = -1.5$ and several values of the nonlinear gain saturation parameter: $\mu = -0.000\,01$ (dashed curves), $\mu = -0.000\,05$ (dashed-dotted curves), and $\mu = -0.0001$ (solid curves). The other parameter values are $\delta = -0.1$, $\beta = 0.2$, $\varepsilon = 0.35$, and $\nu = 0$.

linear pulse compression technique. It might be noted, from Figure 12.11, that there is no limit for the pulse amplitude. However, a limitation could be imposed by some higher-order effects.

12.5 Soliton Bound States

After finding the conditions for the existence of stable solitary-pulse solutions of the CGLE, the next natural step is to consider their interactions. In fact, the problem of soliton interaction is crucial for the transmission of information. In the case of Hamiltonian systems, the interaction between the pulses is inelastic. Energy exchange between the pulses is one of the mechanisms that make the two-soliton solutions of these systems unstable, even when such stationary solutions do exist. The situation is rather different for dissipative systems. In this case, all solutions are a result of a double balance: between nonlinearity and dispersion and also between gain and loss. Moreover, the properties of dissipative solitons are completely determined by the external parameters of the optical system.

The formation of bound states of conventional solitons was first predicted in 1991 by Malomed [85], as mentioned in Section 8.4. Bound states constituted by dissipative solitons were theoretically studied in the following years based on the CGLE model [86–93]. The formation of bound states of solitons in fiber lasers has attracted considerable interest in recent years [94–113].

For given values of the CGLE parameters, the amplitude and width of its soliton solutions are fixed. As a consequence, during the interaction of two solitons, basically, only two parameters may change: their separation r and the phase difference, ϕ, between them. These two parameters provide a two-dimensional plane in which we may analyze pulse interaction, namely the formation of bound states, their stability, and their global dynamics [89, 90, 93]. This reduction in the number of degrees of freedom is a unique feature of systems with gain and loss. In the case of Hamiltonian systems, the amplitudes of the solitons can also change, which can affect the stability of the possible bound states.

In order to analyze numerically the soliton interaction in the 2-D space provided by the separation r, and phase difference, ϕ, between the two solitons, Eq. (12.1) can be solved with an initial condition

$$q(T) = q_0(T - r/2) + q_0(T + r/2)\exp(i\phi) \tag{12.32}$$

where q_0 is the stationary solution obtained numerically from Eq. (12.1) when the values of its parameters are specified. Initial condition (12.32) with arbitrary values for r and ϕ will result in a trajectory on the interaction plane. Bound states will be singular points of this plane.

Figure 12.15 shows an example of a numerical simulation of an interaction between the two plain solitons on the interaction plane, considering the following parameter values: $\delta = -0.01$, $\beta = 0.5$, $\varepsilon = 1.5$, $\nu = 0$, and $\mu = -0.03$. This figure indicates that, for the given set of parameters, there are at least four singular points. The points P_3 and P_4 are saddles and correspond to unstable bound states. In these states, the phase difference between the solitons is zero or π. In addition, there are two symmetrically located stable foci (points P_1 and P_2), which correspond to stable bound states of two solitons with a phase difference $\phi = \pm\pi/2$ between them. The stationary pulse separation in these bound states is $r \approx 1.62$.

Figure 12.16 illustrates the stable propagation of a bound state of two plain pulses with a phase difference $\pi/2$ between them. The bound state moves with a constant velocity and the direction of motion depends on the sign of ϕ. Figure 12.17 shows (a) the initial (dashed curve) and final (full curve) profiles; (b) phase; and (c) spectrum of the bound state.

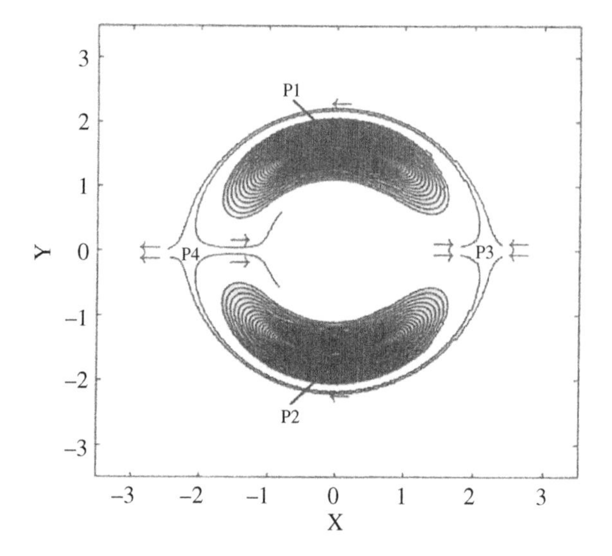

Figure 12.15 Trajectories on the interaction plane showing the evolution of two plain pulses for $\delta = -0.01$, $\beta = 0.5$, $\varepsilon = 1.5$, $\nu = 0$, and $\mu = -0.03$. We have $X = r\cos(\phi)$ and $Y = r\sin(\phi)$.

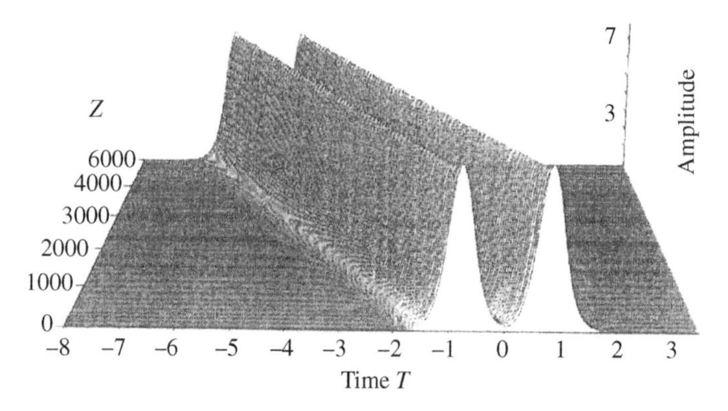

Figure 12.16 Propagation of a bound state of two plain pulses with a phase difference of $\pi/2$ between them.

Figure 12.18 shows the trajectories on the interaction plane showing the evolution of two composite pulses, considering the following parameter values: $\delta = -0.01$, $\beta = 0.5$, $\varepsilon = 2.0$, $\nu = 0$, and $\mu = -0.03$. This figure indicates that, for the given set of parameters, there are at least five singular points. The points P_3, P_4, and P_5 are saddles and correspond to unstable bound states. The points P_1 and P_2 correspond to stable bound states of two solitons with a phase difference $\phi = \pm\pi/2$ between them.

Figure 12.19 shows the stable propagation of a bound state of two composite pulses with a phase difference $\pi/2$. In contrast with the behavior of the PP bound state shown in Figure 12.16, the CP bound state moves at the group velocity. Figure 12.20 shows (a) the initial (dashed-curve) and final (full-curve) profiles; (b) phase; and (c) spectrum of the bound state.

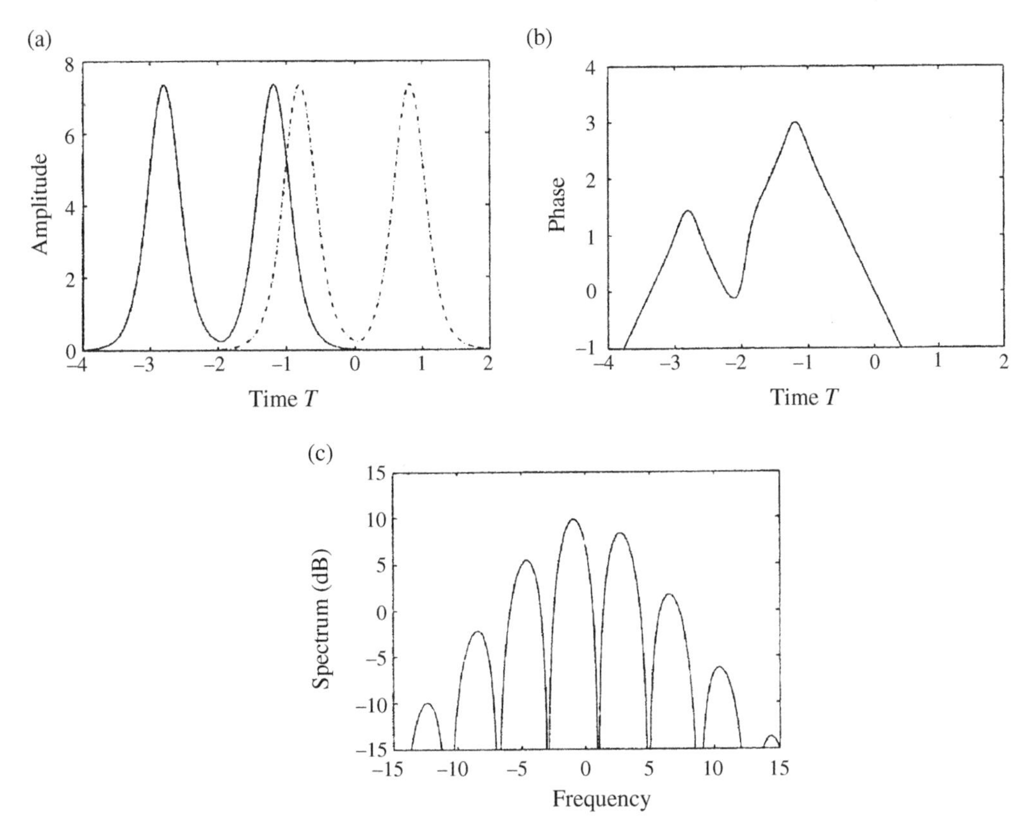

Figure 12.17 (a) Initial (dashed-curve) and final (full-curve) profiles, (b) phase, and (c) spectrum of a bound state of two plain pulses with a phase difference of $\pi/2$ between them.

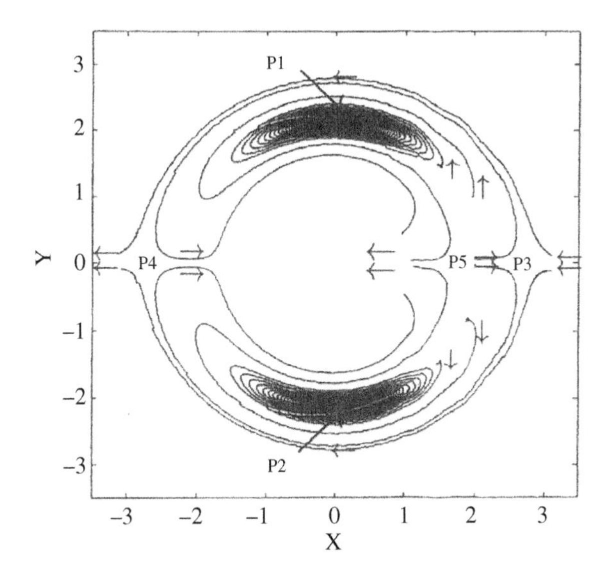

Figure 12.18 Trajectories on the interaction plane showing the evolution of two composite pulses for $\delta = -0.01$, $\beta = 0.5$, $\varepsilon = 2.0$, $\nu = 0$, and $\mu = -0.03$. We have $X = r\cos(\phi)$ and $Y = r\sin(\phi)$.

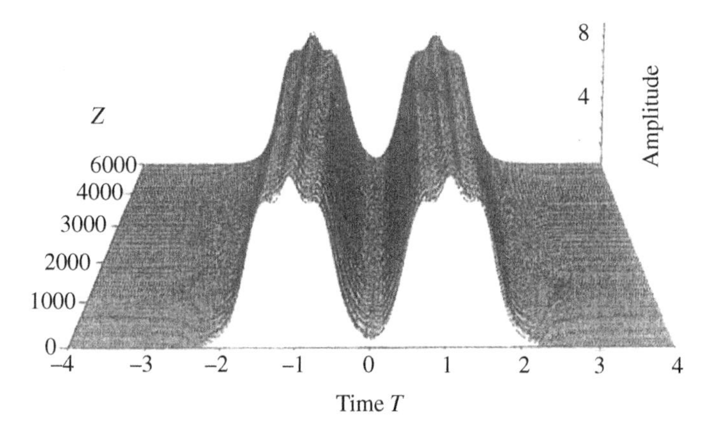

Figure 12.19 Propagation of a bound state of two composite pulses with a phase difference of $\pi/2$ between them.

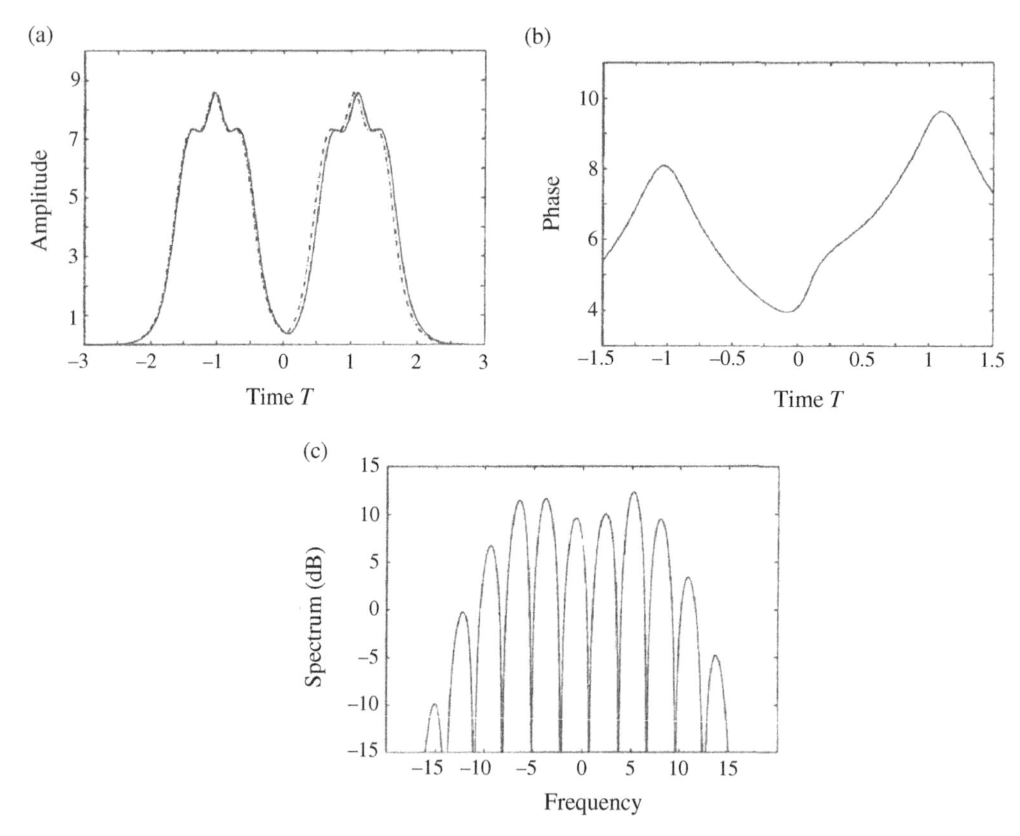

Figure 12.20 (a) Initial (dashed-curve) and final (full-curve) profiles, (b) phase, and (c) spectrum of a bound state of two composite pulses with a phase difference of $\pi/2$ between them.

12.6 Impact of Higher-Order Effects

It has been seen in Chapter 8 that, in the absence of intrapulse Raman scattering (IRS) and considering only the nonlinear gain proportional to the second order of the amplitude, when the pulse is stable, the background is unstable, and vice versa. However, as demonstrated in Chapter 9, the existence of stable pulse propagation in a stable background can be achieved in the presence of IRS when $\delta < 0$, $\varepsilon > 0$, and $\mu = 0$ or $\mu < 0$.

Figure 12.21 shows the contour plot for the amplitude (a), the spectrum (b), as well as the final amplitude profile (c) obtained by numerically solving Eq. (9.35) considering an initial pulse given by the plain pulse represented in Figure 12.7. The following non-small values for the parameters were used: $\beta = 0.5$, $\delta = -0.01$, $\varepsilon = 1.5$, $\mu = -0.03$, and $\tau_R = 0.01$. Since $\delta < 0$, the background is truly stable in this case. The trajectory in Figure 12.21a is a straight line, which indicates a constant

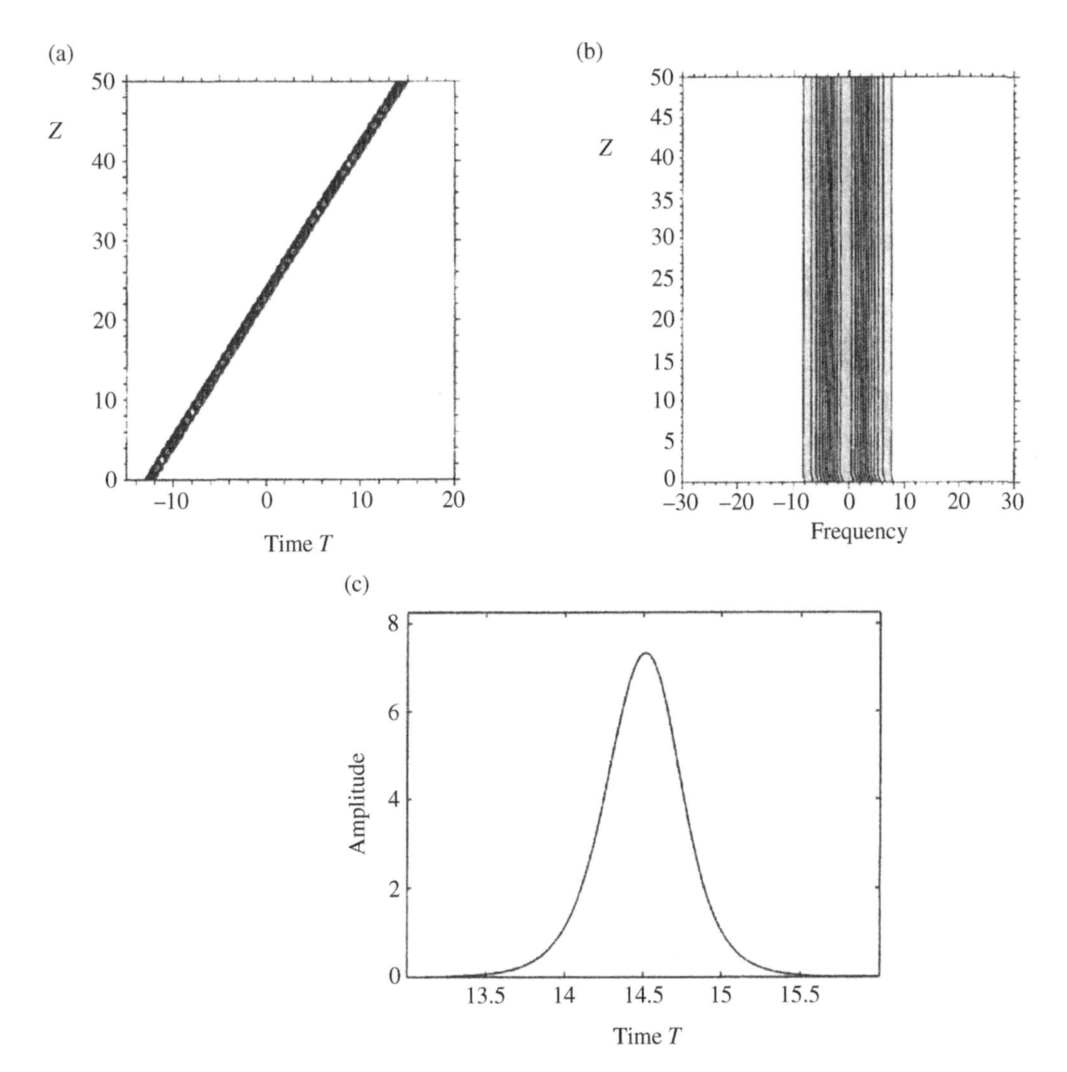

Figure 12.21 Contour maps for (a) the amplitude and (b) the spectrum, as well as (c) the final pulse profile for $\delta = -0.01$, $\beta = 0.5$, $\varepsilon = 1.5$, $\mu = -0.03$, and $\tau_R = 0.01$.

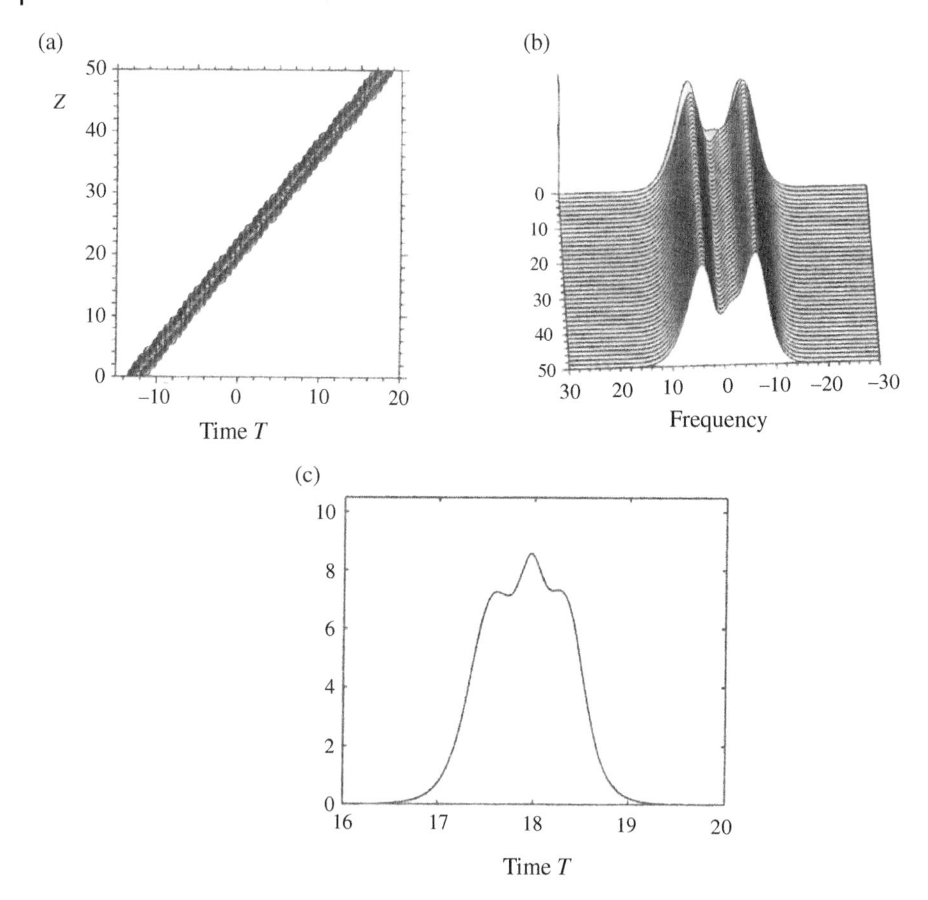

Figure 12.22 Contour maps for (a) the amplitude and (b) the spectrum, as well as (c) the final pulse profile for $\tau_R = 0.025$, $\beta = 0.1$, $\delta = -0.025$, $\varepsilon = 0.06$, and $\mu = -0.006$.

velocity of the pulse. This case clearly demonstrates that both the soliton self-frequency shift (SSFS) and the background instability can be effectively controlled using spectral filters together with nonlinear gain proportional to the second and fourth powers of the amplitude.

As seen in Section 12.3, the cubic-quintic CGLE given by Eq. (12.1) can have stable stationary solutions in the form of composite pulses with a symmetric profile and a dual-frequency but symmetric spectrum. Figure 12.22 shows the impact of the Raman effect in the profile of a narrow CP, assuming the following parameter values: $\tau_R = 0.01$, $\beta = 0.5$, $\delta = -0.01$, $\varepsilon = 2.0$, and $\mu = -0.03$. It can be seen from Figure 12.22a that the pulse trajectory is a straight line, which indicates a constant velocity of the CP. However, due to the IRS effect, the lower frequencies of the CP experience Raman gain at the expense of the higher-frequency components, resulting in a deformation of its spectrum, as seen from Figure 12.22b. As a consequence, the amplitude profile also becomes asymmetric, as observed in Figure 12.22c [114].

Concerning the VHA pulses discussed in Section 12.4, it has been shown in Ref. [78] that the singularity observed as $\mu \to 0^-$ is no longer present when the IRS effect is taken into account. This is illustrated by Figure 12.23, which represents the stationary amplitude against the nonlinear gain saturation parameter, μ, for $\delta = -0.1$, $\varepsilon = 0.35$, $\tau_R = 0.01$, and four values of the filter strength: $\beta = 0.05$, 0.1, 0.2, and 0.5. Similar to the results obtained in the absence of the IRS

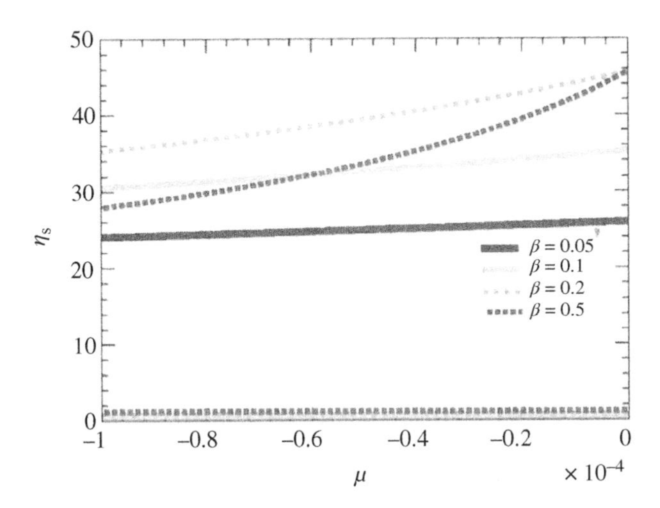

Figure 12.23 Equilibrium amplitude versus nonlinear gain saturation parameter, μ, in the presence of IRS, considering four different values of the spectral filtering: $\beta = 0.05$ (thin solid curve), $\beta = 0.1$ (thick solid curve), $\beta = 0.2$ (dashed-dotted curve) and $\beta = 0.5$ (dashed curve). The other parameter values are: $\delta = -0.1$, $\varepsilon = 0.35$, $\tau_R = 0.01$.

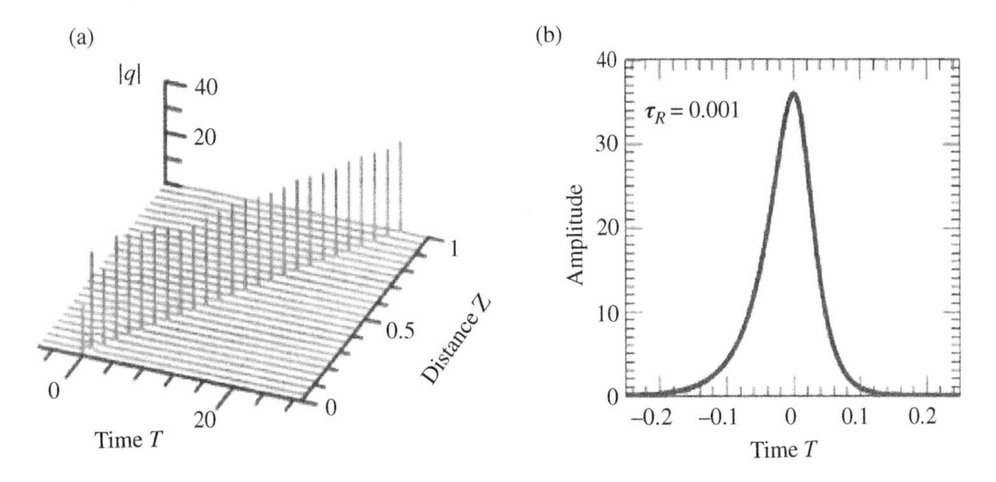

Figure 12.24 (a) Amplitude evolution and (b) stationary pulse profile of a VHA pulse in the presence of IRS ($\tau_R = 0.01$), considering the following parameter values: $\delta = -0.1$, $\beta = 0.2$, $\varepsilon = 0.35$, and $\mu = -0.0001$.

effect, both SA and VHA solutions are obtained. As $\mu \to 0^-$, the amplitude of the VHA solution increases. However, the singularity observed in the absence of IRS is no longer present.

Figure 12.24 shows (a) the evolution in the temporal domain and (b) the stationary pulse profile of the VHA pulse obtained by numerically solving Eq. (9.35) for the following set of parameter values: $\delta = -0.1$, $\beta = 0.2$, $\varepsilon = 0.35$, $\mu = -0.0001$, and $\tau_R = 0.01$. Besides the reduction of the amplitude, Figure 12.24 shows that IRS also determines a deceleration of the pulse.

Figure 12.25 shows the propagation of VHA pulses in the presence of both IRS ($\tau_R = 0.01$) and third-order dispersion (TOD): (a) $\beta_3 = -0.0041$ and (b) $\beta_3 = -0.005\,375$. The other parameter values are: $\delta = -0.1, \beta = 0.2, \varepsilon = 0.35$, and (a) $\mu = -0.000\,01$ and (b) $\mu = -0.0001$, respectively. The values of

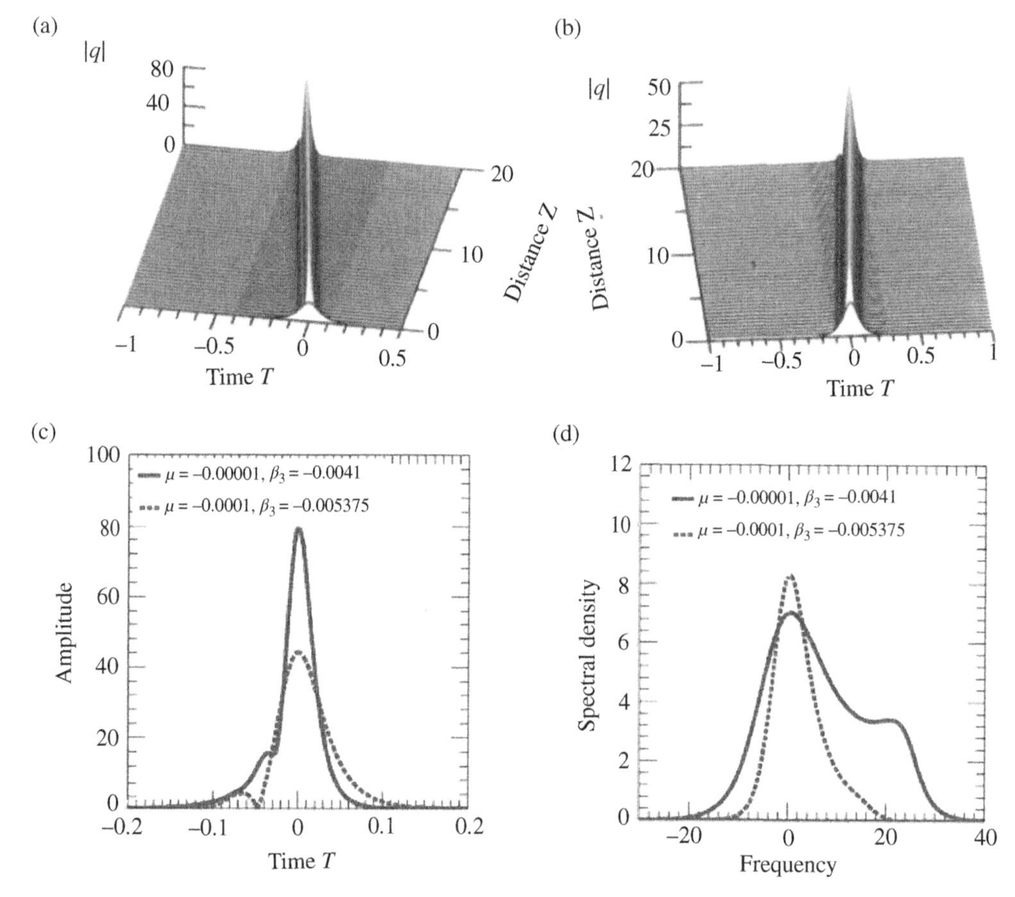

Figure 12.25 Propagation of VHA pulses in the presence of both IRS ($\tau_R = 0.01$) and TOD: (a) $\beta_3 = -0.0041$ and (b) $\beta_3 = -0.005\,375$. The other parameter values are: $\delta = -0.1$, $\beta = 0.2$, $\varepsilon = 0.35$, and (a) $\mu = -0.000\,01$ and (b) $\mu = -0.0001$. (c) and (d) show the amplitude and spectral density profiles for the pulses presented in (a) (solid curves) and in (b) (dashed curves), respectively.

β_3 have been chosen in order to provide a reduced pulse velocity. Figure 12.25c and d shows the pulse amplitude profiles and their spectral densities for both values of μ and β_3. We observe that the amplitude profiles are asymmetric, with a major distortion occurring on the pulses leading edge, which is mainly due to TOD. The spectral profile also becomes asymmetric as a consequence of both higher-order effects.

The IRS effect has also a significant impact on the formation and characteristics of the dissipative soliton bound states discussed in Section 12.5. This is illustrated in Figure 12.26. which shows the interaction plane for the case of two plain pulses considering two values for the IRS coefficient (a) $\tau_R = 0.01$ and (b) $\tau_R = 0.02$, respectively. We observe that the two stationary points remain symmetrically located in the interaction plane, but the line joining them is rotated in the counterclockwise direction by an angle $\Delta\phi$, which depends on the magnitude of the IRS coefficient τ_R. Noticeably, we verify that the point P_1 remains a stable stationary point, whereas P_2 turns out to be an unstable stationary point. The phase difference between the two pulses forming the bound state is $\phi = \pm 2\pi/3$ rad in (a) and $\phi = \pm 11\pi/13$rad in (b). The stationary pulse separation in the bound states is slightly increased in comparison to the value $r \approx 1.62$ in the absence of IRS: $r \approx 1.63$ in (a), and to $r \approx 1.65$ in (b).

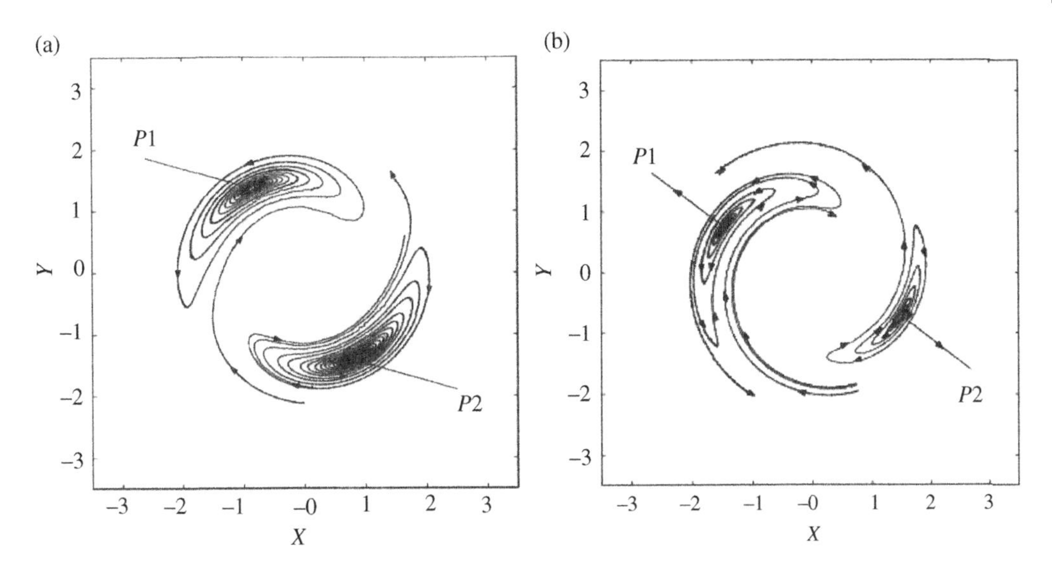

Figure 12.26 Trajectories on the interaction plane showing the evolution of two plain pulses for $\delta = -0.01$, $\beta = 0.5$, $\varepsilon = 1.5$, $\mu = -0.03$, $\nu = 0$, and $\tau_R = 0.01$ (a), or $\tau_R = 0.02$ (b). We have $X = r\cos(\phi)$ and $Y = r\sin(\phi)$.

References

1 Akhmediev, N. and Ankiewicz, A. (ed.) (2005). *Dissipative Solitons*. Berlin: Springer.
2 Akhmediev, N. and Ankiewicz, A. (ed.) (2008). *Dissipative Solitons: From Optics to Biology and Medicine*. Berlin: Springer.
3 Aranson, I. and Kramer, L. (2002). *Rev. Mod. Phys.* **74**: 99.
4 Jian, P.S., Torruellas, W.E., Haelterman, M. et al. (1999). *Opt. Lett.* **24**: 400.
5 Ng, C.S. and Bhattacharjee, A. (1999). *Phys. Rev. Lett.* **82**: 2665.
6 Haus, H.A. (1975). *IEEE J. Quantum Electron.* **QE-11**: 736.
7 Dunlop, A.M., Wright, E.M., and Firth, W.J. (1998). *Optics Commun.* **147**: 393.
8 Akhmediev, N., Rodrigues, A., and Townes, G. (2001). *Opt. Commun.* **187**: 419.
9 Kalashnikov, V.L., Podivilov, E., Chernykh, A., and Apolonski, A. (2006). *Appl. Phys. B* **83**: 503.
10 Komarov, A., Leblond, H., and Sanchez, F. (2005). *Phys. Rev. E* **72**: 025604.
11 Matsumoto, M., Ikeda, H., Uda, T., and Hasegawa, A. (1995). *J. Lightwave Technol.* **13**: 658.
12 Ferreira, M.F., Facão, M.V., Latas, S.V., and Sousa, M.H. (2005). *Fiber Integrated Opt.* **24**: 287.
13 Ferreira, M.F. (2011). *Nonlinear Effects in Optical Fibers*. Hoboken, NJ: John Wiley & Sons.
14 Chang, W., Ankiewicz, A., Akhmediev, N., and Soto-Crespo, J. (2007). *Phys. Rev. E.* **76**: 016607.
15 Mancas, S. and Choudhury, S. (2009). *Chaos, Solitons Fractals* **40**: 91.
16 Pereira, N. and Stenflo, L. (1977). *Phys. Fluids* **20**: 1733.
17 Nozaki, K. and Bekki, N. (1984). *Phys. Soc. Japan* **53**: 1581.
18 Conte, R. and Musette, M. (1993). *Physica D* **69**: 1.
19 Akhmediev, N., Afanasjev, V., and Soto-Crespo, J. (1996). *Phys. Rev. E* **53**: 1190.
20 Akhmediev, N. and Ankiewicz, A. (1997). *Solitons, Nonlinear Pulses and Beams*. Chapman & Hall.
21 Thual, O. and Fauvre, S. (1988). *J. Phys.* **49**: 1829.
22 Soto-Crespo, J., Akhmediev, N., and Ankiewicz, A. (2000). *Phys. Rev. Lett.* **85**: 2937.
23 Akhmediev, N., Soto-Crespo, J., and Town, G. (2001). *Phys. Rev. E.* **63**: 056602.
24 Akhmediev, N. and Soto-Crespo, J. (2003). *Phys. Lett A.* **317**: 287.

25 Akhmediev, N. and Soto-Crespo, J. (2004). *Phys. Rev. E.* **70**: 036613.
26 Soto-Crespo, J. and Akhmediev, N. (2005). *Phys. Rev. Lett.* **95**: 024101.
27 Soto-Crespo, J. and Akhmediev, N. (2005). *Math. Comp. Simulation* **69**: 526.
28 Tsoy, E., Ankiewicz, A., and Akhmediev, N. (2006). *Phys. Rev. E.* **73**: 036621-(1-10).
29 Ankiewicz, A., Akhmediev, N., and Devine, N. (2007). *Opt. Fiber Technol.* **13**: 91.
30 Manousakis, M., Papagiannis, P., Moshonas, N., and Hizanidis, K. (2001). *Opt. Commun.* **198**: 351.
31 Mancas, S. and Choudhury, S. (2007). *Theor. Math. Phys.* **152**: 1160.
32 Ferreira, M.F. (2018). *IET Optoelectron.* **12**: 122.
33 Marcq, P., Chaté, H., and Conte, R. (1994). *Physica D* **73**: 305.
34 Soto-Crespo, J.M., Akhmediev, N.N., and Afanasjev, V.V. (1996). *J. Opt. Soc. Am. B* **13**: 1439.
35 Latas, S.C., Ferreira, M.F., and Rodrigues, A.S. (2005). *Opt. Fiber Technol.* **11**: 292.
36 Serbin, J., Bauer, T., Fallnich, C. et al. (2002). *Appl. Surf. Sci.* **197**: 737.
37 König, K., Krauss, O., and Riemann, I. (2002). *Opt. Express* **10**: 171.
38 Taylor, R.S., Hnatovsky, C., Simova, E. et al. (2003). *Opt. Lett.* **28**: 1043.
39 Cheng, G., Wang, Y., White, J.D. et al. (2003). *J. Appl. Phys.* **94**: 1304.
40 Nishizawa, N., Chen, Y., Hsiung, P. et al. (2004). *Opt. Lett.* **29**: 2846.
41 Maxwell, I., Chung, S., and Mazur, E. (2005). *Med. Laser Appl.* **20**: 193.
42 Suzuki, K., Sharma, V., Fujimoto, J.G., and Ippen, E.P. (2006). *Opt. Express* **14**: 2335.
43 Sakakura, M., Kajiyama, S., Tsutsumi, M. et al. (2007). *Jpn. J. Appl. Phys.* **46** (Part 1): 5859.
44 Diddams, S.A., Hollberg, L., and Mbele, V. (2007). *Nature* **445**: 627.
45 Genty, G., Coen, S., and Dudley, J.M. (2007). *J. Opt. Soc. Am. B* **24**: 1771.
46 Labruyère, A., Tonello, A., Couderc, V. et al. (2012). *Opt. Fiber Technol.* **18**: 375.
47 Kawagoe, H., Ishida, S., Aramaki, M. et al. (2014). *Biomed. Opt. Express* **5**: 932.
48 Hu, S., Yao, J., Liu, M. et al. (2016). *Opt. Express* **24**: 10786.
49 Rodrigues, S.G., Facão, M.V., and Ferreira, M.F. (2017). *J. Nonlinear Opt. Phys. Mat.* **56**: 1750049.
50 Kang, J., Kong, C., Feng, P. et al. (2018). *IEEE Photonics Technol. Lett.* **30**: 311.
51 Akhmediev, N., Soto-Crespo, J.-M., and Grelu, P. (2008). *Phys.Lett. A* **372**: 3124.
52 Chang, W., Ankiewicz, A., Soto-Crespo, J.M., and Akhmediev, N. (2008). *Phys. Rev. A* **78**: 023830.
53 Chang, W., Soto-Crespo, J.M., Ankiewicz, A., and Akhmediev, N. (2009). *Phys. Rev. A* **79**: 033840.
54 Grelu, P., Chang, W., Ankiewicz, A. et al. (2010). *J. Opt. Soc. Am. B* **27**: 2336.
55 Komarov, A., Amrani, F., Dmitriev, A. et al. (2013). *Phys. Rev. A* **87**: 023838.
56 Du, W., Li, H., Li, J. et al. (2019). *Opt Express* **27**: 8059.
57 Grischkowsky, D. and Balant, A.C. (1982). *Appl. Phys. Lett.* **41**: 1.
58 Wu, X., Tang, D.Y., Zhang, H., and Zhao, L.M. (2009). *Opt. Express* **17**: 5580.
59 Liu, X. (2010). *Phys. Rev. A* **81**: 053819.
60 Ding, E., Grelu, P., and Kutz, J.N. (2011). *Opt. Lett.* **36**: 1146.
61 Wang, S.-K., Ning, Q.-Y., Luo, A.-P. et al. (2013). *Opt. Express* **21**: 2402.
62 Li, D., Tang, D., Zhao, L., and She, D. (2015). *J. Lightwave Technol.* **33**: 3781.
63 Lin, W., Wang, S., Xu, S. et al. (2015). *Opt. Express* **23**: 14860.
64 Xu, Y., Song, Y., Du, G. et al. (2015). *IEEE Photonics J.* **7**: 1502007.
65 Krzempek, K., Sotor, J., and Abramski, K. (2016). *Opt. Lett.* **41**: 4995.
66 Semaan, G., Braham, F., Fourmont, J. et al. (2016). *Opt. Lett.* **41**: 4767.
67 Krzempek, K. (2015). *Opt. Express* **23**: 30651.
68 Krzempek, K. and Abramski, K. (2016). *Opt. Express* **24**: 22379.
69 Lee, J.S., Koo, J.H., and Lee, J.H. (2016). *Opt. Eng.* **55**: 081309.
70 Li, D., Li, L., Zhou, J. et al. (2016). *Scientific Rep.* **6**: 23631.
71 Rivera, A., Laborde, C., Carrascosa, A. et al. (2016). *Opt. Express* **24**: 9966.

72 Krzempek, K., Tomaszewska, D., and Abramski, K. (2017). *Opt. Express* **25**: 24853.

73 Lyu, Y., Zou, X., Shi, H. et al. (2017). *Opt. Express* **25**: 13286.

74 Wang, P., Zhao, K., Xiao, X., and Yang, C. (2017). *Opt. Express* **25**: 30708.

75 Zhao, G.-K., Lin, W., Chen, H.-J. et al. (2017). *Opt. Express* **25**: 20923.

76 Du, T., Li, W., Ruan, Q. et al. (2018). *Appl. Phys. Express* **11**: 052701.

77 Wang, N., Cai, J.-H., Qi, X. et al. (2018). *Opt. Express* **26**: 1689.

78 Latas, S.C., Ferreira, M.F.S., and Facão, M. (2017). *J. Opt. Soc. Am. B* **34**: 1033.

79 Latas, S.C. and Ferreira, M.F. (2019). *J. Opt. Soc. of Am. B* **36**: 3016.

80 Soto-Crespo, J.M., Akhmediev, N.N., Afanasjev, V.V., and Wabnitz, S. (1997). *Phys. Rev. E* **55**: 4783.

81 Afanasjev, V.V. (1995). *Opt. Lett.* **20**: 704.

82 Chong, A., Renninger, W.H., and Wise, F.W. (2007). *Opt. Lett.* **32**: 2408.

83 Zhao, L.M., Tang, D.Y., and Wu, J. (2006). *Opt. Lett.* **31**: 1788.

84 Grelu, P. and Akhmediev, N. (2012). *Nat. Photonics* **6**: 84.

85 Malomed, B.A. (1991). *Phys. Rev. A* **44**: 6954.

86 Malomed, B.A. (1993). *Phys. Rev. E* **47**: 2874.

87 Afanasjev, V. and Akhmediev, N. (1996). *Phys. Rev. E* **53**: 6471.

88 Afanasjev, V.V., Malomed, B.A., and Chu, P.L. (1997). *Phys. Rev. E* **56**: 6020.

89 Akhmediev, N., Ankiewicz, A., and SotoCrespo, J.M. (1997). *Phys. Rev. Lett.* **79**: 4047.

90 Akhmediev, N., Ankiewicz, A., and Soto-Crespo, J. (1998). *J. Opt. Soc. Am. B* **15**: 515.

91 Ferreira, M.F. and Latas, S.V. (2001). *SPIE Proc.* **4271**: 268.

92 Ferreira, M.F. and Latas, S.V. (2002). Bound states of plain and composite pulses in optical transmission lines and fiber lasers. In: *Applications of Photonic Technology*, vol. **4833** (ed. R. Lessard, G. Lampropoulos and G. Schinn), 845. SPIE.

93 Latas, S.V. and Ferreira, M.F. (2005). *Opt. Fiber Tech.* **11**: 292.

94 Lee, R.K., Lai, Y.C., and Malomed, B.A. (2005). *Opt. Lett.* **30**: 3084.

95 Zhao, L.M., Tang, D.Y., and Zhao, B. (2005). *Opt. Commun.* **252**: 167.

96 Hsiang, W.W., Lin, C.Y., and Lai, Y.C. (2006). *Opt. Lett.* **31**: 1627.

97 Zhao, L.M., Tang, D.Y., Wu, X. et al. (2007). *Opt. Lett.* **32**: 3191.

98 Komarov, A., Haboucha, A., and Sanchez, F. (2008). *Opt. Lett.* **33**: 2254.

99 Komarov, A., Komarov, K., and Sanchez, F. (2009). *Phys. Rev. A* **79**: 033807.

100 Zhao, L.M., Tang, D.Y., and Liu, D. (2010). *Appl. Phys. B: Lasers Opt.* **99**: 441.

101 Wu, X., Tang, D.Y., Luan, X.N., and Zhang, Q. (2011). *Opt. Commun.* **284**: 3615.

102 Gui, L.L., Xiao, X.S., and Yang, C.X. (2013). *J. Opt. Soc. Am. B* **30**: 158.

103 Gumenyuk, R. and Okhotnikov, O.G. (2013). *IEEE Photonics Technol. Lett.* **25**: 133.

104 Guo, J. (2014). *J. Mod. Opt.* **61**: 980.

105 Yun, L. and Han, D. (2014). *Opt. Commun.* **313**: 70.

106 Komarov, A., Komarov, K., and Sanchez, F. (2015). *Opt. Commun.* **354**: 158.

107 Liu, H.H. and Chow, K.K. (2015). *IEEE Photonics Technol. Lett.* **27**: 867.

108 Zeng, C., Cui, Y.D., and Guo, J. (2015). *Opt. Commun.* **347**: 44.

109 He, X., Hou, L., Li, M. et al. (2016). *IEEE Photonics J.* **8**: 1500706.

110 Li, L., Ruan, Q., Yang, R. et al. (2016). *Opt. Express* **24**: 21020.

111 Li, K.-X., Song, Y.-R., Tian, J.-R. et al. (2017). *IEEE Photonics J.* **9**: 1400209.

112 Li, X., Xia, K., Wu, D. et al. (2017). *IEEE Photonics Technol. Lett.* **29**: 2071.

113 Liang, H., Wang, Z., He, R. et al. (2018). *IEEE Photonics Technol. Lett.* **30**: 1475.

114 Latas, S.V. and Ferreira, M.F. (2007). *J. Math. Comp. Simul. Elsevier* **74**: 379.

13

Pulsating Dissipative Solitons

Besides the stationary solutions, the cubic-quintic complex Ginzburg–Landau equation (CGLE) presents also different types of localized solutions with periodically or quasi-periodically varying amplitude, width, and energy. Among such pulsating solitons, we may refer the plain pulsating, creeping, erupting and chaotic solutions [1–7]. Actually, in order to observe these pulsating solitons, the parameters of the CGLE must be far enough from the nonlinear Schrödinger equation. The regions of existence of pulsating solitons have been presented by Akhmediev et al. [2].

The first experimental observation of an erupting soliton was reported in 2002 by Cundiff et al. in a sapphire laser [8], whereas the pulsating soliton was observed for the first time in 2004 by Soto-Crespo et al. in a mode-locked fiber ring laser [9]. More recently, a new technique, called dispersive Fourier transform (DFT) [10], was proposed that has enabled the experimental observation in real time of transient dynamics of diverse pulsating solitons and other complex ultrafast nonlinear phenomena in fiber lasers [11–29].

In this chapter, we describe several types of pulsating dissipative solitons, namely, plain pulsating, creeping, erupting, and chaotic solutions. The impact of some higher-order effects on the dynamics of such pulsating pulses and the possibility of converting them into stationary pulses will be discussed.

13.1 Dynamic Models for CGLE Solitons

None of the pulsating pulses can be found in analytic form, as it happens for some of the stationary soliton solutions discussed in Chapter 12. However, the existence of such pulsating solitons can be anticipated using some dynamic models based on the variational approach [30–34] or the method of moments [35–38]. Using any of these approaches, one can obtain a dynamical system for the evolution of several variables that correspond to the soliton parameters, such as the maximum amplitude, pulse width and chirp.

The stability and other properties of a dissipative soliton are linked to the stability and characteristics of its attractor in the terminology of nonlinear dynamics. For example, a stationary soliton corresponds to a fixed point, whereas a pulsating soliton is associated with a limit cycle. By changing the system parameters, we can transform one attractor into the other and even produce some type of irregular and chaotic dynamics.

Solitons in Optical Fiber Systems, First Edition. Mario F. S. Ferreira.

13.1.1 The Variational Approach

The standard variational approach, presented in Chapter 6, can be modified to allow for dissipative terms in the CGLE. Using such approach, Eq. (12.1) can be obtained from the Euler-Lagrange equation [34]:

$$\frac{\partial}{\partial \tau}\left(\frac{\partial L_d}{\partial q_T^*}\right) + \frac{\partial}{\partial z}\left(\frac{\partial L_d}{\partial q_Z^*}\right) - \frac{\partial L_d}{\partial q^*} = S \tag{13.1}$$

where L_d is the Lagrangian density, given by

$$L_d = \frac{i}{2}\left(q\frac{\partial q^*}{\partial Z} - q^*\frac{\partial q}{\partial Z}\right) + \frac{1}{2}\left|\frac{\partial q}{\partial T}\right|^2 - \frac{1}{2}|q|^4 - \frac{\nu}{3}|q|^6 \tag{13.2}$$

and S represents the dissipative terms:

$$S = i\delta q + i\beta\frac{\partial^2 q}{\partial T^2} + i\varepsilon|q|^2 q + i\mu|q|^4 q \tag{13.3}$$

Considering the Lagrangian

$$L = \int_{-\infty}^{\infty} L_d dT \tag{13.4}$$

and a trial solution containing parameters $\nu_j, j = 1,2,...$ the modified Euler-Lagrange equations are given by [34]:

$$\frac{d}{dz}\left(\frac{\partial L}{\partial v_z}\right) - \frac{\partial L}{\partial v} = 2\,\mathrm{Re}\left(\int_{-\infty}^{\infty} S\frac{\partial q^*}{\partial \nu}\right)dT \tag{13.5}$$

for each parameter ν.

Let us consider a soliton solution given by:

$$q(Z,T) = B(Z,T)\exp(-i\theta(Z)) \tag{13.6}$$

where $B(Z, T)$ is the complex amplitude and $\theta(Z)$ gives the phase evolution on propagation. Then, using Eq. (13.5), we have for a general parameter ν:

$$\frac{\partial L}{\partial v} - \frac{d}{dz}\left(\frac{\partial L}{\partial v_z}\right) = 2\,\mathrm{Re}\int_{-\infty}^{\infty}\frac{1}{iq^*}\frac{\partial q^*}{\partial \nu}\left(\delta|B|^2 + \varepsilon|B|^4 + \mu|B|^6 + \beta B^*\frac{\partial^2 B}{\partial T^2}\right)dT \tag{13.7}$$

If $\nu = \theta(Z)$ in Eq. (13.7), we have

$$\frac{\partial L}{\partial \theta} = 0, \quad \frac{\partial L}{\partial \theta'} = -E(Z), \quad \frac{1}{iq^*}\frac{\partial q^*}{\partial \theta} = 1 \tag{13.8}$$

where

$$E = \int_{-\infty}^{\infty} |q|^2 dT \tag{13.9}$$

is the pulse energy and

$$\frac{dE}{dZ} = 2\,\mathrm{Re} \int_{-\infty}^{\infty} \left[\delta|B|^2 + \varepsilon|B|^4 + \mu|B|^6 + \beta B^* \frac{\partial^2 B}{\partial T^2} \right] dT \tag{13.10}$$

Equation (13.10) corresponds to the balance equation for the energy E. Obviously, the right-hand side is zero in the case of a stationary soliton solution.

13.1.1.1 Sech Ansatz

Let us assume that

$$B(Z,T) = A\mathrm{sech}\left(\frac{T}{w(Z)}\right) \exp\left(-iC(Z)T^2\right) \tag{13.11}$$

where $A(Z)$, $w(Z)$, and $C(Z)$ are the amplitude, width and chirp parameter, respectively. In this case, we have

$$E = 2A^2 w \tag{13.12}$$

and Eq. (13.10) becomes

$$\frac{dE}{dZ} = 2E\left[\delta - \beta\left(\frac{1}{3w^2} + \frac{\pi^2}{3}C^2w^2\right) + \frac{\varepsilon}{3}\frac{E}{w} + \frac{2\mu}{15}\frac{E^2}{w^2}\right] \tag{13.13}$$

The Lagrangian corresponding to Eqs. (13.6) and (13.11) is given by:

$$L = E\left[\frac{\pi^2C^2w^2}{6} + \frac{1}{6w^2} - \frac{E}{6w} - \frac{2\nu E^2}{45w^2} - \frac{\pi^2w^2}{12}\frac{dC}{dZ} - \frac{d\theta}{dZ}\right] \tag{13.14}$$

If $\nu = C(Z)$ in Eq. (13.7), we obtain the following equation for the pulse width

$$\frac{dw}{dZ} = -2Cw + \beta\left(\frac{8}{\pi^2 w} - \frac{16\pi^2}{15}C^2w^3\right) - \frac{2\varepsilon}{\pi^2}E - \frac{\mu}{\pi^2}\frac{C^2}{w} \tag{13.15}$$

The equation for the chirp parameter $C(Z)$ is obtained using the parameter $\nu = w(Z)$ in Eq. (13.7). We have

$$\frac{dC}{dZ} = -2\left(\frac{1}{\pi^2w^4} - C^2\right) + \frac{1}{\pi^2}\frac{E}{w^3} - 4\left(\frac{1}{3} + \frac{1}{\pi^2}\right)\beta\frac{C}{w^2} + \frac{8\nu}{15\pi^2}\frac{E^2}{w^4} \tag{13.16}$$

Finally, using $\nu = E(Z)$ in Eq. (13.7), we obtain the following equation for the phase evolution

$$\frac{d\theta}{dZ} = \frac{1}{3w^2} + \beta C\left(\frac{1}{3} + \frac{\pi^2}{9}\right) - \frac{5}{12}\frac{E}{w} - \frac{8\nu}{45}\frac{E^2}{w^2} \tag{13.17}$$

13.1.1.2 Gaussian Ansatz

As a second example, let us consider a simple Gaussian-type ansatz:

$$B(Z,T) = A(Z)\exp\left(-\frac{T^2}{w^2(Z)} - iC(Z)T^2\right) \tag{13.18}$$

where $A(Z)$, $w(Z)$, and $C(Z)$ are the amplitude, width and chirp parameter, respectively. In this case, the pulse energy E is given by

$$E(Z) = \sqrt{\frac{\pi}{2}} A^2(Z) w(Z) \tag{13.19}$$

while the Lagrangian can be written as:

$$L = E\left[\frac{C^2 w^2}{2} + \frac{1}{2w^2} - \frac{E}{2\sqrt{\pi} w} - \frac{2\nu E^2}{3\sqrt{3}\pi w^2} - \frac{1}{4} w^2 \frac{dC}{dZ} - \frac{d\theta}{dZ}\right] \tag{13.20}$$

Considering $\nu = \theta, C, w, E$ in Eq. (13.7), the following evolution equations for the pulse parameters are obtained:

$$\frac{dE}{dZ} = 2E\left(\delta - \frac{\beta}{w^2} - \beta C^2 w^2 + \frac{1}{\sqrt{\pi}} \frac{\varepsilon E}{w} + \frac{2}{\pi\sqrt{3}} \frac{\mu E^2}{w^2}\right) \tag{13.21}$$

$$\frac{dw}{dZ} = 2\frac{\beta}{w} - 2Cw - 2\beta C^2 w^3 - \frac{1}{2\sqrt{\pi}} \varepsilon E - \frac{4}{3\pi\sqrt{3}} \frac{\mu E^2}{w} \tag{13.22}$$

$$\frac{dC}{dZ} = -\frac{2}{w^4} + 2C^2 - 8\frac{\beta C}{w^2} + \frac{1}{\sqrt{\pi}} \frac{E}{w^3} + \frac{8}{3\pi\sqrt{3}} \frac{\nu E^2}{w^4} \tag{13.23}$$

$$\frac{d\theta}{dZ} = \frac{1}{w^2} + 2\beta C - \frac{5}{4\sqrt{\pi}} \frac{E}{w} - \frac{8}{3\pi\sqrt{3}} \frac{\nu E^2}{w^2} \tag{13.24}$$

We observe that Eqs. (13.21)–(13.24) and Eqs. (13.13)–(13.17) are similar. The only difference lies in the numerical values of the coefficients of the terms.

13.1.2 The Method of Moments

The method of moments [35] allows a reduction of the complete evolution problem with an infinite number degrees of freedom to the evolution of a finite set of pulse characteristics. The method has been applied to various problems described by the perturbed NLSE [35, 39, 40] and by the cubic-quintic CGLE in [36–38], where simplified trial functions were considered.

For a localized solution with a single maximum, the following characteristics are required: peak amplitude, pulse width, center of mass position and phase parameters. For an arbitrary localized solution, the method of moments makes use of the integrals for the energy, E, and the momentum M, as given by Eqs. (12.2) and (12.3), as well as of the higher-order generalized moments:

$$I_1 = \int_{-\infty}^{+\infty} T|q|^2 dT \tag{13.25}$$

$$I_2 = \int_{-\infty}^{+\infty} (T - T_0)^2 |q|^2 dT \tag{13.26}$$

$$I_3 = \int_{-\infty}^{+\infty} (T - T_0)\left(q^* q_T - q q_T^* \; dT \tag{13.27}$$

where $T_0 = I_1/E$.

The energy and the momentum, as well as the above three higher-order moments, are conserved quantities of the nonlinear Schrödinger equation. For Eq. (12.1), they are not conserved but satisfy the following truncated set of first-order ordinary differential equations [35]:

$$\frac{dE}{dZ} = i \int_{-\infty}^{+\infty} (qR^* - q^*R)dT \tag{13.28}$$

$$\frac{dM}{dZ} = -i \int_{-\infty}^{+\infty} \left(q_T R^* - q_T^* R\right)dT \tag{13.29}$$

$$\frac{dI_1}{dZ} = iM + i \int_{-\infty}^{+\infty} T(qR^* - q^*R)dT \tag{13.30}$$

$$\frac{dI_2}{dZ} = -iI_3 + i \int_{-\infty}^{+\infty} (T - T_0)^2(qR^* - q^*R)dT \tag{13.31}$$

$$\frac{dI_3}{dZ} = 2M\frac{dT_0}{dz} + i \int_{-\infty}^{+\infty} \left(2|q_T|^2 - |q|^4\right)dT + 2i \int_{-\infty}^{+\infty} (T - T_0)\left(q_T R^* + q_T^* R\right)dT + i \int_{-\infty}^{+\infty} (qR^* + q^*R)dT \tag{13.32}$$

where q_T represents the derivative of q relative to T, q^* is the complex conjugate of q and

$$R = i\delta q + i\beta\frac{\partial^2 q}{\partial T^2} + i\varepsilon|q|^2 q + i\mu|q|^4 q - \nu|q|^4 q \tag{13.33}$$

Equations (13.28) and (13.29) correspond to Equations (12.4)–(12.7). Actually, from Equations (13.28)–(13.33), a system of ODEs for the soliton parameters similar to those derived in Section 13.1.1 can be obtained, assuming the same ansatz [38]. The conclusion is that both the method of moments and the variational approach provide the possibility of reducing the cubic-quintic CGLE to the same finite-dimensional dynamical system.

An analysis of the dynamical system given by Equations (13.28)–(13.33) demonstrates a clear correspondence between attractors and various types of localized waves governed by the cubic-quintic CGLE. Both fixed points and limit cycles can be found for such dynamical system, the former corresponding to stationary solitons and the latter to pulsating solitons. A transformation from a stationary soliton to a pulsating soliton appears as the result of a Hopf bifurcation in the reduced dynamical system [38]. Figure 13.1 shows two examples of limit cycles in three-dimensional phase space for two different sets of parameters for the model given by Eqs. (13.13), (13.15), and (13.16).

Figure 13.2 illustrates a pulsating soliton reconstructed from the sech ansatz given by Eqs. (13.6) and (13.11), together with the dynamical system provided by Eqs. (13.13), (13.15), and (13.16). The following system parameters were assumed: $\beta = 0.08$, $\varepsilon = 0.66$, $\mu = -0.1$, and $\nu = -0.1$. It is found that the soliton width varies periodically with Z, while the soliton amplitude remains approximately constant. Pulsating dissipative solitons appear due to a dynamic balance between dissipation and energy supply.

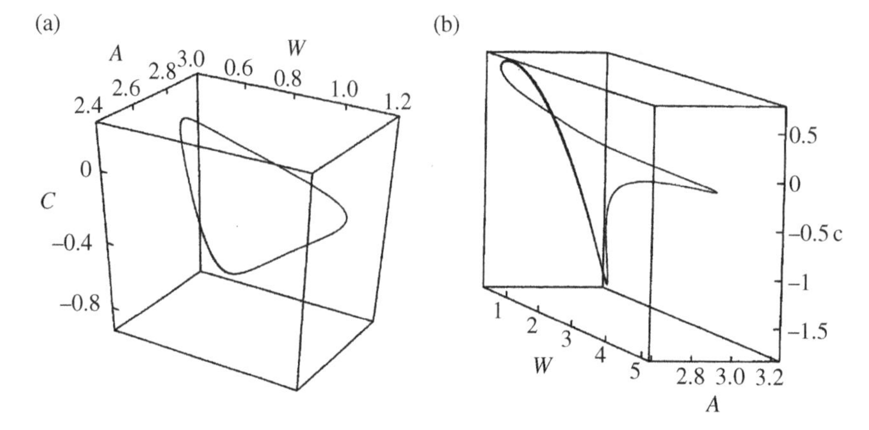

Figure 13.1 Examples of limit cycles in (*A, w, C*)-space for the model given by Eqs. (13.13), (13.15), and (13.16), assuming the following parameter values: $\beta = 0.08$, $\mu = -0.1$ and $\nu = -0.09$, and (a) $\varepsilon = 0.66$ or (b) $\varepsilon = 0.72$. *Source:* After Ref. [38].

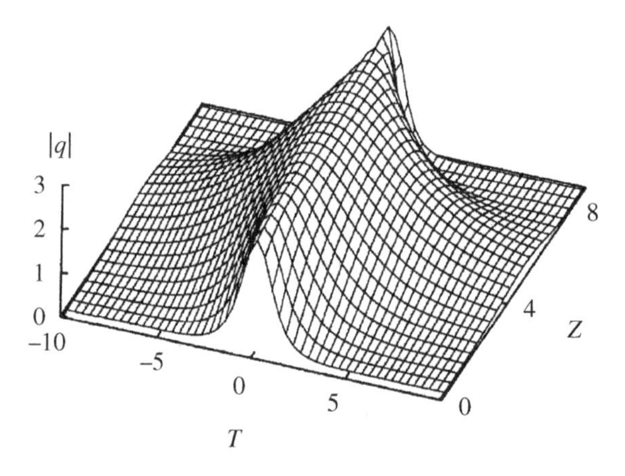

Figure 13.2 Pulsating soliton reconstructed from the model given by Eqs. (13.13), (13.15) and (13.16). The following parameter values are assumed: $\beta = 0.08$, $\varepsilon = 0.66$, $\mu = -0.1$ and $\nu = -0.1$. *Source:* After Ref. [38].

13.2 Plain Pulsating Solitons

A plain pulsating soliton is a localized solution of the CGLE that shows a periodic variation of its width, amplitude, and other pulse-shape parameters. It corresponds to a limit cycle of the infinite-dimensional dynamical system and can be described by a closed loop in the phase space of the system.

Plain pulsating soliton solutions of the cubic-quintic CGLE were first found by Deissler and Brand [41] in the normal dispersion regime. The profile of such solitons shows a periodic variation only in its tails, whereas the energy remains almost constant. In contrast, the plain pulsating solitons found by Akhmediev e al. [1] were observed in the anomalous dispersion regime and both their shape and energy change quite significantly.

Figure 13.3a and b show the amplitude and the spectrum evolution, respectively, of the plain pulsating soliton found in Ref. [1]. The initial condition was given by a unitary sech(T) pulse and the assumed parameters values were the following: $\delta = -0.1$, $\beta = 0.08$, $\varepsilon = 0.66$, $\mu = -0.1$, and $\nu = -0.1$. The pulse shows a periodic behavior, with a period $Z \approx 14$. It has different shapes

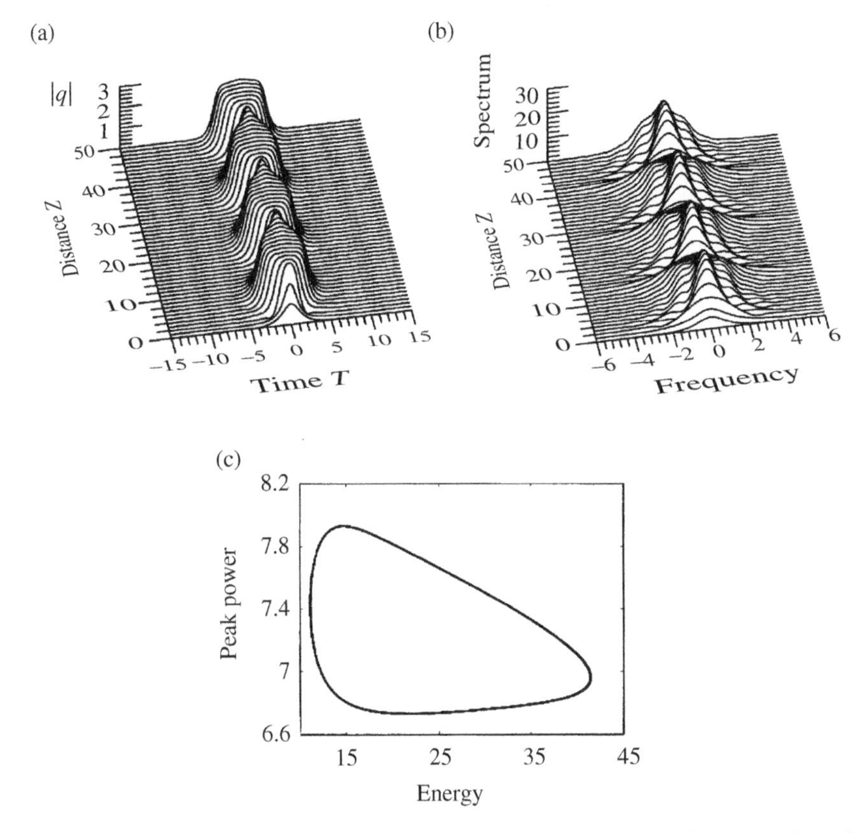

Figure 13.3 (a) Amplitude, (b) spectrum, and (c) peak power against pulse energy for a plain pulsating soliton, for $\delta = -0.1$, $\beta = 0.08$, $\varepsilon = 0.66$, $\mu = -0.1$, and $\nu = -0.1$.

at each position Z, but it recovers exactly its initial shape after a period. The pulse has a single-peak spectrum, which follows a similar periodic behavior, as illustrated in Figure 13.3b). The maximum spectral width occurs at minimum temporal pulse width, and vice-versa. Figure 13.3c shows the soliton peak power, $|q|^2$, against the soliton energy, given by Eq. (13.9). The trajectory corresponds to a cycle repeating itself indefinitely, which means that the pulse energy and peak power also have a periodic behavior.

The first experimental observation of soliton pulsations was reported by Soto-Crespo et al. in a mode-locked fiber ring laser [9]. Nevertheless, the lack of a high-resolution real-time diagnostic method has precluded detailed characterization of pulsating behaviors. This limitation has been broken through by the development of a novel powerful real-time spectra measurement technique called dispersive Fourier transform (DFT) [10], with which it is possible to observe the real-time spectral dynamics of pulsating solitons [22–24, 27, 42, 43].

13.2.1 Impact of Higher-Order Effects

An interesting question concerns the impact on the behavior of a plain pulsating pulse of the higher-order nonlinear and dispersive effects described in Chapter 9, namely, the intrapulse Raman scattering, the self-steepening, and the third-order dispersion effects. With all these effects included, the cubic-quintic CGLE given by Eq. (12.1) is generalized and becomes [44–50]:

$$i\frac{\partial q}{\partial Z}+\frac{D}{2}\frac{\partial^2 q}{\partial T^2}+|q|^2 q=i\delta q+i\beta\frac{\partial^2 q}{\partial T^2}+i\varepsilon|q|^2 q+i\mu|q|^4 q-\nu|q|^4 q+HOEs \tag{13.33a}$$

where

$$HOEs=i\delta_3\frac{\partial^3 q}{\partial T^3}-is\frac{\partial\left(|q|^2 q\right)}{\partial T}+\tau_R q\frac{\partial|q|^2}{\partial T} \tag{13.33b}$$

The parameters δ_3, s, and τ_R, govern, respectively, the effects of third-order dispersion (TOD), self-steepening (SST), and intrapulse Raman scattering (IRS), as defined in Section 9.1. The influence of the TOD, SST, and IRS both on stationary and pulsating CGLE soliton solutions have has been studied by several authors during the recent years [44–72].

The upper row in Figure 13.4 shows the amplitude contour plot for the plain pulsating pulse shown in Figure 13.3 in the presence of a) IRS ($\tau_R = 0.01$), b) negative TOD ($\delta_3 = -0.05$), and c) IRS plus TOD, respectively. The lower row Figure 13.4 shows the peak power, $|q|^2$, against the soliton energy, E, for the same cases, respectively. The other parameter values are the same as for Figure 13.3. Under the IRS effect alone, Figure 13.4a shows that the pulse moves to the right at a constant velocity, whereas the pulse of Figure 13.3a evolves with zero velocity. In the presence of negative TOD, the pulse moves to the left, at constant velocity, as illustrated in Figure 13.4b. In both cases, the oscillations of the pulse shape are slightly reduced, compared with the case of Figure 13.3a. However, if both effects act together, the pulse shape oscillations are reduced more significantly, as illustrated by Figure 13.4c. This behavior can also be observed from Figure 13.4d–f. In particular, when IRS and negative TOD act simultaneously, Figure 13.4f shows that the trajectory slowly spirals inward, converging eventually to an inner limit cycle.

The upper row in Figure 13.5 shows the amplitude contour plot for the plain pulsating pulse in the presence of (a) IRS ($\tau_R = 0.01$) and SST ($s = 0.005$); (b) IRS, SST, and positive TOD ($\delta_3 = +0.05$);

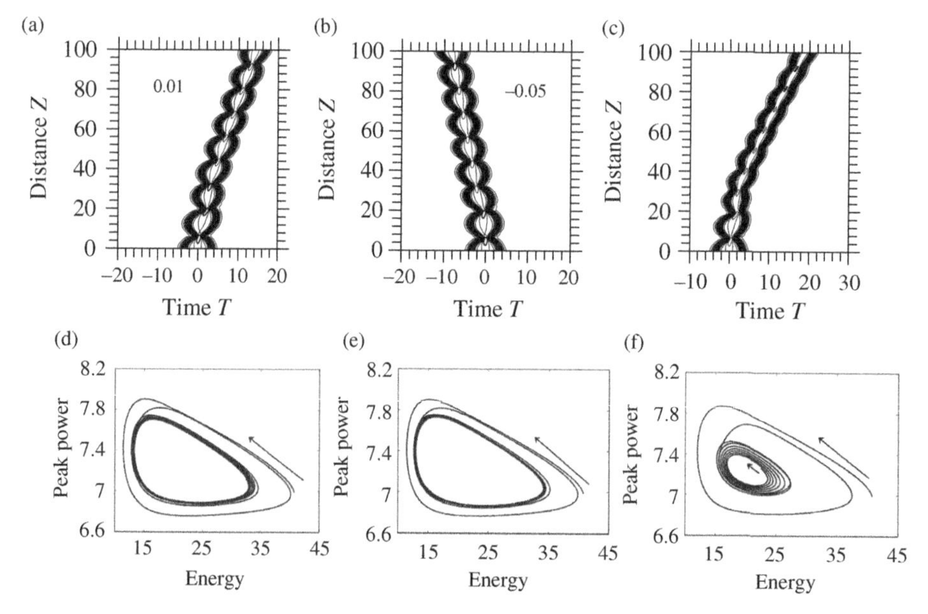

Figure 13.4 Amplitude contour plot (upper row) and pulse peak power against pulse energy (lower row) for a plain pulsating pulse in the presence of (a, d) IRS ($\tau_R = 0.01$), (b, e) negative TOD ($\delta_3 = -0.05$), (c, f) IRS and negative TOD.

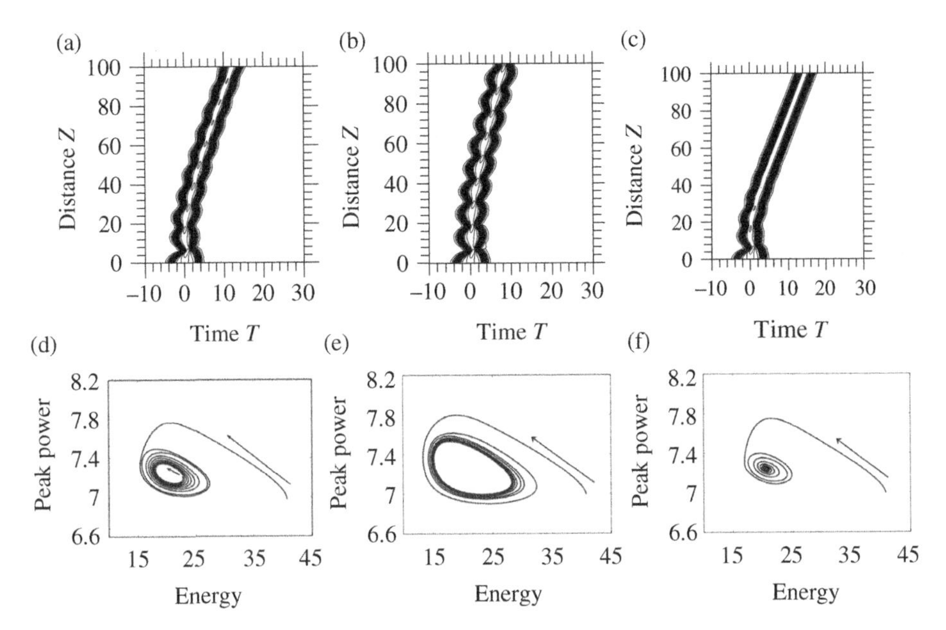

Figure 13.5 Amplitude contour plot for a pulsating pulse in the presence of (a) IRS (τ_R = 0.01) and SST (s = 0.005); (b) IRS, SST, and positive TOD (δ_3 = + 0.05); and (c) IRS, SST, and negative TOD (δ_3 = − 0.05). In (d)–(f), the peak power is shown against the pulse energy for the same cases, respectively. The other parameter values are the same as for Figure 13.3.

and (c) IRS, SST, and negative TOD ($\delta_3 = -0.05$). Figure 13.5a shows that the combination of both nonlinear gradient terms is effective in reducing the amplitude oscillations. However, in the presence of positive TOD the oscillations amplitude is increased, as shown by Figure 13.5b). In contrast, if negative TOD is added to the gradient terms the oscillations amplitude is drastically reduced, as illustrated by Figure 13.5c. In this case, Figure 13.5f shows that the cycle trajectory quickly spirals inward to a fixed point, corresponding to a stationary fixed-shape pulse.

13.3 Creeping Solitons

A creeping soliton is a pulsating localized solution that changes its shape periodically and shifts a finite distance in the transverse direction after each pulsation. It is a rectangular pulse with two fronts and one sink on the top [2, 50], presenting a shape that resembles that of the composite soliton [73, 74]. Creeping solitons were first observed numerically in [1] and their existence was later numerically confirmed for various dissipative systems [2, 44, 50].

Figure 13.6a and b show the amplitude and spectrum evolution, respectively, of a creeping soliton, considering the following set of parameters: $\delta = -0.1$, $\beta = 0.101$, $\varepsilon = 1.3$, $\mu = -0.3$, and $\nu = -0.101$. We observe that the two fronts pulsate back and forth asymmetrically relatively to the sink at the two sides of the pulse. Due to this asymmetry, the center of mass position of the whole soliton shifts a finite distance in the transverse direction after each period of oscillation, which generally determines a nonzero constant velocity. The pulse spectrum oscillates between a single- and a dual-peak spectrum, with three small peaks in the middle, as can be seen from Figure 13.6b. The trajectories represented in Figure 13.6c seem to evolve to a limiting cycle. In this

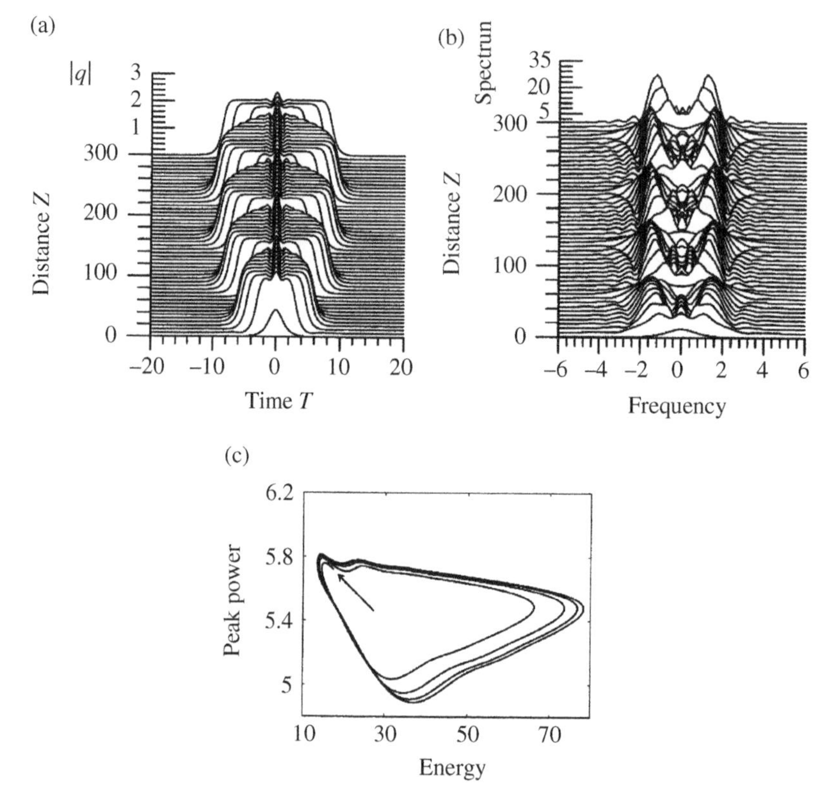

Figure 13.6 (a) Amplitude, (b) spectrum, and (c) peak power against pulse energy for a creeping soliton, for $\delta = -0.1$, $\beta = 0.101$, $\varepsilon = 1.3$, $\mu = -0.3$, and $\nu = -0.101$.

process, the maximum value of the pulse energy grows slightly, while the minimum pulse energy remains almost constant. On the other hand, the minimum value of the pulse peak power tends to decrease, whereas its maximum value remains nearly constant.

Creeping solitons with several frequencies in their motion can also be observed. Figure 13.7 illustrates the case of a creeping soliton having two frequencies. We observe that the pulse moves forth and back around a fixed point, instead of having a constant velocity. Each frequency is related to a particular type of motion. In this case, pulsations occur with the higher frequency, whereas the change of the soliton velocity follows the lower frequency (longer period).

In a 2020 experiment, the motion dynamics of the creeping soliton have been demonstrated in a passively mode-locked fiber laser in the negative-dispersion regime by utilizing a Raman-assisted temporal magnifier system together with the dispersive Fourier transform technique [29]. The periodical variation of pulse width, peak power, and motion range could be observed in real time, while the corresponding spectral evolution exhibited breathing dynamics.

13.3.1 Impact of Higher-Order Effects

Figure 13.8 shows the evolution of the creeping pulse amplitude (upper row) in the presence of (a) IRS ($\tau_R = 0.004$), (b) IRS and positive TOD ($\delta_3 = +0.01$), and (c) IRS and negative TOD ($\delta_3 = -0.01$). The values of the remaining parameters are the same as for Figure 13.6. We observe from Figure 13.8a that even for a relatively low value of the IRS coefficient, the creeping soliton is

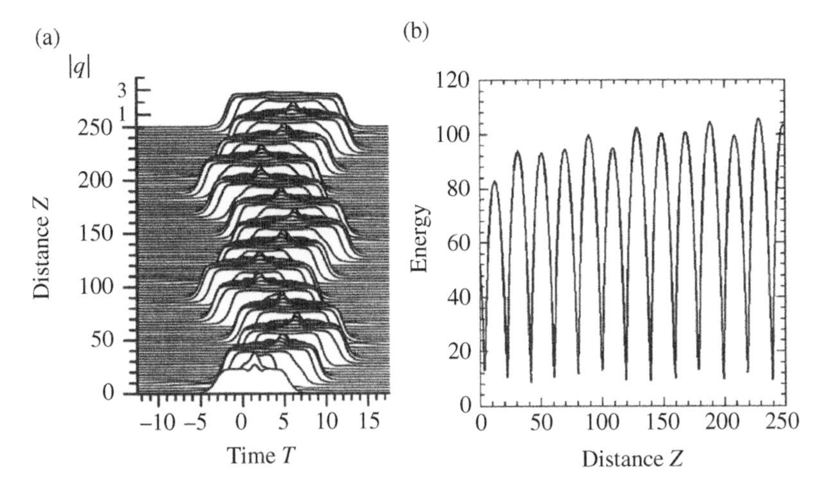

Figure 13.7 Evolution of the (a) profile and (b) energy of a periodic creeping soliton, for the following values of the CGLE parameters: $\delta = -0.1$, $\beta = 0.08$, $\varepsilon = 0.832$, $\mu = -0.11$, and $\nu = -0.08$.

converted into a fixed shape soliton. This is confirmed by Figure 13.8d, in which the cycle trajectory of the peak power versus pulse energy spirals inward to a fixed point. The resulting pulse moves to the right, at constant velocity. Figure 13.8b shows that if both IRS and positive TOD effects act simultaneously, the pulse achieves also a fixed shape, while moving to the right at constant velocity. Figure 13.8e confirms that the pulse quickly evolves to a fixed point in this case. However, a rather different scenario occurs if the pulse propagates under the influence of IRS and negative TOD (δ_3). In this case, Figure 13.8c and f show that the oscillations of both the pulse average energy and peak power increase in amplitude during propagation.

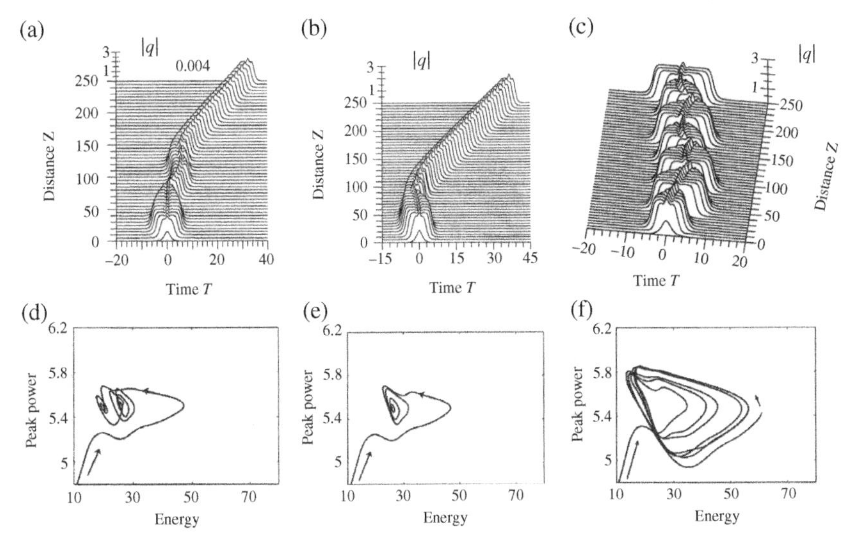

Figure 13.8 Amplitude evolution (upper row) and peak power versus energy (lower row) for a creeping soliton in the presence of (a, d) IRS ($\tau_R = 0.004$), (b, e) IRS and positive TOD ($\delta_3 = +0.01$), and (c, f) IRS and negative TOD ($\delta_3 = -0.01$). The remaining values of the parameters are the same as for Figure 13.6.

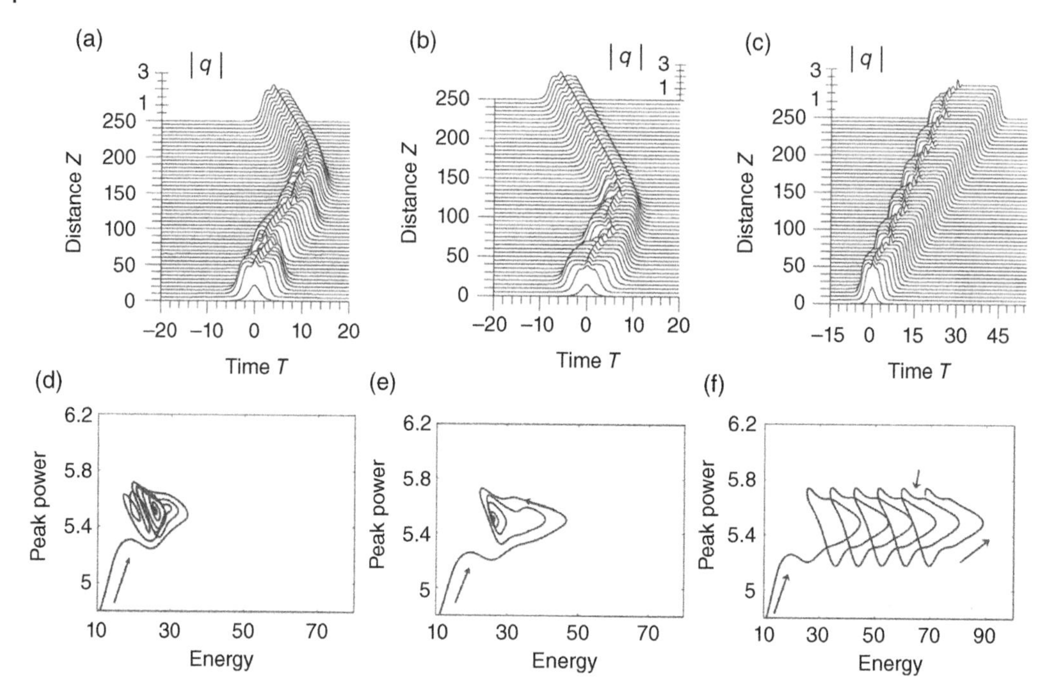

Figure 13.9 Amplitude contour plot for creeping pulse, in the presence of (a) IRS (τ_R = 0.004) and SST (s = 0.002); (b) IRS, SST, and positive TOD (δ_3 = + 0.01); and (c) IRS, SST, and negative TOD (δ_3 = − 0.01). In (d)–(f) the peak power is shown against the pulse energy for the same cases, respectively. The remaining values of the parameters are the same as for Figure 13.6.

Figure 13.9 illustrates the amplitude evolution of the creeping soliton in the presence of (a) IRS ($\tau_R = 0.004$) and SST ($s = 0.002$); (b) IRS, SST, and positive TOD ($\delta_3 = +0.01$); and (c) IRS, SST, and negative TOD ($\delta_3 = -0.01$). It can be seen from Figure 13.9a that the pulse propagating in the presence of IRS and SST converges to a fixed-shape pulse. The same result can be achieved in the presence of IRS, SST, and positive TOD, as illustrated by Figure 13.9b. However, in the presence of negative TOD, Figure 13.9c shows that the front on the left-hand side pulsates back and forth periodically relative to the sink, while the front on the right-hand side spreads with a fixed velocity far away from the sink. Figure 13.9f shows that, in this case, the pulse peak power oscillates, while the mean energy grows without limit, due to the right front expansion.

13.4 Chaotic Solitons

Among the pulsating CGLE solutions, chaotic but localized pulses can also be found [2, 75]. Their profiles evolve during propagation, but never repeat. These pulses can arise as a result of the general dynamics of a passively mode locked fiber laser, namely if there are several periods involved in these dynamics.

An example of a chaotic soliton is illustrated in Figure 13.10a, from which we observe that the pulse shape remains localized and is symmetric. However, in other cases, an asymmetric soliton profile has been observed. In spite of being chaotic, the pulse shape is constrained within certain limits, which means that its parameters remain confined to a given region in a phase space. This is

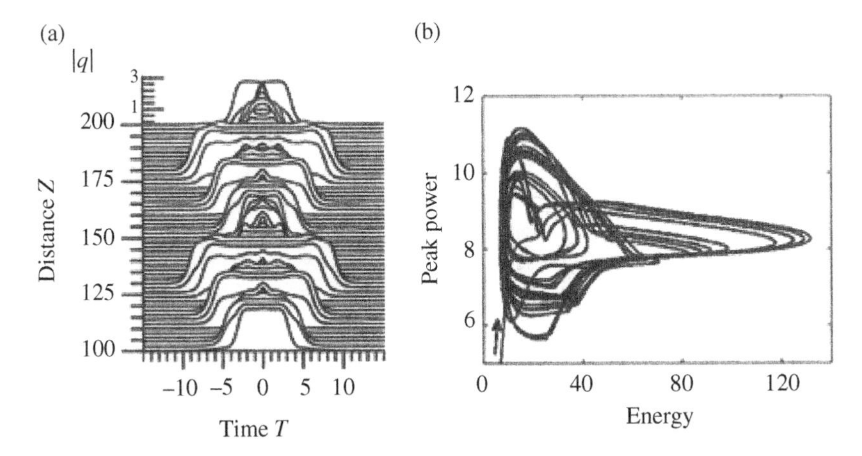

Figure 13.10 (a) Amplitude and (b) peak power against pulse energy of a chaotic soliton, assuming the following parameter values: $\delta = -0.1$, $\beta = 0.04$, $\varepsilon = 0.75$, $\mu = -0.1$, and $\nu = -0.08$.

illustrated by Figure 13.10b, which shows the soliton peak power against the soliton energy. The trajectory was obtained from a simulation over 200 distances, for an initial condition given by a unitary sech profile. It can be seen that such trajectory is not a cycle that repeats itself, as should be expect for a periodic solution. Instead, it occupies a certain region in the plot, which tends to be filled if the propagation distance is increased [51].

The impact of the higher-order effects (HOEs) on the chaotic soliton of Figure 13.10 is illustrated in Figure 13.11, considering the parameter values $s = 0.015$, $\delta_3 = -0.00375$ and $\tau_R = 0.02$. Two different pulses are assumed as the initial condition, as represented in Figure 13.11b: a narrow chaotic pulse (NP) and a wide chaotic pulse (WP), corresponding to the chaotic pulse profiles given by Figure 13.10a at $Z = 140$ and $Z = 130$, respectively. These profiles correspond to extreme values of pulse width and energy. In spite of this difference in the initial chaotic pulses,

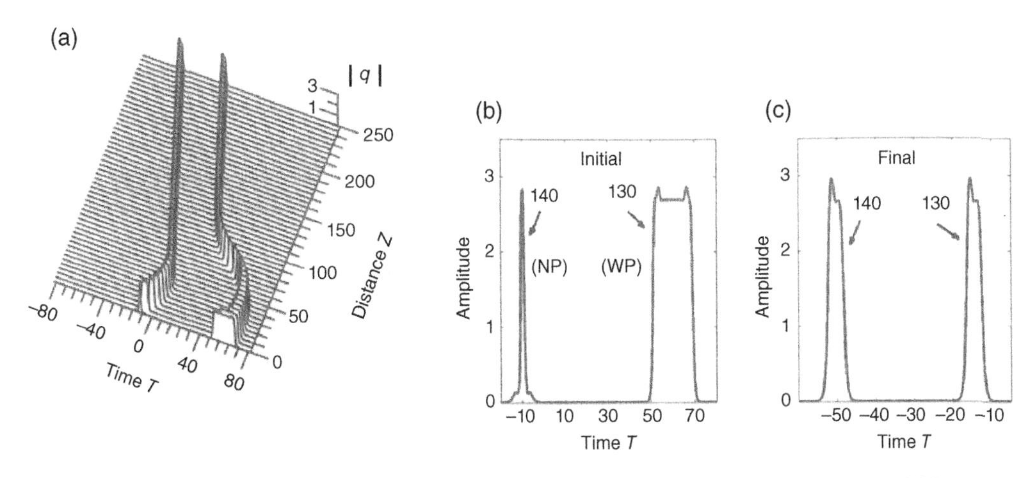

Figure 13.11 (a) Amplitude evolution for the chaotic soliton in the presence of IRS, negative TOD, and SST, considering two different initial conditions. (b) Initial pulses profiles, namely, a narrow chaotic pulse (NP) and a wide chaotic pulse (WP). (c) Final pulse profiles. The parameters for the higher-order effects are: $\tau_R = 0.02$, $\delta_3 = -0.00375$, and $s = 0.015$. The remaining parameter values are the same as for Figure 13.10.

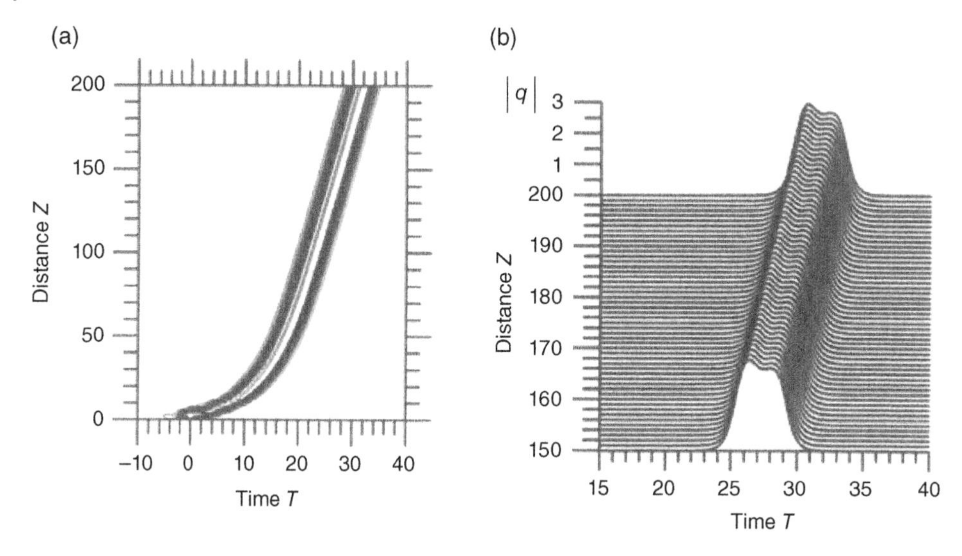

Figure 13.12 (a) Amplitude contour plot and (b) detail of the amplitude evolution, for a chaotic pulsating soliton in the presence of higher-order effects. The higher-order parameter values are $\tau_R = 0.025$, $\delta_3 = 0.03$, and $s = 0.015$. The remaining parameter values are the same as for Figure 13.10.

we observe that the higher-order effects are able to reshape them and make them to converge to the same final fixed-shape solution. However, in this process, the narrow chaotic pulse shows a faster convergence than the wide pulse. The two final pulse shapes are illustrated in by Figure 13.11c.

Figure 13.12 illustrates the evolution of the narrow chaotic pulse towards its final fixed-shape profile, considering the following parameter values for the higher-order effects: $\tau_R = 0.025$, $\delta_3 = 0.03$, and $s = 0.015$. It can be seen that the convergence for the final fixed-shape pulse occurs in an early stage of propagation. The final pulse has a dual-peak profile and moves at a constant velocity, which is smaller than the group velocity. Figure 13.12b shows a detail of the pulse propagation, for $150 < Z < 200$. The pulse shape and peak power are similar to those of the pulses represented in Figure 13.11. However, the pulse velocity is different, which is due to the opposite sign of TOD.

In practical applications, it becomes important to know the regions of existence of fixed-shape solutions. In the absence of higher-order effects, the regions where chaotic pulsating solitons can be found are regions in a five-dimensional parameter space, i.e., in the space defined by δ, β, ε, μ, and ν. However, if higher-order effects are taken into account, the parameter space becomes an eight-dimensional space.

Fixing the five parameters corresponding to the original chaotic soliton, as well as the self-steepening parameter ($s = 0.015$), the region of existence becomes a two-dimensional space, defined on the plane (δ_3 τ_R). Such region of existence is illustrated in Figure 13.13. The curves C1 and C2 define the upper and lower limits of the region of fixed-shape pulses, respectively. The two dots correspond to particular solutions, presented in Figures 13.11 and 13.12.

Figure 13.13 shows that both signs of TOD can prevent the chaotic behavior on a pulsating soliton. The line L corresponds to zero TOD and thus characterizes the impact of the gradient terms on the chaotic pulsating soliton. It can be seen that these terms are sufficient to provide convergence of the chaotic pulse to a fixed-shape pulse. It was found that in all the region where fixed-shape pulses exist, the pulse profiles are similar to the ones presented in Figure 13.11c [51].

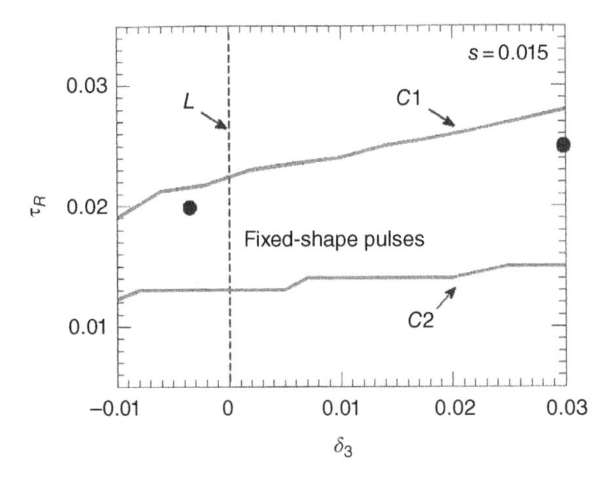

Figure 13.13 Region in the plane (δ_3, τ_R) where fixed-shape solitons can be found for a constant value of SST (s = 0.015). Line L corresponds to zero TOD. The two dots correspond to particular solutions, presented in Figures 13.11 and 13.12.

13.5 Erupting Solitons

One of the most striking forms of pulsating solitons is the erupting soliton, which has been found numerically for the first time in [1], where the soliton eruption manifests itself as a chaotic and quasi-periodic process when the dissipative system is in a meta-stable state. The soliton in such circumstances erupts into pieces in temporal domain abruptly and gradually recover its original state after the eruption, which is similar to exploding behavior and thus regarded as the so-called soliton explosion [1]. Several numerical investigations have been carried out, trying to better understand the intrinsic mechanisms involved in soliton explosions and subsequent recovering of the soliton [3–6, 76–80].

Figure 13.14a–c show the amplitude and the spectrum evolution, respectively, of an erupting soliton, for an initial condition given by a unitary sech profile and considering the following set of parameter values: $\delta = -0.1, \beta = 0.125, \varepsilon = 1.0, \mu = -0.1$, and $\nu = -0.6$. The evolution of the pulse in Figure 13.14a starts from a stationary localized solution. After a while, it becomes covered with small ripples, which seem to move downwards along the two slopes of the soliton, such that after a short distance the pulse becomes covered with this seemingly chaotic structure. When the ripples increase in size, the soliton cracks into pieces, after which it evolves in order to restore its shape. The pulse spectrum exhibits a dual-peak power profile, which evolves as illustrated in Figure 13.14b. Some perturbations appear at the central spectral region, well separated from each of the two main peaks. These perturbations tend to extend over the entire spectrum just before explosions occur.

Despite the return to the same profile, each exploding soliton solution belongs to the class of chaotic solutions. Actually, the erupting process repeats forever, but never exactly in equal periods and none of the explosions repeats the previous one, as observed from Figures 14.14c and 13.14d. In fact, the pulse energy and the pulse peak power have a near periodic behavior, but each trajectory is really different from one explosion to another, having in common only the stationary regime of evolution.

Linear stability analysis was proposed to explain the origin of the soliton explosions [81, 82]. The behavior of the erupting soliton can be studied using such analysis when it is in the laminar

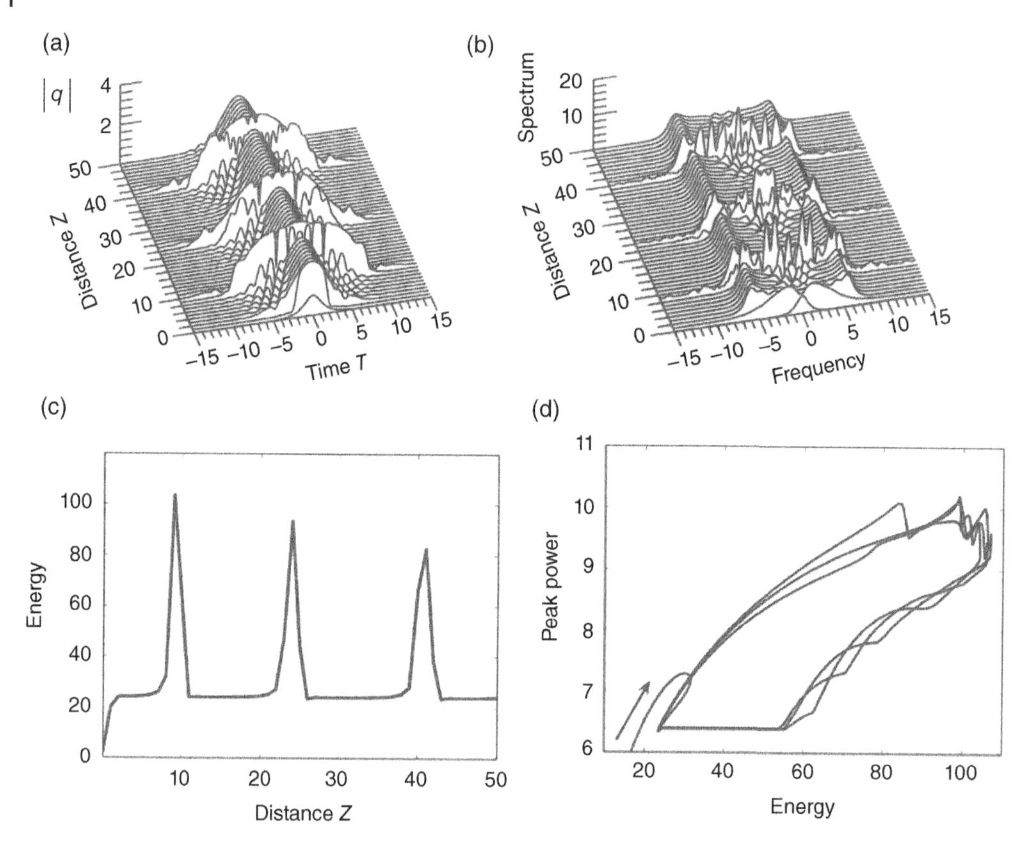

Figure 13.14 (a) Amplitude, (b) spectrum, (c) energy evolution, and (d) peak power against pulse energy of an erupting soliton, for $\delta = -0.1$, $\beta = 0.125$, $\varepsilon = 1.0$, $\mu = -0.1$, and $\nu = -0.6$.

regime [3]. In order to investigate the stability in this regime, we can assume a stationary solution given by $q(Z, T) = q_0 e^{ikZ}$, where $q_0(T)$ is a complex function of T with exponentially decaying tails and k is the propagation constant, assumed to be real. The evolution of the solution in the vicinity of the stationary solution can be described by an expression of the type

$$q(Z, T) = \left[q_0(T) + f(T)e^{\lambda Z} + g(T)e^{\lambda^* Z}\right]e^{ikZ} \tag{13.34}$$

where $f(T)$ and $g(T)$ are small perturbation functions and λ is the associated perturbation growth rate. When the coefficients in the right-hand side of Eq. (12.1) are small, the stability analysis can be done analytically. However, when such coefficients are not small, we must perform numerical calculations in order to obtain the eigenvalues and the eigenfunctions of the linearized problem. It has been found that the full spectrum of an erupting soliton consists of two complex conjugate eigenvalues with positive real parts and a continuous spectrum of complex conjugate eigenvalues, all with negative real parts [3]. It was verified that the eigenfunctions corresponding to the eigenvalues with positive real parts are nonzero mainly in the wings of the soliton [3].

Considering an initial stationary solution with small perturbations, the positive real parts of the eigenvalues will determine an increase of such perturbations. Since the eigenfunctions are almost zero in the central part of the soliton, its center is not affected by the growing instability. When the

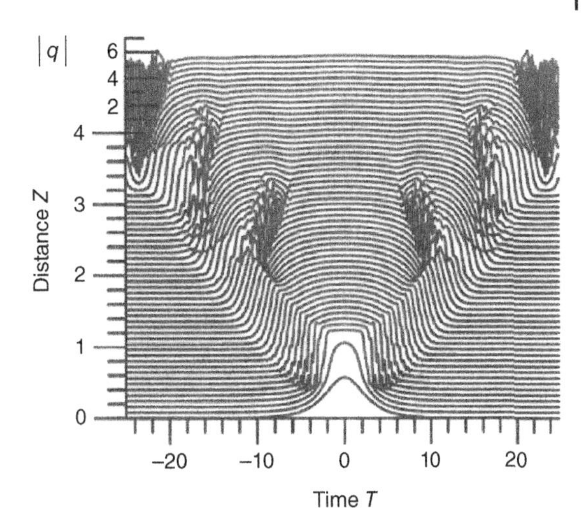

Figure 13.15 Amplitude of an erupting front obtained considering the following parameter values: $\delta = -0.1$, $\beta = 0.08$, $\varepsilon = 1.0$, $\mu = -0.03$, and $\nu = -0.08$.

amplitude of the perturbations becomes similar with the soliton amplitude, the dynamics becomes strongly nonlinear and chaotic. However, the solution remains localized, both in amplitude and in width. Indeed, a positive value of β guarantees that the total width in the frequency domain remains finite, whereas a negative value of μ ensures that the maximum amplitude is limited. As all radiative waves have eigenvalues with negative real parts, they decay and quickly disappear. As a result, the evolution returns to the state of a stationary soliton with a small perturbation that has an eigenvalue with positive real part. As a consequence, the instability will develop again later, thus repeating the whole process.

Besides the erupting solitons, Eq. (12.1) has also erupting front solutions. An example of an erupting front is shown in Figure 13.15. Fronts and solitons can coexist and any of them can be stable or unstable. In general, the transition from solitons to fronts happens along a certain boundary in the parameter space. If explosion existed in the region of solitons, they continue to exist after the transition to fronts. Since the perturbation functions are nonzero mainly at the soliton wings, if the soliton splits into fronts the two parts of the perturbation function move away each attached to the corresponding front.

The possibility of controlling the soliton explosions in passively mode-locked fiber lasers through management of the system parameters has been reported recently [48]. In particular, the important role played by the fiber dispersion, filtering, and nonlinear gain characteristics were demonstrated by the authors. Soliton stabilization can be achieved also using rapid and strong variations of the nonlinearity, the so-called nonlinearity management [83]. This technique can be realized in a fiber ring laser with two-step (focusing and defocusing) variations of nonlinearity.

Figure 13.16 illustrates the filter spectral response, $T(\omega) = \exp(\delta - \beta\omega^2)$, considering the parameters used in Figure 13.14 ($\delta = -0.1$, $\beta = 0.125$) and three other cases: A ($\delta = -0.1$, $\beta = 0.3$), B ($\delta = -0.5$, $\beta = 0.08$), and C ($\delta = -0.5$, $\beta = 0.125$), respectively. Figure 13.16 shows that the peak of the filter response curve decreases when the magnitude of the loss parameter, δ, increases. Furthermore, the width of the curve is reduced when the parameter β increases, meaning a stronger filtering effect.

Changing the filter spectral response has a strong impact on the soliton explosions, as illustrated in Figure 13.17. Taking as reference the case of Figure 13.14, it can be seen from Figure 13.17a that increasing the filter strength β reduces significantly the extension of the laminar stage, whereas the number of explosions remains practically the same. However, increasing

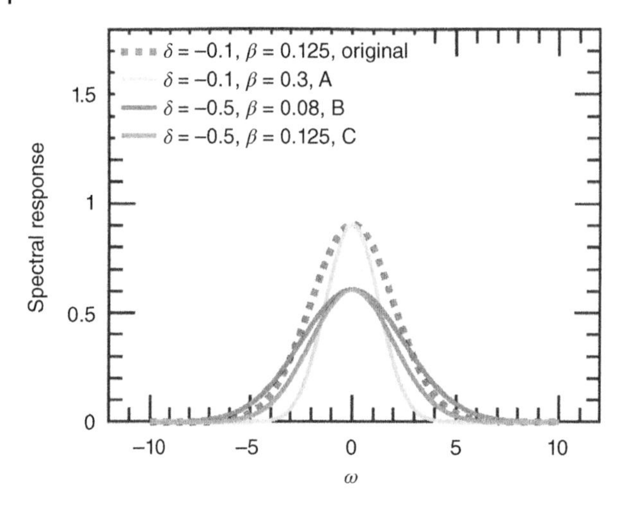

Figure 13.16 Spectral filter profiles corresponding to the original case (dashed curve, $\delta = -0.1, \beta = 0.125$), and three other cases: A ($\delta = -0.1$, $\beta = 0.3$), B ($\delta = -0.5$, $\beta = 0.08$), and C ($\delta = -0.5$, $\beta = 0.125$).

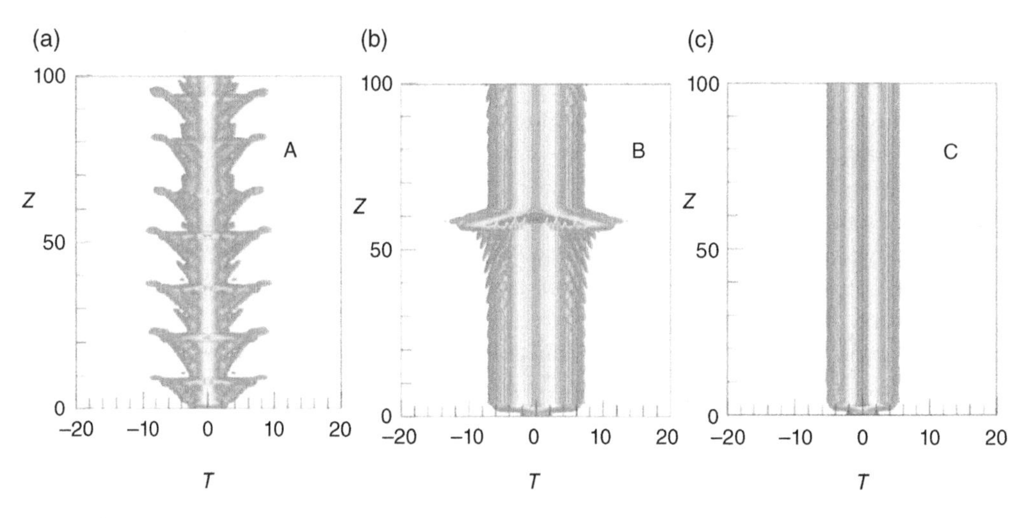

Figure 13.17 Amplitude contour plot of an erupting soliton for three different filter profiles: A ($\delta = -0.1$, $\beta = 0.3$), (b) B ($\delta = -0.5$, $\beta = 0.08$), and (c) C ($\delta = -0.5$, $\beta = 0.125$).

the magnitude of the loss parameter, δ, reduces or eliminates completely such explosions, as observed from Figure 13.17b and c.

Figure 13.18 shows the amplitude contour plot of the erupting soliton for different values of the nonlinear gain parameter. We observe from Figure 13.18a that for $\varepsilon = 0.5$, the number of explosions is reduced in comparison with the case of Figure 13.14 ($\varepsilon = 1$), which is also illustrated by Figure 13.18b. Moreover, both the temporal width of the stationary pulse and the temporal extension of the explosions are enhanced for such lower value of ε. Figure 13.18c shows that increasing the nonlinear gain parameter to $\varepsilon = 1.5$ reduces both the number of explosions and the pulse width. A further increase of the nonlinear gain parameter to $\varepsilon = 2$ suppresses completely the soliton explosions, as shown in Figure 13.18d.

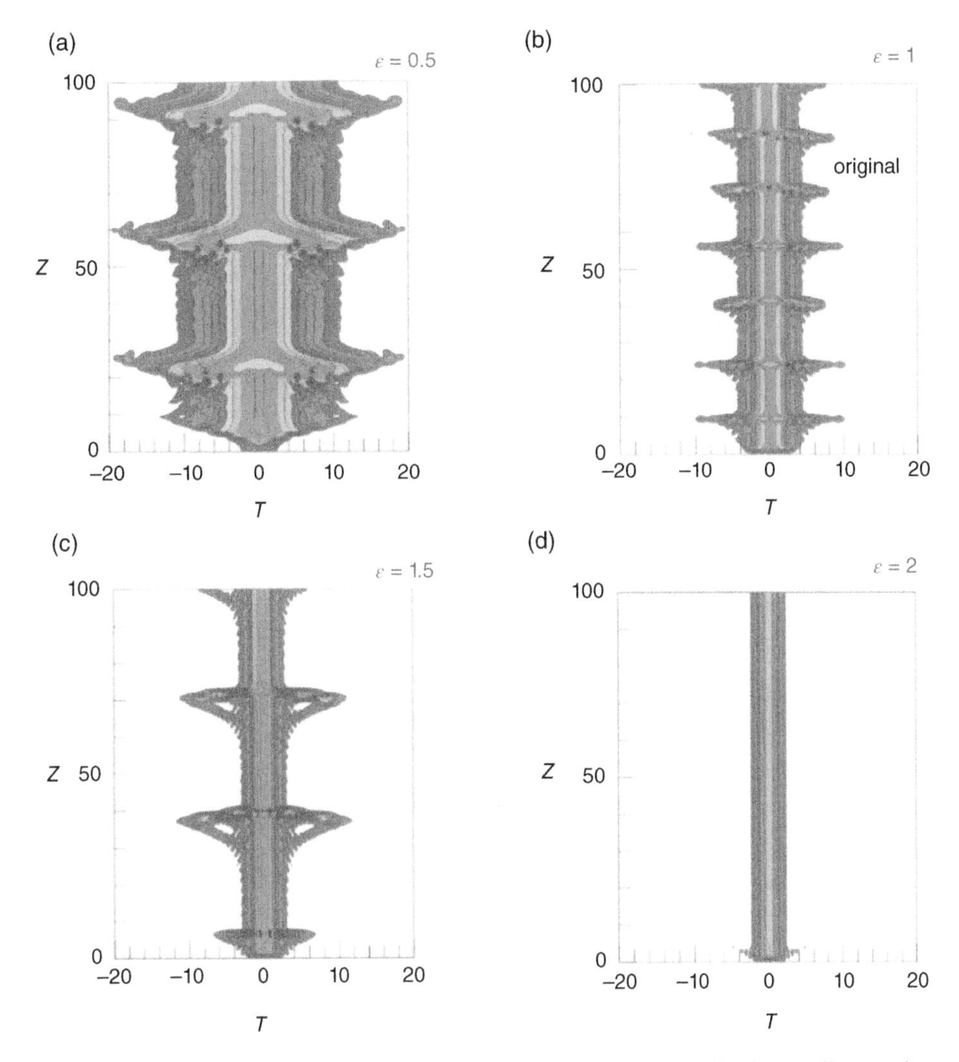

Figure 13.18 Amplitude contour plot of an erupting soliton, for the following nonlinear gain parameter values: (a) $\varepsilon = 0.5$, (b) $\varepsilon = 1.0$, (b) $\varepsilon = 1$, (c) $\varepsilon = 1.5$, and (d) $\varepsilon = 2.0$. The other parameter values are the same as in Figure 13.14.

13.5.1 Impact of Higher-Order Effects

The influence of different higher-order effects on the CGLE erupting solitons was studied numerically in [44–55, 84]. In particular, it was shown that a proper combination of the three HOEs can provide a shape stabilization of an exploding soliton. The transition of erupting solutions under the influence of SST to fixed shape solutions has been characterized by means of subcritical bifurcation controlled by the SST parameter [63]. The existence of a periodic erupting solution has been observed under the influence of IRS and has been related to the sequence of period-halving and period-doubling bifurcations controlled by the IRS parameter [62].

It was found that the higher-order effects can filter in different ways the spectral perturbations that feed the ripples growth, determining the pulse explosions [47]. Figure 13.19 illustrates the impact of each higher-order effect on the dynamics of the erupting soliton. The upper row shows the

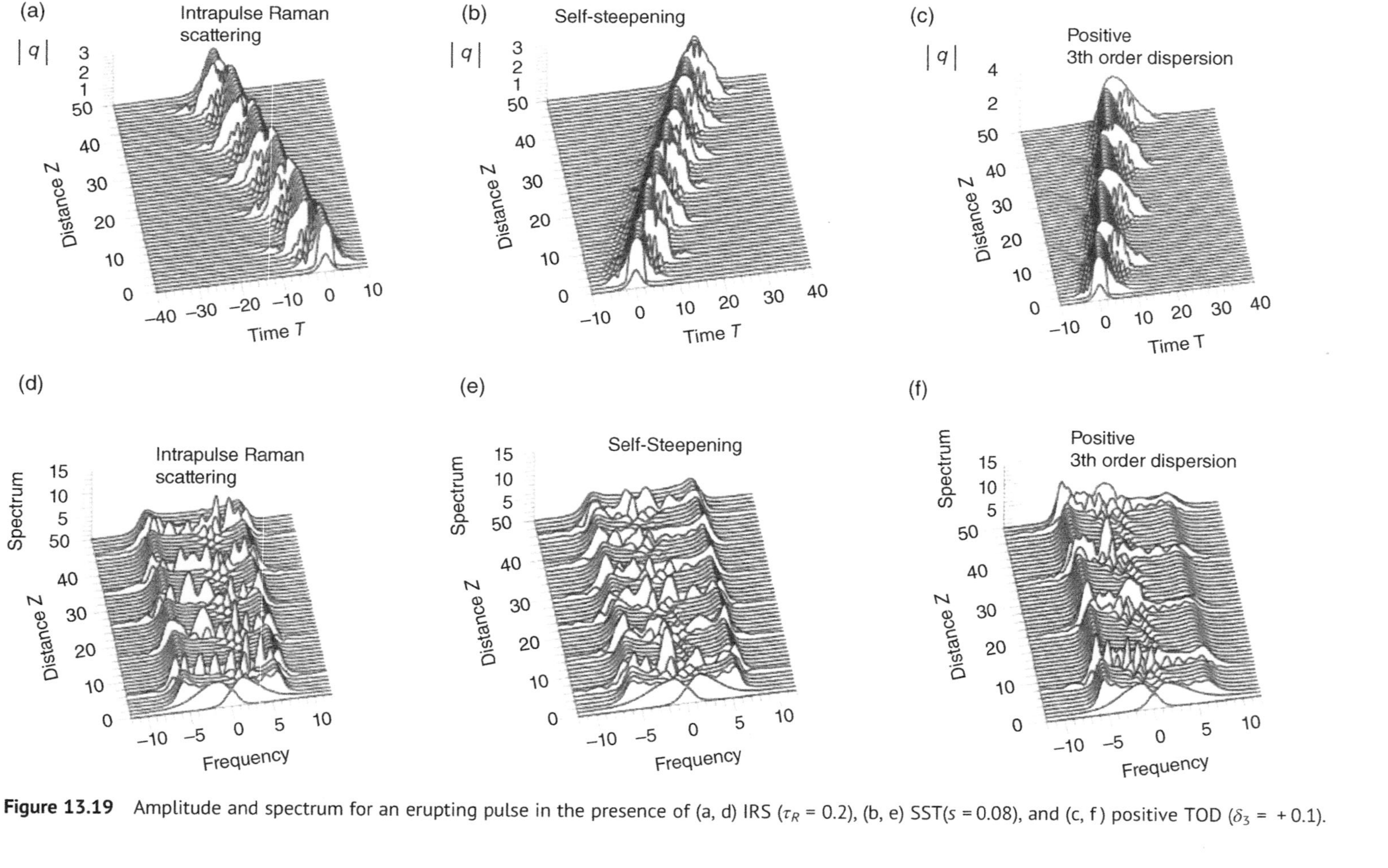

Figure 13.19 Amplitude and spectrum for an erupting pulse in the presence of (a, d) IRS ($\tau_R = 0.2$), (b, e) SST($s = 0.08$), and (c, f) positive TOD ($\delta_3 = +0.1$).

amplitude evolution, whereas the lower row shows the spectrum for an erupting pulse, in the presence of (a, d) IRS ($\tau_R = 0.2$), (b, e) SST($s = 0.08$), and (c, f) positive TOD ($\delta_3 = +0.1$), respectively. The other parameter values and the initial condition are the same as for Figure 13.14.

Figure 13.19a shows that, under the effect of IRS, the pulse moves to the left and only cracks on the leading edge, if compared with the pulse evolution in Figure 13.14a. On the other hand, in the presence of SST or of positive TOD the pulse moves to the right and the explosions only occur on trailing edge, as observed in Figure 13.19b and c, respectively. In the three cases, the pulse moves at constant velocity, and the number of explosions increases in comparison to Figure 13.14a. A dual-pulse spectrum is yet observed in the three cases, but it becomes asymmetric, with one of the peaks higher than the other. In the case of Figure 13.19d, the spectral perturbations just grow on the high frequency region of the spectrum, while explosions occur at the pulse left-hand side, in the time domain. The reverse occurs in the Figure 13.19e and f. Initially, the spectral perturbations appear at the central frequency, but afterwards they spread to the entire spectrum. When they reach a given magnitude, the explosion occurs. After that, the process repeats itself.

The impact of the simultaneous action of the three higher-order effects is illustrated in Figure 13.20a, which shows the evolution of the erupting pulse in the presence of SSFS ($\tau_R = 0.275$), positive TOD ($\delta_3 = 0.1$), and SST ($s = 0.005$). In the temporal domain, it can be seen that the small ripples on both sides of the pulse have been completely removed and a fixed-shape pulse is effectively achieved, moving to the right at a nonzero velocity. In the frequency domain, a clean but asymmetric dual-peak spectrum is obtained, as illustrated in Figure 13.20b. This illustrates the possibility of attenuating the spectral perturbations on both sides of the spectrum and consequently achieving a complete control the soliton explosions by using a proper combination of the three higher-order effects. Numerical simulations have shown that, keeping the remaining parameters as indicated above, a fixed-shape pulse can emerge from the erupting soliton for Raman coefficients in the range $0.1 \leq \tau_R \leq 0.8$. It has been observed that a fixed-shape solution can also be achieved with a negative TOD.

13.5.2 Experimental Observation of Soliton Explosions

Soliton explosions were experimentally observed for the first time by Cundiff et al. [8] in a solid-state, Kerr-lens mode-locked Ti:sapphire laser. The output of the laser was spectrally dispersed

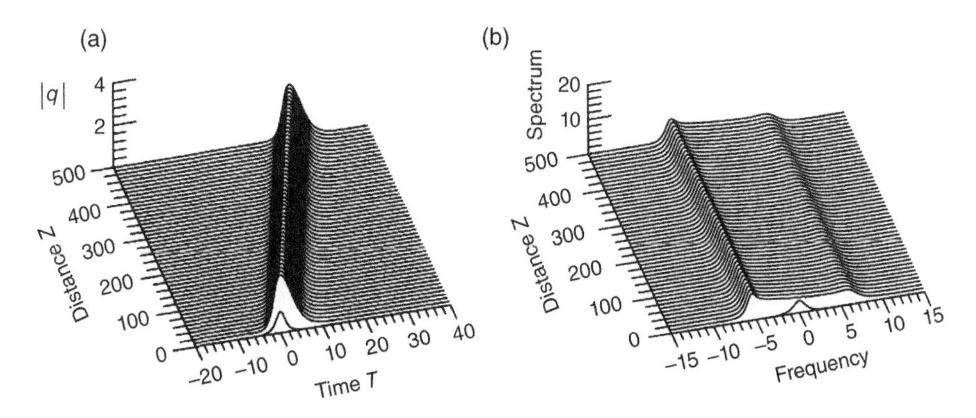

Figure 13.20 (a) Amplitude and (b) spectrum of an erupting soliton in the presence of higher-order effects ($\tau_R = 0.275$, $\delta_3 = 0.1$, $s = 0.005$). The other parameter values are the same as in Figure 13.14.

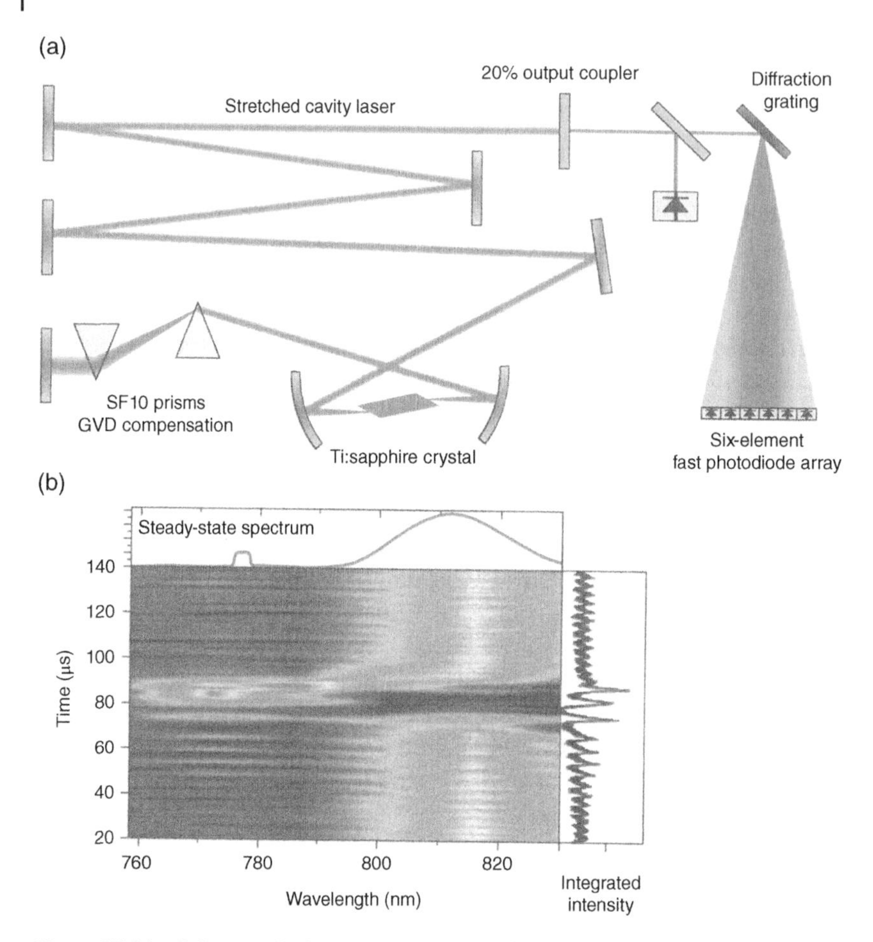

Figure 13.21 Soliton explosions in a mode-locked Ti:sapphire laser. (a) The experimental set-up. GVD, group-velocity dispersion. (b) Fast spectral dynamics displaying a soliton explosion event. *Source:* Figure (a) after Ref. [85]. Figure (b) after Ref. [8].

across a six-element detector array and the corresponding temporally resolved spectrum was measured. Signatures of explosion events were observed even though the detection apparatus was limited to 12 nm spectral resolution and to averaging over approximately five consecutive pulses. Figure 13.21 shows (a) a scheme of the laser and (b) the spectral signature of one soliton explosion.

Later on, with the help of the new DFT technique [10], soliton explosions were also found experimentally in an all-normal dispersion, all-polarization maintaining passively mode-locked Yb-doped fiber laser [11]. Such explosions were observed when the laser was operating in a transition zone between stable mode-locking [86] and noise-like emission [87, 88]. Figure 13.22a concatenates 100 experimentally measured single-shot spectra of consecutive pulses emitted by the laser, from which we can identify seven explosion events [11]. When an explosion occurs, the spectrally broad dissipative soliton collapses into a narrower spectrum with higher amplitude, but after a few roundtrips returns back to its previous state. Figure 13.22b shows four consecutive spectra around a particular explosion event. In addition to the collapse and revival of the pulse at 1028 nm, we can also see how the explosion events trigger the emission of Raman components at 1075 nm. Actually, the

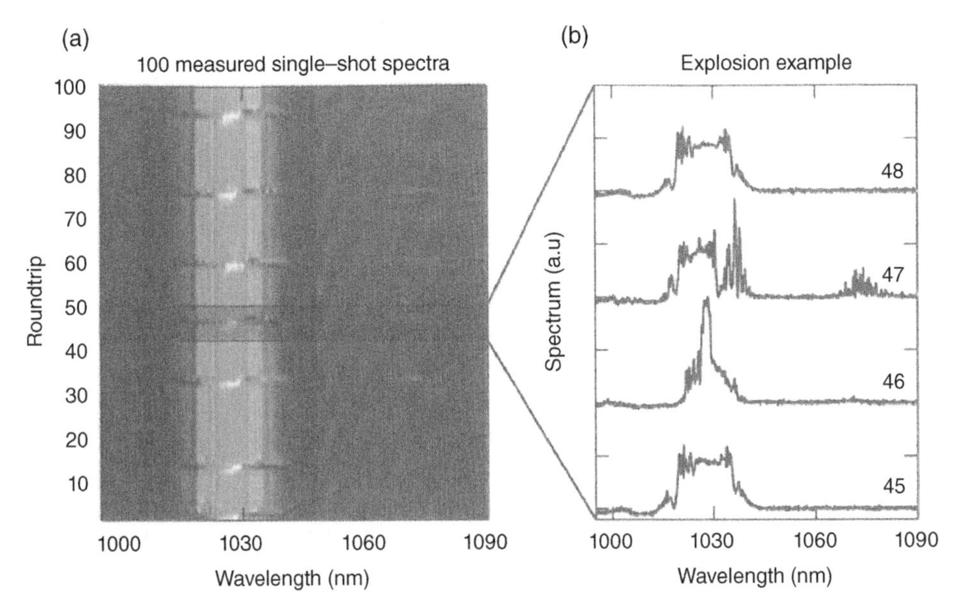

Figure 13.22 (a) Experimentally measured single-shot spectra of 100 consecutive pulses with the laser operating in the transition regime. Seven soliton explosions can clearly be identified. (b) Example spectra at indicated roundtrip numbers around a particular explosion event. *Source:* After Ref. [11].

condensation of energy into a narrow spectral band as a consequence of the explosion constitutes an efficient Raman pump.

The experimental results in [11] have been successfully compared to realistic numerical simulations based on an envelope function approach [89]. The connection between this pulse propagation model and the CGLE with additional higher-order nonlinear and dispersive effects was established in [90].

The use of the DFT technique has boosted the experimental investigations of soliton explosions in mode-locked fiber lasers during the recent years [12, 13, 17, 21, 25, 91, 92]. The results in [11] were expanded in 2016 by the authors considering different cavity configurations and gain levels [12]. Among other observations, it was shown that the probability with which explosions occur depends on the pump power and that the explosion characteristics depend critically on the position in the cavity where the output coupler is located. The evolution of soliton dynamics of an ultrafast fiber laser from steady state to soliton explosions, and to huge explosions by simply adjusting the pump power level has been also observed in another 2016 experiment [13].

The first investigation of soliton explosions in an ultrafast fiber laser in the multi-soliton regime was reported in 2018 [21]. It was demonstrated that explosion of one soliton could be induced by another one through the soliton interactions mediated by the transient gain response of an erbium-doped fiber. The first experimental observation of periodic spectrum changing via soliton explosion in a passively mode-locked fiber laser by a nonlinear polarization evolution was reported also in 2018 [91]. Using time stretch to capture 7220 consecutive single-shot spectra over a 100 μs time window in real time, the soliton explosions were observed in a transition between two different mode-locking states.

In a 2019 experiment, soliton collision-induced explosions were also observed in a mode-locked fiber laser, benefiting from synchronous measurements of the spatio-temporal intensity evolution and the real-time spectra evolution using the DFT technique [25]. Up to seven nonlinear regimes were observed successively in the laser by increasing the pump power, including single-pulse mode locking, standard soliton explosions, noise-like mode locking, stable double pulsing, soliton collision-induced explosions, soliton molecules, and double-pulse noise-like mode locking.

References

1 Soto-Crespo, J., Akhmediev, N., and Ankiewicz, A. (2000). *Phys. Rev. Lett.* **85**: 2937.
2 Akhmediev, N., Soto-Crespo, J., and Town, G. (2001). *Phys. Rev. E.* **63**: 056602.
3 Akhmediev, N. and Soto-Crespo, J. (2003). *Phys. Lett A.* **317**: 287.
4 Akhmediev, N. and Soto-Crespo, J. (2004). *Phys. Rev. E.* **70**: 036613.
5 Soto-Crespo, J. and Akhmediev, N. (2005). *Phys. Rev. Lett.* **95**: 024101.
6 Soto-Crespo, J. and Akhmediev, N. (2005). *Math. Comp. Simulation* **69**: 526.
7 Chang, W., Ankiewicz, A., Akhmediev, N., and Soto-Crespo, J.M. (2007). *Phys. Rev. E* **76**: 016607.
8 Cundiff, S.T., Soto-Crespo, J.M., and Akhmediev, N. (2002). *Phys. Rev. Lett.* **88**: 073903.
9 Soto-Crespo, J.M., Grapinet, M., Grelu, P., and Akmediev, N. (2004). *Phys. Rev. E* **70**: 066612.
10 Goda, K. and Jalali, B. (2013). *Nat. Photonics* **7**: 102.
11 Runge, A.F.J., Broderick, N.G.R., and Erkintalo, M. (2015). *Optica* **2**: 36.
12 Runge, A.F.J., Broderick, N.G.R., and Erkinatalo, M. (2016). *J. Opt. Soc. Am. B* **33**: 46.
13 Liu, M., Luo, A.P., Yan, Y.R. et al. (2016). *Opt. Lett.* **41**: 1181.
14 Herink, G., Jalali, B., Ropers, C., and Solli, D. (2016). *Nat. Photonics* **10**: 321.
15 Wei, X., Li, B., Yu, Y. et al. (2017). *Opt. Express* **25**: 29098.
16 Krupa, K., Nithyanandan, K., and Grelu, P. (2017). *Optica* **4**: 1239.
17 Wang, P., Xiao, X., Zhao, H., and Yang, C. (2017). *IEEE Photonics J.* **9**: 1.
18 Ryczkowski, P., Närhi, M., Billet, C. et al. (2018). *Nat. Photonics* **12**: 221.
19 Chen, H.-J., Liu, M., Yao, J. et al. (2018). *Opt. Express* **26**: 2972.
20 Chen, H.-J., Liu, M., Yao, J. et al. (2018). *IEEE Photonics J.* **10**: 1.
21 Yu, Y., Luo, Z.-C., Kang, J., and Wong, K.K.Y. (2018). *Opt. Lett.* **43**: 4132.
22 Wang, Z., Wang, Z., Liu, Y. et al. (2018). *Opt. Lett.* **43**: 478.
23 Du, Y., Xu, Z., and Shu, X. (2018). *Opt. Lett.* **43**: 3602.
24 Wei, Z.-W., Liu, M., Ming, S.-X. et al. (2018). *Opt. Lett.* **43**: 5965.
25 Peng, J. and Zeng, H. (2019). *Commun. Phys.* **2**: 34.
26 Wang, G., Chen, G., Li, W., and Zeng, C. (2019). *IEEE J. Sel. Top. Quantum Electron.* **25**: 1.
27 Wang, X., Liu, Y.-G., Wang, Z. et al. (2019). *Opt. Express* **27**: 17729.
28 Wei, Z.W., Liu, M., Ming, S.X. et al. (2020). *Opt. Lett.* **45**: 531.
29 Zhang, Y., Cui, Y., Huang, L. et al. (2020). *Op. Lett.* **45**: 6246.
30 Manousakis, M., Papagiannis, P., Moshonas, N., and Hizanidis, K. (2001). *Opt. Commun.* **198**: 351.
31 Ankiewicz, A., Akhmediev, N., and Devine, N. (2007). *Optical Fib. Technol.* **13**: 91.
32 Mancas, S.C. and Choudhury, S.R. (2007). *Theor. Math. Phys.* **152**: 1160.
33 Mancas, S.C. and Choudhury, S.R. (2009). *Chaos, Solitons Fractals* **40**: 91.
34 Ferreira, M.F. (2018). *IET Optoelectronics* **12**: 122.
35 Maimistov, A.I. (1993). *J. Exp. Theor. Phys.* **77**: 727.
36 Zhuravlev, M.N. and Ostrovskaya, N.V. (2004). *J. Exp. Theor. Phys.* **99**: 427.
37 Tsoy, E. and Akhmediev, N. (2005). *Phys. Lett. A* **343**: 417.
38 Tsoy, E., Ankiewicz, A., and Akhmediev, N. (2006). *Phys. Rev. E.* **73**: 036621.
39 Tsoy, E. and de Sterke, C.M. (2000). *Phys. Rev. E* **62**: 2882.
40 Abdullaev, F.K., Navotny, D.V., and Baizakov, B.B. (2004). *Physica D* **192**: 83.

41 Deissler, R. and Brandt, H. (1994). *Phys. Rev. Lett.* **72**: 478.
42 Chen, H.-J., Tan, Y.-J., Long, J.-G. et al. (2019). *Opt. Express* **27**: 28507.
43 Du, W., Li, H., Li, J. et al. (2020). *Opt. Lett.* **45**: 5024.
44 Tian, H., Li, Z., Tian, J. et al. (2004). *Appl. Phys. B* **78**: 199.
45 Song, L., Li, L., Li, Z., and Zhou, G. (2005). *Opt. Commun.* **249**: 301.
46 Latas, S.C. and Ferreira, M.F. (2010). *Opt. Lett.* **35**: 1771.
47 Latas, S.C. and Ferreira, M.F. (2011). *Opt. Lett.* **36**: 3085.
48 Latas, S.C. and Ferreira, M.F. (2020). *Fiber Integr. Opt.* **39**.
49 Ferreira, M.F. (2011). *Nonlinear Effects in Optical Fibers*. Hoboken, NJ: John Wiley & Sons.
50 Latas, S.C., Ferreira, M.F., and Facão, M.V. (2011). *Appl. Phys. B* **104**: 131.
51 Latas, S.C. and Ferreira, M.F. (2012). *Opt. Lett.* **37**: 3897.
52 Facão, M.V. and Carvalho, M.I. (2012). *Phys. Lett. A* **376**: 950.
53 Latas, V. and Ferreira, M.F. (2012). *SPIE Proc.* **8434**: 84340L.
54 Latas, S.C. and Ferreira, M.F. (2018). *J. Nonlinear Opt. Phys. Mat.* **27**: 1850008.
55 Uzunov, I., Georgiev, Z.D., and Arabadzhiev, T.N. (2018). *Phys. Rev. E* **97**: 052215.
56 Horikis, T.P. and Ablowitz, M.J. (2014). *J. Opt. Soc. Am. B* **31**: 2748.
57 Tian, H., Li, Z., Hu, Z. et al. (2003). *J. Opt. Soc. Am. B* **20**: 59.
58 Latas, S.C. and Ferreira, M.F. (2005). *Opt. Commun.* **251**: 415.
59 Latas, S.C. and Ferreira, M.F. (2007). *J. Math. Comp. Simul.* **74**: 379–387.
60 Facao, M.V. and Carvalho, M.I. (2011). *Phys. Lett. A* **375**: 2327.
61 Uzunov, I.M., Georgiev, Z.D., and Arabadzhiev, T.N. (2014). *Phys. Rev. E* **90**: 0429067.
62 Cartes, C. and Descalzi, O. (2015). *Fiber Integrated. Opt.* **34**: 14.
63 Cartes, C. and Descalzi, O. (2014). *Eur. Phys. J.: Spec. Top.* **223**: 91.
64 Gagnon, L. and Belanger, P.A. (1991). *Phys. Rev. A* **43**: 6187.
65 Khatri, F.I., Moores, J.D., Lenz, G., and Haus, H.A. (1995). *Opt. Commun.* **114**: 447.
66 Li, Z., Li, L., Tian, H. et al. (2002). *Phys. Rev. Lett.* **89**: 263901.
67 Bélanger, P.A. (2006). *Opt. Express* **14**: 12174.
68 Facão, M., Carvalho, I., Latas, S., and Ferreira, M.F. (2010). *Phys. Lett. A* **374**: 4844.
69 Uzunov, I.M., Arabadzhiev, T.N., and Georgiev, Z.D. (2015). *Proc. SPIE* **9447**: 94471G.
70 Uzunov, I.M., Arabadzhiev, T.N., and Georgiev, Z.D. (2015). *Opt. Fiber Technol.* **24**: 15.
71 Achilleos, V., Bishop, A.R., Diamantidis, S. et al. (2016). *Phys. Rev. E* **94**: 012210.
72 Latas, S.C., Ferreira, M.F., and Facão, M. (2017). *J. Opt. Soc. Am. B* **34**: 1033.
73 Akhmediev, N. and Ankiewicz, A. (1997). *Solitons, Nonlinear Pulses and Beams*. Chapman & Hall.
74 Latas, S.C. and Ferreira, M.F. (2005). *Opt. Fiber Technol.* **11**: 292.
75 Akhmediev, N. and Ankiewicz, A. (ed.) (2005). *Dissipative Solitons*. Berlin Heidelberg: Springer-Verlag.
76 Descalzi, O. and Brand, H.R. (2010). *Phys. Rev. E* **82**: 026203.
77 Descalzi, O., Cartes, C., Cisternas, J., and Brand, H.R. (2011). *Phys. Rev. E* **83**: 056214.
78 Cartes, C., Descalzi, O., and Brand, H. (2012). *Phys. Rev. E* **85**: 015205.
79 Cisternas, J. and Descalzi, O. (2013). *Phys. Rev. E* **88**: 022903.
80 Chang, W. and Akhmediev, N. (2014). *Advanced Photonics*, NM3A.6. Barcelona: Optical Society of America.
81 Kalashnikov, V.L. (2012). *Solid State Lasers* (ed. A. Al-Kursan), 145. Intech Open.
82 Rozanov, N. (1997). *Optical Bistability and Hysteresis in Distributed Nonlinear Systems*. Moscow, Nauka: Physical and Mathematical Literature Publishing Company.
83 Abdullaev, F., Tadjimuratov, S., and Abdumalikov, A. (2021). *Optik* **228**: 166213.
84 Gurevich, S.V., Schelte, C., and Javaloyes, J. (2019). *Phys. Rev. A.* **99**: 0618013.

85 Grelu, P. and Akhmediev, N. (2012). *Nat. Photonics* **6** (2): 84–92.
86 Erkintalo, M., Aguergaray, C., Runge, A., and Broderick, N. (2012). *Opt. Express* **20**: 22669.
87 Aguergaray, C., Runge, A., Erkintalo, M., and Broderick, N. (2013). *Opt. Lett.* **38**: 2644.
88 Runge, A., Aguergaray, C., Broderick, N., and Erkintalo, M. (2014). *Opt. Lett.* **39**: 319.
89 Runge, A.F.J., Aguergaray, C., Provo, R. et al. (2014). *Opt. Fiber Technol.* **20**: 657.
90 Cartes, C. and Descalzi, O. (2016). *Phys. Rev. A* **93**: 031801.
91 Suzuki, M., Boyraz, O., Asghari, H. et al. (2018). *Opt. Lett.* **43**: 1862.
92 Wei, Z.-W., Liu, M., Ming, S.-X. et al. (2019). *Opt. Lett.* **45**: 531.

14

Soliton Fiber Lasers

A soliton fiber laser is defined as any nonlinear optical oscillator containing optical fibers, and which generates optical pulses due to a balance between various sources of linear and nonlinear gain, loss, self-phase modulation, and dispersion. The variety of such lasers is very large, and so for the most part of this chapter we shall consider mainly lasers in which the gain medium is an optical fiber amplifier. Actually, owing to the simplicity, easy, and precise control on the experimental parameters, optical fiber lasers are ideal systems in which to generate and observe solitons. Various kinds of optical solitons have been experimentally observed in fiber laser systems, namely, vector solitons, dissipative solitons, breathers and dispersion-managed solitons, as well as soliton molecules. Ultrashort pulses can be generated directly by optical fiber soliton lasers using different mode-locking techniques [1–12].

This chapter provides an overview of fundamental matters relating to solitons fiber lasers. Typical cavity configurations, the main mode-locking techniques, and modeling of soliton fiber lasers are reviewed. Different techniques to achieve high-energy soliton pulses, the formation of soliton molecules, and recent experimental results concerning pulsating solitons are also discussed.

14.1 The First Soliton Laser

The first soliton laser used the passive mode-locking technique and was demonstrated by Mollenauer and Stolen [13] in 1984. Figure 14.1 shows the schematic diagram of such soliton laser. The output signal of the mode-locked laser is injected into the fiber with anomalous dispersion through the beam splitter S. The light that is reflected by the mirror M_3 is fed back through the same fiber into the laser cavity. After propagating back and forth in this feedback loop, the pulse comes out as a short soliton. A minimum pulse width of 60 fs was obtained with this system. However, the soliton laser action started and stopped sporadically as vibration and thermal drift caused the relative lengths of the two cavities to vary in and out of the proper interference condition. Servo stabilization of the cavity lengths enabled the laser to achieve a very stable mode of operation [14].

A theoretical interpretation of the soliton laser has been given in terms of the soliton theory [15]. Shortening of the laser output pulses using fibers with normal dispersion instead of anomalous dispersion has been also demonstrated [16, 17]. In such operation, generally called additive-pulse mode-locking [18], self-phase modulation from a purely nonlinear element in the external cavity causes the returned pulses to interfere constructively with the main laser pulses near their peaks, while interfering destructively with them in their wings, thereby producing a narrower pulse.

Solitons in Optical Fiber Systems, First Edition. Mario F. S. Ferreira.

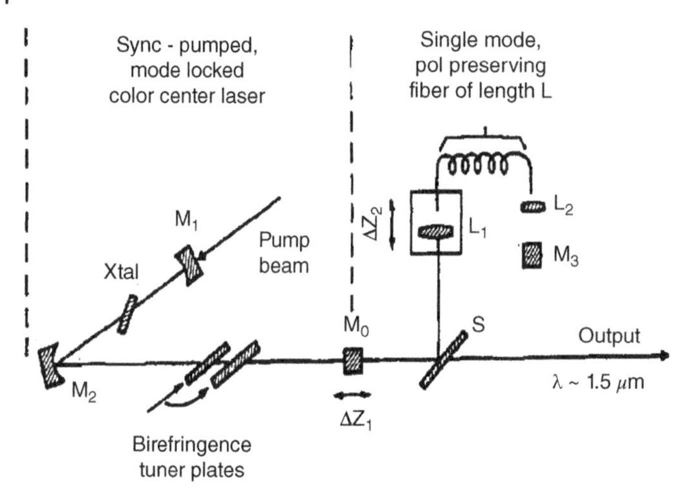

Figure 14.1 Schematic diagram of the first soliton laser. *Source:* After Ref. [13].

The pulse-shortening process continues until it is balanced by temporal broadening mechanisms, such as bandwidth limiting and group velocity dispersion.

14.2 Fundamentals of Fiber Soliton Lasers

The advent and rapid progress of the optical fiber amplifiers have enabled one to use them as a gain medium to fabricate fiber lasers. In fact, a fiber laser can be realized simply by placing a fiber amplifier inside a cavity in order to provide optical feedback. An optical fiber Raman amplifier has been used in one of the first all-fiber soliton lasers reported [19] but rare-earth-doped fiber amplifiers are now more commonly used.

In addition to the single-mode optical fiber amplifier, a typical fiber laser contains several other basic elements. For example, a wavelength selective coupler (WSC) is required to couple the pump into the fiber amplifier without coupling the signal out. An optical filter is useful for wavelength selection and tuning, reject noise, or modify the cavity dispersion. An output coupler is required to out-couple a portion of the signal in the laser cavity for external use. A polarization controller is often also needed to compensate for unwanted birefringence.

The optical feedback can be provided considering different configurations for the laser cavity [20]. In the linear configuration, optical feedback can be achieved simply by butt-coupling high-reflecting mirrors to the ends of a doped fiber, thus providing a *Fabry-Perot cavity*. However, the alignment of such a cavity is not easy, which causes an increase of the cavity losses. Another solution consists in depositing adequate dielectric coatings directly onto the polished ends of the fiber [21]. One problem of this solution concerns with the possibility of damage of the dielectric coatings when high-power pump light is coupled into the fiber. Alternatively, fiber Bragg gratings (FBGs) can be used as mirrors at the ends of the doped fiber, which provides an all-fiber Fabry–Perot cavity [22]. Fiber Bragg gratings will be described in Chapter 15. Actually, besides acting as a high-reflectivity mirror for the laser light, the FBG can be designed in order to be transparent to the pump light. FBGs are also commonly used for dispersion and spectral control in fiber lasers, providing the possibility of laser operation in a single longitudinal mode. Another approach to provide optical feedback makes use of fiber-loop mirrors [23].

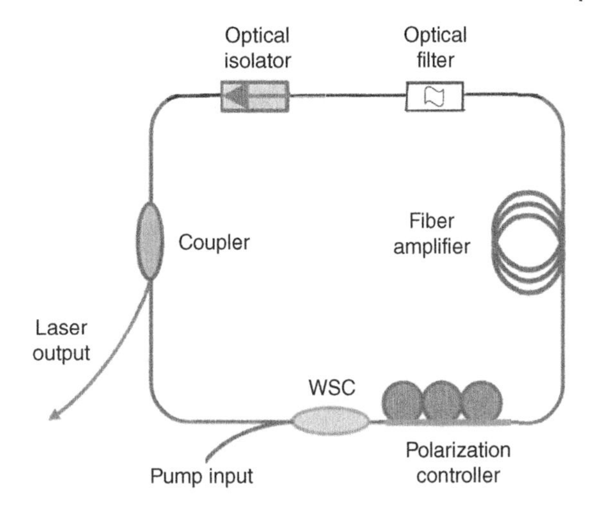

Figure 14.2 Schematic of a ring fiber laser. WSC: wavelength selective coupler.

In the ring configuration, the fiber is sliced into a loop, as represented in Figure 14.2. This is an all-fiber laser cavity, which does not use any mirrors, and can be implemented simply by connecting two ports of a wavelength-selective coupler. The inclusion of an isolator within the loop is necessary to force unidirectional operation. A polarization controller is used in the case of a conventional doped fiber that does not preserve polarization. An optical filter can also be included to help define or tune the lasing wavelength, besides rejecting noise.

Fiber lasers can be used to generate short pulses based on two techniques: Q-switching and mode-locking [1, 24, 25]. Q-switching makes use of a device that increases the cavity losses but opens periodically for a short time, during which such losses are reduced. Nanosecond pulses with high peak powers (>1 kW) can be generated in this way. Moreover, tuning the pulse wavelengths over the entire gain spectrum is often possible [26–31]. In a 1999 experiment, 2 kW-peak power Q-switched pulses from a double-clad Yb-fiber laser have been tuned from 1060 to 1100 nm [30].

Q-switching produces relatively broad optical pulses (~100 ns). In contrast, mode-locking generates ultrashort pulses, with durations from hundreds of picoseconds down to a few femtoseconds. However, the finite gain-bandwidth of optical fiber amplifiers usually limits the minimum possible pulse duration that can be generated in soliton fiber lasers to tens of femtoseconds.

Besides the pulse duration, other main performance characteristics of pulsed fiber lasers are pulse bandwidth and energy, repetition rate, output power, laser stability and self-starting, noise, and jitter. In general, the performance tradeoffs found in soliton transmission systems also apply in soliton lasers. For example, assuming a given pulse duration, the interaction between adjacent solitons limits the maximum repetition rate achievable in soliton lasers [32].

The Kerr nonlinearity and the group-velocity dispersion play a major role in determining the properties of optical fiber soliton lasers. Both effects are often not distributed but lumped into discrete sections, thus acting as perturbations to the solitons. However, such perturbations are not significant if the cavity length is much shorter than the soliton period, $z_0 = \pi L_D/2$. In this case, the soliton behavior is mainly determined by the average dispersion and nonlinearity in the system.

Actually, the dynamics of soliton fiber lasers and the variety of solitons which may be produced in a real system is considerably richer than the simple picture provided by a simple lossless system. Such solitons emerge from a balance between nonlinearity and dispersion, which can vary along the laser cavity, as well as between various sources of gain and loss. In order to improve the performance of soliton lasers, soliton control techniques similar to those discussed in Chapter 8 for soliton transmission systems can be used. These control techniques are necessary to limit the effects of soliton interactions [33], soliton self-frequency shift [34], timing jitter [35], and interaction with dispersive waves [36].

Mode-locked lasers can be considered as an ideal playground for exploring experimentally nonlinear dissipative dynamics. Actually, a mode-locked laser offers essentially an infinite distance of propagation and, since a pulse is emitted once per round trip, it allows the observation of the pulse evolution. Moreover, it provides the possibility of studying some parameter regimes that are not easily accessible in soliton transmission systems.

14.3 Mode-Locking Techniques

The longitudinal modes of a cavity laser occur at frequencies $\nu_j = jc/2nl$, where j is an integer, c is the speed of light in vacuum, n is the refractive index in the cavity, and l is the cavity length. A train of short pulses, spaced by the cavity round-trip time, can be obtained by phase-locking several of these modes. This can be achieved using an adequate active or a passive element in the laser cavity. Since optical fiber lasers are relatively long, the spacing between longitudinal modes is relatively small (typically between 1 GHz and 1 MHz). To generate ultrashort optical pulses in these lasers at high repetition rates therefore requires mode-locking of many harmonically related modes.

14.3.1 Active Mode-Locking

Active mode-locking is typically achieved by inserting a modulator into the cavity to modulate either the amplitude or the phase of the intracavity optical field at a frequency equal to the mode spacing. In both cases, modulation sidebands are generated, spaced apart by the modulation frequency. These sidebands overlap with the neighboring modes, leading to a phase synchronization.

Harmonically, mode-locking occurs when the modulation frequency is an integer multiple of the mode spacing [1]. Both acousto-optic and electro-optic modulators have been used for this purpose. In the presence of nonlinearity and anomalous dispersion, spectral broadening and temporal shortening of the pulses occur, resulting in the generation of short solitons [3, 4]. Active mode-locking is an attractive technique for producing ultralow-jitter pulses in the picosecond range at repetition rates as high as tens of GHz. Actually, the generation of picosecond pulses at repetition rates above 100 GHz has been demonstrated using this technique [37].

In practice, $LiNbO_3$modulators are commonly used for mode-locking fiber lasers, not only because they can be integrated within the laser cavity with relatively low insertion losses but also because they can be modulated at speeds of some tens of GHz [38]. In a 1990 experiment, 2-ps pulses were generated by phase modulation realized using a $LiNbO_3$ modulator inserted in a Fabry-Perot cavity that included a 10 m-long fiber [39]. This laser was referred as the *fiber-soliton laser*, since generated pulses presented a "sech" shape and a time-bandwidth product of only 0.3. In another experiment, a high-speed $LiNbO_3$modulator was used in a ring cavity laser to achieve a pulse repetition rate of 30 GHz [40]. Several other experiments have demonstrated the improved performance of harmonically mode-locked fiber lasers [41–57].

14.3.2 Passive Mode-Locking

Passive mode-locking is a widely used technique, which does not require any externally modulated media or devices [10–12] but only makes use of a nonlinear device whose response depends on the intensity of the input pulse. To promote pulse generation and shortening, they use a fast saturable absorber, the function of which is to reject low-intensity light from the cavity, but to pass high intensities. This idea can be implemented using different approaches, namely, using a saturable absorber medium, a nonlinear fiber-loop mirror, or a nonlinear polarization rotation scheme. Nonlinear effects such as self-phase modulation (SPM) and cross-phase modulation (XPM), as well as fiber dispersion, play an important role in the operation of most passively mode-locked fiber lasers.

In the case of a fast saturable absorber, the wings of an optical pulse experience more loss than its central part, which is intense enough to saturate the absorber. As a consequence, the exiting pulse is narrower than the input pulse. A semiconductor absorbing medium is often used to realize such fast saturable absorber [58–65]. Several other types of mode-lockers have been reported recently, like carbon nanotubes [66–75], graphene [76–88], black phosphorus [89–98], and other graphene-like two-dimensional materials [99–102].

With the inclusion of the above-mentioned saturable absorbers, fiber lasers lose their all-fiber nature. However, this undesirable feature can be avoided if the fast saturable absorber function is implemented in optical fiber, using one of two methods: a nonlinear loop mirror [103, 104], or nonlinear polarization rotation and a polarizer [105]. Actually, the power-dependent transmission achieved with these techniques can shorten an optical pulse just as a saturable absorber material does. The corresponding passive mode-locking mechanisms are called interferometric or additive-pulse mode-locked. The first soliton laser, reported by Mollenauer and Stolen [13] and described in Section 14.1, made use of this pulse-forming mechanism.

14.3.3 Nonlinear Optical Loop Mirrors

The nonlinear optical loop mirror (NOLM) is basically an antiresonant nonlinear Sagnac interferometer, with an intensity-dependent transmission. The Sagnac interferometer is represented in Figure 14.3 and is made by connecting the two outputs of a fiber coupler through a piece of long fiber. The input beam is divided in the fiber coupler into two beams counter-propagating in the fiber loop. After one round trip, both beams arrive at the same time at the coupler, where they interfere coherently, according to their relative phase difference.

Considering an input beam with power $P_0 = |A_0|^2$ incident at one input port, the amplitudes of the optical fields at the two output ports of the coupler are given by:

$$A_1 = \sqrt{f}A_0, \quad A_2 = i\sqrt{1-f}A_0 \tag{14.1}$$

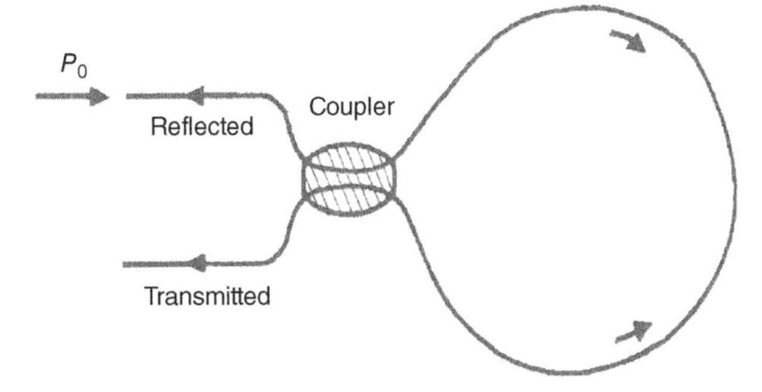

Figure 14.3 Schematic illustration of an all-fiber Sagnac interferometer.

where f is the coupler power-splitting fraction. After one round trip, both the forward- and backward-propagating fields acquire linear and nonlinear phase shifts, taking the form:

$$A'_f = A_f \exp\left[i\beta L + i\gamma\left(|A_f|^2 + 2|A_b|^2\right)L\right] \tag{14.2}$$

$$A'_b = A_b \exp\left[i\beta L + i\gamma\left(|A_b|^2 + 2|A_f|^2\right)L\right] \tag{14.3}$$

where L is the loop length, γ is the fiber nonlinear parameter, and β is the propagation constant.

The reflected and transmitted fields can be obtained by using the transfer matrix of the fiber coupler and are given by:

$$\begin{pmatrix} A_t \\ A_r \end{pmatrix} = \begin{pmatrix} \sqrt{f} & i\sqrt{1-f} \\ i\sqrt{1-f} & \sqrt{f} \end{pmatrix} \begin{pmatrix} A'_f \\ A'_b \end{pmatrix} \tag{14.4}$$

From Eqs. (14.1)–(14.4) we obtain the following result for the transmittivity $T = |A_t|^2/|A_0|^2$ of the Sagnac loop:

$$T = 1 - 4f(1-f)\cos^2\left[\left(f - \frac{1}{2}\right)\gamma P_0 L\right] \tag{14.5}$$

From Eq. (14.5) we see that $T = 0$ for a 3-dB coupler. In this case, the field is totally reflected and the Sagnac interferometer serves as an ideal mirror. For this reason, it is also known as a nonlinear optical loop mirror.

An asymmetry between the clockwise and anticlockwise waves propagating around the loop in Figure 14.3 can be created by using a coupler with an uneven coupling ratio ($f \neq 0.5$). In the case of a nonlinear amplifying loop mirror (NALM), the asymmetry is provided by a fiber amplifier, which is placed closer to one of the coupler arms than the other. In either case, light traveling clockwise around the loop has a different intensity to light traveling anticlockwise and hence experience different amounts of self-phase modulation in the loop due to fiber nonlinearity. The interference at the coupler, constructive or destructive depends on the phase difference between the two waves, which is proportional to the power of the incoming signal. If the phase shift is equal to π for the central part of the pulse, this part is transmitted, but its wings are reflected because of their lower-phase shifts. As a consequence, the exiting pulse is narrower than the input pulse. Apart from providing short pulse generation in optical fiber lasers, nonlinear loop mirrors may also be used for ultrafast all-optical switching [106, 107] and pulse regeneration in soliton transmission systems [108].

14.3.4 Figure-Eight Laser

Mode-locked fiber lasers using NALMs were reported firstly in several 1991 experiments [6, 109–111], in which soliton pulses shorter than 400 fs were generated. Its use provides passively mode-locked fiber lasers with a cavity geometry, which makes them known as figure-eight lasers [112–120]. A schematic of this laser is illustrated in Figure 14.4. It can be regarded as a ring laser (to the left of the 50 : 50 coupler), with a NALM on the right, which provides fast saturable gain. The NALM causes additive-pulse mode-locking of the laser and consists of a 3-dB fiber coupler, a doped fiber, and a wavelength division multiplexer for pumping. Mode-locking is favored since the loop becomes fully transmissive for pulses whose power attains a critical value. An isolator in the left cavity ensures unidirectional operation. The laser output is provided by a fiber coupler with low transmission to minimize the cavity losses. A further reduction of the cavity losses can be achieved if the central coupler

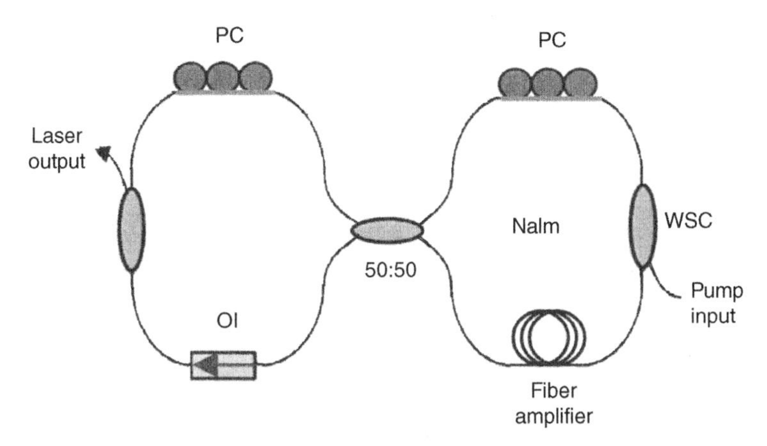

Figure 14.4 Schematic of a figure-eight laser. PC: polarization controller; WSC: wavelength selective coupler; OI: optical isolator.

in Figure 14.4 is unbalanced. This was verified in a 1994 experiment, in which 94% of the intracavity power was propagating in the direction in which laser emission occurred [115]. Optical pulses shorter than 1 ps were generated with such fiber laser.

14.3.5 Nonlinear Polarization Rotation

Passive mode-locking in fiber lasers can also be achieved using the nonlinear polarization rotation technique, which occurs during propagation in a birefringent nonlinear medium [121, 122]. Mode-locking occurs in this case due to intensity-dependent changes in the state of polarization resulting from a combination of SPM and XPM of the two orthogonally polarization components of a single propagating pulse. A polarizer placed after the fiber can be adjusted such that it lets the central intense part of the pulse pass but rejects the low-intensity pulse wings. The resulting intensity-dependent transmissivity is similar to that of a NOLM or NALM and provides a fast saturable absorber action.

The technique of nonlinear polarization rotation has been used for the first time in 1992 for passive mode-locking of fiber lasers and has resulted in considerable improvement of such lasers [123–126]. The shortest (38 fs) pulse has been generated in a fiber laser using this mode-locking technique [125]. Further improvements occurred when it was realized that fiber lasers could operate not only in the presence of anomalous GVD in the ring cavity but also when the average GVD is normal. This was observed in a 1993 experiment, in which 76-fs pulses with 90-pJ energy and 1 kW of peak power were generated [126].

14.3.6 Hybrid Mode-Locking

Another mode-locking technique is called hybrid mode-locking, since it combines various mode-locking techniques mentioned above to improve the laser performance. For example, an amplitude or phase modulator can be used inside a passively mode-locked fiber lasers to promote ultrashort pulse generation at a stable repetition rate [127]. In particular, the performance of fiber lasers has been considerably improved by combining the active and passive mode-locking techniques through the use of a phase modulator [127, 128].

14.4 High-Energy Soliton Fiber Lasers

Mode-locked fiber laser configurations implemented in early 1990s used mainly anomalous dispersion fibers, and the generated pulses presented standard sech-like profiles. However, in the mid-1990s, Haus et al. begun studying the dispersion-managed mode-locked fiber laser, which combines fibers of opposite dispersion signs, thus providing the formation of highly stretched pulses with higher energies. Moreover, the use of a dispersion-managed cavity with a total dispersion close to zero provides also the possibility of suppressing the sideband instability, which is a problem affecting usually the fiber soliton lasers operating with pulses in the sub-picosecond regime [111, 126, 129].

Figure 14.5 shows an experimental configuration of a dispersion-managed fiber laser, also called a stretched-pulse mode-locked laser [126]. In fact, pulses circulating inside the cavity stretch considerably in the section with normal GVD. All the characteristics of dispersion-managed soliton transmission systems, as discussed in Chapter 10, can also be observed in a stretched-pulse laser, namely, the enhanced pulse energy, reduced pulse interactions, and reduced timing jitter [126, 129–134].

A fiber laser with a constant dispersion close to zero would normally generate very short pulses, but with very low energy in each pulse. In the case of stretched-pulse fiber lasers, however, this is not the case. As seen in Chapter 10, dispersion-managed solitons experience large changes in their temporal width, returning to its minimum temporal width (and maximum peak power) at the midpoint of each fiber section. Due to this behavior, the location of the output coupler plays an important role. Mode-locked pulses as short as 63 fs have been generated with proper optimization [135]. Moreover, since these pulses have a high peak power at only two points in the laser cavity, the average nonlinearity is reduced, and the pulse energy can reach the nanojoule range — an order of magnitude higher than in all-anomalous-dispersion lasers. Dispersion-managed soliton mode-locked fiber lasers has been intensively studied during recent years [136–145].

Mode-locked operation in all-normal fiber lasers has been also realized in recent years [146–148]. In these cases, the cavity dynamics is dominated by the balance between spectral filtering on one side, and nonlinearity, dispersion and gain on the other side. Figure 14.6 shows an example of such a laser cavity and its output features [148]. These devices are usually called "dissipative soliton

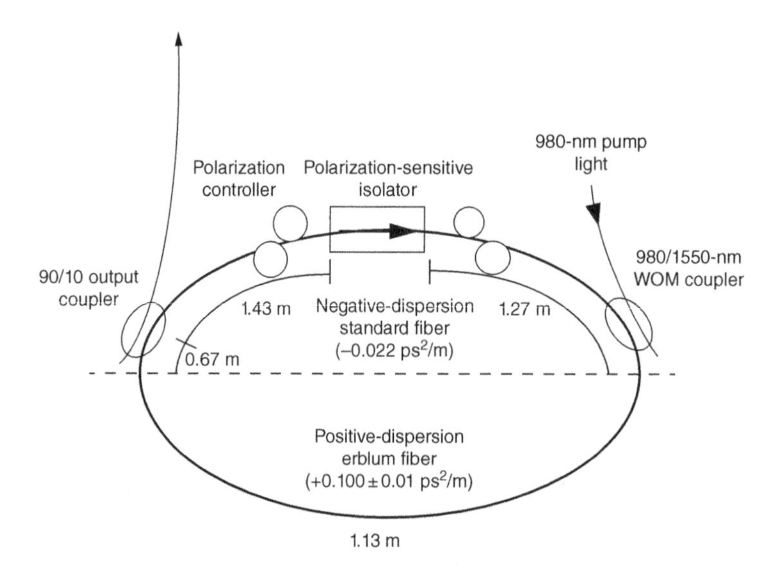

Figure 14.5 Schematic of a stretched-pulse mode-locked laser. *Source:* After Ref [126].

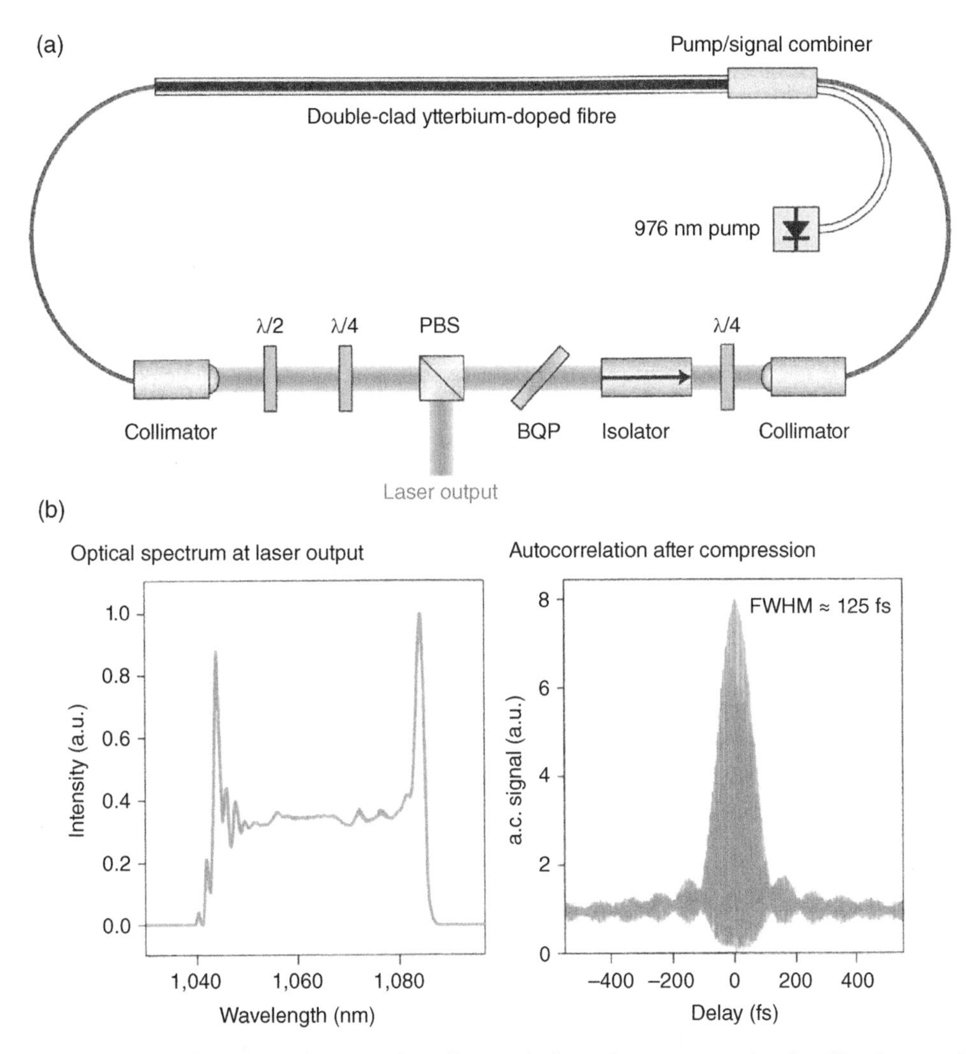

Figure 14.6 (a) Experimental set-up of an all-normal, dispersion-compensation-free fiber laser cavity operating at a 70 MHz repetition rate and generating 31 nJ pulses. PBS, polarizing beam splitter; BQP, birefringent quartz plate; $\lambda/4$, quarter waveplate, $\lambda/2$, half waveplate; (b) Characterization of the output pulses. Autocorrelation is performed after dechirping outside the laser cavity, achieving 80 fs pulses with 200 kW of peak power. *Source:* After Ref. [148].

lasers" [148–150] and can produce pulses with energies higher than 20 nJ, often presenting M-shaped optical spectra. Actually, dissipative solitons can be found in mode-locked fiber lasers using either all normal or anomalous dispersion fibers. In any case, the dissipation due to several kinds of loss is balanced by gain. Besides this, another balance must be verified between nonlinearity and dispersion, which are the key ingredients for any soliton system.

More recently, several techniques to achieve high pulse energy mode-locked fiber lasers have been implemented. One of them relies on the use of microstructured optical fibers where micro-joule-level pulse energies have been achieved [151]. Another technique consists in shifting the mode-locking operation further into the normal dispersion regime, thereby increasing the pulse chirping effect [152, 153].

An alternative approach to achieve high-pulse energies in fiber lasers is based on the dissipative soliton resonance (DSR) phenomenon. This effect was described in Section 12.4 and has been frequently observed and reported [154–163]. DSR can occur in both anomalous and normal dispersion regimes [154, 155] of a mode-locked fiber laser and enables an almost infinite boost in the pulse energy without wave breaking [158]. The energy of a DSR pulse increases mainly due to the increase of the pulse width, while keeping the amplitude at a constant level. In order to obtain high-energy ultrashort pulses, a linear pulse compression technique has to be used outside the laser cavity.

In a 2009 experiment, the DSR phenomenon was observed with the generation of 281.2 nJ mode-locked pulses directly from an erbium-doped fiber laser mode-locked with the nonlinear polarization rotation technique [156]. By increasing the pump power, the pulse width of the observed flat top solitons was also increased. A figure-9 Er:Yb double-clad fiber laser has been used in 2016 to obtain a pulse energy of 2.3 μJ and a pulse width of 455 ns [160]. The DSR pulse energy was further increased to 10 μJ with a pulse width of 416 ns in another 2016 experiment using a figure-8 double-clad Er:Yb fiber laser [161]. More recently, a pulse energy of 353 nJ and a maximum peak power of 84 W was reported for a dissipative soliton resonance in a 2 μm thulium-doped double-clad fiber laser [163].

A different kind of high-energy pulses corresponds to the very high amplitude (VHA) solitons also discussed in Section 12.4 [164, 165]. Such VHA solitons can occur in both anomalous and normal dispersion regimes when the nonlinear gain saturation effect tends to vanish. The increase in energy of these pulses is mainly due to the increase of the pulse amplitude, whereas the pulse width becomes narrower. Using this approach, high-energy ultrashort pulses can be obtained without using any additional pulse compression technique.

14.5 Modeling of Soliton Fiber Lasers

The master equation approach to mode-locked lasers was first proposed by H. Haus in mid-1970s [166, 167]. Such master equation is similar to the cubic complex Ginzburg-Landau equation (CGLE), which is commonly used to describe systems in the vicinity of bifurcations [168–170]. The coefficients that appear in the master equation were related to the physical parameters in a phenomenological way by Haus et al. [171]. Later, such relation has been derived more rigorously for the case of a fiber laser mode-locked by means of the nonlinear polarization rotation technique [172–174].

As seen in Section 8.5, the cubic CGLC is never stable, neither for anomalous dispersion [175] nor in the case of normal dispersion [176]. To achieve the stabilization of the solitons in the model, higher-order nonlinear quintic terms must be introduced. The resulting cubic-quintic CGLE and can be written as in Eq. (12.1) and includes all the main physical effects present in any mode-locked laser. In particular, linear loss and nonlinear gain account for the required saturable absorber effect. On the other hand, even if the correction to the nonlinear refractive index is not imposed by most laser materials under common operation conditions, it can be justified by the discrete nature of the laser cavity [177]. An explicit relation of the coefficients on the cubic-quintic CGLE with regard to the physical parameters has been obtained for the case of a fiber ring laser mode-locked through nonlinear polarization rotation [178]. Besides passively mode-locked lasers, this model has been used to describe a wide range of nonlinear optical systems, such as parametric oscillators, wide-aperture lasers, and nonlinear optical transmission lines [179–181].

Actually, Eq. (12.1) models the fiber laser as a distributed system, which is reasonable if the pulse shape changes only slightly during each round trip. There are many advantages in using such distributed model, governed by a continuous equation, since it allows, to some extent, an analytic study. However, when the discrete nature of the laser cavity cannot be ignored, the cubic-quintic CGLE given by Eq. (12.1) can be used such that the various parameters vary periodically with Z, the period corresponding to a cavity round trip.

The study of pulsed dynamics in the framework of the cubic-quintic CGLE has provided an adequate explanation of several unusual single- and multiple-pulse phenomena that have been observed in mode-locked lasers. Such studies have been also of fundamental importance in the development of the concepts of dissipative soliton [180, 182] and dissipative soliton resonance (DSR) [154, 155, 183–185].

As seen in Chapters 12 and 13, many different types of soliton solutions of the cubic-quintic CGLE, both stationary and nonstationary, can be found. However, stable pulses can be generated in a very narrow range of the laser parameters, which must be carefully adjusted for such purpose. In general, the pulses may change their shape from one round trip to another and have complicated dynamics in time [186]. In some circumstances, the pulse-shape evolution in time may even become chaotic. Soliton explosions are among the most striking phenomena, which can be observed in soliton lasers [187–195].

The cubic-quintic CGLE model assumes the existence of a second-order spectral filtering effect, characterized by a spectral response with a single maximum. A more realistic model would require the inclusion of higher-order spectral filtering terms, taking into account that the gain curve is usually wide and can have several maxima. Such more general filtering profile becomes particularly convenient when considering the generation of ultrashort pulses. The inclusion of such effect transforms the CGLE into the complex Swift–Hohenberg equation (CSHE) [196, 197]:

$$i\frac{\partial q}{\partial Z} + \frac{1}{2}\frac{\partial^2 q}{\partial T^2} + |q|^2 q = i\delta q + i\beta\frac{\partial^2 q}{\partial T^2} + i\varepsilon|q|^2 q + i\mu|q|^4 q - \nu|q|^4 q + i\sigma\frac{\partial^4 q}{\partial T^4} \tag{14.6}$$

where the parameter σ in the last term of the right-hand side accounts for fourth-order spectral filtering.

The CSHE has a diversity of known analytical and numerical solutions [198–201]. In particular, several types of stationary and moving composite solitons of this equation were found to be stable and to have a wider range of existence than those of the complex quintic Ginzburg–Landau equation [196]. Figure 14.7 illustrates three types of stationary solutions, including one flat top pulse (SP1) and two sech-like pulses (SP2, SP3).

Similar to the CGLE, it has been shown that the CSHE also supports erupting pulse solutions [201]. Figure 14.8a shows a CSHE erupting soliton evolving along Z, for the following set of parameter values: $\beta = -0.03$, $\delta = -0.2$, $\varepsilon = 1$, $\mu = -0.1$, $\nu = -0.7$, and $\sigma = 0.0025$. The filter spectral response, given by $T(\omega) = \exp(\delta - \beta\omega^2 - \sigma\omega^4)$ [199], is represented in Figure 14.8b and exhibits a dual-peak structure.

At first sight, there is no significant differences in the temporal evolution of this erupting soliton when compared with the erupting solitons of the CGLE, discussed in Chapter 13. However, we may notice that the explosion itself has a bigger temporal extension, whereas the pulse profile becomes almost rectangular. The consecutive explosions are similar, but not exactly equal.

Figure 14.9 illustrates the propagation of CGLE and CSHE erupting solitons in the presence of different filter profiles. Four filter spectral responses are represented in Figure 14.9a: dashed curve, ($\delta = -0.1$, $\beta = 0.125$, $\sigma = 0.00$), thick solid curve, ($\delta = -0.3$, $\beta = -0.055$, $\sigma = 0.004$), medium solid curve, ($\delta = -0.4$, $\beta = -0.075$, $\sigma = 0.005$), and thin solid curve, ($\delta = -0.65$, $\beta = -0.1$, $\sigma = 0.005$).

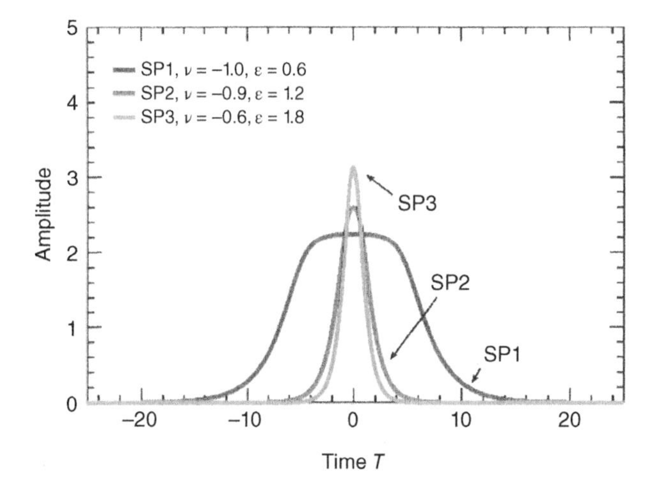

Figure 14.7 Three kind of stationary pulse (SP) solutions of the CSHE, assuming the parameter values in the figure and $D = +1$, $\delta = -0.2$, $\beta = -0.03$, $\mu = -0.1$, and $\sigma = 0.0025$.

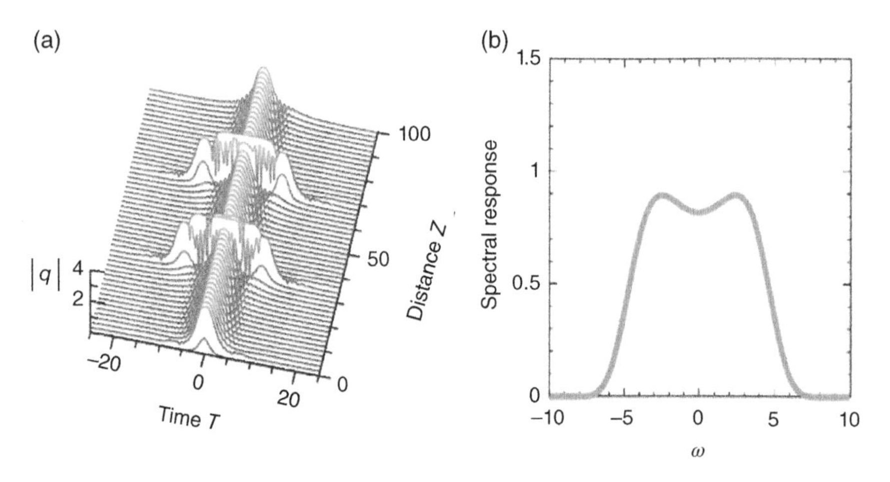

Figure 14.8 (a) Erupting soliton amplitude evolution in the presence of the filter spectral response represented in (b), for the following set of parameter values: $\delta = -0.2$, $\beta = -0.03$, $\varepsilon = 1$, $\mu = -0.1$, $\nu = -0.7$, $\sigma = 0.0025$.

The other parameter values in Figure 14.9(b–d) are $\varepsilon = 1$, $\mu = -0.1$, and $\nu = -0.6$. The pulse in Figure 14.9b corresponds to a CGLE erupting soliton, obtained for the filter spectral response 1 (dashed curve) in Figure 14.9a.

The CSHE erupting pulse evolution in Figure 14.9c, corresponding to the filter with the smallest deep in Figure 14.9a, shows only one explosion, whereas the CGLE erupting soliton in Figure 14.9b shows three explosions in the same distance span. Moreover, the pulse explosions are suppressed in the case of the filter represented by curve 3 in Figure 14.9a, in spite of the ripples that are still visible in Figure 14.9d, both in the pulse leading and trailing edges. A further increase of the filter deep provides a full suppression of such ripples and the formation of a clean stationary pulse, as illustrated in Figure 14.9e.

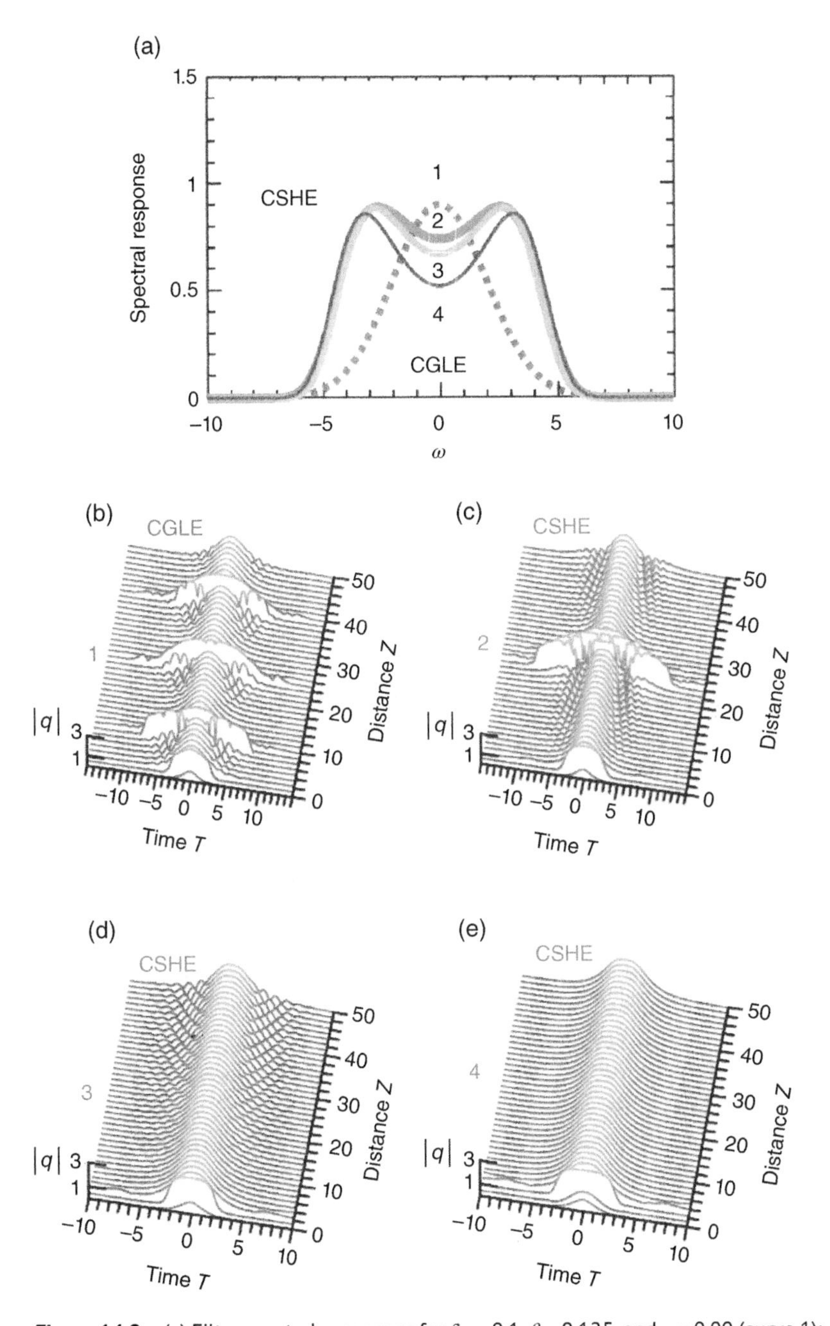

Figure 14.9 (a) Filter spectral responses for $\delta = -0.1, \beta = 0.125$, and $\sigma = 0.00$ (curve 1); $\delta = -0.3, \beta = -0.055$, and $\sigma = 0.004$ (curve 2); $\delta = -0.4, \beta = -0.075$, and $\sigma = 0.005$ (curve 3); and $\delta = -0.65, \beta = -0.1$, and $\sigma = 0.005$ (curve 4). (b) CGLE erupting soliton, and (c)–(e) CSHE erupting solitons, assuming the parameter values of curves 1, 2, 3, and 4 in (a), respectively. The other parameter values are $\varepsilon = 1$, $\mu = -0.1$, and $\nu = -0.6$.

14.6 Polarization Effects

As discussed in Chapter 11, common optical fibers do not preserve polarization. As a result, the state of polarization of output light may change from pulse to pulse or even over the duration of a single pulse. As a consequence, taking into account the polarization evolution is important for a proper modeling of mode-locked fiber lasers [202–205].

The light propagation in weakly birefringent cavity fiber laser where the laser gain and losses are included is described by the following coupled Ginzburg–Landau equations (GLEs):

$$\frac{\partial U_x}{\partial z} + \beta_{1x}\frac{\partial U_x}{\partial t} + \frac{i}{2}\beta_{2x}\frac{\partial^2 U_x}{\partial t^2} = i\gamma\left(|U_x|^2 + \frac{2}{3}|U_y|^2\right)U_x + \frac{i\gamma}{3}U_x^* U_y^2 \exp\left[-2i(\beta_{0x} - \beta_{0y})z\right] + \frac{g}{2}U_x + \frac{g}{2\Omega_g^2}\frac{\partial^2 U_x}{\partial t^2} \tag{14.7}$$

$$\frac{\partial U_y}{\partial z} + \beta_{1y}\frac{\partial U_y}{\partial t} + \frac{i}{2}\beta_{2y}\frac{\partial^2 U_y}{\partial t^2} = i\gamma\left(|U_y|^2 + \frac{2}{3}|U_x|^2\right)U_y + \frac{i\gamma}{3}U_y^* U_x^2 \exp\left[2i(\beta_{0x} - \beta_{0y})z\right] + \frac{g}{2}U_y + \frac{g}{2\Omega_g^2}\frac{\partial^2 U_y}{\partial t^2} \tag{14.8}$$

where U_j ($j = x, y$) are the slowly varying amplitudes along the slow and the fast axes, γ is the nonlinear coefficient, β_{0j} ($j = x, y$) represent the propagation constants of the two orthogonal linearly polarized waves, $\beta_{1j} = d\beta_j/d\omega|_{\omega=\omega_0}$, and $\beta_{2j} = d^2\beta_j/d\omega^2|_{\omega=\omega_0}$. The first, second, and third terms on the right-hand side of these equations correspond to the self-phase modulation, cross-phase modulation, and four-wave mixing effects, respectively. In the fourth and fifth terms of Eqs. (14.7) and (14.8), g is the laser gain and Ω_g represents its bandwidth.

The coupled GLEs given by Eqs. (14.7) and (14.8) are commonly used to model the vector soliton formation and dynamics in dissipative nonlinear systems. These equations can also be used to study the vector soliton formation in fiber lasers. The experimental generation of such vector solitons in passively mode-locked fiber lasers has been intensively studied [206–217]. To generate the vector solitons, all the fibers and passive components of the mode-locked fiber lasers have to be polarization insensitive. The main challenge for achieving this objective is to find an appropriate saturable absorber (SA) that has polarization insensitive saturable absorption. In fiber lasers, semiconductor saturable absorber mirrors (SESAMs), carbon nanotubes, graphene, and graphene-like 2D materials are SAs, which have a polarization-independent saturable absorption. So far, vector solitons have been observed in ultrafast fiber lasers passively mode-locked both by carbon nanotubes [210] and graphene [211, 212].

In weak linear birefringence cavity, vector solitons will experience coherent energy exchange caused by the four-wave mixing, as observed by Zhang et al. [206]. Besides, depending on the cavity birefringence and cross-polarization coupling strength, vector solitons formed can be classified as polarization-locked vector solitons (PLVS) [202, 203, 208, 210], polarization rotation vector solitons (PRVS) [208, 209, 214], group-velocity-locked vector solitons (GVLVS) [207], dark-bright vector solitons [217], and so on.

Polarization-locked vector solitons have been investigated by Cundiff et al. in a 1997 experiment using a SESAM mode-locked fiber laser [202]. They presented a comprehensive study on polarization locked vector solitons in fiber lasers in 2000 [204, 205]. It was found that the polarization state of a PLVS can be linear or elliptical. The relative phase difference is fixed at $\pi/2$ in all cases but the amplitude difference depends on the linear birefringence within the cavity.

A theoretical model based on the coupled GLEs (14.7) and (14.8) is able to explain most of the experimental data [205].

14.7 Dissipative Soliton Molecules

Dissipative soliton molecules correspond to bound states arising from the interaction of initially separate single solitons. They are truly stationary and robust structures, whose energy is proportional to the number of involved solitons. Such structures differ significantly from higher-order soliton solutions of the nonlinear Schrödinger equation, which are easily destroyed by external perturbations or higher-order effects, such as third-order dispersion or intrapulse Raman scattering.

Two-soliton molecules formed by plain pulses or composite pules were already discussed in Section 12.5 and are characterized by a quadrature-phase difference. Phase-locked two-soliton molecules were experimentally observed in 2002 using a stretched-pulse fiber laser [218]. After that, they have been found numerically and experimentally considering different laser configurations and dispersion regimes [219–223].

The two-soliton molecule can be assumed as the building block to construct various multi-soliton molecules [223]. Examples are given in Figure 14.10 of the four- and five-soliton solutions with a phase difference of $\pi/2$ between adjacent plain pulses. As observed in the case of the two-soliton solution, multi-soliton solutions formed by plain pulses also move with the same constant velocity along the T axis. This is in contrast with the behavior of similar structures based on composite pulses. Figure 14.11 shows an example of the seven-soliton bound state, with a phase difference of $\pi/2$ between adjacent composite pulses. We observe that, in spite of the asymmetric phase profile of the structure, it has zero velocity along the T axis. Soliton molecules like those illustrated in Figures 14.10 and 14.11 may increase the capacity of telecommunication and can be also attractive in all-optical information storage [224, 225].

An interesting question concerns with the possibility to achieve a multi-soliton molecule formed by plain pulses with zero velocity, by choosing appropriately its (symmetric) phase profile. Figure 14.12a illustrates the evolution of a four-soliton molecule, whose initial phase profile is given by the dash-dotted line in Figure 14.12b. This phase profile evolves during the propagation and achieves the aspect of the full line in Figure 14.12b at $Z = 1000$. We observe from Figure 14.12a that the four-soliton solution separates in two pairs, one propagating to lower and the other to higher values of T. A different but symmetric initial phase profile of the four-soliton molecule is considered in Figures 14.12c and d. In this case, we observe indeed a stable propagation with zero velocity, in spite of the fact that the separation between the two inner pulses is somewhat greater than the separation of the two-soliton molecule.

Soliton fiber lasers constitute an ideal testbed for the study on soliton molecules. Actually, there have been during the recent years many reports on these structures in fiber lasers mode-locked with the carbon nanotubes, graphene, black phosphorus, and so on [226–231]. Since these saturable absorbers may be polarization independent, soliton molecules formed by vector solitons were also reported. The experimental observation of tightly and loosely bound states of vector solitons in a carbon nanotube mode-locked fiber laser in the anomalous dispersion regime was reported in 2013 [232]. In a latter experiment, localized interactions between vector solitons, vector soliton with bound vector solitons, and vector soliton with a bunch of vector solitons were observed in a fiber laser passively mode-locked by graphene [212]. Owing to the bound soliton interactions,

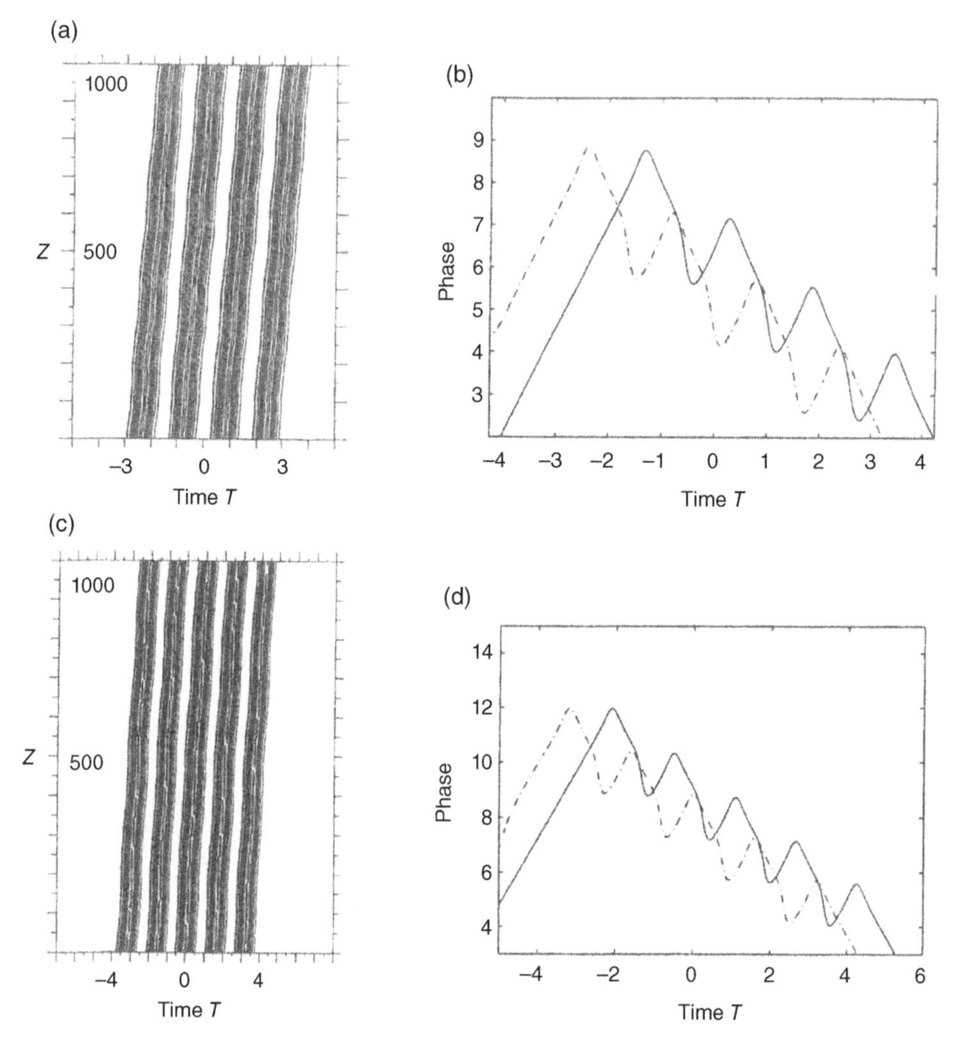

Figure 14.10 Four (a,b) and five (c,d) plain pulse solutions with a phase difference of $\pi/2$ between adjacent pulses. The dash-dotted (full) lines in (b) and (d) correspond to the initial (final) phase profiles.

bound-bound solitons were also found experimentally [233]. The first experimental observation of group-velocity-locked vector soliton molecules has been reported in 2017 in a SESAM mode-locked fiber laser using a birefringence-enhanced fiber [234].

14.8 Experimental Observation of Pulsating Solitons

Conservative optical solitons resulting from a balance between dispersive and nonlinear effects in single-mode fibers were first observed by Mollenauer et al. in 1980 [235]. In the case of soliton pulses appearing in fiber lasers, the single balance between nonlinearity and dispersion is replaced by a double balance: between nonlinearity and dispersion and also between gain and loss. These soliton pulses are named as dissipative solitons [197], and they exhibit a number of unique features

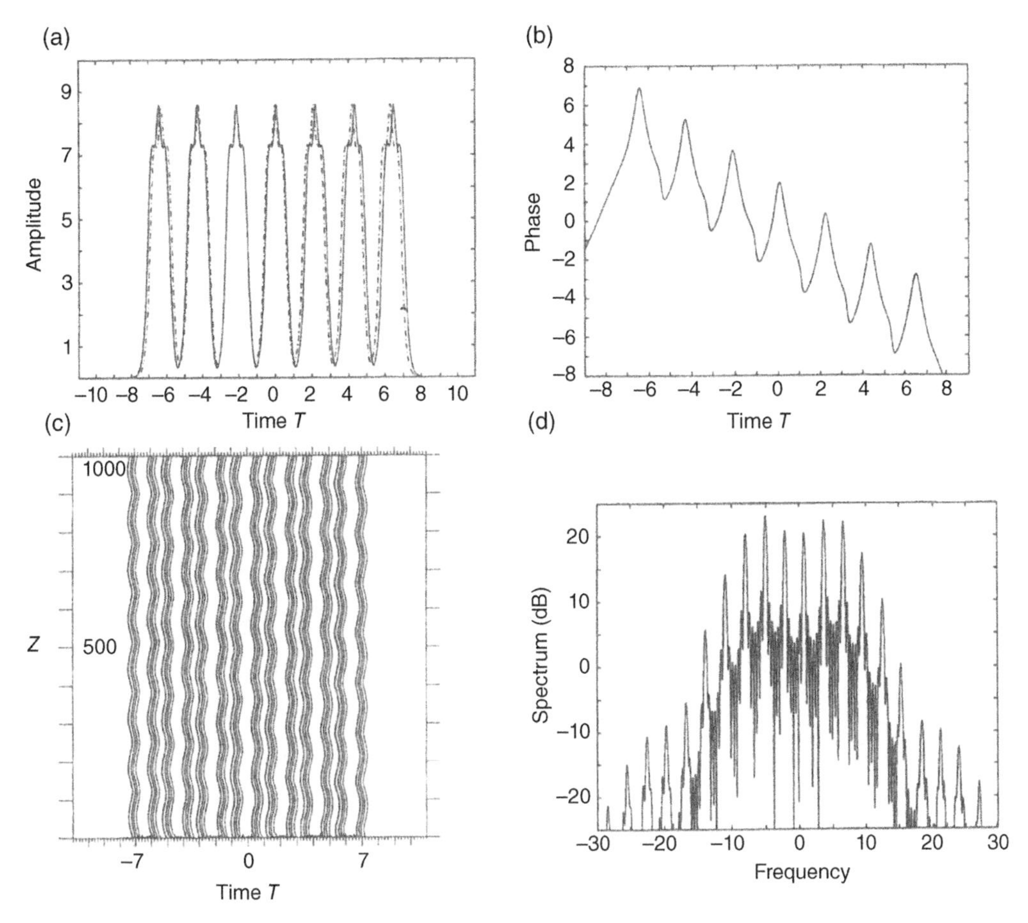

Figure 14.11 Seven-composite pulse solution with a phase difference of $\pi/2$ between adjacent pulses. (a) Initial (dashed curve) and final (full curve) amplitude profiles, (b) phase, (c) contour plot, and (d) spectrum.

that differ from those of conservative solitons. Several types of stationary and pulsating dissipative solitons have been observed both numerically and experimentally, as discussed in Chapters 12 and 13. The mode-locked fiber laser constitutes an ideal test bed for investigating the complex nonlinear dynamics associated by these pulses.

The first experimental observation of an erupting soliton was reported in 2002 by Cundiff et al. in a sapphire laser [236], whereas the plain pulsating soliton was observed for the first time in 2004 by Soto-Crespo et al. in a mode-locked fiber ring laser [237]. Nevertheless, the lack of a high-resolution real-time diagnostic method has precluded a detailed characterization of the pulsating behavior of such pulses. This limitation has been broken recently through the development of a novel powerful real-time spectra measurement technique called dispersive Fourier transform (DFT) [238]. Using this technique, the spectrum of the soliton could be mapped into a temporal waveform by using dispersive element with enough group-velocity dispersion. Thus, the ultrafast spectral signals can be captured by a real-time oscilloscope with a high-speed photodetector. Actually, the DFT technique has enabled the experimental observation in real time of transient dynamics of diverse pulsating solitons and other complex ultrafast nonlinear phenomena in fiber lasers [191, 192, 239–253].

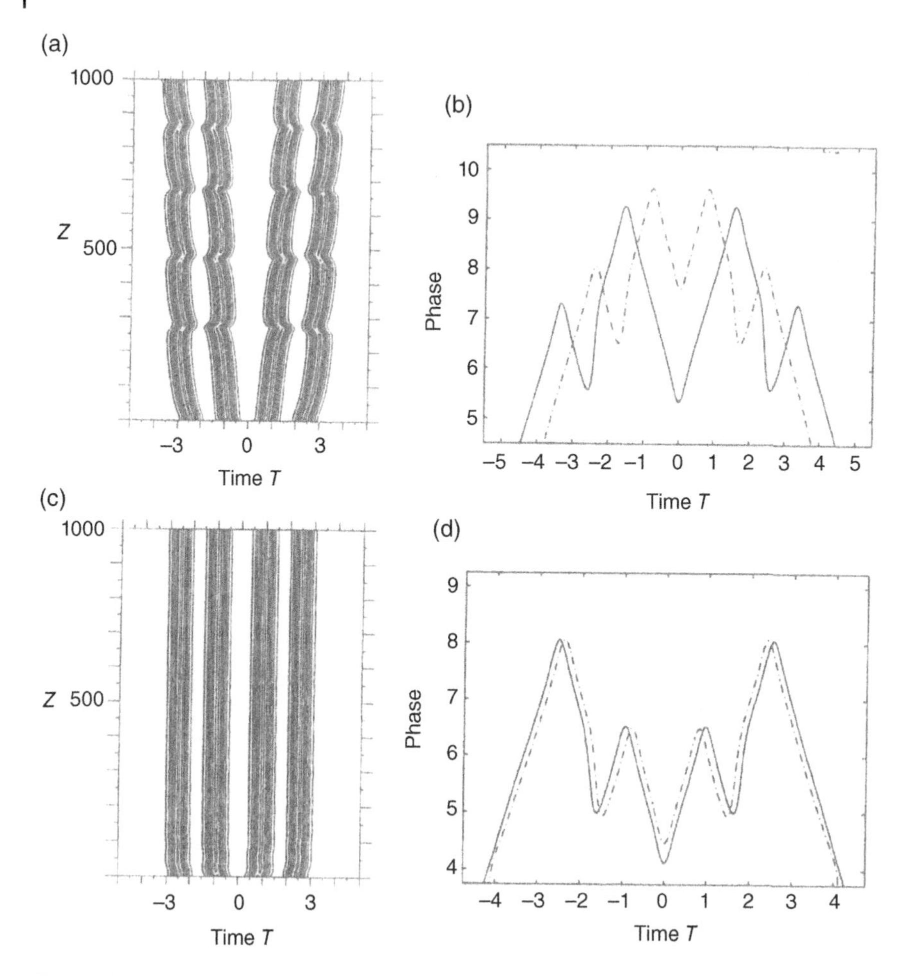

Figure 14.12 Four-plain pulse solutions and the correspondent phase profiles. The dash-dotted (full) lines in (b) and (d) correspond to the initial (final) phase profiles of the pulses in (a) and (c), respectively.

Soliton self-organization and pulsation in a passively mode-locked fiber laser was observed experimentally in 2018 [247]. The soliton pulsation process showed a period corresponding to tens of the cavity round-trip time. In another 2018 experiment, pulsating dissipative solitons in a mode-locked fiber laser at normal dispersion were observed for the first time using the DFT technique [248]. In that experiment, the artificial saturable absorbers, as well as the birefringent filter formed by the nonlinear polarization rotation, made the polarization controller an effective component to adjust the laser state from stationary to pulsating. The pulsating dissipative solitons were accompanied with the spectrum breathing and oscillating structures due to the nonlinear pulse propagation. Also in 2018, the first experimental evidence of the pulsating soliton with chaotic behavior in an ultrafast fiber laser was reported [249]. Using the DFT technique, the chaotic behavior of soliton pulsation was visualized by the fact that the mode-locked spectrum collapsed abruptly in an unpredictable way during the pulsating process.

In a 2019 experiment, three types of soliton pulsations were observed in an L-band normal-dispersion mode-locked fiber laser via the DFT technique [251]. They were classified as single-periodic pulsating soliton, double-periodic pulsating soliton, and soliton explosion.

These pulsations exhibited common features such as energy oscillation, bandwidth breathing, and temporal shift. However, the pulse was repeated every two oscillations for double-periodic pulsating soliton. In the case of soliton explosions, the spectrum was observed to crack into pieces at a periodic manner. The motion dynamics of a creeping soliton in a passively mode-locked fiber laser was also observed in 2020 using the DFT technique [253]. The periodical variation of pulse width, peak power, and motion range could be observed in real time, while the corresponding spectral evolution exhibited breathing dynamics.

Most phenomena related to soliton pulsations have been observed in mode-locked fiber lasers based on the nonlinear polarization rotation (NPR). Since a polarizer is required for the implementation of the NPR technique, the polarization degree of freedom is frozen, and pulsating solitons formed in these lasers are scalar ones. In a 2020 experiment, the generation of pulsating group-velocity-locked vector solitons in a net-normal dispersion fiber laser mode-locked by nonlinear multimode interference has been reported [254]. In another 2020 experiment, the vector nature of various pulsating solitons in an ultrafast fiber laser with single-wall carbon nanotubes has been investigated [255]. By virtue of the DFT technique, the polarization-resolved spectral evolution of pulsating vector solitons was measured in real time. Double-periodic pulsation in the cavity was also observed. Also in 2020, the vector features of pulsating solitons were studied in an erbium-doped fiber (EDF) laser [256]. Three categories of vector solitons with different polarization evolution characteristics could be obtained by adjusting the pump power and polarization controller, namely, pulsating polarization-locked vector soliton (PLVS), pulsating polarization-rotation vector soliton (PRVS), and progressive pulsating PRVS.

Using the DFT technique, distinct dynamical diversity of pulsating solitons in a fiber laser were revealed in a 2019 experiment [257]. Weak to strong explosive behaviors of pulsating solitons, as well as the rogue wave generation during explosions were observed. Moreover, the concept of soliton pulsation was extended to the multi-soliton case. It was found that the simultaneous pulsations of energy, separation, and relative phase difference could be observed for solitons inside the molecule, while the pulsations of each individual in a multi-soliton bunch could be regular or irregular. Other recent experiments have enabled the real-time observation of internal motion within ultrafast optical soliton molecules [258–264].

Figure 14.13 shows the characteristics of a periodically pulsating dual-soliton molecule [257]. Figure 14.13a shows the spatial-spectral dynamical evolution, from which one can see that the modulated spectra breathe periodically. The periodic variation of the energy of the soliton molecule is presented in the inset of Figure 14.13a, whereas the interference fringes of the soliton molecule spectrum can be clearly observed in Figure 14.13b. The temporal evolution of the soliton molecule is demonstrated in Figure 14.13c, from which one can see that two pulsating solitons bound as a unit and evolves stably. Figure 14.13d shows the RF spectrum and Figure 14.13e demonstrate that the separation and phase difference between the two solitons simultaneously vibrate with the round trips. Figure 14.13f shows the evolution trajectory of the two interacting solitons in the phase plane, indicating that the oscillations possess stably periodic behavior. ~

As mentioned in Chapter 13, the use of the DFT technique has boosted also the experimental investigations of soliton explosions in mode-locked fiber lasers [191, 192, 194, 195, 246, 265]. The first investigation of soliton explosions in an ultrafast fiber laser in the multi-soliton regime was reported in 2018 using the DFT technique [246]. In another 2018 experiment, the periodic spectrum changing via soliton explosion in a passively mode-locked fiber laser by a nonlinear polarization evolution was observed for the first time [194]. Soliton collision induced explosions in a mode-locked fiber laser were also reported in 2019 [195]. Up to seven nonlinear regimes were observed successively in the laser by increasing the pump power.

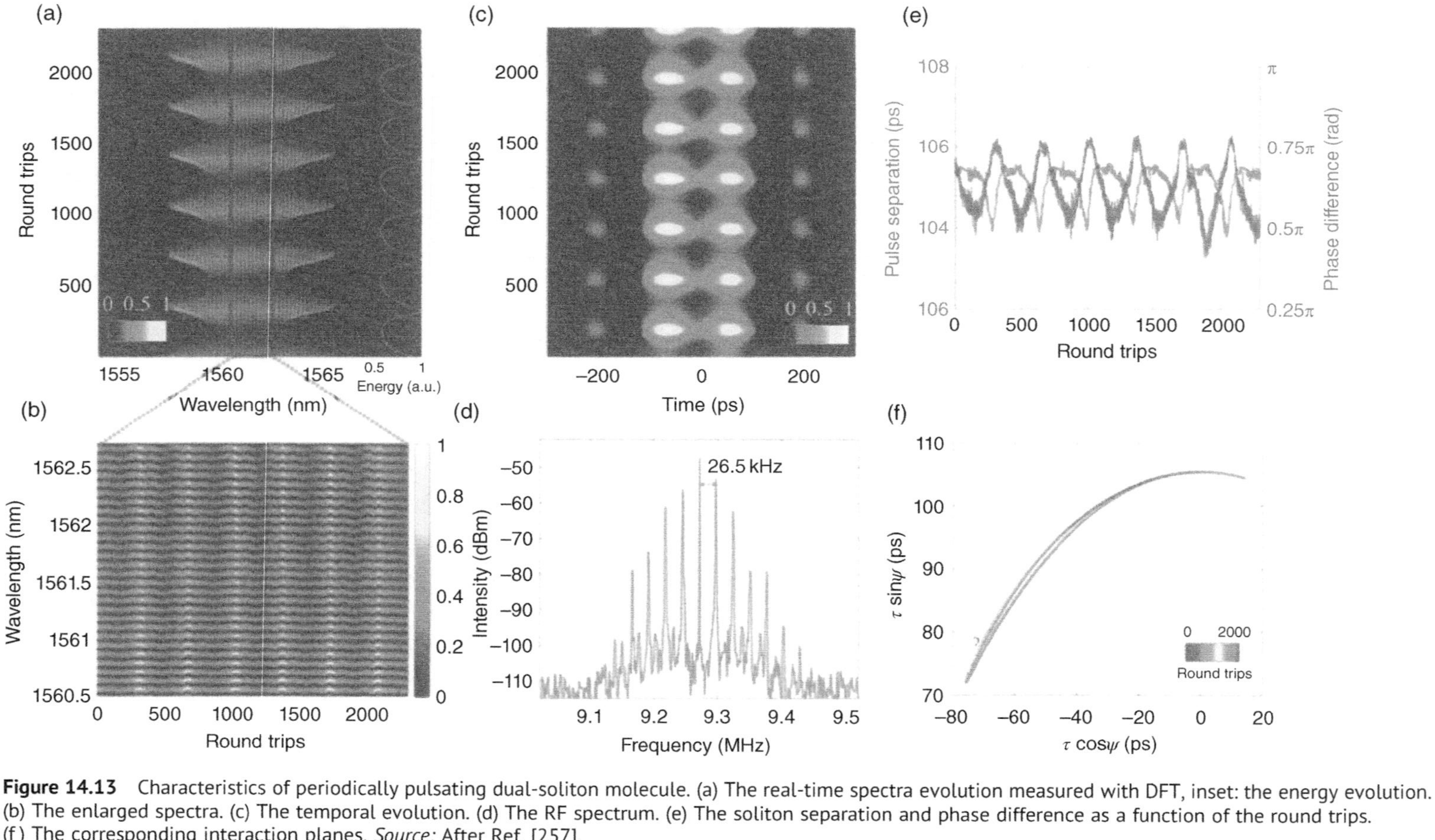

Figure 14.13 Characteristics of periodically pulsating dual-soliton molecule. (a) The real-time spectra evolution measured with DFT, inset: the energy evolution. (b) The enlarged spectra. (c) The temporal evolution. (d) The RF spectrum. (e) The soliton separation and phase difference as a function of the round trips. (f) The corresponding interaction planes. *Source:* After Ref. [257].

References

1 Siegman, A.E. (1986). *Lasers*. Mill Valley: University Science Books.
2 Haus, H. (2000). *IEEE J. Sel. Topics Quant. Electron.* **6**: 1173.
3 Kafka, J., Baer, T., and Hall, D. (1989). *Opt. Lett.* **14**: 1269.
4 Kärtner, F., Kopf, D., and Keller, U. (1995). *J. Opt. Soc. Am B* **12**: 486.
5 Richardson, D., Laming, R., Payne, D. et al. (1991). *Electron. Lett.* **27**: 542.
6 Duling, I. III (1991). *Opt. Lett.* **16**: 539.
7 Hofer, M., Ferman, M., Haberl, F. et al. (1991). *Opt. Lett.* **16**: 502.
8 Matsas, V., Newson, T., Richardson, D., and Payne, D. (1992). *Electron. Lett.* **28**: 1391.
9 Hargrove, L., Fork, R., and Pollack, M. (1964). *Appl. Phys. Lett.* **5**: 4.
10 Clobes, A. and Brienza, M. (1969). *Appl. Phys. Lett.* **14**: 287.
11 Wood, O. and Schwarz, S. (1968). *Appl. Phys. Lett.* **12**: 263.
12 Bulushev, A., Dianov, E., and Okhotnikov, O. (1990). *Opt. Lett.* **15**: 968.
13 Mollenauer, L. and Stolen, R. (1984). *Opt. Lett.* **9**: 13.
14 Mitschke, F. and Mollenauer, L. (1986). *IEEE J. Quantum Electron.* **22**: 2242.
15 Haus, H. and Islam, M. (1985). *IEEE J. Quantum Electron.* **21**: 1172.
16 Blow, K. and Wood, D. (1988). *J. Opt. Soc. Am B* **5**: 629.
17 Blow, K. and Nelson, B. (1988). *Opt. Lett.* **13**: 1026.
18 Ippen, E., Haus, H., and Liu, L. (1991). *J. Opt. Soc. Am B* **8**: 2077.
19 Kafka, J. and Baer, T. (1987). *Opt. Lett.* **12**: 181.
20 Digonnet, M.J. (ed.) (1993). *Rare Earth Doped Fiber Lasers and Amplifiers*. New York: Marcel Dekker.
21 Shimizu, M., Suda, H., and Horiguchi, M. (1987). *Electron. Lett.* **23**: 768.
22 Ball, G., Morey, W., and Glenn, W. (1991). *IEEE Photon. Technol. Lett.* **3**: 613.
23 Mortimore, D. (1988). *J. Lightwave Technol.* **6**: 1217.
24 Silfvast, W.T. (1996). *Laser Fundamentals*. New York: Cambridge University Press.
25 Svelto, O. (1998). *Principles of Lasers*, 4e. New York: Plenum.
26 Morkel, P., Jedrzejewski, K., and Taylor, E. (1993). *IEEE J. Quantum Electron.* **29**: 2178.
27 Abdulhalim, I., Pannell, C., Reekie, L. et al. (1993). *Opt. Commun.* **99**: 355.
28 Okhotnikov, O. and Salcedo, J. (1994). *Electron. Lett.* **30**: 702.
29 Paschotta, R., Haring, R., Gini, E. et al. (1999). *Opt. Lett.* **24**: 388.
30 Renaud, C., Selvas-Aguilar, R., Nilsson, J. et al. (1999). *IEEE Photon. Technol. Lett.* **8**: 976.
31 Alvarez-Chavez, J., Offerhaus, H., Nilsson, J. et al. (2000). *Opt. Lett.* **25**: 37.
32 Romagnoli, M., Midrio, M., Franco, P., and Fontana, F. (1995). *J. Opt. Soc. Am. B* **12**: 1732.
33 Gordon, J. (1983). *Opt. Lett.* **8**: 596.
34 Gordon, J. (1986). *Opt. Lett.* **11**: 662.
35 Gordon, J. and Haus, H. (1986). *Opt. Lett.* **11**: 665.
36 Gordon, J. (1992). *J. Opt. Soc. Am. B* **9**: 91.
37 Abedin, K., Hyodo, M., and Onodera, N. (2000). *Electron. Lett.* **36**: 1185.
38 Noguchi, K., Mitomi, O., and Miyazawa, H. (1998). *J. Lightwave Technol.* **16**: 615.
39 Smith, K., Armitage, J., Wyatt, R. et al. (1990). *Electron. Lett.* **26**: 1149.
40 Takada, A. and Miyazawa, H. (1990). *Electron. Lett.* **26**: 216.
41 Takara, H., Kawanishi, S., Saruwatari, M., and Noguchi, K. (1992). *Electron. Lett.* **28**: 2095.
42 Harvey, G. and Mollenauer, L. (1993). *Opt. Lett.* **18**: 107.
43 Pfeiffer, T. and Veith, G. (1993). *Electron. Lett.* **29**: 1849.
44 Pataca, D., Rocha, M., Smith, K. et al. (1994). *Electron. Lett.* **30**: 964.

45 Marti-Panameno, E., Sanchez-Mondragon, J., and Vysloukh, V. (1994). *IEEE J. Quantum Electron.* **30**: 822.
46 Yoshida, E., Kimura, Y., and Nakazawa, M. (1995). *Electron. Lett.* **31**: 377.
47 Haus, H., Jones, D., Ippen, E., and Wong, W. (1996). *J. Lightwave Technol.* **14**: 622.
48 Jones, D., Haus, H., and Ippen, E. (1996). *Opt. Lett.* **21**: 1818.
49 Baldo, M., Town, G., and Romagnoli, M. (1997). *Opt. Commun.* **140**: 19.
50 Wey, J., Goldhar, J., and Burdge, G. (1997). *J. Lightwave Technol.* **15**: 1171.
51 Bakshi, B., Andrekson, P., and Zhang, X. (1998). *Electron. Lett.* **34**: 884.
52 Jeon, M., Lee, H., Kim, K. et al. (1998). *Opt. Commun.* **149**: 312.
53 Abedin, K., Onodera, N., and Hyodo, M. (1999). *Opt. Lett.* **24**: 1564.
54 Yoshida, E., Shimizu, N., and Nakazawa, M. (1999). *IEEE Photon. Technol. Lett.* **11**: 1587.
55 Carruthers, T., Duling, I. III, Horowitz, M., and Menyuk, C. (2000). *Opt. Lett.* **25**: 153.
56 Horowitz, M., Menyuk, C., Carruthers, T., and Duling, I. III (2000). *IEEE Photon. Technol. Lett.* **12**: 266.
57 Thoen, E., Grein, M., Koontz, E. et al. (2000). *Opt. Lett.* **25**: 948.
58 Loh, W.H., Atkinson, D., Morkel, P.R. et al. (1993). *IEEE Photon. Technol. Lett.* **5**: 35.
59 Loh, W.H., Atkinson, D., Morkel, P.R. et al. (1993). *Appl. Phys. Lett.* **63**: 4.
60 Barnett, B., Rahman, L., Islam, M. et al. (1995). *Opt. Lett.* **20**: 471.
61 Keller, U., Weingarten, K., Kartner, F. et al. (1996). *IEEE J. Sel. Top. Quantum Electron.* **2**: 435.
62 Collings, B., Bergman, K., Cundiff, S. et al. (1997). *IEEE J. Sel. Topics Quantum Electron.* **3**: 1065.
63 Hofer, M., Fermann, M., and Goldberg, L. (1998). *IEEE Photon. Technol. Lett.* **10**: 1247.
64 Jiang, M., Sucha, G., Fermann, M. et al. (1999). *Opt. Lett.* **24**: 1074.
65 Haus, J.W., Hapduk, M., Kaechele, W. et al. (2000). *Opt. Commun.* **174**: 204.
66 Yamashita, S., Inoue, Y., Maruyama, S. et al. (2004). *Opt. Lett.* **29**: 1581.
67 Song, Y., Set, S., Goh, C., and Kotake, T. (2005). *IEEE Photonics Technol. Lett.* **17**: 1623.
68 Song, Y., Yamashita, S., Einarsson, E., and Maruyama, S. (2007). *Opt. Lett.* **32**: 1399.
69 Sun, Z., Rozhin, A., Wang, F. et al. (2008). *Appl. Phys. Lett.* **93**: 061114.
70 Nicholson, J. and DiGiovanni, D. (2008). *IEEE Photonics Technol. Lett.* **20**: 2123.
71 Solodyankin, M., Obraztsova, E., Lobach, A. et al. (2008). *Opt. Lett.* **33**: 1336.
72 Choi, S., Rotermund, F., Jung, H. et al. (2009). *Opt. Express* **17**: 21788.
73 Ouyang, C., Shum, P., Wang, H. et al. (2010). *Opt. Lett.* **35**: 2320.
74 Jiang, K., Fu, S., Shum, P., and Lin, C. (2010). *IEEE Photonics Technol. Lett.* **22**: 754.
75 Kieu, K., Jones, R., and Peyghambarian, N. (2010). *IEEE Photonics Technol. Lett.* **22**: 1521.
76 Zhang, H., Bao, Q., Tang, D. et al. (2009). *Appl. Phys. Lett.* **95**: 141103.
77 Zhao, L., Tang, D., Zhang, H. et al. (2010). *Opt. Lett.* **35**: 3622.
78 Xu, J., Wu, S., Li, H. et al. (2012). *Opt. Express* **20**: 23653.
79 Sotor, J., Sobon, G., and Abramski, K. (2012). *Opt. Lett.* **37**: 2166.
80 Ismail, M., Ahmad, H., and Harun, S. (2014). *IEEE J. Quantum Electron.* **50**: 85.
81 He, X., Chen, T., and Wang, D. (2014). *Laser Phys.* **24**: 085109.
82 Cui, Y., Liu, X., and Zeng, C. (2014). *Laser Phys. Lett.* **11**: 055106.
83 Cheng, Z., Li, H., Shi, H. et al. (2015). *Opt. Express* **23**: 7000.
84 Jeong, H., Choi, S., Kim, M. et al. (2016). *Opt. Express* **24**: 14152.
85 Song, Y., Liang, Z., Zhang, H. et al. (2017). *IEEE Photonics J.* **9**: 4502308.
86 Gui, L. and Yang, C. (2018). *IEEE Photonics J.* **10**: 1502609.
87 Wang, Z., Wang, Z., Liu, Y. et al. (2018). *Laser Phys. Lett.* **15**: 055101.
88 Yang, G., Liu, Y., Wang, Z. et al. (2018). *Optics Laser Technol.* **105**: 76.
89 Sotor, J., Sobon, G., Kowalczyk, M. et al. (2015). *Opt. Lett.* **40**: 3885.

90 Chen, Y., Chen, S., Liu, J. et al. (2016). *Opt. Express* **24**: 13316.
91 Song, Y., Chen, S., Zhang, Q. et al. (2016). *Opt. Express* **24**: 25933.
92 Hisyam, M., Rusdi, M., Latiff, A., and Harun, S. (2017). *IEEE J. Sel. Top. Quantum Electron.* **23**: 1100205.
93 Song, H., Wang, Q., Zhang, Y., and Li, L. (2017). *Opt. Commun.* **394**: 157.
94 Gao, B., Ma, C., Huo, J. et al. (2018). *Opt. Commun.* **410**: 191.
95 Jin, X., Hu, G., Zhang, M. et al. (2018). *Opt. Express* **26**: 12506.
96 Wu, K., Chen, B., Zhang, X. et al. (2018). *Opt. Commun.* **406**: 214.
97 Gao, B., Guo, W., Huo, J. et al. (2018). *Opt. Commun.* **406**: 192.
98 Ma, C., Tian, X., Gao, B., and Wu, G. (2018). *Opt. Commun.* **406**: 177.
99 Wang, Y., Mao, D., Gan, X. et al. (2015). *Opt. Express* **23**: 205.
100 Jiang, Z., Chen, H., Li, J. et al. (2017). *Appl. Phys. Express* **10**: 122702.
101 Wang, Z., He, R., Liu, Y. et al. (2018). *Appl. Phys. Express* **11**: 072504.
102 Wang, P., Zhao, K., Gui, L. et al. (2018). *IEEE Photonics Technol. Lett.* **30**: 1210.
103 Doran, N. and Wood, D. (1988). *Opt. Lett.* **13**: 56.
104 Fermann, M., Haberl, F., Hofer, M., and Hochreiter, H. (1990). *Opt. Lett.* **15**: 752.
105 Stolen, R., Botineau, J., and Ashkin, A. (1982). *Opt. Lett.* **7**: 512.
106 Doran, N., Forrester, D., and Nayar, B. (1989). *Electron. Lett.* **25**: 267.
107 Blow, K., Doran, N., and Nayar, B. (1989). *Opt. Lett.* **14**: 754.
108 Yamada, E. and Nakazawa, M. (1994). *IEEE J. Quantum Electron.* **30**: 1842.
109 Duling, I. III (1991). *Electron. Lett.* **27**: 544.
110 Richardson, D., Laming, R., Payne, D. et al. (1991). *Electron. Lett.* **27**: 730.
111 Richardson, D., Laming, R., Payne, D. et al. (1991). *Electron. Lett.* **27**: 1451.
112 Chernikov, S. and Mamyshev, P. (1991). *J. Opt. Soc. Am. B* **8**: 1633.
113 Guy, M., Noske, D., and Taylor, J. (1993). *Opt. Lett.* **18**: 1447.
114 Noske, D. and Taylor, J. (1993). *Electron. Lett.* **29**: 2200.
115 Stentz, A. and Boyd, R. (1994). *Opt. Lett.* **19**: 1462.
116 Noske, D., Guy, M., Rottwitt, K. et al. (1994). *Opt. Commun.* **108**: 297.
117 Noske, D., Pandit, N., and Taylor, J. (1995). *Opt. Commun.* **115**: 105.
118 Town, G., Chow, J., and Romagnoli, M. (1995). *Electron. Lett.* **31**: 1452.
119 Oh, W., Kim, B., and Lee, H. (1996). *IEEE J. Quantum Electron.* **32**: 333.
120 Tsun, T., Islam, M., and Chu, P. (1997). *Opt. Commun.* **141**: 65.
121 Winful, H. (1985). *Appl. Phys. Lett.* **47**: 213.
122 Menyuk, C. (1989). *IEEE J. Quantum Electron.* **25**: 2674.
123 Matsas, V., Newson, T., and Zervas, M. (1992). *Opt. Commun.* **92**: 61.
124 Tamura, K., Haus, H., and Ippen, E. (1992). *Electron. Lett.* **28**: 2226.
125 Hofer, M., Ober, M., Haberl, F., and Fermann, M. (1992). *IEEE J. Quantum Electron.* **28**: 720.
126 Tamura, K., Ippen, E., Haus, H., and Nelson, L. (1993). *Opt. Lett.* **18**: 1080.
127 Carruthers, T., Duling, I., and Dennis, M. (1994). *Electron. Lett.* **30**: 1051.
128 Davey, R., Fleming, R., Smith, K. et al. (1991). *Electron. Lett.* **27**: 2087.
129 Haus, H., Tamura, K., Nelson, L., and Ippen, E. (1995). *IEEE J. Quantum Electron.* **31**: 591.
130 Smith, N., Knox, F., Doran, N. et al. (1996). *Electron. Lett.* **32**: 54.
131 Namiki, S. and Haus, H. (1997). *IEEE J. Quantum Electron.* **33**: 649.
132 Bakhshi, B., Andrekson, P., and Zhang, X. (1998). *Opt. Fiber Technol.* **4**: 293.
133 Waiyapot, S. and Matsumoto, M. (2001). *Opt. Commun.* **188**: 167.
134 Clark, T., Carruthers, T., Matthews, P., and Duling, I. III (1999). *Electron. Lett.* **35**: 720.
135 Tamura, K., Ippen, E., and Haus, H. (1995). *Appl. Phys. Lett.* **67**: 158.

136 Ablowitz, M., Horikis, T., and Ilan, B. (2008). *Phys. Rev. A* **77**: 033814.
137 Bale, B., Boscolo, S., and Turitsyn, S. (2009). *Opt. Lett.* **34**: 3286.
138 Lei, D.J. and Dong, H. (2013). *Optik* **124**: 2544.
139 Han, X. (2014). *J. Lightwave Technol.* **32**: 1472.
140 Wang, S., Fan, X., Zhao, L. et al. (2014). *Appl. Opt.* **53**: 8216.
141 Lin, J., Chan, C., Lee, H., and Chen, Y. (2015). *IEEE Photonics J.* **7**: 1.
142 Wang, W., Wang, F., Yu, Q. et al. (2016). *Opt. Laser Technol.* **85**: 41.
143 Dvoretskiy, D.A., Sazonkin, S., Negin, M. et al. (2017). *IEEE Photonics Technol. Lett.* **29**: 1588.
144 Hou, L., Guo, H., Wang, Y. et al. (2018). *Opt. Express* **26**: 9063.
145 Luo, Y., Xiang, Y., Liu, B. et al. (2018). *IEEE Photonics J.* **10**: 7105210.
146 Chong, A., Renninger, W.H., and Wise, F.W. (2007). *Opt. Lett.* **32**: 2408.
147 An, J., Kim, D., Dawson, J. et al. (2007). *Opt. Lett.* **32**: 2010.
148 Kieu, K., Renninger, W., Chong, A., and Wise, F. (2009). *Opt. Lett.* **34**: 593.
149 Lecaplain, C., Ortac, B., and Hideur, A. (2009). *Opt. Lett.* **34**: 3731.
150 Zhang, H., Tang, D., Knize, R. et al. (2010). *Appl. Phys. Lett.* **96**: 111112.
151 Ortaç, B., Baumgartl, M., Limpert, J., and Tünnermann, A. (2009). *Opt. Lett.* **34**: 1585.
152 Kalashnikov, V., Podivilov, E., Chernykh, A., and Apolonski, A. (2006). *Appl. Phys. B* **83**: 503.
153 Kelleher, E., Travers, J., Ippen, E. et al. (2009). *Opt. Lett.* **34**: 3526.
154 Chang, W., Ankiewicz, A., Soto-Crespo, J., and Akhmediev, N. (2008). *J. Opt. Soc. Am. B* **25**: 1972.
155 Chang, W., Soto-Crespo, J., Ankiewicz, A., and Akhmediev, N. (2009). *Phys. Rev. A* **79**: 033840.
156 Wu, X., Tang, D., Zhang, H., and Zhao, L. (2009). *Opt. Express* **17**: 5580.
157 Ding, E., Grelu, P., and Kutz, J.N. (2011). *Opt. Lett.* **36**: 1146.
158 Komarov, A., Amrani, F., Dmitriev, A. et al. (2013). *Phys. Rev. A* **87**: 023838.
159 Xu, Y., Song, Y., Du, G. et al. (2015). *IEEE Photonics J.* **7**: 1502007.
160 Krzempek, K., Sotor, J., and Abramski, K. (2016). *Opt. Lett.* **41**: 4995.
161 Semaan, G., Braham, F., Fourmont, J. et al. (2016). *Opt. Lett.* **41**: 4767.
162 Krzempek, K., Tomaszewska, D., and Abramski, K. (2017). *Opt. Express* **25**: 24853.
163 Du, T., Li, W., Ruan, Q. et al. (2018). *Appl. Phys. Express* **11**: 052701.
164 Latas, S.C., Ferreira, M.F.S., and Facão, M. (2017). *J. Opt. Soc. Am. B* **34**: 1033.
165 Latas, S.C. and Ferreira, M.F. (2019). *J. Opt. Soc. of Am. B* **36**: 3016.
166 Haus, H. (1975). *J. Appl. Phys.* **46**: 3049.
167 Haus, H., Fujimoto, J., and Ippen, E. (1991). *J. Opt. Soc. Am. B* **8**: 2068.
168 Bekki, N. and Nozaki, K. (1985). *Phys. Lett. A* **110**: 133.
169 Van Saarloos, W. and Hohenberg, P. (1990). *Phys. Rev. Lett.* **64**: 749.
170 Aranson, I. and Kramer, L. (2002). *Rev. Mod. Phys.* **74**: 99.
171 Haus, H., Ippen, E., and Tamura, K. (1994). *IEEE J. Quantum Electron.* **30**: 200.
172 Leblond, H., Salhi, M., Hideur, A. et al. (2002). *Phys. Rev. A* **65**: 063811.
173 Salhi, M., Leblond, H., and Sanchez, F. (2003). *Phys. Rev. A* **68**: 033815.
174 Salhi, M., Leblond, H., Sanchez, F. et al. (2004). *J. Opt. A: Pure Appl. Opt.* **6**: 774.
175 Akhmediev, N., Afanasjev, V., and Soto-Crespo, J. (1996). *Phys. Rev. E* **53**: 1190.
176 Soto-Crespo, J., Akhmediev, N., Afanasjev, V., and Wabnitz, S. (1997). *Phys. Rev. E* **55**: 4783.
177 Ding, E. and Kutz, J. (2009). *Opt. Soc. Am. B* **26**: 2290.
178 Komarov, A., Leblond, H., and Sanchez, F. (2005). *Phys. Rev. E* **72**: 025604R.
179 Akhmediev, N. and Ankiewicz, A. (1997). *Solitons, Nonlinear Pulses and Beams*. Chapman and Hall.
180 Akhmediev, N. and Ankiewicz, A. (ed.) (2008). *Dissipative Solitons: From Optics to Biology and Medicine*. Berlin: Springer.
181 Ferreira, M.F. (2011). *Nonlinear Effects in Optical Fibers*. Hoboken, NJ: John Wiley & Sons.

182 Grelu, P. and Akhmediev, N. (2012). *Nat. Photonics* **6**: 84.
183 Akhmediev, N., Soto-Crespo, J., and Grelu, P. (2008). *Phys.Lett. A* **372**: 3124.
184 Grelu, P., Chang, W., Ankiewicz, A. et al. (2010). *J. Opt. Soc. Am. B* **27**: 2336.
185 Du, W., Li, H., Li, J. et al. (2019). *Opt Express* **27**: 8059.
186 Akhmediev, N., Soto-Crespo, J.M., and Town, G. (2001). *Phys. Rev. E* **63**: 056602.
187 Soto-Crespo, J., Akhmediev, N., and Ankiewicz, A. (2000). *Phys. Rev. Lett.* **85**: 2937.
188 Latas, S. and Ferreira, M. (2010). *Opt. Lett.* **35**: 1771.
189 Latas, S. and Ferreira, M. (2011). *Opt. Lett.* **36**: 3085.
190 Latas, S., Ferreira, M., and Facão, M. (2011). *Appl. Phys. B* **104**: 131.
191 Runge, A., Broderick, N., and Erkintalo, M. (2015). *Optica* **2**: 36.
192 Runge, A., Broderick, N., and Erkintalo, M. (2016). *J. Opt. Soc. Am. B* **33**: 46.
193 Liu, M., Luo, A., Yan, Y. et al. (2016). *Opt. Lett.* **41**: 1181.
194 Suzuki, M., Boyraz, O., Asghari, H. et al. (2018). *Opt. Lett.* **43**: 1862.
195 Peng, J. and Zeng, H. (2019). *Comm Phys.* **2**: 34.
196 Soto-Crespo, J. and Akhmediev, N. (2002). *Phy. Rev. E* **66**: 066610.
197 Akhmediev, N. and Ankiewicz, A. (ed.) (2005). *Dissipative Solitons*. Berlin: Springer.
198 Lega, J., Moloney, J., and Newell, A. (1994). *Phys. Rev. Lett.* **73**: 2978.
199 Sakaguchi, H. and Brand, H. (1998). *Phys. D* **117**: 95.
200 Wang, H. and Yanti, L. (2011). *Numer. Math.Theor. Appl.* **4**: 237.
201 Latas, S., Ferreira, M., and Facão, M. (2018). *J. Opt. Soc. Am. B* **35**: 2266.
202 Cundiff, S., Collings, B., and Knox, W. (1997). *Opt. Express* **1**: 12.
203 Cundiff, S., Collings, B., Akhmediev, N. et al. (1999). *Phys. Rev. Lett.* **82**: 3988.
204 Collings, B., Cundiff, S., Akhmediev, N. et al. (2000). *J. Opt. Soc. Am. B* **17**: 354.
205 Soto-Crespo, J., Akhmediev, N., Collings, B. et al. (2000). *J. Opt. Soc. Am. B* **17**: 366.
206 Zhang, H., Tang, D., Zhao, L., and Xiang, N. (2008). *Opt. Express* **16**: 12618.
207 Zhao, L., Tang, D., Zhang, H. et al. (2008). *Opt. Express* **16**: 9528.
208 Zhao, L., Tang, D., Zhang, H., and Wu, X. (2008). *Opt. Express* **16**: 10053.
209 Zhao, L., Tang, D., Wu, X. et al. (2009). *Opt. Lett.* **34**: 3059.
210 Mou, C., Sergeyev, S., Rozhin, A., and Turistyn, S. (2011). *Opt. Lett.* **36**: 3831.
211 Song, Y., Li, L., Zhang, H. et al. (2013). *Opt. Express* **21**: 10010.
212 Song, Y., Zhang, H., Zhao, L. et al. (2016). *Opt. Express* **24**: 1814.
213 Luo, Y., Li, L., Liu, D. et al. (2016). *Opt. Express* **24**: 18718.
214 Liu, M., Luo, A., Luo, Z., and Xu, W. (2017). *Opt. Lett.* **42**: 330.
215 Akosman, A., Zeng, J., Samolis, P., and Sander, M. (2018). *IEEE J. Sel. Top. Quantum Electron.* **24**: 1101107.
216 Li, D., Shen, D., Li, L. et al. (2018). *Opt. Lett.* **43**: 1222.
217 Ma, J., Shao, G., Song, Y. et al. (2019). *Opt Lett.* **44**: 2185.
218 Grelu, P., Belhache, F., Gutty, F., and Soto-Crespo, J. (2002). *Opt. Lett.* **27**: 966.
219 Tang, D., Man, W., Tam, H., and Drummond, P. (2001). *Phys. Rev. A* **64**: 033814.
220 Seong, N. and Kim, D. (2002). *Opt. Lett.* **27**: 1321.
221 Grelu, P., Beal, J., and Soto-Crespo, J. (2003). *Opt. Express* **11**: 2238.
222 Martel, G., Chédot, C., Réglier, V. et al. (2007). *Opt. Lett.* **32**: 343.
223 Latas, S. and Ferreira, M. (2005). *Opt. Fiber Technol.* **11**: 292.
224 Stratmann, M., Pagel, T., and Mitschke, F. (2005). *Phys. Rev. Lett.* **95**: 143902.
225 Pang, M., He, W., Jiang, X., and Russell, P. (2016). *Nat. Photonics* **10**: 454.
226 Luo, A., Zheng, X., Liu, M. et al. (2015). *Appl. Phys. Express* **8**: 042702.
227 Wang, P., Bao, C., Fu, B. et al. (2016). *Opt. Lett.* **41**: 2254.

228 Ma, C., Tian, X., Gao, B., and Wu, G. (2017). *Laser Phys.* **27**: 065102.
229 Liu, B., Xiang, Y., Luo, Y. et al. (2018). *Appl. Phys. B: Lasers Opt.* **124**: 151.
230 Niknafs, A., Rooholamininejad, H., and Bahrampour, A. (2018). *Laser Phys.* **28**: 045406.
231 Qin, H.Q., Xiao, X., Wang, P., and Yang, C. (2018). *Opt. Lett.* **43**: 1982.
232 Mou, C., Sergeyev, S., Rozhin, A., and Turitsyn, S. (2013). *Opt. Express* **21**: 26868.
233 Gui, L., Zhang, W., Li, X. et al. (2011). *IEEE Photonics Technol. Lett.* **23**: 1790.
234 Luo, Y., Cheng, J., Liu, B. et al. (2017). *Sci. Rep.* **7**: 2369.
235 Mollenauer, L., Stolen, R., and Gordon, J. (1980). *Phys. Rev. Lett.* **45**: 1095.
236 Cundiff, S., Soto-Crespo, J., and Akhmediev, N. (2002). *Phys. Rev. Lett.* **88**: 073903.
237 Soto-Crespo, J., Grapinet, M., Grelu, P., and Akmediev, N. (2004). *Phys. Rev. E* **70**: 066612.
238 Goda, K. and Jalali, B. (2013). *Nat. Photonics* **7**: 102.
239 Herink, G., Jalali, B., Ropers, C., and Solli, D. (2016). *Nat. Photonics* **10**: 321.
240 Wei, X., Li, B., Yu, Y. et al. (2017). *Opt. Express* **25**: 29098.
241 Krupa, K., Nithyanandan, K., and Grelu, P. (2017). *Optica* **4**: 1239.
242 Wang, P., Xiao, X., Zhao, H., and Yang, C. (2017). *IEEE Photonics J.* **9**: 1.
243 Ryczkowski, P., Närhi, M., Billet, C. et al. (2018). *Nat. Photonics* **12**: 221.
244 Chen, H., Liu, M., Yao, J. et al. (2018). *Opt. Express* **26**: 2972.
245 Chen, H., Liu, M., Yao, J. et al. (2018). *IEEE Photonics J.* **10**: 1.
246 Yu, Y., Luo, Z., Kang, J., and Wong, K. (2018). *Opt. Lett.* **43**: 4132.
247 Wang, Z., Wang, Z., Liu, Y. et al. (2018). *Opt. Lett.* **43**: 478.
248 Du, Y., Xu, Z., and Shu, X. (2018). *Opt. Lett.* **43**: 3602.
249 Wei, Z., Liu, M., Ming, S. et al. (2018). *Opt. Lett.* **43**: 5965.
250 Wang, G., Chen, G., Li, W., and Zeng, C. (2019). *IEEE J. Sel. Top. Quantum Electron.* **25**: 1.
251 Wang, X., Liu, Y., Wang, Z. et al. (2019). *Opt. Express* **27**: 17729.
252 Wei, Z., Liu, M., Ming, S. et al. (2020). *Opt. Lett.* **45**: 531.
253 Zhang, Y., Cui, Y., Huang, L. et al. (2020). *Op. Lett.* **45**: 6246.
254 Luo, Y., Xiang, Y., Shum, P.P. et al. (2020). *Opt. Express* **28**: 4216.
255 Du, W., Li, H., Li, J. et al. (2020). *Opt. Lett.* **45**: 5024.
256 Li, T., Liu, M., Luo, A. et al. (2020). *Opt. Express* **28**: 32010.
257 Chen, H., Tan, Y., Long, J. et al. (2019). *Opt. Express* **27**: 28507.
258 Herink, G., Kurtz, F., Jalali, B. et al. (2017). *Science* **356**: 50.
259 Krupa, K., Nithyanandan, K., Andral, U. et al. (2017). *Phys. Rev. Lett.* **118**: 243901.
260 Liu, X., Yao, X., and Cui, Y. (2018). *Phys. Rev. Lett.* **121**: 023905.
261 Peng, J. and Zeng, H. (2018). *Laser Photonics Rev.* **12**: 1800009.
262 Hamdi, S., Coillet, A., and Grelu, P. (2018). *Opt. Lett.* **43**: 4965.
263 Liu, M., Li, H., Luo, A. et al. (2018). *J. Opt.* **20**: 034010.
264 Wang, Z., Nithyanandan, K., Coillet, A. et al. (2019). *Nat. Commun.* **10**: 830.
265 Wei, Z., Liu, M., Ming, S. et al. (2019). *Opt. Lett.* **45**: 531.

15

Other Applications of Optical Solitons

Besides deserving much interest in the areas of optical communications and fiber laser systems, optical fiber solitons are also useful in many other applications, namely all-optical switching, optical regeneration, and pulse compression. Actually, two widely adopted techniques to produce ultrashort pulses are the higher-order soliton-effect [1–3] and adiabatic pulse-compression techniques [4–6]. Fiber gratings can also be used for pulse compression. Ultrashort pulses have great potential in several areas of research and application, such as measurements of ultrafast processes, time-resolved spectroscopy and sampling systems.

15.1 All-Optical Switching

An optical switch with fully transparent features in both time and wavelength domains is a key device for providing several functions required in optical signal processing. Ultrafast switching operations with speeds beyond the limits of electrical devices can only be achieved through an all-optical approach. The third-order nonlinearity in optical fibers can provide such features [7, 8].

15.1.1 The Fiber Coupler

Fiber couplers, also known as directional couplers, are four-port devices that are commonly used in many optical fiber systems [9–12]. Their function is to split coherently an optical field into two physically separated parts. Figure 15.1 shows the schematic of a fiber coupler in which the cores of two single-mode fibers become conveniently close in the central region. In these conditions, the fundamental modes propagating in each core overlap partially in the cladding region between the two cores, which can lead to the transfer of optical power from one core to the other. One important parameter of a fiber coupler is the *coupling length*, L_c, which corresponds to the shortest distance at which maximum power is transferred from one core to the other core. A fiber coupler with a length $L = L_c$ transfers all the input power to the second core, whereas all of the launched power returns to the same core when $L = 2L_c$.

The nonlinear effects in fiber couplers were studied since 1982 [13–22] and can be used to achieve all-optical switching. Actually, an input optical pulse can be directed toward different output ports depending on its peak power. An important parameter in this case is the *switching power*, P_c, which corresponds to the power of a continuous wave (CW) signal launched in one input port which becomes equally divided between the two output port cores of a fiber coupler of length $L = L_c$.

Solitons in Optical Fiber Systems, First Edition. Mario F. S. Ferreira.

Figure 15.1 Schematic of a fiber coupler.

Nonlinear pulse switching in fiber couplers can be described by the following normalized coupled field equations [11, 12]:

$$i\frac{\partial q_1}{\partial Z} - \frac{s}{2}\frac{\partial^2 q_1}{\partial T^2} + |q_1|^2 q_1 + Kq_2 = 0 \tag{15.1}$$

$$i\frac{\partial q_2}{\partial Z} - \frac{s}{2}\frac{\partial^2 q_2}{\partial T^2} + |q_2|^2 q_2 + Kq_1 = 0 \tag{15.2}$$

where q_1 and q_2 are the normalized amplitudes of the optical fields propagating inside the two cores, $s = \text{sgn}(\beta_2)$ and K is the normalized coupling coefficient between the two fibers. For $K = 0$, these equations reduce to two uncoupled NLS equations.

In general, the coupled NLS equations, Eqs. (15.1) and (15.2), must be solved numerically using the split-step Fourier method described in Section 6.8. The switching behavior depends on whether the group-velocity dispersion (GVD) is normal or anomalous. It has been shown that, in the anomalous dispersion regime, soliton pulses switch between the cores as an entire pulse [23]. However, pulse switching does not occur in the normal-dispersion regime if the dispersion length L_D becomes comparable to the coupling length L_c [24]. Figure 15.2 illustrates the superior switching characteristics of solitons (full curve) compared with quasi-CW pulses (dashed curve). These results were obtained by solving Eqs. (15.1) and (15.2) in the case of a fiber coupler of length $L = L_c$ with the initial conditions:

$$q_1(0, T) = N\text{sech}(T) \quad q_2(0, T) = 0 \tag{15.3}$$

15.1.2 The Sagnac Interferometer

All-optical switching can be realized using the Sagnac interferometer, as described in Section 14.3. Actually, if the coupler in Figure 14.3 has a power-splitting fraction $f \neq 0.5$, the fiber loop exhibits different behavior at low and high powers and can act as an optical switch. At low powers, we verify

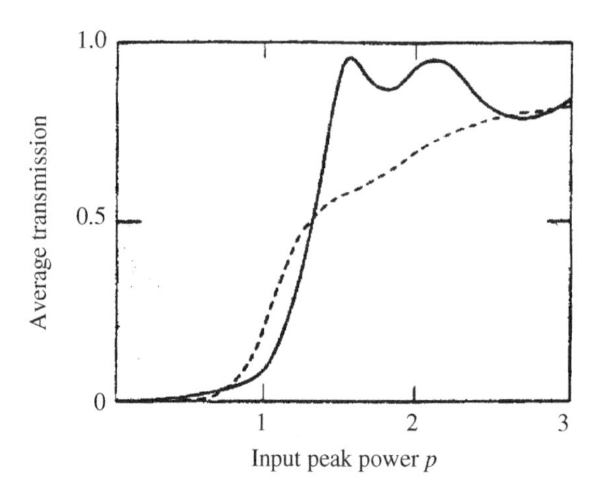

Figure 15.2 Transmitted pulse energy as a function of normalized pulse peak power $p = P_0/P_c$ in the case of solitons (solid curve) and quasi-CW pulses (dashed curve) for a coupler of length. $L = L_c$. *Source:* After Ref. [23].

from Eq. (14.5) that the transmittivity of a nonlinear optical loop mirror (NOLM) has a minimum, given by

$$T_{\min} = 1 - 4f(1-f) \tag{15.4}$$

This shows that the NOLM can be used to suppress the zero-level noise, which becomes more effective for a coupling ratio near 0.5 [25, 26]. At high input powers, we verify from Eq. (14.5) that a maximum transmittivity $T = 1$ is achieved if the condition

$$\left|f - \frac{1}{2}\right|\gamma P_0 L = (2N-1)\frac{\pi}{2}, \quad (N \text{ integer}) \tag{15.5}$$

is satisfied. For $N = 1$, a fiber loop 50-m-long will need a switching power of 63 W when $f = 0.45$ and $\gamma = 10\ \ \mathrm{W^{-1}/km}$.

The switching power of a Sagnac interferometer can be significantly reduced if a fiber amplifier is included within the loop. This device is referred to as the *nonlinear amplifying loop mirror* (NALM) and it can operate as a switch even if a 3-dB coupler ($f = 0.5$) is used. The asymmetry of an NALM is provided by the fiber amplifier which is placed closer to one of the coupler arms than the other. In this case, light traveling clockwise around the loop has a different intensity to light traveling anticlockwise, and hence experience different amounts of self-phase modulation (SPM) in the loop due to fiber nonlinearity.

Using short pulses in Sagnac interferometers, the power dependence of loop transmittivity, as given by Eq. (14.5), can lead to significant distortion and pulse narrowing [27]. This happens since only the central part of the pulse is sufficiently intense to undergo switching. The same feature can be used for pulse shaping and pulse cleanup. For example, a low-intensity pedestal accompanying a short optical pulse can be removed by passing it through a Sagnac interferometer [28].

Nonlinear switching without the deformation effect can be achieved by using optical solitons since they have a uniform nonlinear phase across the entire pulse. Any low-energy dispersive radiation accompanying a train of optical solitons is reflected back, while they are transmitted by the loop. Figure 15.3 shows the switching efficiency of a NOLM with and without the soliton effect [29]. We observe that the switching efficiency is almost 100% when the soliton effect is considered. Soliton switching in Sagnac interferometers has been observed experimentally by several authors [29–31].

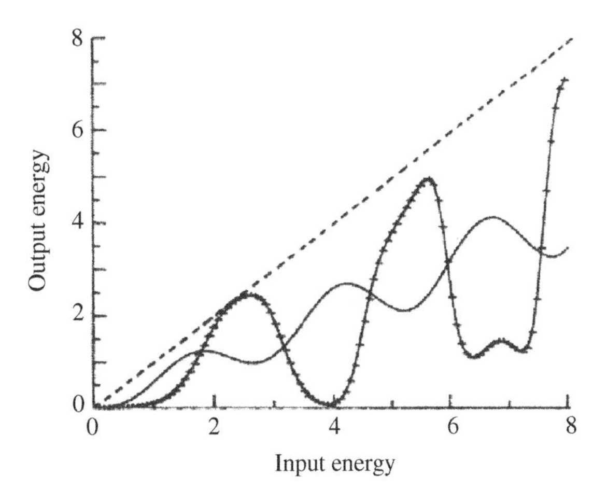

Figure 15.3 Energy transmission of a nonlinear optical loop mirror (NOLM) self-switch with (crossed curve) and without (solid curve) the soliton effect. The dashed line indicates 100% switching. The energy is normalized to that of the fundamental soliton. *Source:* After Ref. [29].

As discussed in Section 4.2, the effect of cross-phase modulation (XPM) produces an alteration in the phase of one pulse due to the intensity of another pulse propagating at the same time in the fiber. This effect can be used in a Sagnac interferometer, as shown in Figure 14.3, to also realize an optical switch. Let us consider a control pulse that is injected into the Sagnac loop such that it propagates in one direction only. Due to the XPM effect in the fiber, the signal pulse that propagates in that direction experiences a phase shift while the counter-propagating pulse remains unaffected. The XPM-induced phase shift is given by $\Delta\varphi_{NL} = 2\gamma P L_{eff}$, where P is the power of the control pulse. If this power is such that $\Delta\varphi_{NL} = \pi$, the signal pulse will be transmitted instead of being reflected. The potential of XPM-induced switching in all-fiber Sagnac interferometers was demonstrated in several experiments [32–35].

Sagnac interferometers can be used for demultiplexing optical time-division multiplexed (OTDM) signals. In this case, the control signal consists of a train of optical pulses that is injected into the loop such that it propagates in a given direction. If the control (clock) signal is timed such that it overlaps with pulses belonging to a specific OTDM channel, the XPM-induced phase shift allows the transmission of this channel, whereas all the remaining channels are reflected. Different channels can be demultiplexed simultaneously by using several Sagnac loops [36].

Using a standard fiber, a length of several kilometers is necessary for a control pulse power in the mW range. In these circumstances, if the signal and control pulses have different wavelengths, the walk-off effects due to the group-velocity mismatch must be taken into consideration. The control pulse width and the walk-off between the signal and control pulses determine the switching speed as well as the maximum bit rate for demultiplexing with an NOLM [37].

One possibility to suppress the pulse walk-off is to use an orthogonally polarized pump at the same wavelength as that of the signal. In this case, the group-velocity mismatch due to the polarization-mode dispersion is relatively low. Moreover, it can be used to advantage by constructing a Sagnac loop consisting of multiple sections of polarization-maintaining fibers that are spliced together in such a way that the slow and fast axes are interchanged periodically. As a result, the pump and signal pulses are forced to collide multiple times inside the Sagnac loop, and the XPM-induced phase shift is enhanced significantly. This idea has been verified using two orthogonally polarized pumps and signal soliton pulses, which were launched in a 10.2-m loop constituted by 11 sections [38].

15.2 2R Optical Regeneration

Reamplification and reshaping (2R) optical pulse regeneration using a highly nonlinear fiber with subsequent spectral filtering was first proposed by Mamyshev [39]. This scheme relies on SPM to broaden the spectrum of a return-to-zero (RZ) input signal and has been the object of a lot of attention since it presents a number of key features [40–47]. Actually, an SPM-based regenerator has the important characteristic of discriminating pulses of different widths. This capacity is enabled from the slope dependence of the spectral broadening in SPM, which allows to discriminate logical ones from logical zeros and thereby provides a bit error rate (BER) improvement. Up to 1 million kilometers of unrepeated transmission was reported in a 2002 experiment using such a regenerator [48].

Figure 15.4 illustrates the operation of a Mamyshev regenerator. A noisy input RZ signal is phase modulated by its own waveform (SPM), while it propagates through a nonlinear fiber. The temporal phase variation induces the instantaneous frequency shift and the signal spectrum is broadened.

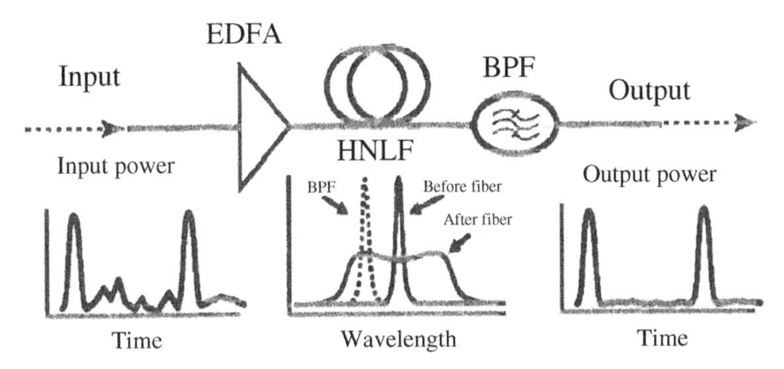

Figure 15.4 An SPM-based 2R regenerator (top) and its action on a bitstream (bottom). HNLF stands for a highly nonlinear fiber. *Source:* After Ref. [47].

The intensity slope point of the original signal is spectrally shifted and can be selected using a bandpass filter (BPF) offset from the input center wavelength. Furthermore, above a certain level, any variation in the signal intensity will translate into a change in the width of the SPM-induced broadened spectrum. Since the bandwidth of the optical filter is fixed, the intensity of the filtered output is insensitive to those variations of the signal power, which results in effective suppression of the amplitude noise at bit "one" level. On the other hand, optical noise at bit "zero" is low and does not induce any spectral shift, being filtered out by the BPF. This results in an improvement of the optical signal-to-noise ratio (SNR).

2R optical regenerators based on SPM with subsequent spectral filtering have been demonstrated using different types of highly nonlinear fibers [40–51]. Since logical "ones" (e.g. pulses) and logical "zeros" (e.g. ASE noise) are processed with distinct transfer functions, such regenerators are able to directly improve the BER of a noisy signal. In a 2006 experiment, a BER improvement from 5×10^{-9} to 4×10^{-12} for an optical SNR of 23 dB was achieved using an 851-m-long highly nonlinear silica fiber [47].

SPM-based 2R regenerators can also be used when soliton-like pulses propagate in the anomalous dispersion regime of optical fiber [52–56]. In such a case, the input pulse is compressed or decompressed, depending on whether its amplitude is larger or smaller than that of the fundamental soliton in the fiber. As a consequence, the spectral width of the pulse is broadened or narrowed, respectively. By using a bandpass optical filter at the fiber output, larger (smaller) energy loss is induced to the pulse having a larger (smaller) initial amplitude, which provides a mechanism of self-amplitude stabilization.

To compensate for the filter-induced loss, some excess gain must be provided to the soliton. However, this excess gain also amplifies the small-amplitude waves coexistent with soliton. Such amplification results in a background instability which can affect significantly and even destroy the soliton itself, especially when a number of regenerators are cascaded. An approach to avoid the background instability consists in using filters whose central frequency is gradually shifted along the cascaded regenerators. This is similar to the soliton control technique based on the use of sliding-frequency guiding filters in long-distance transmission [57], as discussed in Chapter 8. Another way to suppress the noise growth is to use saturable absorbing elements [58, 59]. This technique has been used in a 2002 experiment in which soliton pulses were transmitted at 40 Gb/s along 7600 km, with a regenerator spacing of 240 km [60]. Scaling guidelines of a soliton-based power limiter for 2R optical regeneration applications have been reported in 2010 [56].

15.3 Pulse Compression

Pulse compression can be achieved using different techniques. Some of these techniques utilize dispersion, possibly from gratings or prisms, simultaneously to compensate for chirp and achieve pulse compression. As seen in Section 3.1, when an optical pulse propagates in a linear dispersive medium, it acquires a dispersion-induced chirp. If the pulse has an initial chirp in the opposite direction to that imposed by the dispersive medium, the two tend to cancel each other, resulting in the compression of the optical pulse.

Actually, early work on optical pulse compression did not make use of any nonlinear optical effects. Only during the 1980s, after having understood the evolution of optical pulses in silica fibers, the SPM effect was used to achieve pulse compression. Using such an approach, an optical pulse of 50-fs width from a colliding pulse-mode-locked dye laser oscillating at 620 nm was compressed down to 6 fs [61].

15.3.1 Grating-Fiber Compression

One scheme for pulse compression uses the so-called *fiber-grating compressor* [62, 63]. In this scheme, which is generally used at wavelengths $\lambda < 1.3$ μm [61, 64–68], the input pulse is propagated in the normal-dispersion regime of the fiber, which imposes a nearly linear, positive chirp on the pulse through a combination of SPM and GVD. The output pulse is then sent through a grating pair, which provides the anomalous (or negative) GVD required to get the pulse compression.

Different spectral components of an optical pulse incident at one grating are diffracted at slightly different angles. As a consequence, they experience different time delays during their passage through the grating pair, the blue-shifted components arriving early than the red-shifted ones. In the case of an optical pulse with a positive chirp, the blue-shifted (red-shifted) components occur near the trailing (leading) edge of the pulse and the passage through the grating pair then provides its compression.

A limitation of the grating pair is that the spectral components of the pulse are dispersed not only temporally but also spatially. As a consequence, the optical beam becomes deformed, which is undesirable. This problem can be avoided by simply using a mirror to reflect the pulse back through the grating pair. Reversing the direction of propagation not only allows the beam to recover its original cross section, but also doubles the amount of GVD, thereby reducing the grating separation by a factor of 2. Such double-pass configuration was used in a 1984 experiment in which 33-ps input pulses at 532 nm from a frequency-doubled Nd : YAG laser were propagated through a 105-m-long fiber and were compressed to 0.42 ps after passing through the grating pair [62]. A schematic of the optical pulse compressor used in this experiment is shown in Figure 15.5. In this configuration, the mirror M_3 is used to send the beam back through the grating pair, whereas the mirror M_4 is used to deflect the compressed pulse out of the compressor.

To achieve optimum performance from a grating-fiber compressor, it is necessary to optimize the fiber length, as well as the grating separation. Concerning the first aspect, the effects of both GVD and SPM during the propagation of the pulse inside the fiber must be considered. SPM alone determines a linear chirp only over the central part of an optical pulse. Since the grating pair compresses only this region, while a significant amount of energy remains in the wings, the compressed pulse is not of high quality in this case. The effect of GVD turns out to be important since it broadens and reshapes the pulse, which develops a nearly linear chirp across its entire width. In these circumstances, the grating pair can compress most of the pulse energy into a narrow pulse.

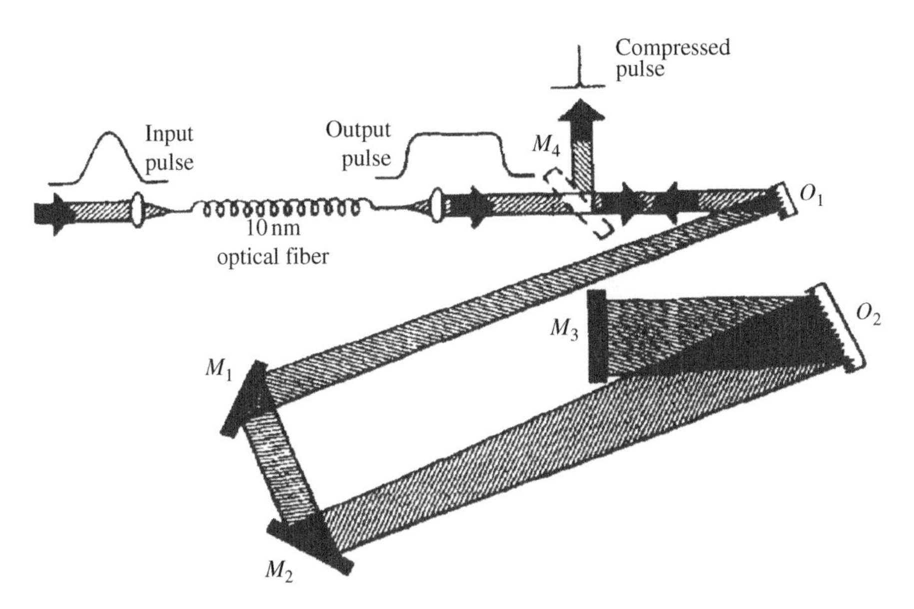

Figure 15.5 Schematic representation of a grating-fiber compressor in the double-pass configuration. *Source:* After Ref. [62].

15.3.2 Higher-Order Soliton-Effect Compression

In another pulse compression scheme, known as a *soliton-effect compressor*, the fiber itself acts as a compressor without the need for an external grating pair [69, 70]. The input pulse propagates in the anomalous-GVD regime of the fiber and is compressed through an interplay between SPM and GVD. This compression mechanism is related to a fundamental property of the higher-order solitons. As seen in Chapter 6, these solitons follow a periodic evolution pattern such that they go through an initial narrowing phase at the beginning of each period. If the fiber length is suitably chosen, the input pulses can be compressed by a factor that depends on the soliton order. Obviously, this compression technique is applicable only in the case of optical pulses with wavelengths exceeding 1.3 μm when propagating in standard SMFs.

The evolution of a soliton of order N inside an optical fiber is governed by the nonlinear Schrödinger equation (NLSE), given by Eq. (3.59), where the soliton order N is given by Eq. (3.60). Even though higher-order solitons follow an exact periodic evolution only for integer values of N, Eq. (3.59) can be used to describe pulse evolution for arbitrary values of N.

In practice, soliton effect compression is carried out by initially amplifying optical pulses up to the power level required for the formation of higher-order solitons. The peak optical power of the initial pulse required for the formation of the Nth-order soliton can be obtained from Eq. (3.60) and is given by

$$P_0 = 3.11 \frac{|\beta_2| N^2}{\gamma t_{FWHM}^2} \tag{15.6}$$

These Nth-order solitons are then passed through the appropriate length of optical fiber to achieve a highly compressed pulse. The optimum fiber length, z_{opt}, and the optimum pulse compression factor, F_{opt}, of a soliton-effect compressor can be estimated from the following empirical relations [70]:

$$z_{opt} \approx z_0 \left(\frac{0.32}{N} + \frac{1.1}{N^2} \right) \tag{15.7}$$

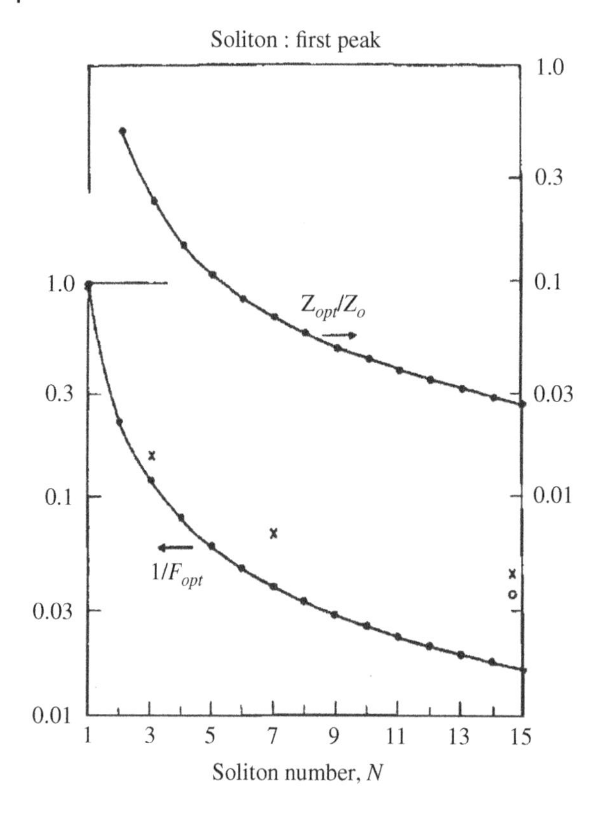

Figure 15.6 Inverse of the compression factor (left scale) and fiber length (right scale) for which the compression is maximum in the higher-order soliton compression. The fiber length is expressed as a fraction of the soliton period. Data points correspond to experimental results. *Source:* After Ref. [69].

$$F_{opt} \approx 4.1N \tag{15.8}$$

where z_0 is the soliton period, given by Eq. (6.25). These relations are accurate within a few percent for $N > 10$. Figure 15.6 shows the numerically and theoretically obtained compression factor and the fiber length for which the compression is maximum as a function of the initial pulse amplitude [69].

Compression factors as large as 500 have been attained using soliton-effect compressors [71]. However, the pulse quality is relatively poor since the compressed pulse carries only a fraction of the input energy, while the remaining energy appears in the form of a broad pedestal. From a practical point of view, the pedestal is deleterious since it causes the compressed pulse to be unstable, making it unsuitable for some applications. Despite this, soliton-effect-compressed pulses can still be useful because there are some techniques that can eliminate the pedestal. One of them consists in the use of an NOLM [72, 73]. Since the transmission of an NOLM is intensity-dependent, the loop length can be chosen such that the low-intensity pedestal is reflected while the higher-intensity pulse peak is transmitted, resulting in a pedestal-free transmitted pulse.

One difficulty faced when using the soliton-effect compressor is that pulses with high peak power are required for the formation of high-order solitons in conventional fibers. As suggested by Eq. (15.6), the use of dispersion-shifted fibers (DSFs) with small values of β_2 at the operating wavelength can reduce the peak power required for soliton generation by an order of magnitude. However, because the soliton period z_0 is inversely proportional to $|\beta_2|$, Eq. (15.7) indicates that longer lengths of DSFs will be required for solitons to achieve optimum compression. As a result, the total fiber loss experienced by those solitons will be larger and the loss-induced pulse broadening will have a significant impact on the compressor global performance.

Another problem introduced by the use of DSFs in soliton-effect compressors is the third-order dispersion, which generally degrades the quality of the compressed pulse [74]. In the case of soliton-effect compressors using conventional optical fibers, the effects of third-order dispersion (TOD) become significant only when the widths of the compressed pulse become very short ($\ll$1 ps). In DSFs, however, the relative importance of TOD is increased and those effects become more pronounced because the GVD parameter β_2 is small. In this case, TOD is detrimental in the compression process even for pulses' widths of a few picoseconds, resulting in serious degradation of the optimum compression factor F_{opt}.

The result given by Eq. (15.8) for the optimum compression factor holds only for the case of an ideal soliton-effect compressor, when high-order nonlinear and dispersive effects, such as intrapulse Raman scattering (IRS), TOD, and self-steepening, are neglected. However, these effects cannot be neglected for the highly compressed pulses because of their sub-picosecond widths and high intensities.

For optical pulses propagating not too close to the zero-dispersion wavelength of the fiber, the dominant higher-order effect is IRS, which manifests as a shift of the pulse spectrum toward the red side – the so-called soliton self-frequency shift, which was discussed in Chapter 9. Such effect can be used with advantage to improve the quality of the compressed pulse, by removing the pedestal mentioned above [75]. As a consequence of the change in the group velocity induced by the IRS effect, the sharp narrow spike corresponding to the compressed pulse travels more slowly than the pedestal and separates from it. Moreover, the pedestal can be removed by spectral filtering and a red-shifted, pedestal-free, and highly compressed pulse is then produced. Other studies revealed that the combined effects of negative TOD and IRS, together with the use of dispersion-decreasing fibers (DDFs), can improve further the performance of a soliton-effect compressor [76, 77].

15.3.3 Compression of Fundamental Solitons

As described in the last section, the propagation of higher-order solitons provides a rapid compression method, but it suffers from the existence of residual pedestals. Some techniques can help to reduce or even eliminate those pedestals, but energy is always wasted in this process. A less rapid technique that provides better pulse quality is the adiabatic compression of fundamental solitons, which are of primary importance in the domain of optical communications. Such adiabatic compression can be achieved using slow amplification [78–80], a proper selection of the initial conditions [81], or by slowly decreasing dispersion [82] along the fiber.

An optical pulse in a fiber having distributed gain obeys the perturbed NLSE:

$$i\frac{\partial q}{\partial Z} + \frac{1}{2}\frac{\partial^2 q}{\partial T^2} + |q|^2 q = igq \tag{15.9}$$

The gain term with a gain coefficient g in the right-hand side leads to the amplification of the pulse energy. Solitons have a fixed area. So, the increased energy from amplification is accommodated by an increase in power and a decrease in width. To avoid distortion, the amplification per soliton period cannot be too great. According to the perturbation theory presented in Section 7.1, an approximate solution of Eq. (15.9) is given by

$$q(T, Z) = \eta(Z)\text{sech}[\eta(Z)T]\exp[i\sigma(Z)] \tag{15.10}$$

where

$$\eta(Z) = \eta_0 \exp(2gZ) \tag{15.11}$$

and

$$\sigma(Z) = \frac{\eta_0^2}{8g}[\exp(4gZ) - 1] \tag{15.12}$$

From Eqs. (15.10) and (15.11), we verify that the pulse width decreases exponentially as the pulse amplitude increases.

The same effect as adiabatic amplification can be achieved using an optical fiber with dispersion that decreases along the length of the fiber. In fact, for optical solitons, a small variation in the dispersion has a similar perturbative effect as amplification or loss: such a variation perturbs the equilibrium between the dispersion and nonlinearity in such a way that when, for example, the dispersion decreases, the soliton pulse is compressed. It can be seen from Eq. (3.60) that if the value of $|\beta_2|$ decreases along the fiber and to keep the soliton order N ($N = 1$ for the fundamental soliton), the pulse width must indeed decrease as $|\beta_2|^{1/2}$. Hence, the use of fibers with variable dispersion is viewed as a passive and effective method to control optical solitons in soliton communication systems. DDFs, in particular, have been recognized to be very useful for high-quality, stable, polarization-insensitive, adiabatic soliton pulse compression and soliton train generation [6, 76, 81, 82]. These fibers can be made by tapering the core diameter of a single-mode fiber during the drawing process, and hence changing the waveguide contribution to the second-order dispersion. Provided the dispersion variation in the DDF is sufficiently gradual, soliton compression can be an adiabatic process where an input fundamental soliton pulse can be ideally compressed as it propagates, while retaining its soliton character and conserving the energy.

The evolution of the fundamental soliton in a DDF can be described by the following NLSE:

$$i\frac{\partial q}{\partial Z} + \frac{1}{2}p(Z)\frac{\partial^2 q}{\partial T^2} + |q|^2 q = 0 \tag{15.13}$$

where the variable coefficient $p(Z) = |\beta_2(Z)/\beta_2(0)|$ takes into account the variation of dispersion along the fiber. Using the transformations $s = \int_0^Z p(y)dy$ and $u = q/\sqrt{p}$, Eq. (15.13) assumes the form:

$$i\frac{\partial u}{\partial s} + \frac{1}{2}\frac{\partial^2 u}{\partial T^2} + |u|^2 u = -i\frac{1}{2p}\frac{dp}{ds}u \tag{15.14}$$

Equation (15.14) clearly shows that the effect of decreasing dispersion is mathematically equivalent to the effect of distributed amplification, adding a gain term to the NLSE. The effective gain coefficient is related to the rate at which GVD decreases along the fiber.

Since decreasing dispersion is equivalent to an effective gain, a DDF can be used in place of a conventional fiber amplifier to generate a train of ultrashort pulses. To achieve such an objective, a CW beam with a weak sinusoidal modulation imposed on it is injected into the DDF. The sinusoidal modulation can be imposed, for example, by beating two optical signals with different wavelengths. As a result of the combined effect of GVD, SPM, and decreasing GVD, the nearly CW beam is converted into a high-quality train of ultrashort solitons, whose repetition rate is governed by the frequency of the initial sinusoidal modulation.

If the input to the DDF is a fundamental soliton given by $q(0, T) = A\,\mathrm{sech}(AT)$ and provided that the dispersion variation is sufficiently adiabatic, the pulse after a length L of a lossless DDF will be compressed to a soliton-like pulse of the form [6, 81]:

$$q(L, T) = \sqrt{G_{eff}}A\mathrm{sech}\left[G_{eff}AT\right]\exp\left\{\frac{iA^2L}{2}\right\} \tag{15.15}$$

where

$$G_{eff} = \frac{\beta_2(0)}{\beta_2(L)} \tag{15.16}$$

is commonly called the effective amplification of the fiber. Adiabatic compression means that all the energy of the input pulse remains localized in the compressed pulse. In such a case, the compression factor is given by

$$F_c = \frac{t_{FWHM}}{t_{comp}} = \frac{|q(L,\ 0)|^2}{|q(0,\ 0)|^2} = G_{eff} \tag{15.17}$$

where energy has been defined as the product of pulse intensity and pulse-width. In the case of a real fiber, the loss must be taken into account, which leads to pulse broadening. As a consequence, the final compression factor is smaller than in the ideal lossless case and becomes:

$$F_c = G_{eff} \exp(-2\Gamma L) \tag{15.18}$$

where Γ is the normalized fiber attenuation coefficient.

Different approaches have been developed to determine the optimum GVD profile and its dependence on the width and peak power of input pulses. In the case of picosecond soliton pulse compression, direct numerical simulations of the NLSE show that linear, Gaussian, and exponential dispersion profiles may all be used effectively to provide ideal, adiabatic compression, where the input pulse energy is conserved and remains localized within the pulse [81]. Compression factors larger than 50 are possible launching input pulses with peak powers corresponding to the fundamental soliton into a DDF whose length is about one soliton period. This technique takes advantage of soliton-effect compression but requires lower peak powers and produces compressed pulses of better quality than those obtained using higher-order solitons.

In the case of sub-picosecond soliton pulses, the influence of higher-order nonlinear and dispersive effects must be accounted for. Some studies show that, in this case, the linear and Gaussian dispersion profiles are the most suitable to achieve high-quality, pedestal-free, adiabatic compression of fundamental solitons. However, in the presence of higher-order effects, the final compression factor is generally lower, and after reaching a maximum at a particular length of DDF, it decreases steadily [6, 81]. This can be explained in terms of pulse compression stabilization, which originates from a competition between the rate of dispersion decrease in the DDF and the rate of dispersion increase due to the combined effects of soliton self-frequency shift and third-order dispersion. In contrast to the degradation occurring for a positive TOD, the combination of a negative TOD and IRS can significantly enhance the compression of fundamental solitons [6].

15.3.4 Dissipative Soliton Compression

Pulse compression can also be achieved in the case of dissipative solitons, corresponding to pulse solutions of the complex Ginzburg–Landau equation (CGLE). Such compression effect can be observed both with the fixed-amplitude and with the arbitrary-amplitude solitons discussed in Section 12.2. In the first case, the final pulse width can be set by an appropriate choice of the various system parameters, whereas in the second case the filtering strength plays a decisive role [83].

An analytical expression for the fixed-amplitude soliton solution of the cubic CGLE is given by Eqs. (12.17) through (12.20). Figure 15.7a illustrates the evolution of the soliton peak power for an initial pulse amplitude $A = 1$ but different values of the initial inverse pulse width B, considering a filter strength $\beta = 0.15$ and a nonlinear gain $\varepsilon = 0.07$. In all cases, the pulse evolves toward the

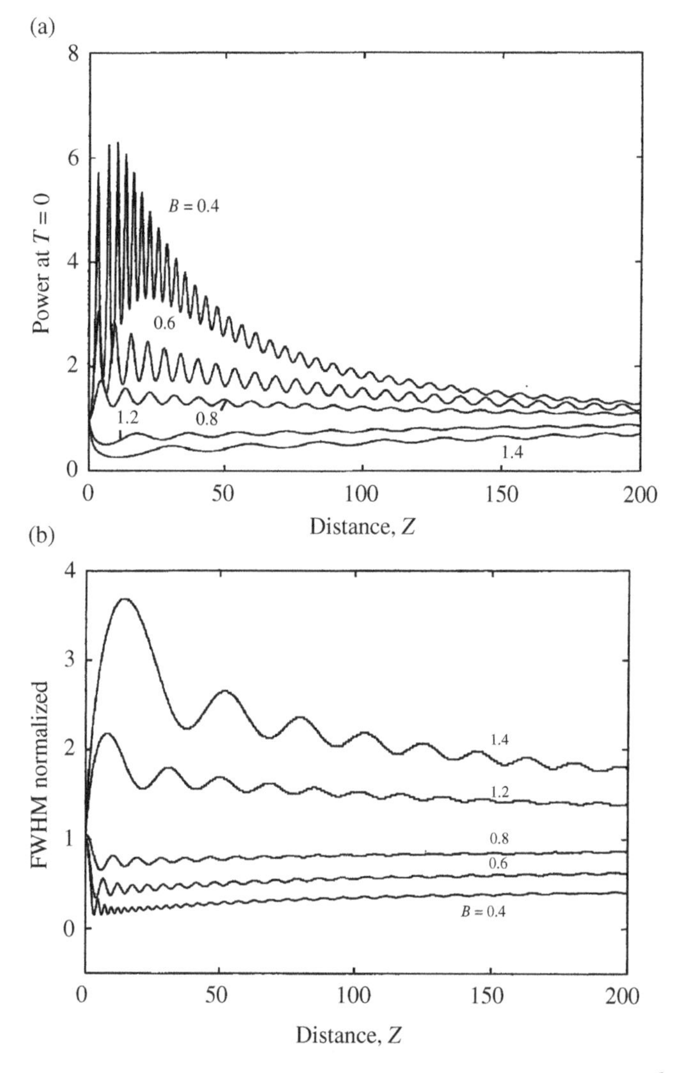

Figure 15.7 Evolution of (a) the soliton peak power $P_0 = |q(Z, T = 0)|^2$ and (b) pulse FWHM, normalized by its initial value, for an initial pulse amplitude $A = 1$ but different values of the initial inverse pulse width B. The parameters $\beta = 0.15$ and $\varepsilon = 0.07$ are assumed.

stationary solution given by Eq. (12.17). This behavior can be used to get soliton compression. This possibility is illustrated in Figure 15.7b, where the evolution of the pulse full width at half-maximum (FWHM), normalized by its initial value, is shown for the same cases of Figure 15.7a. In general, pulse compression is achieved when the input pulse width $1/B$ is larger than its stationary value, which is determined by the system parameters. In the case of Figure 15.7, the stationary pulse width is $1/B = 1.03$ and pulse compression is observed for $B = 0.4$, 0.6, and 0.8.

A maximum compression factor of about seven is observed in Figure 15.7b for the case $B = 0.4$ at a normalized distance $Z = 3$, corresponding to the first minimum of the curve. After some initial oscillations, the pulse width increases monotonously, tending to the stationary value. The excess linear gain corresponding to the parameters $\beta = 0.15$ and $\varepsilon = 0.07$ is positive but relatively small, $\delta = 0.001735$, which makes the background instability develop very slowly. However, to guarantee a truly stable propagation, the linear gain must vanish. In such a case, if the nonlinear gain coefficient is given by Eq. (12.21), the system supports the propagation of arbitrary-amplitude solitons, described by Eqs. (12.22) through (12.25). The amplitude-width product C/D for this type of solitons depends uniquely on the filtering parameter β, as seen from Eq. (12.23). An interesting point concerns with a system with a purely nonlinear gain when the initial amplitude-width product of the pulse does not satisfy such relation.

Figure 15.8a illustrates the evolution of the peak power for $\beta = 0.15$, $\delta = 0$, and $\varepsilon = \varepsilon_s$ (given by Eq. (12.21)), assuming that $C = 1$ and $D = 0.4$, 0.6, and 0.8. We observe that the power P_0 increases from its initial value $P_0 = 1$ and shows an oscillatory behavior, which is more significant in the initial stage of propagation and for lower values of D at the input. On the other hand, the pulse width is reduced, as shown in Figure 15.8b. In case $D = 0.4$, a soliton compression factor of about 6.5 is achieved

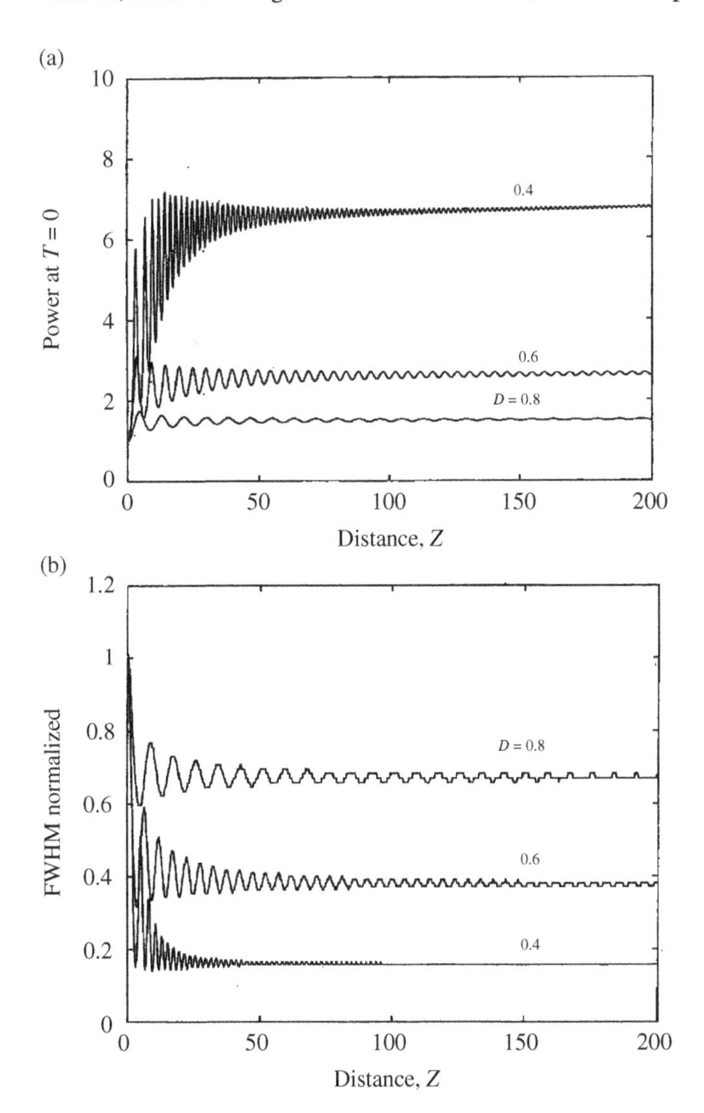

Figure 15.8 Evolution of (a) the soliton peak power $P_0 = |q(Z, T = 0)|^2$ and (b) pulse FWHM, normalized by its initial value, for $\beta = 0.15$, $\delta = 0$, and $\varepsilon = \varepsilon_s$ (given by Eq. (12.21)). The parameters C = 1 and D = 0.4, 0.6, and 0.8 are assumed.

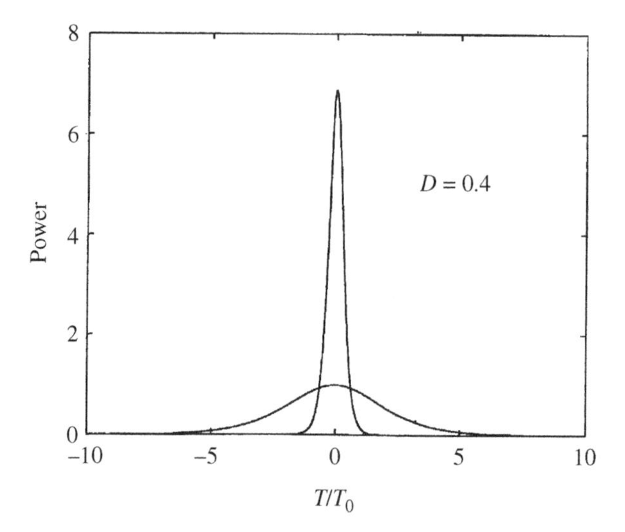

Figure 15.9 Initial and final pulse profiles for $\beta = 0.15$, $\varepsilon = \varepsilon_S$ (given by Eq. (12.21)), $C = 1$ and $D = 0.4$. A soliton compression factor of 6.5 is achieved in this case.

for $Z \approx 3$, a distance similar to that observed in Figure 15.7b. However, while the pulse width in Figure 15.7b for the cases $B = 0.8$, 0.6, and 0.4 increases monotonously with the transmission distance, in the case of Figure 15.8b, the pulse width remains practically constant (curve for $D = 0.4$) or oscillates slightly around a given value (curves for $D = 0.6$ and 0.8). The final soliton shape shows a perfect $\text{sech}^2(DT)$ profile, while the amplitude-width product satisfies the relation (12.23). Figure 15.9 shows the initial and the final pulse profiles corresponding to case $D = 0.4$.

Considering the quintic CGLE, the result given by Eq. (8.69) for the stationary pulse amplitude shows that, for $\varepsilon > \beta/2$, the higher amplitude solution grows to infinity when the nonlinear gain saturation μ tends to zero. Such singularity was predicted in Refs. [84, 85] and provides the possibility of observing ultrashort very high-amplitude pulses [86, 87], as discussed in Section 12.4.

The pulse compression effect occurring as $\mu \to 0-$ is illustrated in Figure 15.10, which shows the numerical simulation of pulse propagation for $\delta = -0.1$, $\beta = 0.2$, $\varepsilon = 0.35$, and (a) $\mu = -0.00001$ or (b) $\mu = -0.0001$. Figure 15.10 also shows (c) the temporal and (d) the spectral density profiles for the stationary pulses presented in (a) (solid curves) and (b) (dashed curves). For μ closer to zero, the pulse amplitude becomes higher and its width is reduced, which means the possibility of achieving an effective pulse compression.

In case $\mu = -0.00001$, the eigenvalues computed using the linear stability analysis described in Section 9.4 are $\lambda_1 = -8.3 \times 10^3$ and $\lambda_2 = -2.1 \times 10^4$, whereas, for $\mu = -0.0001$, the eigenvalues are $\lambda_1 = -8.3 \times 10^2$ and $\lambda_2 = -2.1 \times 10^3$. Since in both cases the eigenvalues are real and negative, we conclude that such stationary solutions are stable, as confirmed by the numerical simulation of the pulse propagation shown in Figure 15.10.

15.4 Solitons in Fiber Gratings

A fiber Bragg grating (FBG) is obtained when the refractive index varies periodically along the fiber length. The realization of such index gratings is possible due to the photosensitivity of optical fibers, which change permanently their optical properties when subject to intense ultraviolet radiation.

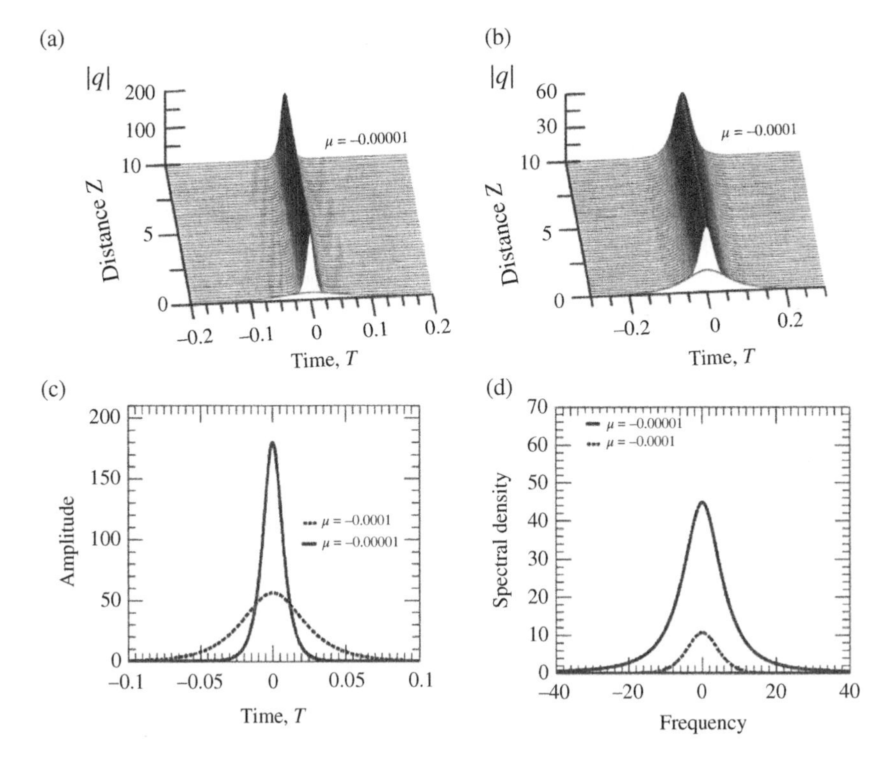

Figure 15.10 Pulse evolution for two different values of the nonlinear gain saturation parameter, namely (a) $\mu = -0.00001$ and (b) $\mu = -0.0001$. (c) Temporal and (d) spectral density profiles for the cases presented in (a) (solid curves) and (b) (dashed curves). The other parameter values are: $\delta = -0.1$, $\beta = 0.2$, and $\varepsilon = 0.35$.

Index changes occur only in the fiber core and are typically of the order of $\sim 10^{-4}$ in the 1550 nm spectral region. However, the amount of index change induced by ultraviolet absorption can be enhanced by more than one order of magnitude in fibers with high Ge concentration [88].

The holographic technique is now routinely used to fabricate such fiber gratings with a controllable period [89]. This technique uses two optical beams, obtained from the same laser and making an angle 2θ between them, which are made to interfere at the exposed core of an optical fiber. The grating period, Λ, is determined by that angle and by the wavelength of the ultraviolet radiation, λ_{uv}, according to the relation

$$\Lambda = \frac{\lambda_{uv}}{2\sin\theta} \tag{15.19}$$

Taking into account both the frequency and the intensity dependences, in addition to its periodic variation along the fiber length, the refractive index of an FBG can be written in the form

$$n(\omega, z, I) = \overline{n}(\omega) + n_2 I + \delta n_g(z) \tag{15.20}$$

where n_2 is the Kerr coefficient and $\delta n_g(z)$ describes the periodic variation of the grating refractive index. In the following, we will consider a sinusoidal grating of the form

$$\delta n_g = n_a \cos\left(\frac{2\pi z}{\Lambda}\right) \tag{15.21}$$

where n_a is the refractive index modulation depth.

A fiber grating acts as a reflector for a given wavelength of light λ, such that

$$\lambda \equiv \lambda_B = 2\overline{n}\Lambda \tag{15.22}$$

The Bragg wavenumber and the Bragg frequency are given by $\beta_B = \pi/\Lambda$ and $\omega_B = \pi c/(\overline{n}\Lambda)$, respectively. Equation (15.22) is known as the *Bragg-resonance condition* and it ensures that the phase of the reflected light adds up constructively, producing a strong reflection of the light wave.

In order to study the wave propagation along the fiber grating, we assume that the nonlinear effects are relatively weak and solve the Helmholtz equation in the frequency domain:

$$\nabla^2\widetilde{E} + \overline{n}^2(\omega,\ z)\left(\frac{\omega}{c}\right)^2\ \widetilde{E} = 0 \tag{15.23}$$

The field $\widetilde{E}$ in Eq. (15.23) includes both forward- and backward-propagating waves and can be written in the form

$$\widetilde{E}(r,\ \omega) = F(r)\left[\widetilde{A}_f(z,\ \omega)\exp\left(i\beta_B z\right) + \widetilde{A}_b(z,\ \omega)\exp\left(-i\beta_B z\right)\right] \tag{15.24}$$

where $F(r)$ describes the transverse modal distribution in a single-mode fiber andβ_B is the Bragg wave number. Using Eqs. (15.20) through (15.24) and using the slowly varying approximation for the amplitudes $\widetilde{A}_f$ and $\widetilde{A}_b$, the following coupled-mode equations are obtained [90]:

$$\frac{d\widetilde{A}_f}{dz} = i[\delta(\omega) + \Delta\beta]\widetilde{A}_f + i\kappa_g\widetilde{A}_b \tag{15.25}$$

$$-\frac{d\widetilde{A}_b}{dz} = i[\delta(\omega) + \Delta\beta]\widetilde{A}_b + i\kappa_g\widetilde{A}_f \tag{15.26}$$

where

$$\delta = \left(\frac{\overline{n}}{c}\right)(\omega - \omega_B) \equiv \beta(\omega) - \beta_B \tag{15.27}$$

is the detuning from the Bragg frequency ω_B and

$$\kappa_g = \frac{\pi n_a}{\lambda_B} \tag{15.28}$$

is the coupling coefficient. In Eqs. (15.25) and (15.26) the nonlinear effects are included through $\Delta\beta$.

15.4.1 Pulse Compression Using Fiber Gratings

If the input intensity is sufficiently low, the nonlinear effects can be neglected, considering $\Delta\beta = 0$. In these circumstances, we can obtain from Eqs. (15.23) and (15.24) the following result for the reflection coefficient of an FBG of length L_g[89]:

$$r_g = \frac{A_b(0)}{A_f(0)} = \frac{i\kappa_g\sin\left(QL_g\right)}{Q\cos\left(QL_g\right) - i\delta\sin\left(QL_g\right)} \tag{15.29}$$

where Q satisfies the linear dispersion relation

$$\delta = \pm\sqrt{\kappa_g^2 + Q^2}. \tag{15.30}$$

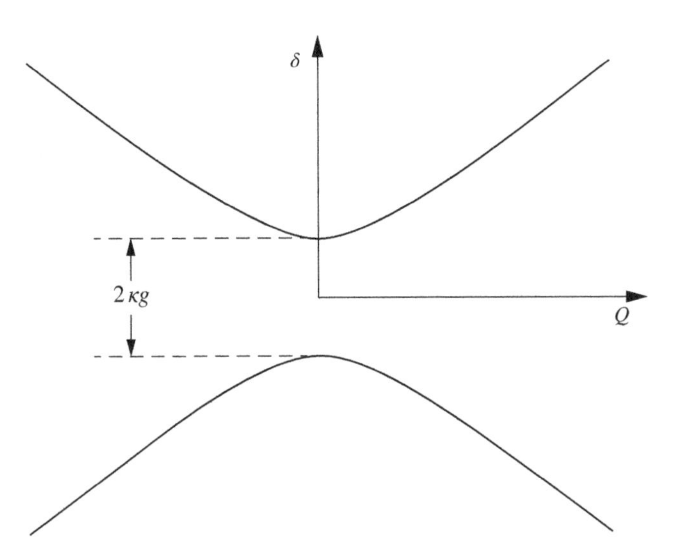

Figure 15.11 Dispersion relation of a uniform fiber grating.

Maximum reflectivity occurs at the center of the stopband and can be obtained from Eqs. (15.29) and (15.30) making $\delta = 0$:

$$R = |r_g|^2 = \tanh^2 \kappa_g L \tag{15.31}$$

We have $R = 0.93$ for $\kappa_g L = 2$ and $R \approx 1$ for $\kappa_g L \geq 3$. The bandwidth of the reflection spectrum, which is defined as the wavelength spacing between the two reflection minima on either side of the central peak, is approximately given by [91]

$$\Delta\lambda = \frac{\lambda_B^2}{\pi n_{eff} L} \left(\kappa_g^2 L^2 + \pi^2\right)^{1/2} \tag{15.32}$$

Figure 15.11 illustrates the relation between Q and δ given by Eq. (15.30). The most interesting feature in this figure is the existence of a photonic bandgap for detunings such that $-\kappa_g < \delta < \kappa_g$. Light with frequencies within this bandgap cannot propagate and is reflected. Frequencies outside the gap, however, can propagate, but at velocities that can be significantly reduced in comparison with the speed of light in the uniform medium. According to Figure 15.11, the group velocity vanishes at the band edge and asymptotically approaches $c/\overline{n}$ far from the Bragg resonance.

Outside but close to the stop-band edges, the grating provides large dispersion. Noting that the effective propagation constant of the forward- and backward-propagating waves is $\beta_e = \beta_B \pm Q$, where the choice of a sign depends on the sign of δ, we obtain the following result for the group velocity inside the grating:

$$V_G = \left[\frac{dQ}{d\omega}\right]^{-1} = \pm v_g \sqrt{1 - x^2} \tag{15.33}$$

where $x = \kappa_g/\delta$. On the other hand, we can obtain the following results for the dispersion parameters of a fiber grating [92]:

$$\beta_2^g = \frac{d^2Q}{d\omega^2} = -\frac{1}{v_g^2}\frac{x^2}{\delta(1-x^2)^{3/2}}, \quad \beta_3^g = \frac{d^3Q}{d\omega^3} = \frac{3}{v_g^3}\frac{x^2}{\delta^2(1-x^2)^{5/2}} \tag{15.34}$$

The grating-induced GVD depends on the sign of the detuning δ. The GVD is anomalous ($\beta_2^g < 0$) on the high-frequency side of the stop-band and becomes normal ($\beta_2^g > 0$) on the low-frequency side. Near the stop-band edges, the grating exhibits large GVD. Since an FBG acts as a dispersive delay line, it can be used for dispersion compensation in fiber transmission systems [93]. It can also be used in place of the bulk-grating pair in a grating-fiber compressor, providing a compact all-fiber device. In fact, since β_2^g can typically exceed 10^7 ps^2/cm, a 1-cm-long fiber grating may provide as much dispersion as 10 km of silica fiber or a bulk-grating pair with more than one-meter spacing. The main limitation in this case comes from the grating-induced TOD, which can significantly affect the quality of the pulses when the optical frequency falls close to the edges of the stop-band.

Both the compression factor and the pulse quality can be significantly enhanced by using an FBG in which the grating period varies linearly with position – the so-called *chirped fiber grating* [92]. In this case, the grating reflects different frequency components of the pulse at different points along its length. Such a device has been used to compensate for dispersion-induced broadening of pulses in fiber transmission systems as well as for pulse compression, where compression factors above 100 have been achieved [94, 95]. The only disadvantage of a chirped fiber grating is that the compressed pulse is reflected instead of being transmitted.

15.4.2 Fiber Bragg Solitons

If the frequency of the lightwave propagating in the fiber grating approaches the band edge, Figure 15.11 shows that the group velocity tends to zero and the small group velocity in this region creates a larger local intensity. Due to the Kerr nonlinearity, the refractive index increases in proportion to this intensity, which enables the propagation of a soliton-like pulse within the stopband. Such soliton was reported for the first time in 1981 [96] and it is often referred to as *Bragg soliton*, or *grating soliton*. It is also called *gap soliton* because it exists only inside the bandgap of the Bragg resonance.

Expanding $\beta(\omega)$ in Eq. (15.27) in a Taylor series, we have

$$\beta(\omega) = \beta_0 + \beta_1(\omega-\omega_0) + \frac{1}{2}\beta_2(\omega-\omega_0)^2 + \frac{1}{6}\beta_3(\omega-\omega_0)^3 + \cdots \tag{15.35}$$

In most practical situations, the GVD dispersion can be neglected in FBGs. Keeping terms in Eq. (15.35) only to first-order and replacing $(\omega-\omega_0)$ with the differential operator $i(\partial/\partial t)$, Eqs. (15.25) and (15.26) can be converted to the time domain, assuming the form

$$i\frac{\partial A_f}{\partial z} + i\beta_1\frac{\partial A_f}{\partial t} + \delta A_f + \kappa_g A_b + \gamma\left(|A_f|^2 + 2|A_b|^2\right)A_f = 0 \tag{15.36}$$

$$-i\frac{\partial A_b}{\partial z} + i\beta_1\frac{\partial A_b}{\partial t} + \delta A_b + \kappa_g A_f + \gamma\left(|A_b|^2 + 2|A_f|^2\right)A_b = 0 \tag{15.37}$$

In Eqs. (15.36) and (15.37), $\beta_1 = 1/v_g$, where v_g is the group velocity and $\gamma = n_2\omega_0/(cA_{eff})$ is the nonlinear parameter.

Unlike the NLSE, the system of equations (15.36) and (15.37) is not exactly integrable. Nevertheless, a family of exact soliton solutions of these equations was found following the pattern of the previously known exact solutions in the so-called massive Thirring model [97]. The Bragg soliton solution is given by [98–100]:

$$A_f(z,\ t) = \sqrt{\frac{\kappa_g(1+\nu)}{\gamma(3-\nu^2)}}\left(1-\nu^2\right)^{1/4} W(X)\exp\left(i\psi\right) \tag{15.38}$$

$$A_b(z,\ t) = -\sqrt{\frac{\kappa_g(1-\nu)}{\gamma(3-\nu^2)}}\left(1-\nu^2\right)^{1/4} W^*(X)\exp\left(i\psi\right) \tag{15.39}$$

where

$$W(X) = (\sin\theta)\text{sech}\left(\frac{X\sin\theta - i\theta}{2}\right),\quad X = \frac{z - V_G t}{\sqrt{1-v^2}}\kappa \tag{15.40a}$$

$$\psi = vX\cos\theta - \frac{4v}{(3-v^2)}\arctan\left[\left|\cot\left(\frac{\theta}{2}\right)\right|\coth(\varsigma)\right] \tag{15.40b}$$

The above solution is a *two-parameter family* of Bragg solitons. The parameter $v = V_G/v_g$ is in the range $-1 < v < 1$ while the parameter θ can be chosen anywhere in the interval $[0,\ \pi]$. Bragg solitons correspond to specific combinations of counter-propagating waves. Depending on the relative amplitudes of the two waves, the soliton can move forward or backward at the reduced speed V_G if the counterpropagating waves have equal amplitudes, we have $V_G = 0$, which corresponds to a *stationary gap soliton.*

The first experimental observation of Bragg solitons occurred in 1996, using a 6-cm-long fiber grating [101]. Since then, other experiments were performed which allows a better understanding of these nonlinear pulses. For example, in a 1997 experiment, soliton formation at frequencies slightly above the gap frequency was reported using 80-ps pulses obtained from a Q-switched, mode-locked Nd:YLF laser [102]. In a later experiment, Bragg solitons were also observed in a 7.5-cm-long apodized fiber grating using the same type of input pulses [103]. Figure 15.12 shows the experimental results for three values of the detuning of input frequencies from the bandgap [102]. The solid curve corresponds to the input pulse shape, which has a width of 80 ps and a peak intensity of 18GW/cm^2. Figure 15.12a shows the transmitted intensity when the pulse was centered on 1053.00 nm, where the linear grating transmission is approximately 90%. As a result of the dispersive and nonlinear effects, the transmitted pulse was compressed to 40 ps, whereas its peak intensity is enhanced by approximately 40%. The compressed pulse also leaves the grating ~40 ps later than the pulse propagating without dispersion, which is due to the smaller group velocity at this frequency, corresponding to 81% of the speed of light in the uniform fiber. Figure 15.12b shows the transmitted intensity when the pulse was tuned closer to the photonic bandgap, such that linear transmissivity is 80%. The transmitted pulse now has a width of approximately 25 ps and is delayed by ~45 ps, corresponding to an average velocity of 79% of the speed of light in the uniform medium. The nonlinear losses are responsible for the reduced intensity of the compressed pulse. Finally, Figure 15.12c shows the transmitted intensity when the frequency is further tuned closer to the bandgap, where the linear transmissivity is 50%. The transmitted pulse now has a width of approximately 23 ps and is retarded by approximately 55 ps, corresponding to an average velocity of 76% of the speed of light in the uniform medium. These results clearly show the slowdown and the compression of light as a result of the reduced group velocity and the soliton effect. However, gap solitons that form within the stopband of a fiber grating have not been

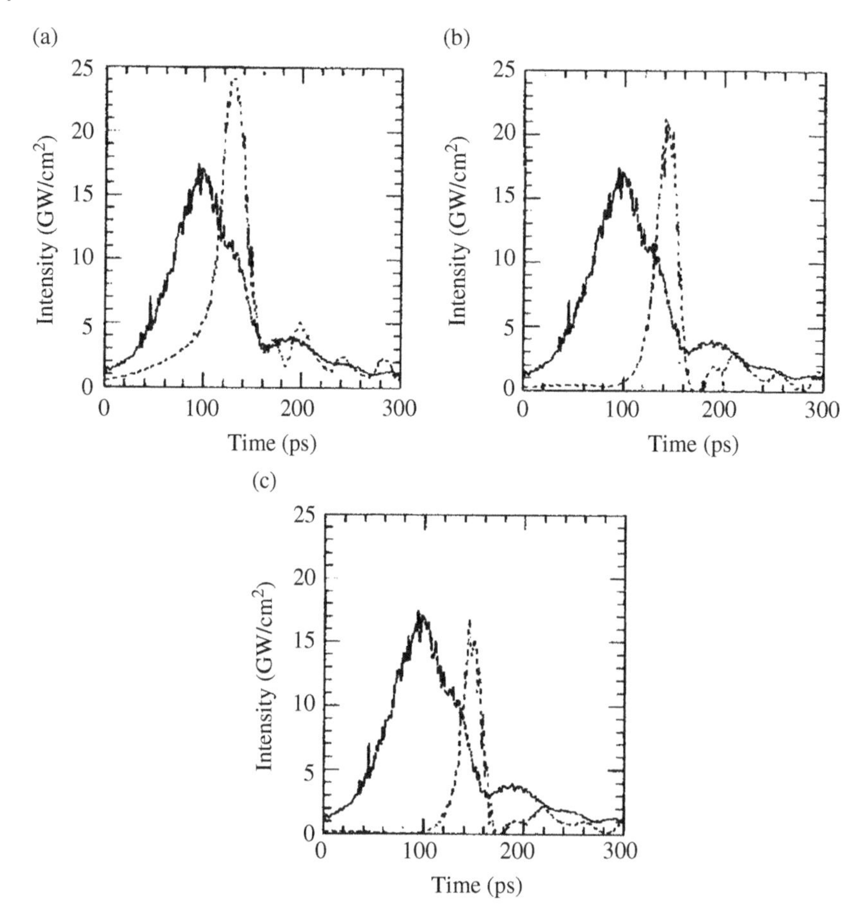

Figure 15.12 Transmitted intensities (dotted lines) and input intensities (solid lines) of light pulse in a fiber grating. Figures (a), (b), and (c) correspond to grating detunings of 15, 13.2, and 11.4 cm^{-1}, respectively. *Source:* After Ref. [102].

observed because of a practical difficulty: the light whose wavelength falls inside the stopband of a Bragg grating is reflected.

It can be shown that if $\kappa_g L_{LN} \gg 1$, where $L_{NL} = (\gamma P_0)^{-1}$ is the nonlinear length, the coupled-mode equations (15.36) and (15.37) reduce to the following effective NLSE [100]:

$$i\frac{\partial U}{\partial z} - \frac{\beta_2^g}{2}\frac{\partial^2 U}{\partial T^2} + \gamma_g |U|^2 U = 0 \tag{15.41}$$

where U corresponds to the envelope associated with Bloch wave formed by a linear combination of A_f and A_b, $z = V_G t$, $T = t - z/V_G$, β_2^g is given by Eq. (15.34) and

$$\gamma_g = \frac{3 - v^2}{2v}\gamma \tag{15.42}$$

Given the above correspondence, fiber Bragg solitons can be used for pulse compression in a way similar to that described in Section 15.2 for higher-order solitons in fibers without grating. The relations given there for the compression factor and for the optimum fiber length also apply in this case as long as the GVD parameter β_2 and the nonlinear parameter γ are replaced by the corresponding fiber-grating parameters, given by Eqs. (15.34) and (15.42), respectively.

When a beam is launched at one end of the grating, both its intensity and wavelength with respect to the stopband will play an important role concerning the beam transmission along the FBG. At low input powers, if the beam wavelength is located near the Bragg wavelength, the transmissivity will be small. However, increasing the input power above a given level will result in the transmission of most of that power. In fact, due to the power dependence of the refractive index, the effective detuning δ in Eqs. (15.30) and (15.32) also changes. As a consequence, light whose wavelength was inside the stopband can be tuned out of the stopband and get transmitted through the FBG, a behavior known as *SPM-induced switching*. This phenomenon was observed in a 1998 experiment [104], using an 8-cm-long grating with its Bragg wavelength centered near 1536 nm. In such an experiment, switching from 3 to 40% of the pulse energy was obtained for nanosecond pulses at internal field strengths of the order of 15 GW/cm^2. At higher intensities, multiple Bragg solitons were obtained with durations in the range of 100–500 ps.

References

1 Mollenauer, L., Stolen, R., and Gordon, J. (1980). *Phys. Rev. Lett.* **45**: 1095.
2 Ahmed, K., Chan, K., and Liu, H. (1995). *IEEE J. Sel. Top. Quantum Electron.* **1**: 592.
3 Chan, K. and Liu, H. (1995). *IEEE J. Quantum Electron.* **31**: 2226.
4 Chernikov, S. and Mamyshev, P. (1991). *J. Opt. Soc. Am. B* **8**: 1633.
5 Pelusi, M. and Liu, H. (1997). *IEEE J. Quantum Electron.* **33**: 1430.
6 Chan, K. and Cao, W. (2000). *Opt. Commun.* **184**: 463.
7 Ferreira, M.F. (2011). *Nonlinear Effects in Optical Fibers.* Hoboken, NJ: John Wiley & Sons.
8 Ferreira, M.F. (2020). *Optical Signal Processing in Highly Nonlinear Fibers.* Oxon, UK: CRC Press.
9 Green, P.E. Jr. (1993). *Fiber-Optic Networks.* Upper Saddle River, NJ: Prentice-Hall Chap. 3.
10 Hecht, J. (1999). *Understanding Fiber Optics.* Upper Saddle River, NJ: Prentice-Hall Chap. 15.
11 Ghatak, A.K. and Thyagarajan, K. (1999). *Introduction to Fiber Optics.* New York: Cambridge University Press Chap. 17.
12 Agrawal, G.P. (2001). *Applications of Nonlinear Fiber Optics.* San Diego, CA: Academic Press Chap. 2c.
13 Jenson, S. (1982). *IEEE J. Quantum Electron.* **QE-18**: 1580.
14 Daino, B., Gregori, G., and Wabnitz, S. (1985). *J. Appl. Phys.* **58**: 4512.
15 Wabnitz, S., Wright, E., Seaton, C., and Stegeman, G. (1986). *Appl. Phys. Lett.* **49**: 838.
16 Friberg, S., Weiner, A., Silberberg, Y. et al. (1988). *Opt. Lett.* **13**: 904.
17 Trillo, S., Wabnitz, S., Banyai, W. et al. (1989). *IEEE J. Quantum Electron.* **25**: 104.
18 Snyder, A., Mitchell, D., Poladian, L. et al. (1991). *J. Opt. Soc. Am. B* **8**: 2102.
19 Samir, W., Pask, C., and Garth, S. (1994). *J. Opt. Soc. Am. B* **11**: 2193.
20 Yasumoto, K., Maeda, H., and Maekawa, N. (1996). *J. Lightwave Technol.* **14**: 628.
21 Artigas, D., Dios, F., and Canal, F. (1997). *J. Mod. Opt.* **44**: 1207.
22 Vigil, S., Zhou, Z., Canfield, B. et al. (1998). *J. Opt. Soc. Am. B* **15**: 895.
23 Trillo, S., Wabnitz, S., Wright, E., and Stegeman, G. (1988). *Opt. Lett.* **13**: 672.
24 Trillo, S., Wabnitz, S., and Stegeman, G. (1989). *IEEE J. Quantum Electron.* **25**: 1907.
25 Smith, K., Doran, N., and Wigley, P. (1990). *Opt. Lett.* **15**: 1294.
26 Yamada, E. and Nakazawa, M. (1994). *IEEE J. Quantum Electron.* **30**: 1842.
27 Doran, N., Forrester, D., and Nayar, B. (1989). *Electron. Lett.* **25**: 267.

28 Smith, K., Doran, N., and Wigley, P. (1990). *Opt. Lett.* **15**: 1294.
29 Doran, N. and Wood, D. (1988). *Opt. Lett.* **13**: 56.
30 Blow, K., Doran, N., and Nayar, B. (1989). *Opt. Lett.* **14**: 754.
31 Islam, M., Sunderman, E., Stolen, R. et al. (1989). *Opt. Lett.* **14**: 811.
32 Blow, K., Doran, N., Nayar, B., and Nelson, B. (1990). *Opt. Lett.* **15**: 248.
33 Jino, M. and Matsumoto, T. (1991). *Electron. Lett.* **27**: 75.
34 Ellis, A. and Cleland, D. (1992). *Electron. Lett.* **28**: 405.
35 Bülow, H. and Veith, G. (1993). *Electron. Lett.* **29**: 588.
36 Bodtkrer, E. and Bowers, J. (1995). *J. Lightwave Technol.* **13**: 1809.
37 Uchiyama, K., Takara, H., Morioka, T. et al. (1993). *Electron. Lett.* **29**: 1313.
38 Moores, J., Bergman, K., Haus, H., and Ippen, E. (1991). *Opt. Lett.* **16**: 138.
39 Mamyshev, P.V. (1998). *European Conference on Optical Communications (ECOC98)*, Madrid, Spain (20–24 September 1998), 475–477.
40 Matsumoto, M. (2002). *IEEE Photon. Technol. Lett* **14**: 319.
41 Matsumoto, M. (2004). *J. Light. Technol* **23**: 1472.
42 Her, T., Raybon, G., and Headley, C. (2004). *IEEE Photon. Technol. Lett.* **16**: 200.
43 Miyazaki, T. and Kubota, F. (2004). *IEEE Photon. Technol. Lett* **16**: 1909.
44 Johannisson, P. and Karlsson, M. (2005). *IEEE Photon. Technol. Lett* **17**: 2667.
45 Striegler, A. and Schmauss, B. (2006). *J. Lightwave Technol.* **24**: 2835.
46 Rochette, M., Kutz, J., Blows, J. et al. (2005). *IEEE Photon. Technol. Lett.* **17**: 908.
47 Rochette, M., Libin, F., Ta'eed, V. et al. (2006). *IEEE J. Select. Top. Quantum Electron.* **12**: 736.
48 Raybon, G., Su, Y., Leuthold, J. et al. (2002). *Optical Fiber Communication Conference (OFC2002)*, Anaheim, CA, PD10-1.
49 Igarashi, K. and Kikuchi, K. (2008). *IEEE J. Sel. Top. Quantum Electron.* **14**: 551.
50 Fok, M. and Shu, C. (2008). *IEEE J. Sel. Top. Quantum Electron.* **14**: 587.
51 Radic, S., Moss, D.J., and Eggleton, B.J. (2008). Nonlinear optics in communications: from crippling impairment to ultrafast tools. In: *Optical Fiber Telecommunications VA, Components and Subsystems* (ed. I. Kaminow, T. Li and A. Willner). London: Academic Press.
52 Leclerc, O., Lavigne, B., Balmefrezol, E. et al. (2003). *J. Lightwave Technol.* **21**: 2779.
53 Ohara, T., Takara, H., Hirano, A. et al. (2003). *IEEE Photon. Technol. Lett.* **15**: 763.
54 Ohara, T., Takara, H., Kawanishi, S. et al. (2004). *IEEE Photon. Technol. Lett.* **16**: 2311.
55 Matsumoto, M. and Leclerc, O. (2002). *Electron. Lett.* **38**: 576.
56 Fatome, J. and Finot, C. (2010). *J. Lightwave Technol.* **28**: 2552.
57 Mollenauer, L., Gordon, J., and Evangelides, S. (1992). *Opt. Lett.* **17**: 1575.
58 Matsumoto, M. and Leclerc, O. (2002). *Electron. Lett.* **38**: 576.
59 Gay, M., Silva, M., Nguyen, T. et al. (2010). *IEEE Photon. Technol. Lett.* **22**: 158.
60 D. Rouvillain, F. Seguineau, L. Pierre et al. (2002). *Proceedings of the Optical Fiber Communication Conference*, Paper FD11, Anaheim, California, USA (17–22 March 2002).
61 Fork, R., Cruz, C., Becker, P., and Shank, C. (1987). *Opt. Lett.* **12**: 483.
62 Johnson, A., Stolen, R., and Simpson, W. (1984). *Appl. Phys. Lett.* **44**: 729.
63 Tomlinson, W., Stolen, R., and Shank, C. (1984). *J. Opt. Soc. Am B* **1**: 139.
64 Blow, K., Doran, N., and Nelson, B. (1985). *Opt. Lett.* **10**: 393.
65 Strickland, B. and Mourou, G. (1985). *Opt. Commun.* **55**: 447.
66 Tai, K. and Tomita, A. (1986). *Appl. Phys. Lett.* **48**: 309.
67 Valk, B., Vilhelmsson, Z., and Salour, M. (1987). *Appl. Phys. Lett.* **50**: 656.
68 Dianov, E., Karasik, A., Mamyshev, P. et al. (1987). *Sov. J. Quantum Electron.* **17**: 415.
69 Mollenauer, L., Stolen, R., Gordon, J., and Tomlinson, W. (1983). *Opt. Lett.* **8**: 289.

70 Dianov, E., Nikonova, Z., Prokhorov, A., and Serkin, V. (1986). *Sov. Tech. Phys. Lett.* **12**: 311.
71 Gouveia-Neto, A., Gomes, A., and Taylor, J. (1988). *J. Mod. Opt.* **35**: 7.
72 Tamura, K. and Nakazawa, M. (1999). *IEEE Photon. Technol. Lett.* **11**: 230.
73 Pelusi, M., Matsui, Y., and Suzuki, A. (1999). *IEEE J. Quantum Electron.* **35**: 867.
74 Chan, K. and Liu, H. (1994). *Opt. Lett.* **19**: 49.
75 Agrawal, G. (1990). *Opt. Lett.* **15**: 224.
76 Wai, P. and Cao, W. (2003). *J. Opt. Soc. Am. B* **20**: 1346.
77 Shumin, Z., Fuyun, L., Wencheng, X. et al. (2004). *Opt. Commun.* **237**: 1.
78 Nakazawa, M., Kurokawa, K., Kubota, H., and Yamada, E. (1990). *Phys. Rev. Lett.* **65**: 1881.
79 Nabiev, R., Melnikov, I., and Nazarkin, A. (1990). *Opt. Lett.* **15**: 1348.
80 Quiroga-Teixeiro, M., Anderson, D., Andrekson, P. et al. (1996). *J. Opt. Soc. Am. B* **13**: 687.
81 Mostofi, A., Hanza, H., and Chu, P. (1997). *IEEE J. Quantum Electron.* **33**: 620.
82 Chernikov, S., Dianov, E., Richardson, D., and Payne, D. (1993). *Opt. Lett.* **18**: 476.
83 Ferreira, M. and Latas, S. (2002). *Opt. Eng.* **41**: 1696.
84 Afanasjev, V. (1995). *Opt. Lett.* **20**: 704.
85 Soto-Crespo, J., Akhmediev, N., Afanasjev, V., and Wabnitz, S. (1997). *Phys. Rev. E* **55**: 4783.
86 Latas, S.C., Ferreira, M.F.S., and Facão, M. (2017). *J. Opt. Soc. Am. B* **34**: 1033.
87 Latas, S.C. and Ferreira, M.F. (2019). *J. Opt. Soc. Am. B* **36**: 3016.
88 Archambault, J., Reekie, L., and Russel, P. (1993). *Electron. Lett.* **29**: 453.
89 Othonos, A. and Kalli, K. (1999). *Fiber Bragg Gratings*. Boston, MA: Artech Howse.
90 Haus, H.A. (1984). *Waves and Fields in Optoelectronics*. Englewood Cliffs, NJ: Prentice-Hall.
91 Ghatak, A.K. and Thyagarajan, K. (1989). *Optical Electronics*. Cambridge, UK: Cambridge University Press.
92 Agrawal, G.P. (1997). *Fiber-Optic Communication Systems*, 2ee. New York: John Wiley & Sons.
93 Litchinitser, N., Eggleton, B., and Patterson, D. (1977). *J. Lightwave Technol.* **15**: 1303.
94 William, J., Bennion, I., Sugden, K., and Doran, N. (1994). *Electron. Lett.* **30**: 985.
95 Kashyap, R., Chernikov, S., McKee, P., and Taylor, J. (1994). *Electron. Lett.* **30**: 1078.
96 Voloshchenko, Y., Ryzhov, Y., and Sotin, V. (1981). *Sov. Phys. Tech. Phys.* **26**: 541.
97 Thirring, W. (1958). *Ann. Phys.* **3**: 91.
98 Aceves, A. and Wabnitz, S. (1989). *Phys. Lett. A* **141**: 37.
99 Christodoulides, D.N. and Joseph, R.I. (1989). *Phys. Rev. Lett.* **62**: 1746.
100 Sterke, C. (1998). *J. Opt. Soc. Am. B* **15**: 2660.
101 Eggleton, B., Slusher, R., Sterke, C. et al. (1996). *Phys. Rev. Lett.* **76**: 1627.
102 Eggleton, B., Sterke, C., and Slusher, R. (1997). *J. Opt. Soc. Am. B* **14**: 2980.
103 Eggleton, B., Sterke, C., and Slusher, R. (1999). *J. Opt. Soc. Am B* **16**: 587.
104 Traverner, D., Broderick, N., Richardson, D. et al. (1998). *Opt. Lett.* **23**: 328.

16

Highly Nonlinear Optical Fibers

In standard single-mode silica fibers, the fiber nonlinear parameter has a typical value of only $\gamma \approx 1.3\ W^{-1}/km$ [1]. Such value is too small for some applications requiring highly efficient nonlinear processes. Highly nonlinear fibers (HNLFs) made with silica and presenting a standard structure but with a smaller effective mode area, and hence a larger nonlinear coefficient ($\gamma \approx 10-20\ W^{-1}/km$) [2], have been widely used. The nonlinear parameter can be further increased by tailoring appropriately the fiber structure. Different types of silica-based microstructured optical fibers have been designed to address this purpose [3], providing a nonlinear coefficient $\gamma \approx 70\ W^{-1}/km$. Significantly higher values of γ can be achieved by combining tight mode confinement with the use of glasses with a greater intrinsic material nonlinearity coefficient than that of silica. Examples of suitable glasses that have been used include bismuth oxide ($\gamma \approx 1100\ W^{-1}/km$) [4], lead-silicate ($\gamma \approx 1860\ W^{-1}/km$) [5], and chalcogenide ($\gamma \approx 2450\ W^{-1}/km$) [6] glasses. Such HNLFs exhibit peculiar dispersive characteristics, which strongly affect the soliton dynamics.

16.1 Highly Nonlinear Silica Fibers

The fiber nonlinearity is commonly characterized by the nonlinear parameter γ, given by [7]:

$$\gamma = \frac{2\pi}{\lambda} \frac{n_2}{A_{eff}} \tag{16.1}$$

where λ is the light wavelength, n_2 is the nonlinear-index coefficient of the fiber core, and A_{eff} is the effective mode area. Equation (16.1) shows that, for a fixed wavelength, since n_2 is determined by the material from which the fiber is made, the most practical way of increasing the nonlinear parameter γ is to reduce the effective mode area A_{eff}. However, the nonlinear parameter γ can also be enhanced using the dopant dependence of the nonlinear refractive index coefficient n_2. This possibility is illustrated in Figure 16.1, which shows the relationship between relative index difference Δ_d and n_2 in bulk glass [8]. The relative index difference is defined as $\Delta_d = \left(n_d^2 - n_0^2\right)/2n_d^2$, where n_0 and n_d denote the refractive indexes of pure silica and doped glass, respectively. The closed and open circles correspond to GeO_2- and F-doped bulk glasses, respectively, whereas, the squares show the n_2 of pure silica glasses. The dopant dependence can be approximated using a linear relation, as shown in Figure 16.1. It can be observed that n_2 increases with the relative index difference for the GeO_2-doped bulk glass. On the other hand, the dopant dependence of n_2 for the F-doped bulk glass is the inverse and about 2.3 times smaller than that for the GeO_2-doped case.

In order to enhance the nonlinear parameter γ, the best option will be provided by a fiber heavily doped with GeO_2 in the core and having a large refractive index difference between the core and the

Solitons in Optical Fiber Systems, First Edition. Mario F. S. Ferreira.

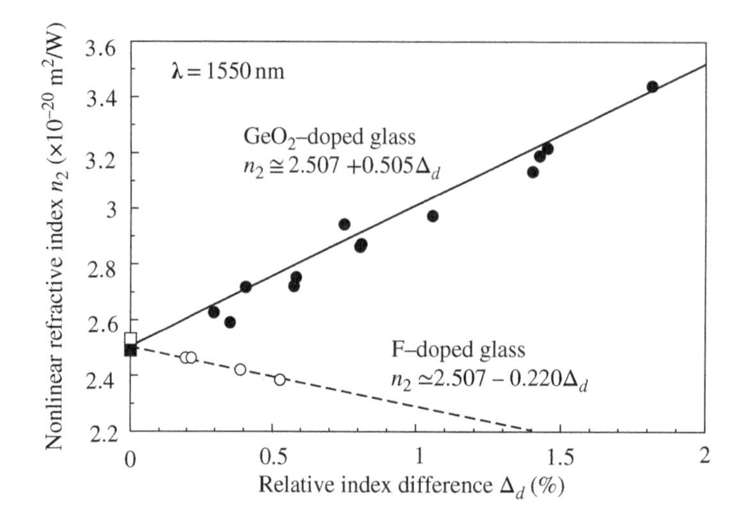

Figure 16.1 Relationship between the relative index difference Δ_d and the nonlinear refractive index n_2 of GeO_2- (closed circles) and F- (open circles) doped bulk-glass. *Source:* After Ref. [8].

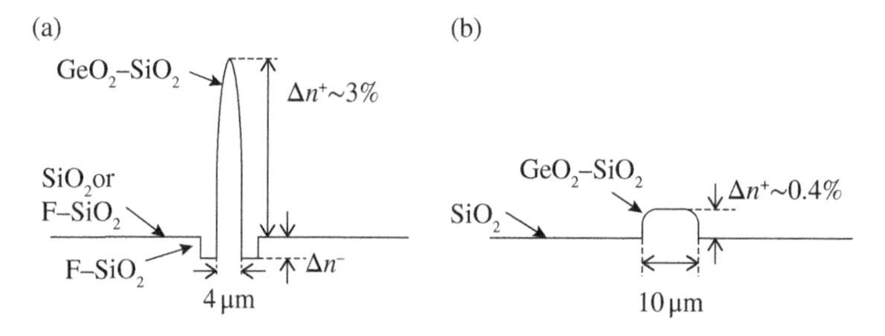

Figure 16.2 Schematic refractive index profiles of (a) an HNLF and (b) an SSMF. *Source:* After Ref. [2].

cladding because such fibers exhibit a large n_2 as well as a small effective area A_{eff} [9]. Figure 16.2 shows the schematic refractive index profiles of a silica HNLF and a standard single-mode fiber (SMF) [2]. In typical HNLFs, the relative refractive index difference of the core to the outer cladding (Δn^+) is around 3%, while the core diameter is around 4 μm. The W-cladding profile with a fluorine-doped depressed cladding having a refractive index difference of Δn^- allows the single-mode operation in the communication bands as well as flexibility in the designing of chromatic dispersion [10]. Using such approaches, new kinds of HNLFs with specific dispersive properties, namely dispersion-shifted fibers, dispersion-compensating fibers, dispersion-decreasing fibers, and dispersion-flattened fibers, have been developed [9, 11, 12]. The values of the nonlinear parameter γ for most of these fibers are in the range of 10–20 W^{-1}/km [8]. Increasing γ much above these values is not possible in these types of fibers since the optical mode confinement is lost when the core diameter is further reduced.

16.1.1 Tapered Fibers

The fiber-normalized frequency V is given by

$$V = k_0 a n_1 \sqrt{2\Delta} \tag{16.2}$$

where k_0 is the vacuum wavenumber, a is the core radius, n_1 is the core refractive index, and

$$\Delta = \frac{(n_1 - n_c)}{n_1} \tag{16.3}$$

is the relative index difference between the core and the cladding of the fiber, n_c being the cladding refractive index. An SMF is achieved when $V < 2.405$. If Δ is kept constant and the core radius a is reduced, the normalized frequency V decreases and the mode confinement is lost. Equation (16.2) shows that in order to maintain a given value of V when decreasing a, Δ must increase such that $a^2\Delta$ remains constant.

The maximum value of Δ is achieved when the cladding material is replaced by air ($n_c \approx 1$). In such a case, the mode remains confined to the core even if the core diameter is close to 1 μm. However, such confinement is reduced for smaller values of the core diameter. Figure 16.3 shows the fundamental mode of a silica fiber surrounded by air, for two different fiber diameters: (a) $d = 2$ μm and (b) $d = 0.5$ μm. In the case of Figure 16.3a, the field is completely confined by the fiber. However, in the case of Figure 16.3b, the fiber cannot completely confine the propagating wave, leading to a relatively large evanescent field.

Narrow-core fibers with air cladding have been produced by tapering standard optical fibers, with an original cladding diameter of 125 μm [13–15]. Tapering a fiber involves heating and stretching it to form a narrow waist connected to the untapered fiber by taper transitions. If the transitions are gradual, light propagating along the fiber suffers a very little loss.

Numerical results have shown that the mean-field diameter of the optical mode of a silica-tapered fiber achieves a minimum value when the fiber diameter is 0.74λ [16]. The nonlinear parameter γ then attains a maximum value, which scales with λ^{-3}. Using $n_2 = 2.6 \times 10^{-20}$ m^2/W, a value $\gamma = 662$ W^{-1}/km is obtained at $\lambda = 0.8$ μm.

As seen in Chapter 3, the overall GVD of a guided mode in an optical fiber depends not only on the material dispersion but also on the waveguide dispersion. The last contribution becomes especially important in the case of fibers using a large difference in refractive index between the core and cladding materials. As seen above, increasing such refractive index difference also provides an enhancement of the nonlinear parameter of the fiber.

The dispersive properties of a tapered fiber are very sensitive to its size and can be adjusted by changing appropriately the fiber diameter [16, 17]. This is illustrated in Figure 16.4, which shows the dispersion curves for the fundamental mode of a tapered fiber considering different fiber diameters. By decreasing such diameter, the zero-dispersion wavelength (ZDW) is shifted toward shorter wavelengths. Moreover, the strongly normal waveguide GVD causes a decrease of the overall GVD at long wavelengths and a second ZDW appears, defining a wavelength window with anomalous GVD. The GVD values are quite large in the anomalous dispersion regime, ranging in the order of 200–300 ps/km/nm. Meanwhile, the first zero-dispersion points do not exceed a wavelength of 800 nm for a fiber diameter below 3 μm.

16.2 Microstructured Optical Fibers

Microstructured optical fibers (MOFs) represent a new class of optical fibers that are characterized by the fact that the silica cladding presents an array of embedded air holes. They are also referred to as photonic crystal fibers (PCFs) since they were first realized in 1996 in the form of a photonic crystal cladding with a periodic array of air holes [18].

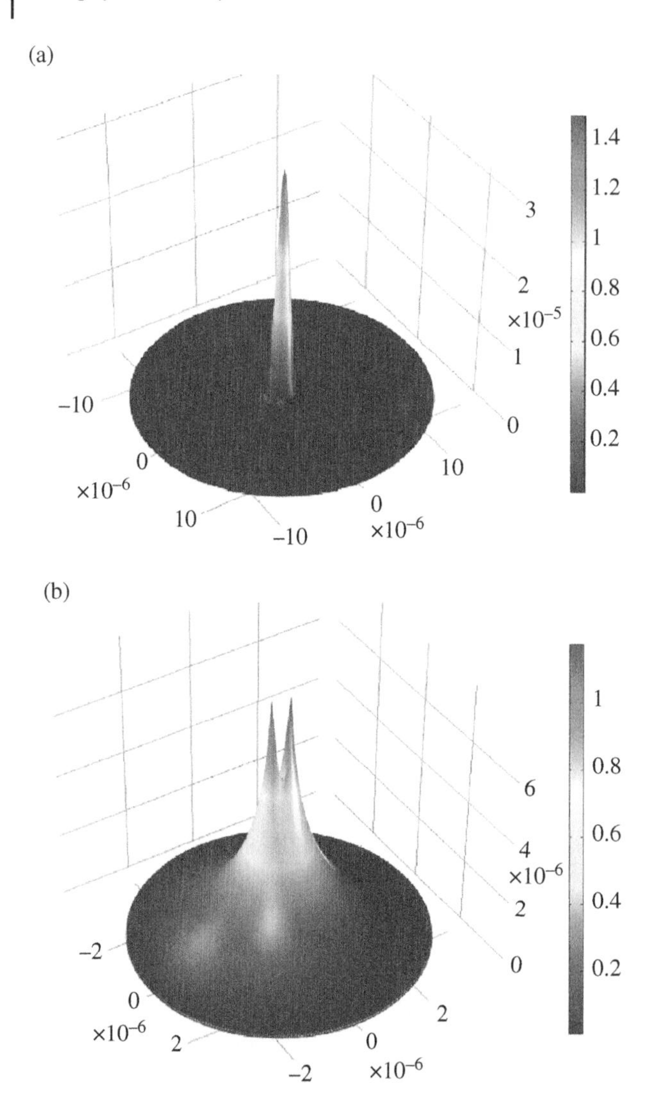

Figure 16.3 Fundamental mode of a tapered silica fiber surrounded by air with a diameter: (a) $d = 2\ \mu\text{m}$ and (b) $d = 0.5\ \mu\text{m}$.

MOFs can be divided into two main types. One class of fibers, first proposed in 1999 [19], has a central region containing air and can confine light by means of a full two-dimensional photonic bandgap (PBG) created by the periodic structure of the cladding. These fibers are usually called hollow-core (HC) MOFs and they will be further discussed in Section 16.6. The second type of microstructured fibers has a solid core in which the light is guided mainly due to total internal reflection since the core has a higher refractive index than the cladding. In such fibers, the periodic nature of air holes in the cladding is not important as long as they provide an effective reduction of its refractive index below that of the silica core [20, 21]. This helps concentrate the mode field in a very small area, which is particularly the case in MOFs with small-scale cladding features and large air-filling fractions.

Figure 16.5 illustrates a solid-core (SC) MOF with a hexagonal pattern of holes in the cladding. This kind of fiber is characterized by two main structural parameters: the hair-hole diameter d and the hole-to-hole spacing Λ.

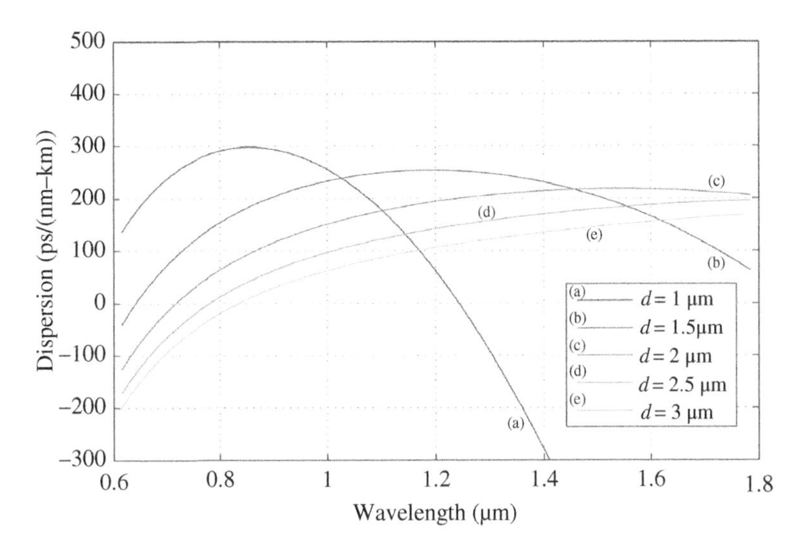

Figure 16.4 Dispersion properties of silica-tapered fiber, for several values of the fiber diameter.

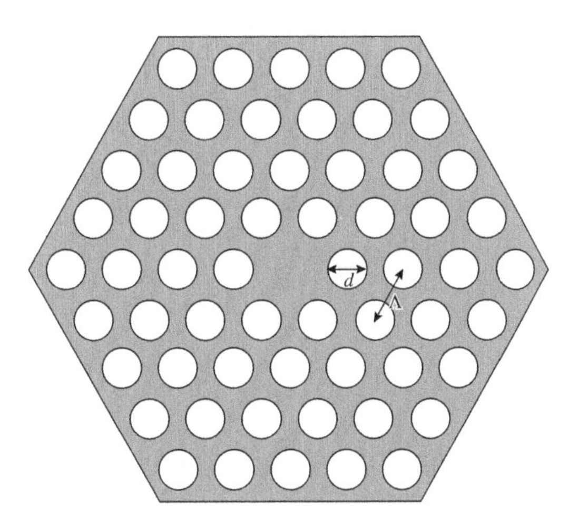

Figure 16.5 Schematic of an solid core microstructured optical fiber (MOF) with a hexagonal pattern of holes.

The number of guided modes in a conventional fiber is determined by the normalized frequency V given by Eq. (16.2), which depends on the wavelength. In conventional fibers, the single-mode condition $V < 2.405$ is only satisfied if the wavelength is in the infrared region. For lower values of the wavelength, the fiber becomes multimode. However, in a MOF the normalized frequency is given by a slightly different expression from Eq. (16.2), with a replaced by Λ and n_c replaced by the effective cladding index of the microstructured cladding. Such an effective refractive index depends significantly on the ratios d/Λ and λ/Λ [22] and asymptotically approaches the core index in the short wavelength limit. As a consequence, both the effective index difference and the numerical aperture go to zero in this limit.

Surprisingly, in an MOF, it is verified that the behavior of the cladding-effective index cancels out the wavelength dependence of the normalized frequency V, such that it approaches a constant value as $\Lambda/\lambda \rightarrow \infty$. As a consequence, it was observed that an MOF shows a single-mode behavior at all wavelengths if $d/\Lambda < 0.43$; such fiber is known as the *endlessly single-mode* (ESM) fiber

[23, 24]. Furthermore, because of the scale invariance of MOFs, the ESM behavior can be achieved for all core diameters.

The dispersive properties of MOFs are quite sensitive to the structural parameters d and Λ and can be tailored by changing appropriately each of them. Figure 16.6 shows the dispersion characteristics of a microstructured fiber with a hexagonal pattern of holes spaced by (a) $\Lambda = 3$ μm, (b) $\Lambda = 2$ μm, and (c) $\Lambda = 1$ μm, considering a wavelength $\lambda = 1.55$ μm, for different hole-pitch ratios.

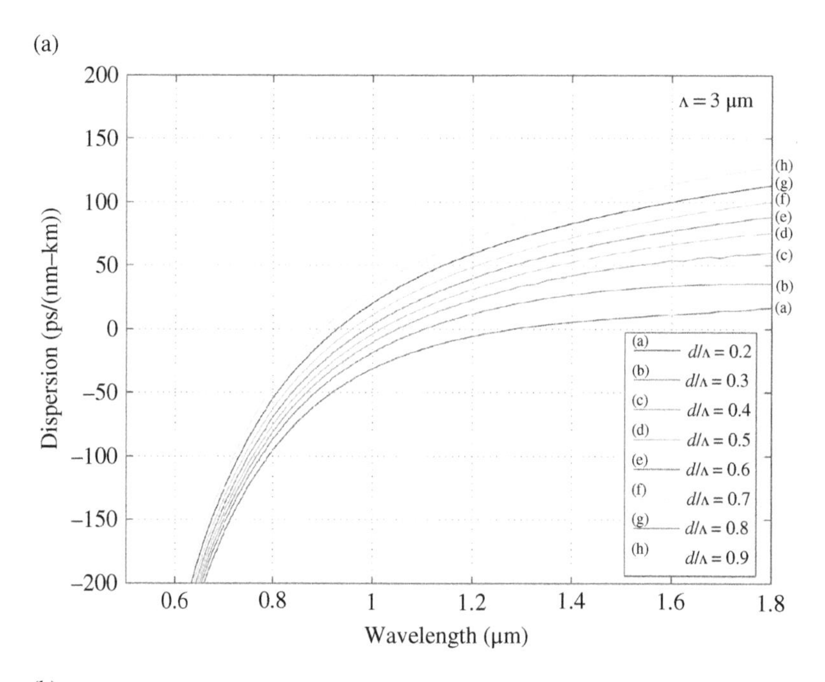

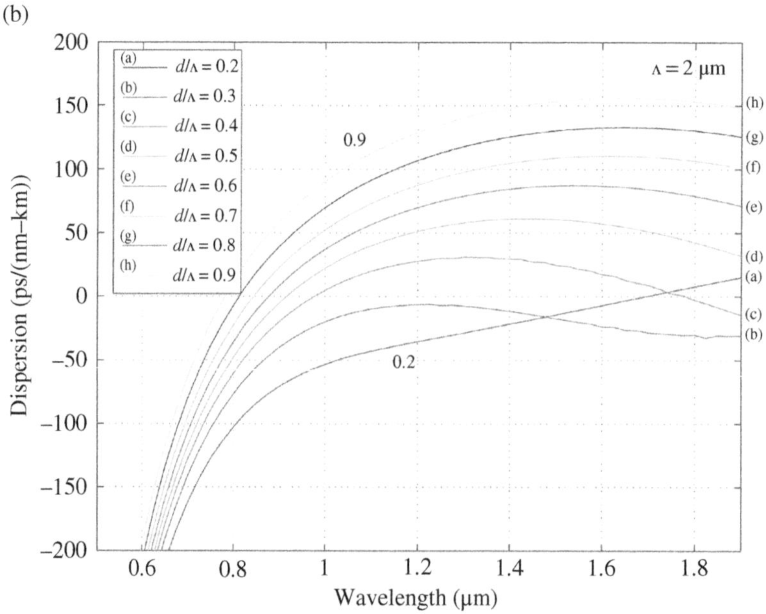

Figure 16.6 Dispersion properties of a microstructured fiber with a hexagonal pattern of holes spaced by (a) $\Lambda = 3$ μm, (b) $\Lambda = 2$ μm, and (c) $\Lambda = 1$ μm, at $\lambda = 1.55$ μm for different hole-pitch ratios.

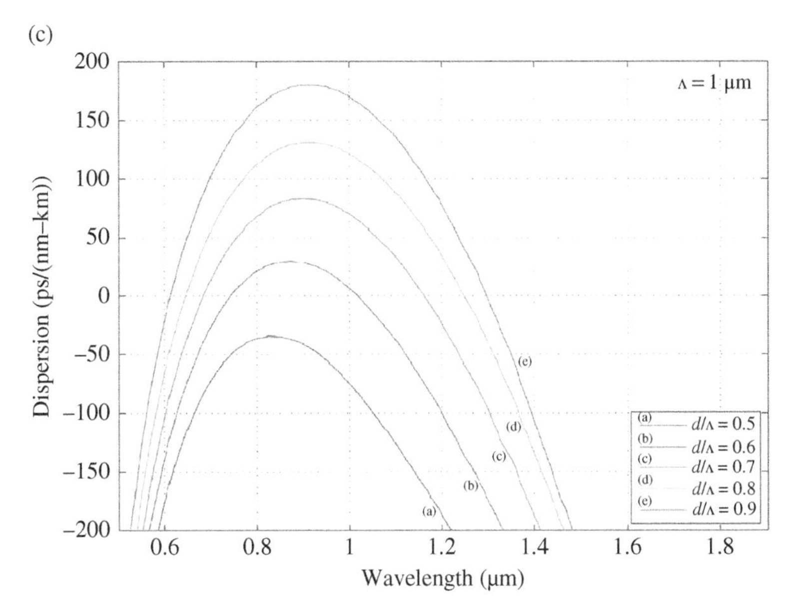

Figure 16.6 (Continued)

In an SC MOF, as the holes get larger, the core becomes more and more isolated, until it resembles an isolated strand of silica glass, also known as jacketed air-suspended rod (JASR). The core diameter of an MOF is given by $d_{core} = 2\Lambda - d$. As observed for tapered fibers, MOFs with larger cores exhibit semi-infinite anomalous dispersion above the ZDW. By decreasing the core size, the ZDW tends to be shifted to a shorter wavelength, leading to the anomalous dispersion at near infrared and visible wavelengths. When the core size is decreased further, a second ZDW arises in the longer wavelength side, such that the GVD is anomalous in the spectral window between the two ZDWs and normal outside it. This situation can be seen in some cases of Figure 16.6b and c. The existence of two ZDWs has a great influence on the nonlinear effects in MOFs [25–28]. In the four-wave mixing (FWM) effect, two phase-matching sidebands can be generated in the fiber with one ZDW [29–31], and four phase-matching sidebands can be generated in an MOF with two ZDWs [32, 33]. These have opened new opportunities for multiwavelength FWM, optical switches, optical parametric amplification, and supercontinuum generation [34–37].

A nearly zero dispersion and flattened behavior is observed in some cases of Figure 16.6. This characteristic can be further controlled by adjusting the number of hole rings around the core of an MOF, as illustrated in Figure 16.7 for the case $\Lambda = 2$ μm and $d = 0.6$ μm. The availability of fibers with nearly zero dispersion is particularly important for supercontinuum generation and other telecom applications making use of the FWM effect.

Figure 16.8 shows the effective mode area for SC-MOFs with a hexagonal pattern of holes spaced by $\Lambda = 1$ μm and $\Lambda = 2.5$ μm, considering a wavelength $\lambda = 1.55$ μm, for different hole-pitch ratios. We observe that the effective area is significantly smaller in the case $\Lambda = 1$ μm, compared with the case $\Lambda = 2.5$ μm. In both cases, increasing the hole-pitch ratio provides a reduced effective area. Consequently, the nonlinear parameter is increased, according to Eq. (16.1).

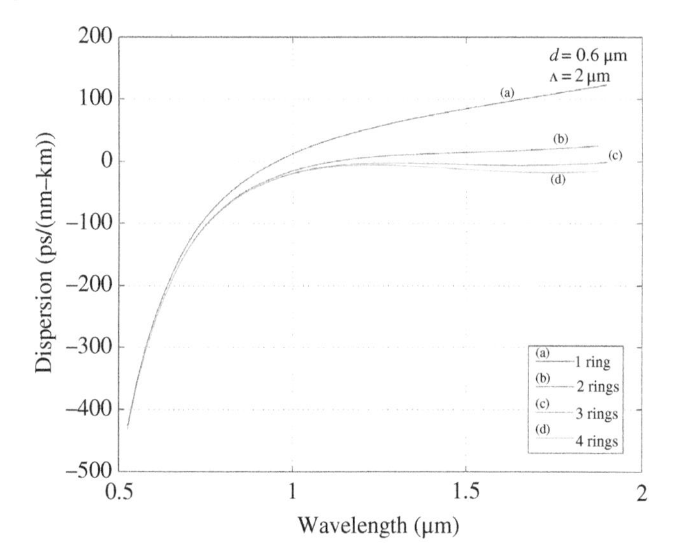

Figure 16.7 Dispersion properties of a microstructured fiber with a hexagonal pattern of holes spaced by $\Lambda = 2\ \mu m$ and hole diameter $d = 0.6\ \mu m$ at $\lambda = 1.55\ \mu m$, for a different number of hole rings around the fiber core.

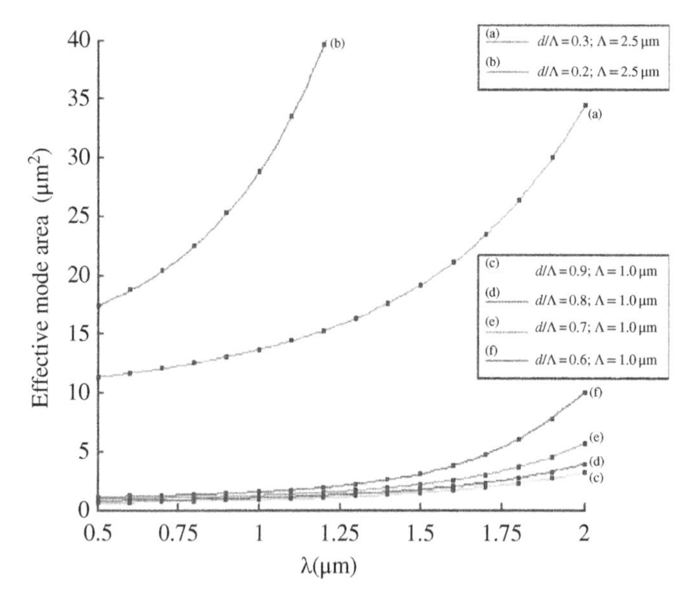

Figure 16.8 Effective mode area for SC-MOFs with a hexagonal pattern of holes spaced by $\Lambda = 1\ \mu m$ and $\Lambda = 2.5\ \mu m$, considering a wavelength $\lambda = 1.55\ \mu m$, for different hole-pitch ratios.

Figure 16.9 shows the effective mode area of an MOF against the hole-to-hole spacing Λ, for different hole-pitch ratios d/Λ (full curves). The dashed curve corresponds to the case of a JASR and shows a minimum mode area for a rod diameter around 1.1–1.2 μm. Moreover, Figure 16.9 shows that SC MOFs also exhibit a minimum effective mode area for a given value

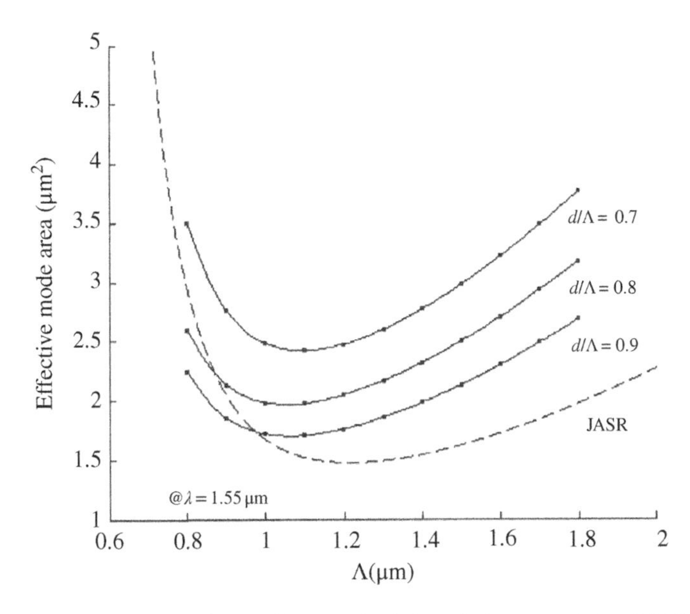

Figure 16.9 Effective mode area of a jacketed air-suspended rod (JASR) (dashed curve) and of an MOF (full curves) as a function of the rod diameter or the hole-to-hole spacing, respectively.

of the hole-to-hole spacing. The smallest effective mode area corresponds to the largest air-filling fraction, which is achieved in Figure 16.9 for the case $d/\Lambda = 0.9$. A minimum effective mode area $A_{eff} = 1.7\ \mu m^2$ is obtained in this case, which occurs for a hole-to-hole spacing $\Lambda \approx 1\ \mu m$.

In the case of microstructured fibers and to take into account the different proportions of light in glass and air, the nonlinear coefficient γ must be redefined as follows [38]:

$$\gamma = k_0 \sum_i \frac{n_2^i}{A_{eff}^i} = k_0 \frac{n_2^{eff}}{A_{core}} \tag{16.4}$$

In Eq. (16.4), n_2^i is the nonlinear refractive index of material i ($2.9 \times 10^{-23}\ m^2/W$ for air, $2.6 \times 10^{-20}\ m^2/W$ for silica), A_{eff}^i is the effective area for the light in material i, and n_2^{eff} is the effective nonlinear index for fiber, with a core area A_{core}. A nonlinear coefficient $\gamma = 240 W^{-1}/km$ at 850 nm was measured for an SC MOF with a core diameter of 1 μm [3], which must be compared with the highest values ~$20\ W^{-1}/km$ obtainable in conventional highly nonlinear (HNL) silica fibers.

A peak power of $P_0 = 5$ kW launched into a silica MOF with a nonlinear coefficient $\gamma = 240\ W^{-1}/km$ provides a nonlinear length $L_{NL} = (\gamma P_0)^{-1} < 1$ mm. For dispersion values in the range of $-400 < \beta_2 < 400\ ps^2/km$ and pulse duration of $t_0 = 300$ fs, we have a dispersion length $L_D = t_0^2/|\beta_2| > 0.2$ m. Considering also typical values of fiber loss ($\alpha \sim 1$–100 dB/km), we conclude that both the effective fiber length $L_{eff} = (1 - \exp(-\alpha L))/\alpha$ and the dispersion length L_D are much longer than the nonlinear length L_{NL}, which means that strong nonlinear effects will be readily observable in SC MOFs.

16.3 Non-Silica Fibers

The value of the nonlinear parameter γ can be enhanced if the optical fiber is made using glasses with higher nonlinearities than silica, such as lead silicate, tellurite, bismuth oxide, and chalcogenide glasses. Such possibility is illustrated in Figure 16.10a, which shows the nonlinear-index coefficient n_2 against the linear refractive index, n, for several optical glasses [39, 40]. High intrinsic nonlinearity is predicted for glasses with a high linear refractive index according to the empirical Miller's rule. Introducing heavy-metal compounds, or introducing the chalcogen elements S, Se, and Te to replace oxygen, acts to increase the polarizability of the components in the glass matrix and also increases the nonlinear index n_2.

Figure 16.10a shows that the nonlinear index can be increased by a factor of almost 1000 compared with its value for silica fibers using some types of highly nonlinear (HNL) glasses. Heavy metal oxide glasses (lead silicate, bismuth oxide, and tellurite) have linear indices in the range of 1.8–2.0, nonlinear indices ~10 times higher than silica, and material ZDWs of 2–3 μm, as observed from Figure 16.10b. In the particular case of bismuth oxide fibers, values for n_2 in the range of 30×10^{-20}–11×10^{-18} m^2/W have been measured [4, 41, 42]. Chalcogenide glasses (GLS, As_2 S_3) have linear indices of 2.2–2.4, nonlinear indices significantly higher than those of the oxide glasses, and material ZDWs larger than 4 μm. A nonlinear refractive index $n_2 = 2.4 \times 10^{-17}$ m^2/W has been measured, which is larger by a factor of about 1000 relatively to the silica case [6]. In general, the ZDW of a glass shifts to longer wavelengths with an increasing linear refractive index, as shown in Figure 16.10b.

The calculated multiphonon absorption losses of several glasses are shown in Figure 16.10c. The multiphonon absorption edge determines the long-wavelength transmission limit and can be moved to longer wavelengths if the lattice vibration frequencies are reduced by using compositions containing heavier atoms or having weaker chemical bonds in the glass network.

Compared with silica, chalcogenide glasses have a wider transmission window, which extends from the visible up to the infrared region of 10 or 15 μm depending on the composition. They also have higher Raman gain coefficients: $4.3 - 5.7 \times 10^{-12}$ m/W for As_2 S_3 and $2 - 5 \times 10^{-11}$ mW for As_2 Se_3, both at 1.5 μm [43, 44]. As a result, optical fibers made of chalcogenide glasses are particularly suitable for generating cascaded stimulated Raman scattering (SRS) in the mid-infrared region [45–47].

Several techniques have been used to fabricate compound glass MOFs [48]. One approach is to manually stack capillary tubes to produce the structured preform. This is the approach that is routinely used to produce silica-microstructured fibers, and the resulting fibers typically consist of a hexagonal lattice of air holes surrounding the fiber core. Soft glass MOFs fabricated using this method have been reported by several groups [48, 49]. In addition, tellurite glass MOFs prepared by stacking have been reported with rare earth dopants [50].

Typical compound glass softening temperatures are 500° C, as opposed to 2000° C for silica. In these circumstances, the extrusion technique for fiber performs manufacture is particularly suitable [40, 51]. Besides being simpler than the stacking technique, extrusion allows the realization of cladding structures consisting of mostly air, as required to increase the fiber nonlinearity.

A nonlinear parameter $\gamma = 1860$ W^{-1}/km at 1.55 μm was measured in a 144-cm-long MOF made of lead-silicate glass (SF57) and presenting an effective mode area $A_{eff} = 1.1$ μm^2 [5]. Moreover, a value $\gamma \approx 1100$ W^{-1}/km at 1550 nm has been measured in a bismuth oxide fiber with a mode field diameter of 1.97 μm [4]. Only a 35-cm length of such an HNLF was needed to demonstrate several applications on nonlinear processing of optical signals.

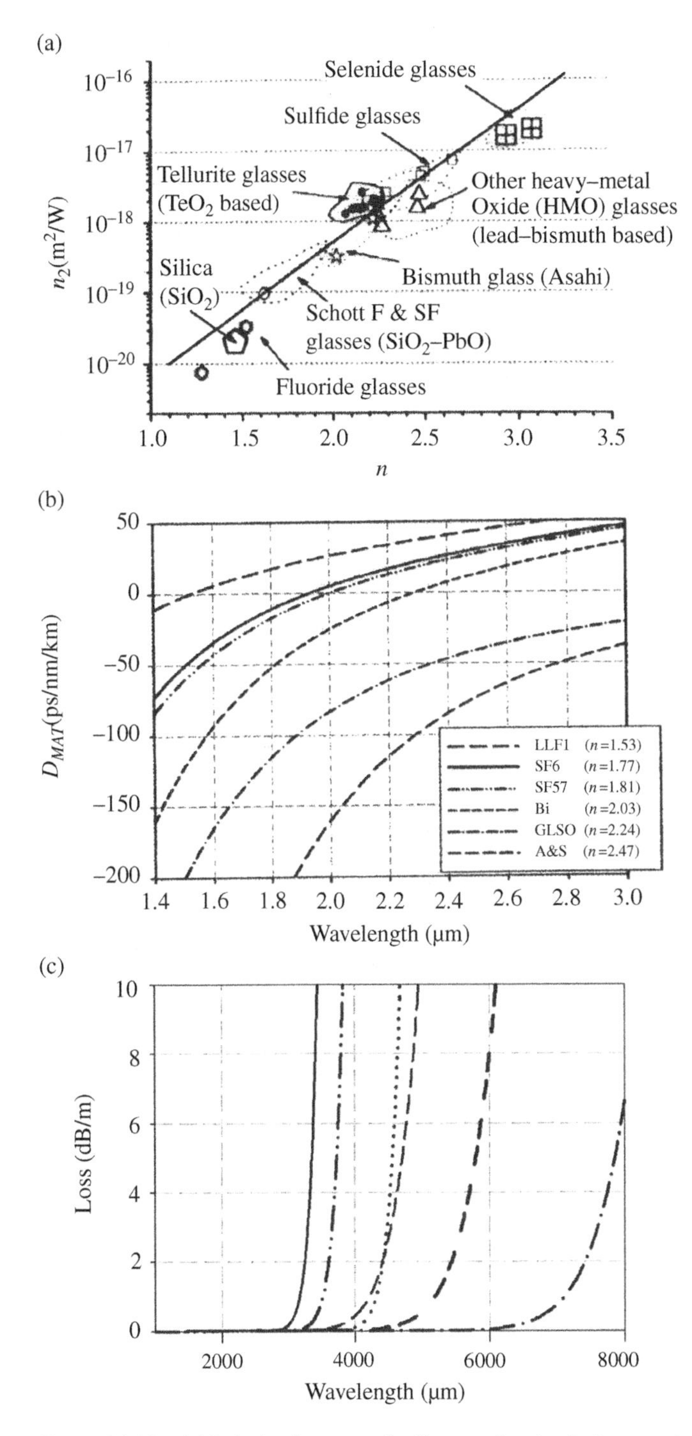

Figure 16.10 (a) Relation between the linear refractive index n and nonlinear refractive index n_2 in various glasses; (b) Material dispersive curves of Schott glasses (LLF1, SF6, SF57), bismuth oxide glass (Bi), and chalcogenide glass (gallium lanthanum sulphide oxide glass [GLSO]). The number in the legend indicates the linear index at 1.06 μm; (c) Multiphonon edge of the glasses (theoretical fit). *Source:* After Ref. [39].

A nonlinear parameter $\gamma = 2450\ \mathrm{W}^{-1}/\mathrm{km}$ was measured in a 2004 experiment for an 85-cm-long chalcogenide fiber with a core diameter of 7 μm [6]. The same fiber also exhibited a Raman gain parameter which was nearly 800 larger than that of silica fibers [6]. This allows the use of either a smaller fiber length or a lower pump power to attain amplifier gain characteristics similar to those of silica-based devices. Due to their high nonlinearity, chalcogenide fibers have been considered as a good option to enhance several nonlinear effects.

MOFs have been fabricated in various types of chalcogenide glasses for several applications. Employment of photonic crystal geometry in the cladding of As_2Se_3 fibers can give rise to a strong overlap between pump and signal, providing higher Raman gain efficiency than conventional fibers, as demonstrated for silica-based MOF structures [52, 53]. It was shown that the Raman gain efficiency in As_2Se_3 MOFs can be improved by a factor of more than four compared with conventional As_2Se_3 fibers [54].

16.4 Soliton Fission and Dispersive Waves

As seen in Chapter 9, the propagation of femtosecond pulses in optical fibers can be perturbed by several higher-order dispersive and nonlinear effects, namely the third-order dispersion, self-steepening, and intrapulse Raman scattering (IRS). In the case of a higher-order soliton, these perturbations can determine the fission of the pulse into its constituent fundamental solitons, a phenomenon called *soliton fission*. This phenomenon was reported for the first time in a 1987 experiment, by observing the spectra at the output of a fiber with a 1 km-length of 830-fs input pulses with different values of the peak power [55].

Modeling the ultrashort pulse propagation and supercontinuum generation in optical fibers can be realized considering a generalized NLSE that includes higher-order dispersive and nonlinear effects [7, 56, 57]. Such an equation can be written as:

$$\frac{\partial U}{\partial z} - i\sum_{k\geq 2}\frac{i^k\beta_k}{k!}\frac{\partial^k U}{\partial \tau^k} + \frac{\alpha(\omega)}{2}U = i\gamma\left(1 + \frac{i}{\omega_0}\frac{\partial}{\partial \tau}\right)\left(U(z,\tau)\int_{-\infty}^{+\tau} R(t')|U(z,\tau-t')|^2 dt'\right) \tag{16.5}$$

where $U(z, t)$ is the electric field envelope, ω_0 is the center frequency, β_k are the dispersion coefficients at the center frequency, $\alpha(\omega)$ is the frequency-dependent fiber loss, and γ is the nonlinear parameter.

The nonlinear response function $R(t)$ in Eq. (16.5) can be written as

$$R(t) = (1 - f_R)\delta(t) + f_R h(t), \tag{16.6}$$

where the δ-function represents the instantaneous electron response (responsible for the Kerr effect), $h(t)$ represents the delayed ionic response (responsible for the Raman scattering), and f_R is the fractional contribution of the delayed Raman response to the nonlinear polarization, in which a value $f_R = 0.18$ is often assumed for silica fibers [58]. It is common to approximate $h(t)$ in the form [57, 59]:

$$h(t) = \frac{\tau_1^2 + \tau_2^2}{\tau_1\tau_2^2}\exp\left(\frac{-t}{\tau_2}\right)\sin\left(\frac{t}{\tau_1}\right) \tag{16.7}$$

where $\tau_1 = 12.2$ fs and $\tau_2 = 32$ fs. More accurate forms of the response function $h(t)$ have also recently been investigated [60].

Equation (16.5) can be used to describe the propagation of femtosecond pulses in optical fibers, in both the normal and anomalous dispersion regimes. When such pulses have enough power, their spectra undergo extreme broadening. In the anomalous dispersion regime, this process is mainly influenced by the phenomenon of soliton fission.

The soliton order is given by $N = \sqrt{L_D/L_{NL}}$, where $L_D = t_0^2/|\beta_2|$ is the dispersion distance and $L_{NL} = 1/\gamma P_0$ is the nonlinear length as defined in Chapter 3. In the presence of higher-order dispersion, an Nth-order soliton gives origin to N fundamental solitons whose widths and peak powers are given by [61]:

$$t_k = \frac{t_0}{2N + 1 - 2k} \tag{16.8}$$

$$P_k = \frac{(2N + 1 - 2k)^2}{N^2} P_0 \tag{16.9}$$

where $k = 1$ to $\tilde{N}$, where $\tilde{N}$ is the integer closest to N when N is not an integer. Soliton fission occurs generally after a propagation distance $L_{fiss} \sim L_D/N$, at which the injected higher-order soliton attains its maximum bandwidth. The fission distance L_{fiss} is a particularly significant parameter in the context of higher-order soliton effect compression, as described in Chapter 15 [62, 63].

Besides higher-order dispersion, another main effect affecting the dynamics of higher-order soliton fission is the intrapulse Raman scattering (IRS). As discussed in Chapter 9, this phenomenon leads to a continuous downshift of the soliton carrier frequency, an effect known as the *soliton self-frequency shift* (SSFS) [64]. Since its discovery in conventional SMFs [65], the SSFS effect has also been observed in other types of fibers, including tapered fibers [66], SC MOFs [67, 68], and air-core MOFs [69]. The SSFS effect can be significantly enhanced in some HNFs, where it has been used during recent years for producing femtosecond pulses [70–73], as well as to realize several signal processing functions [74–76].

The rate of frequency shift per propagation length is given from Eq. (9.14) as

$$\frac{df}{dz} = -\frac{4t_R|\beta_2|}{15\pi t_0^4} = -\frac{4t_R(\gamma P_0)^2}{15\pi|\beta_2|} \tag{16.10}$$

where $t_R \equiv f_R \int_{-\infty}^{+\infty} th(t)dt \approx 5$ the Raman parameter and P_0 is the soliton peak power. Since the SSFS effect is proportional to $(\gamma P_0)^2$, it will be enhanced if short pulses with high peak pulses are propagated in HNLFs.

Figure 16.11 shows the optical spectra of the output pulses from a 60-cm-long MOF, obtained in a 2002 experiment [70]. In this experiment, 70-fs pulses with a 48-MHz repetition rate at a wavelength of 782 nm were launched at the fiber input. As the fiber input power is increased above a given value, the center wavelength of the generated Raman soliton is shifted toward the longer wavelength side continuously due to the SSFS effect. The total shift at the fiber output achieves a maximum value of 900 nm when the fiber input power is 3.6 mW. The spectral shape of the Raman pulses is almost sech^2 and their spectral width has an almost constant value of 18 nm.

It can be observed from Figure 16.11 that a second Raman soliton is generated when the input power is 3.6 mW. In fact, for this pump pulse, the corresponding soliton order exceeds 2. Two fundamental solitons are then created through the fission process experienced by the higher-order soliton initially propagating inside the fiber.

The Raman solitons created through the fission process are generally perturbed by the third- and higher-order dispersion of the fiber, emitting dispersive waves which constitute the so-called

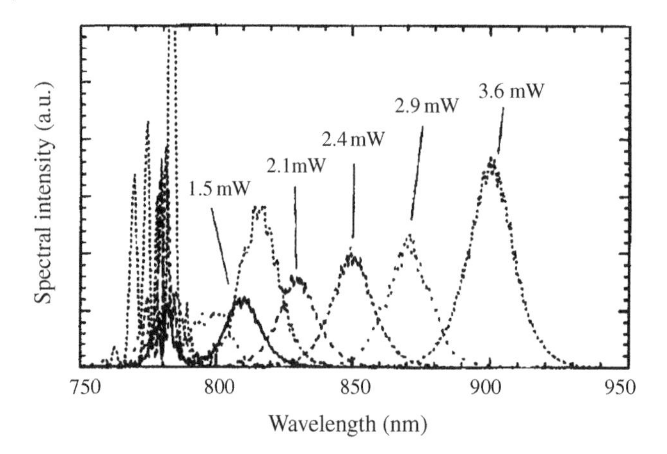

Figure 16.11 Optical spectra of generated wavelength-tunable femtosecond soliton pulse as the fiber input power was changed. *Source:* After Ref. [70].

nonsoliton radiation (NSR) [77] or *Cherenkov radiation* [78]. Such radiation was observed in a 2001 experiment using 110-fs pulses from a laser operating at 1556 nm, which were launched into a polarization-maintaining dispersion-shifted fiber with a GVD $\beta_2 = -0.1\ \text{ps}^2/\text{km}$ at the laser wavelength [79].

The NSR is emitted at a frequency such that its phase velocity matches that of the soliton. The two phases are equal when the following phase-matching condition is satisfied:

$$\beta(\omega) = \beta(\omega_s) + \beta_1(\omega - \omega_s) + \frac{1}{2}\gamma P_s \tag{16.11}$$

In Eq. (16.11), P_s and ω_s are the peak power and the frequency, respectively, of the fundamental soliton formed after the fission process. Expanding $\beta(\omega)$ around ω_s, this phase-matching condition reduces to

$$\sum_{n\geq 2} \frac{\Omega^n}{n!}\beta_n(\omega_s) = \frac{\gamma P_s}{2} \tag{16.12}$$

where $\Omega = \omega - \omega_s$ is the frequency shift between the dispersive wave and the soliton and β_n $(n = 2, 3)$ are the dispersion parameters, calculated at the soliton central frequency ω_s.

Neglecting the fourth- and higher-order dispersion terms, the frequency shift between the frequency of dispersive waves and that of the soliton is given approximately by [78]:

$$\Omega_d \approx -\frac{3\beta_2}{\beta_3} + \frac{\gamma P_0 \beta_3}{3\beta_2^2} \tag{16.13}$$

For solitons propagating in the anomalous-GVD regime, we have $\beta_2 < 0$. In these circumstances, Eq. (16.13) shows that $\Omega > 0$ when $\beta_3 > 0$, in which case the NSR is emitted at wavelengths shorter than that of the soliton. On the contrary, if $\beta_3 < 0$, the NSR is emitted at a longer wavelength than that of the soliton and it can result in the suppression of the SSFS effect [80, 81].

Figure 16.12 shows the dispersion curve of a microstructured fiber with a hexagonal pattern of holes with a diameter $d = 0.8\ \mu\text{m}$, spaced by $\Lambda = 1\ \mu\text{m}$. This MOF has two ZDWs, such that the GVD is anomalous in the spectral window between them. The third-order dispersion parameter β_3 is positive near the first ZDW and negative near the second one.

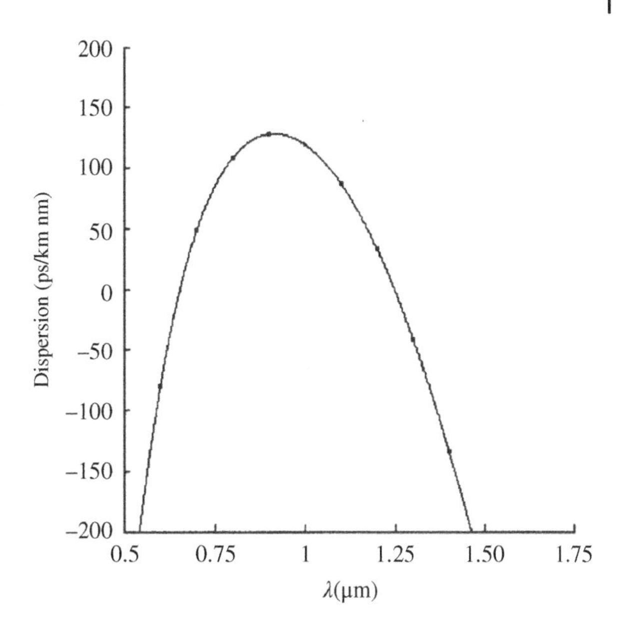

Figure 16.12 Dispersion curve for a microstructured fiber with a hexagonal pattern of holes with a diameter $d = 0.8$ μm, spaced by $\Lambda = 1$ μm.

Figure 16.13 shows the spectral (a) and the temporal (b) evolution of an optical pulse along the MOF whose dispersion is given in Figure 16.12. The fiber is pumped at 860 nm using 100-fs pulses with a peak power of 1 kW. The GVD at the pump wavelength is $\beta_2 = -0.0487$ ps^2/m and the nonlinear parameter is $\gamma = 196.2$ W^{-1}/km, which result in a soliton order $N = 3.6$. As a consequence of the SSFS induced by the IRS effect, the solitons arising from the fission process separate from each other. Since the SSFS is the largest for the shortest soliton, its spectrum shifts the most toward the red side in Figure 16.13a. Near the first ZDW, where $\beta_3 > 0$, the NSR is always blue-shifted so that the recoil and SSFS act in the same spectral direction. However, the opposite behavior occurs near the second ZDW, where $\beta_3 < 0$. As the soliton loses energy to the red-shifted NSR, which is emitted in the normal-GVD regime, it becomes slightly wider in the time domain, as a result of energy conservation. On the other hand, since the detuning between the radiation and soliton gets smaller with propagation, the radiation is exponentially amplified. This amplification leads to the strong spectral recoil of the soliton into the anomalous-GVD regime, providing the suppression of SSFS [80, 81]. These effects are clearly visible in Figure 16.13, regarding the evolution of the first ejected soliton. We observe that the frequency shift of this soliton is suppressed after $Z \sim 0.4$ m.

The cancellation of the SSFS was observed in a 2003 experiment [80], using a silica-core MOF with two ZDWs: one at 600 and the other at 1300 nm. 200-fs pulses at a wavelength of 860 nm and with a peak power of 230 W were launched in it. These experimental conditions provided the initial formation of a pulse corresponding to the fourth-order soliton solution of the NLSE. It was observed that the frequency of the more intense Raman soliton resulting from the fourth-order soliton fission was stabilized at ≈1270 nm, through the recoil mechanism.

The effect of SSFS cancellation described above cannot be observed in conventional optical fibers since, for the commonly used frequencies, the radiation is always blue-shifted so that the recoil and Raman effects act in the same spectral direction. Such cancellation is an example of the unexpected

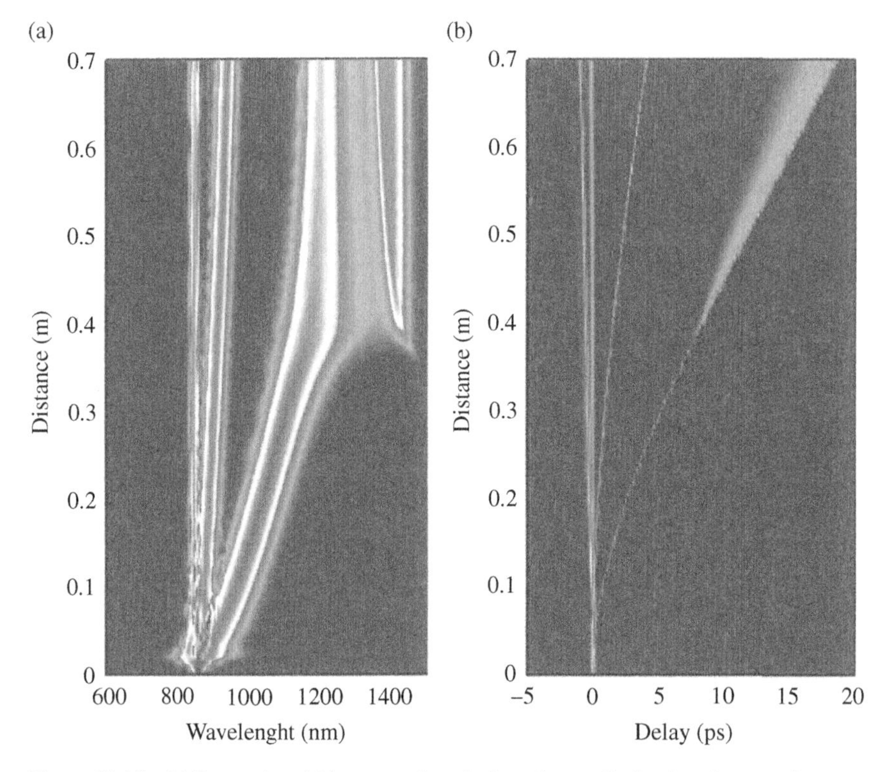

Figure 16.13 (a) Spectral and (b) temporal evolution of an optical pulse along a microstructured fiber whose dispersion is given by curve B in Figure 16.12. The fiber is pumped at 860 nm using 100-fs pulses with a peak power of 1 kW.

effects which can be observed in MOFs due to the unique combination of their dispersive and nonlinear properties.

16.5 Four-Wave Mixing

As discussed in Chapter 4, new frequencies can be generated in conventional optical fibers by parametric FWM when they are pumped close to the ZDW. In this case, the phase-matching condition is sensitive to the exact shape of the dispersion curve. Thus, we must expect that the peculiar dispersive properties of some HNLFs have a profound impact on the FWM process through the phase-matching condition.

The phase-matching condition for the FWM process requires that dispersive effects compensate nonlinear ones as obtained from Eq. (4.48):

$$\beta_2\Omega_s^2 + \frac{\beta_4}{12}\Omega_s^4 + 2\gamma P_p = 0 \tag{16.14}$$

where β_2 and β_4 are, respectively, the second- and fourth-order dispersion terms at the pump frequency, Ω_s is the frequency shift between the signal or idler frequencies and the pump frequency, γ is the nonlinear parameter, and P_p is the pump peak power. The third and all-odd dispersion orders play no role in Eq. (16.14) since they cancel out from the degenerate phase-matching condition $2\beta(\omega_p) - \beta(+\Omega_s) - \beta(-\Omega_s) = 0$, where $\beta(\omega)$ is the exact wave vector.

The impact of the fourth-order dispersion term (β_4) on parametric processes was experimentally studied some years ago in standard telecommunications fibers [82] and in MOFs [83]. In the anomalous dispersion regime ($\beta_2 < 0$), the β_2 term dominates in Eq. (16.14) and the frequency shift is given by Eq. (4.46). However, modulation instability can also occur in the normal-GVD regime ($\beta_2 > 0$). In this case, the positive nonlinear phase mismatch $2\gamma P$ is compensated by the negative value of the linear phase mismatch due to β_4.

In general, the frequency shift between the pump and the FWM sidebands is much larger than the Raman gain band (of 13.2 THz) when the pump wavelength is below the ZDW. Therefore, no significant spectral overlap is possible between Raman and FWM gain bands in this situation. Such feature is particularly important to reduce the Raman-related noise and allows the use of the MOF as a compact bright tunable single-mode source of entangled photon pairs, with wide applications in quantum communications [84–86].

FWM in HNL MOFs can be used to realize both fiber-optic parametric amplifiers (FOPAs) and optical parametric oscillators (OPOs) [87–91]. The physics of parametric amplification in HNLFs is similar to that of standard optical fibers. Differences arise, first, from the enhanced nonlinear parameter γ. Moreover, some MOFs can exhibit single transverse mode propagation over a wide range of wavelengths, which leads to excellent spatial overlap between propagating modes at widely different wavelengths. Finally, the peculiar dispersive characteristics of HNLFs can enhance the phase-matching condition of the FWM process.

FWM in HNLFs can also be used to stabilize the output of multiwavelength erbium-doped fiber lasers, through the continuous annihilation and creation of photons at the wavelengths of interest [92]. A stable tunable dual-wavelength output over 21.4 nm was demonstrated using a 35-cm-long HNL bismuth-oxide fiber [93]. The number of oscillating wavelengths could be increased by simply applying a stronger FWM in the fiber. In such a case, more wavelength components would be involved in the energy exchange process to stabilize the laser output. Using the same 35-cm-long HNL bismuth-oxide fiber and a 30-dBm EDFA output power, up to four different wavelengths were produced in a 2008 experiment [4].

In addition to its use in generating CW multiwavelength laser output, FWM in HNL bismuth-oxide fiber has also been used in several other applications, namely to produce wavelength and width-tunable optical pulses [94, 95], frequency multiplication of a microwave photonic carrier [96], and wavelength conversion of 40 Gb/s polarization-multiplexed ASK-DPSK data signals [4].

16.6 Hollow-Core Microstructured Fibers

As mentioned in Section 16.2, there is one kind of MOF that presents a central region containing air, usually called HC-MOFs. Compared with SC-MOFs, HC-MOFs can offer much higher power delivery capability with a considerably higher damage threshold and significantly lower dispersion and nonlinearity [97–99]. Among the different types of HC-MOFs, PBG MOFs have shown the lowest loss [99]. The main drawback of PBG MOFs is their intrinsically narrow transmission bandwidth determined by the bandgaps, which excludes its implementation in a large number of applications in ultrafast nonlinear optics requiring broadband guidance or guidance in the visible and ultraviolet (UV). Moreover, PBG MOFs present a relatively low-damage threshold due to the significant overlap between the air-guided field and the cladding glass [100]. On the other hand, recently introduced HC-MOFs with a negative-curvature core boundary (hypocycloid), namely the kagomé MOFs [3, 101, 102], provide a particularly lower field overlap with the surrounding

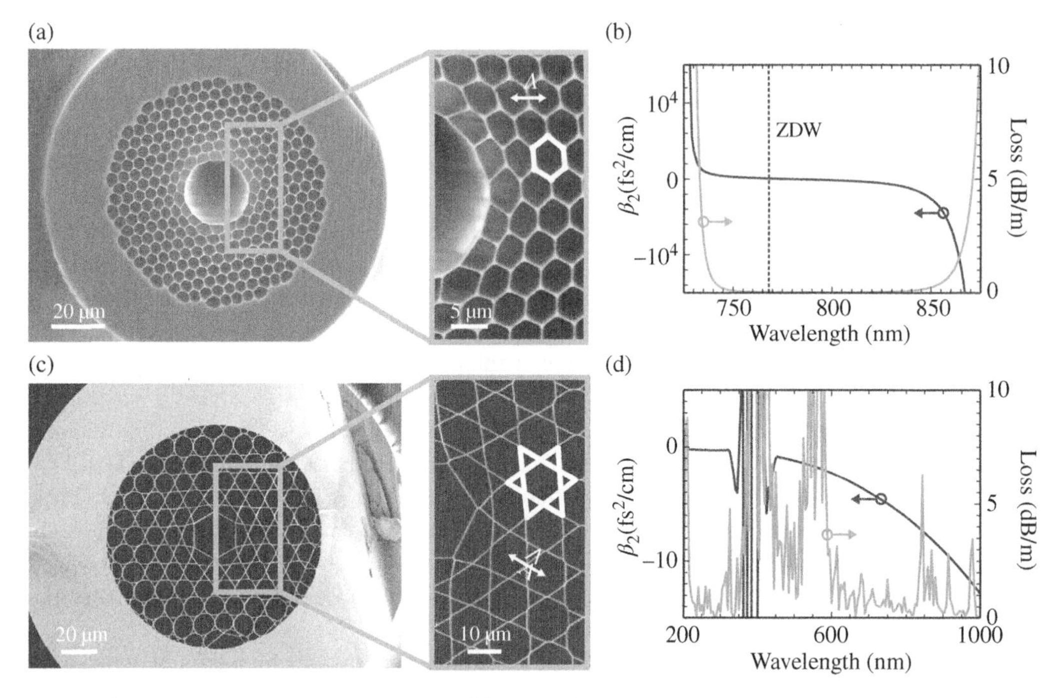

Figure 16.14 Scanning electron micrographs (SEMs) and finite-element modeling (FEM) of the two main types of HC-MOFs: (a) SEM of PBG MOF; (b) GVD and loss calculated using FEM of an idealized PBG MOF structure, designed for operation around 800 nm, with 11 μm core diameter and 2.1 μm pitch (Λ); (c) SEM of a kagomé MOF designed for operation in the UV and around 800 nm; and (d) GVD and loss calculated using the FEM of an idealized kagomé PCF structure with 30 μm core diameter, 15 μm pitch (Λ), and 0.23 μm web thickness. *Source:* After Ref. [103].

cladding structure [3], leading to a considerable increase in damage threshold. Furthermore, at wavelengths around 1 μm and below, kagomé MOFs can now achieve a loss comparable to, or even lower than, PBG MOFs.

Figure 16.14 shows the scanning electron micrographs (SEMs) of the cross-section of the two main varieties of HC-MOFs: (a) PBG MOF; and (c) kagomé MOF [103]. The guidance mechanism in the PBG MOF is based on a full two-dimensional PBG effect, whereas in the second case it has been associated with antiresonance effects from glass membranes adjacent to air-core [103–106]. The GVD and loss calculated using finite-element modeling (FEM) for each fiber are shown in Figure 16.14b and d, respectively. The PBG MOF shows a minimum loss of 0.01 dB/m around 800 nm. Kagomé MOFs provide ultrabroadband (several hundred nanometers) guidance at loss levels of ~1 dB/m, and it displays weak anomalous GVD over the entire transmission window.

The properties exhibited by kagomé MOFs not only help support ultrafast soliton dynamics, but also allow the guidance of any UV light that is subsequently generated at a relatively low loss [103]. The possibility of filling the fiber core with a suitable gas and of adjusting conveniently its pressure also offers a new degree of freedom over conventional fibers, providing a perfect environment for demonstrating many different nonlinear effects [103, 107–111]. Using noble gases in the fiber core, the Raman effects are avoided, allowing one to study Kerr-related phenomena in the absence of perturbations such as the SSFS or Raman-induced noise.

In order to find the modes supported by a kagomé MOF, we can solve the following master equation [110–112]:

$$\nabla \times \left(\frac{1}{n^2(\lambda, r)} \nabla \times \mathbf{H}(\lambda, r) \right) = \frac{4\pi^2}{\lambda} \mathbf{H}(\lambda, r); \tag{16.14a}$$

$$\mathbf{E}(\lambda, r) = \frac{i}{\omega \varepsilon_0 n^2(\lambda, r)} \nabla \times \mathbf{H}(\lambda, r); \tag{16.14b}$$

where $n(\lambda, r)$ is the refractive index in each point of the fiber cross section, λ is the wavelength of the light, ω is its frequency, **H** is the magnetic field and **E** is the electric field. Eq. (16.14a) can be solved on two different domains: (i) considering the whole cross section of the MOF; or (ii) only on a cell of its cladding and applying periodic boundary conditions. With these two approaches, we obtain different, but consistent, information about the fibers: with the former method, we obtain the propagation modes in the core, and optical properties such as dispersion and the nonlinear parameter; with the latter method, we obtain the properties of the cladding structure such as the bandgap regions.

The effective modal refractive index, n_{eff}, of a gas-filled HC-MOF can be approximated by that of a glass capillary, which is given by [113]:

$$n_{eff} = n_{gas} - \frac{1}{2} \left(\frac{u_{01}\lambda}{\pi d_f} \right)^2, \tag{16.15}$$

where d_f is the core diameter, u_{01} is the first zero of the J_0 Bessel function, and n_{gas} is the refractive index of the core-filling gas, which is given by [114]:

$$n_{gas} = \sqrt{1 + \frac{p}{p_0} \left(\frac{b_1 \lambda^2}{\lambda^2 - c_1} + \frac{b_2 \lambda^2}{\lambda^2 - c_2} \right)}. \tag{16.16}$$

In Eq. (16.16), p is the gas pressure, $p_0 = 1$ bar, and the Sellmeier coefficients b_i and c_i depend on the considered gas. The values of these coefficients for various noble gases can be found in Ref. [114].

In kagomé MOFs, the optical field overlap with the surrounding silica cladding is very small due to the fact that good confinement of the field is generally achieved in the guiding region. This is illustrated in Figure 16.15, which presents the spatial distribution of the optical field for a kagomé

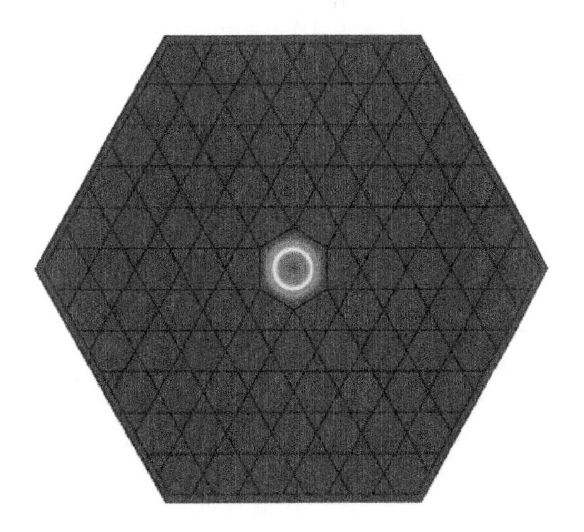

Figure 16.15 Optical field of the fundamental mode of a kagomé MOF with a core diameter d = 40.0 μm and strut thickness t = 0.1 μm, filled with helium at a pressure p = 10 bar and pump wavelength λ = 790 nm.

MOF with a core diameter $d = 40.0$ μm and a strut thickness of $t = 0.1$ μm, filled with helium at a pressure $p = 10$ bar and pump wavelength $\lambda = 790$ nm. The fundamental propagation mode was obtained by solving the master equation (16.14a) for the full draw of the fiber cross section and using the COMSOL Multiphysics software [110, 112, 115].

The propagation constant β of an optical pulse is related with the effective modal refractive index in the form:

$$\beta(\omega) = \frac{\omega n_{eff}}{c} \tag{16.17}$$

The kagomé MOF provides ultra-broadband guidance at low loss levels and it presents, when evacuated, weak anomalous GVD over the entire transmission window. However, when filled with a noble gas, the normal GVD of the gas can be balanced against the anomalous GVD of the fiber, allowing the ZDW to be tuned across the ultraviolet, visible and near-infrared spectral regions. This can be observed from Figure 16.16, which shows (a) the dependence of the GVD parameter $\beta_2 = d^2\beta/d\omega^2$ on the wavelength for a kagomé MOF with a 40 μm core diameter filled with helium at various pressures; and (b) the GVD parameter for helium, argon, and xenon at 10 bar. At pumping wavelengths lower than the ZDW, the dispersion regime of these fibers is normal, whereas at pumping wavelengths higher than λ_{ZD}, the dispersion regime is anomalous. Therefore, we can change the dispersion profile and the position of the ZDW by adjusting the pressure or by changing the filling gas. Compared with an SC MOF, we observe that the dispersion magnitude is significantly smaller in a kagomé HC-MOF, which means that ultrafast pulses broaden much less quickly in this case.

Figure 16.17 shows the ZDW as a function of pressure for the fundamental mode of a kagomé MOF with a core diameter of 40 μm filled with four different gases: xenon, krypton, argon, and helium. In the case of helium and for pressures up to 150 bar, the dispersion is always anomalous at $\lambda_{pump} = 790$ nm; however, for the other gases, the anomalous regime at 790 nm only happens for relatively small pressures. When the pressure varies from 1.0 to 150 bar, the ZDW varies from 215.7 to 659.2 nm for helium, from 406.2 to 1355.9 nm for argon, from 482.6 to 1609.5 nm for krypton, and from 693.2 to 1973.5 nm for xenon.

The Kerr coefficient of the gas filling the fiber depends on the gas pressure as $n_2(p)=pn_2^0$, n_2^0 being the Kerr coefficient at 1 atm [116]. Table 16.1 shows the values of n_2^0 for different noble gases [117]. It can be verified that at 150 bar, the nonlinear refractive index coefficient n_2 of xenon becomes nearly half that of bulk silica.

Pulse propagation in a gas-filled HC-MOF can be described by Eq. (16.5). However, in the case of a noble gas, the Raman contribution must be ignored and the propagation equation becomes:

$$\frac{\partial U}{\partial z} - i\sum_{k\geq 2}\frac{i^k\beta_k}{k!}\frac{\partial^k U}{\partial \tau^k} + \frac{\alpha(\omega)}{2}U = i\gamma\left(1 + \frac{i}{\omega_0}\frac{\partial}{\partial \tau}\right)|U|^2U \tag{16.18}$$

The nonlinear fiber parameter is given by Eq. (16.1), where the Kerr coefficient depends linearly on the gas pressure as mentioned above.

When the input pulse propagates in the anomalous-GVD regime of the fiber, it becomes compressed through an interplay between self-phase modulation (SPM) and GVD. This compression mechanism is related to the periodic evolution of the higher-order solitons, which go through an initial narrowing phase at the beginning of each period. As seen in Chapter 15, if the fiber length is suitably chosen, the input pulses can be compressed by a factor [63]:

$$F_{opt} \approx 4.1N \tag{16.19}$$

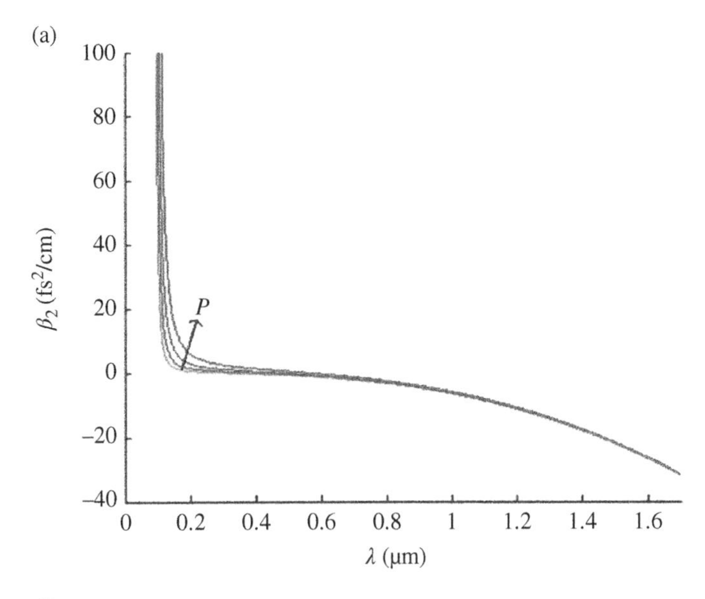

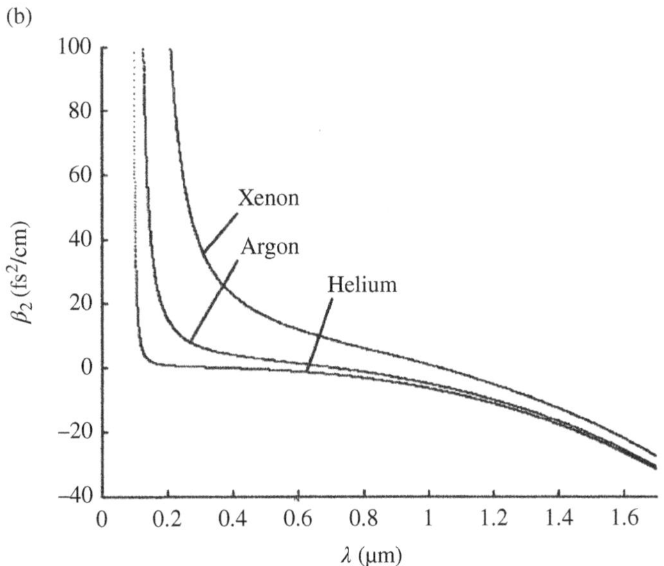

Figure 16.16 (a) GVD for the helium-filled kagomé MOF for $p = 10.0, 20.0, 40.0$, and 80.0 bar; (b) GVD at $p = 10$ bar, for helium, argon, and xenon.

where N is the soliton order. In practice, extreme pulse compression is limited by higher-order effects. However, this limitation becomes less significant in the case of kagomé MOFs filled with noble gases since they have a relatively smaller dispersion slope and they do not present a Raman response. Compressed pulses with a duration of a few fs can be achieved. Soliton-effect self-compression of ∼30 fs pulses to shorter than 10 fs was achieved in an Ar-filled kagomé MOF in a 2011 experiment [118].

Figure 16.18 shows the spectral and temporal evolution of a 30-fs pulse through a kagomé MOF with a 30 μm diameter core filled with Ar, presenting a ZDW at 500 nm. The pumping is

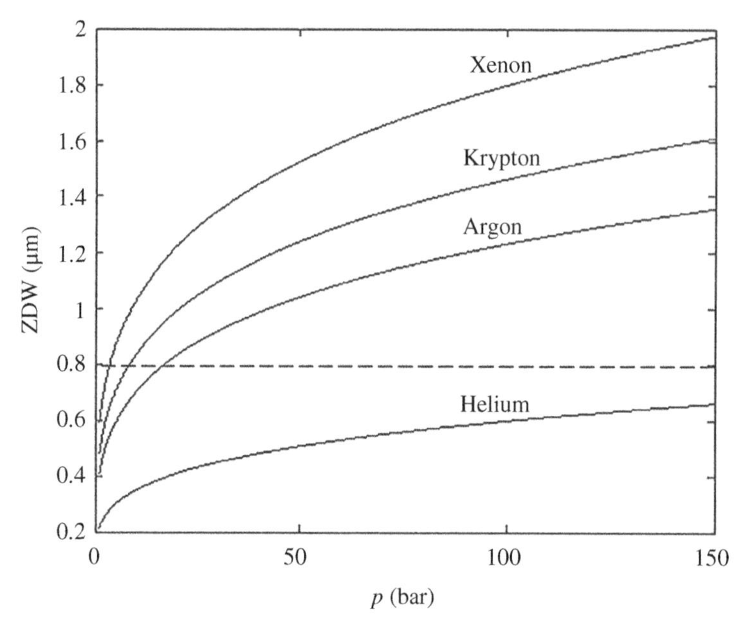

Figure 16.17 Zero dispersion wavelength (ZDW) as a function of the pressure for various noble gases: helium, argon, krypton, and xenon.

Table 16.1 Kerr coefficients for various noble gases at pressure $p = 1$ atm.

		$n_2 \times 10^{16}$ (m^2/W)		
Helium	**Neon**	**Argon**	**Krypton**	**Xenon**
5.21×10^{-9}	1.31×10^{-8}	1.27×10^{-7}	3.07×10^{-7}	9.16×10^{-7}

Source: After Ref. [117].

realized at 800 nm, situated in the anomalous dispersion region, and the corresponding soliton order is $N = 3.5$. As a consequence of the soliton self-compression effect, the pulse temporal profile is dramatically sharpened, producing a ~2 fs pulse. For higher values of N, the self-compressed pulses can achieve subcycle durations, but the corresponding quality factor is reduced [103].

The use of air-filled HC-MOFs has also been considered in recent years [69, 119–122]. Actually, the ambient air has almost two orders of magnitude lower nonlinearity than silica glass [60, 123, 124]. However, its nonlinearity can still become significant in the case of high-peak-power-propagating pulses. The involved nonlinear processes can be exploited for pulse modification [125, 126], new frequency generation [127], or even broad supercontinuum generation with a very high-power density due to the very high-damage threshold of HC-MOFs and the confined gas within the core [100, 128]. The simplicity of air-filled HC-MOFs combined with their very long light-matter interactions, motivated the exploration of potential applications in recent years [119].

The dispersion of an HC-MOF filled with ambient air is almost the same if the air is replaced with 0.8 bar of Ar [114, 122]. On the other hand, the Kerr coefficients at 400 nm in both cases are also very similar [129, 130]. The main difference comes from the Raman effect, which is absent in the

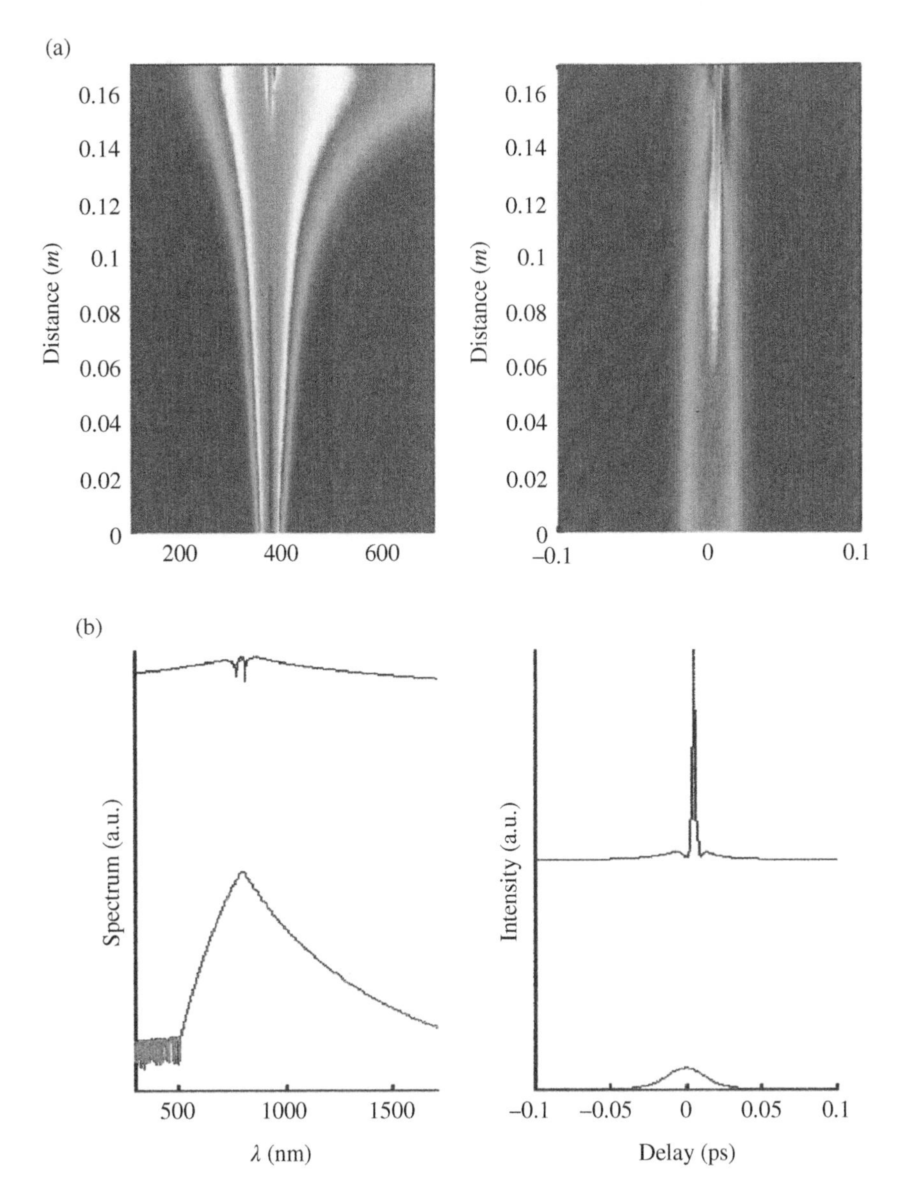

Figure 16.18 (a) Spectral and temporal evolution of self-compression of a 30 fs input pulse at 800 nm through a kagomé MOF with a 30 μm diameter core filled with Ar (ZDW at 500 nm). (b) Initial (lower) and final (upper) pulse profiles in spectral and time domains.

case of Ar and that is beneficial for spectral broadening and pulse compression in the case of ambient air [131, 132].

In a 2021 experiment, 54-fs pulses at 400 nm were launched into an 80-cm-long air-filled kagomé MOF fiber with a 22 μm core diameter [122]. Figure 16.19 shows (a) the experimental and (b) the simulated dependence on the pumping energy of the output spectrum. To model the experimental results, Eq. (16.5) was used in conjunction with a semi-quantum model for air that includes both the rotational Raman scattering and the vibrational Raman scattering responses [120]. As the

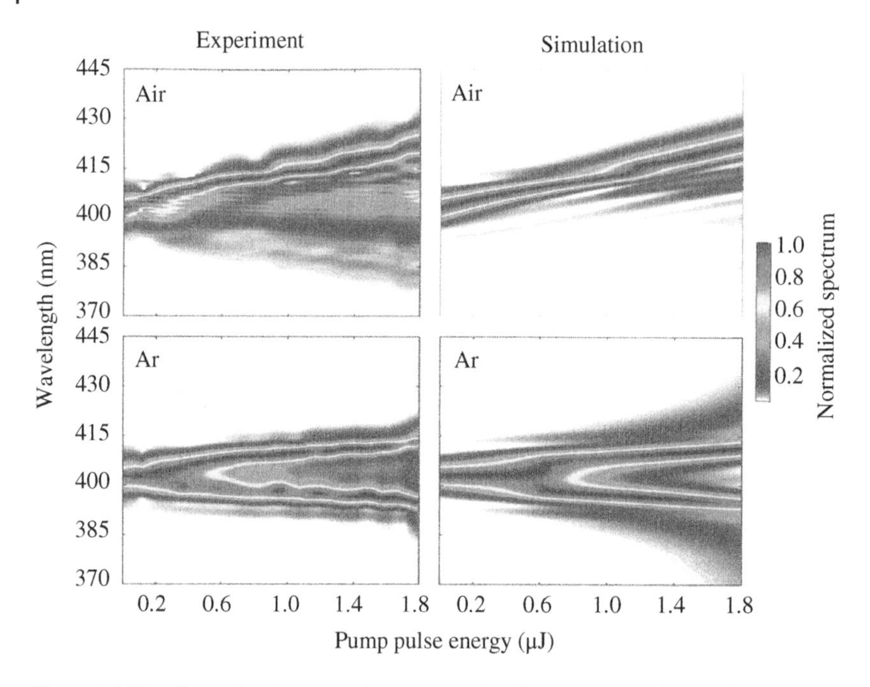

Figure 16.19 Normalized spectral intensity at the fiber output for increasing pump energy. The left-hand column shows the experimental results and the right-hand column shows the simulations. The upper row is for ambient air and the lower for Ar at 0.8 bar. *Source:* After Ref. [122].

launched pump energy is increased from 0.1 to 1.8 μJ, the experimental and simulated spectra broaden and shift toward lower frequency, due to the IRS. In contrast, the experimental and simulated spectra for fiber filled with 0.8 bar of Ar show symmetric spectral broadening, caused by self-phase modulation (lower two panels in Figure 16.19).

In the same experiment, the generation of ultrashort near-UV pulses by soliton self-compression was reported [122]. Pump pulses with the energy of 2.6 μJ and duration of 54 fs were compressed temporally by a factor of 5, to a duration of ~11 fs. The experimental results demonstrated that both Raman and Kerr effects play a role in the compression dynamics.

References

1 Kato, T., Suetsugu, Y., Takagi, M. et al. (1995). *Opt. Lett.* **20**: 988.
2 Hirano, M., Nakanishi, T., Okuno, T., and Onishi, M. (2009). *J. Sel. Top. Quantum Electron.* **15**: 103.
3 Russel, P. (2006). *J. Lightwave Technol.* **24**: 4729.
4 Fok, M. and Shu, C. (2008). *IEEE J. Sel. Top. Quantum Electron.* **14**: 587.
5 Leong, J., Petropoulos, P., Price, J. et al. (2006). *J. Lightwave Technol.* **24**: 183.
6 Slusher, R., Lenz, G., Hodelin, J. et al. (2004). *J. Opt. Soc. Am. B* **21**: 1146.
7 Ferreira, M.F. (2011). *Nonlinear Effects in Optical Fibers*. Hoboken, NJ: John Wiley & Sons.
8 Nakajima, K. and Ohashi, M. (2002). *IEEE Photon. Technol. Lett.* **14**: 492.
9 Okuno, T., Onishi, M., Kashiwada, T. et al. (1999). *IEEE J. Sel. Top. Quantum Electron.* **5**: 1385.
10 Takahashi, M., Sugizaki, R., Hiroishi, J. et al. (2005). *J. Lightwave Technol.* **23**: 3615.
11 Mori, K., Takara, H., and Kawanishi, S. (2001). *J. Opt. Soc. Am. B* **18**: 1780.

12 Okuno, T., Hirano, M., Kato, T. et al. (2003). *Electron. Lett.* **39**: 972.

13 Harbold, J.M., Ilday, F.O., Wise, F.W. et al. (2000). *Opt. Lett.* **27**: 1558.

14 Lu, F. and Knox, W.H. (2004). *Opt. Express* **12**: 347.

15 Foster, M. and Gaeta, A. (2004). *Opt. Express* **12**: 3137.

16 Foster, M., Moll, K., and Gaeta, A. (2004). *Opt. Express* **12**: 2880.

17 Leon-Saval, S., Birks, T., Wadsworth, W., and Russel, P. (2004). *Opt. Express* **12**: 2864.

18 Knight, J., Birks, T., Russel, P., and Atkin, D. (1996). *Opt. Lett.* **21**: 1547.

19 Cregan, R., Mangan, B., and Knight, J. (1999). *Science* **285**: 1537.

20 Broeng, J., Mogilevstev, D., Barkou, S.E., and Bjarklev, A. (1999). *Opt. Fiber Technol.* **5**: 305.

21 Eggleton, B., Kerbage, C., Westbrook, P. et al. (2001). *Opt. Express* **9**: 698.

22 Saito, K. and Koshiba, M. (2005). *J. Lightwave Technol.* **23**: 3580.

23 Birks, T., Knight, J., and Russel, P. (1997). *Opt. Lett.* **22**: 961.

24 Dong, L., McKay, H., and Fu, L. (2008). *Opt. Lett.* **33**: 2440.

25 Reeves, W., Skryabin, D., Biancalana, F. et al. (2003). *Nature* **424**: 511.

26 Benabid, F., Knight, J., Antonopoulos, G., and Russell, P. (2002). *Science* **298**: 399.

27 Dudley, J., Genty, G., and Coen, S. (2006). *Rev. Mod. Phys.* **78**: 1135.

28 Yuan, J., Sang, X., Yu, C. et al. (2011). *J. Lightwave Technol.* **29**: 2920.

29 Petersen, S., Alkeskjold, T., and Lægsgaard, J. (2013). *Opt. Express* **21**: 18111.

30 Yuan, J., Sang, X., Wu, Q. et al. (2015). *Opt. Lett.* **40**: 1338.

31 Yuan, J., Sang, X., Wu, Q. et al. (2013). *Opt. Lett.* **38**: 5288.

32 Hilligsøe, K., Andersen, T., Paulsen, H. et al. (2004). *Opt. Express* **12**: 1045.

33 Andersen, T., Hilligsøe, K., Nielsen, C. et al. (2004). *Opt. Express* **12**: 4113.

34 Petersen, S., Alkeskjold, T., Olausson, C., and Lægsgaard, J. (2015). *Opt. Lett.* **40**: 487.

35 Sévigny, B., Cassez, A., Vanvincq, O. et al. (2015). *Opt. Lett.* **40**: 2389.

36 Zlobina, E., Kablukov, S., and Babin, S. (2015). *Opt. Express* **23**: 833.

37 Gong, Y., Huang, J., Li, K. et al. (2012). *Opt. Express* **20**: 24030.

38 Laegsgaard, J., Mortenson, N., Riishede, J., and Bjarklev, A. (2003). *J. Opt. Soc. Am. B* **20**: 2046.

39 Price, J., Monro, T., Ebendorff-Heidepriem, H. et al. (2007). *IEEE J. Sel. Top. Quantum Electron.* **23**: 738.

40 Feng, X., Mairaj, A., Hewak, D., and Monro, T. (2005). *J. Lightwave Technol.* **23**: 2046.

41 Lee, J., Kikuchi, K., Nagashime, T. et al. (2005). *Opt. Express* **13**: 3144.

42 Lee, J., Nagashima, T., Hasegawa, T. et al. (2006). *J. Lightwave Technol.* **24**: 22.

43 Monro, T. (2011). *Opt. Lett.* **36**: 2351.

44 Tuniz, A., Brawley, G., Moss, D., and Eggleton, B. (2008). *Opt. Express* **16**: 18524.

45 Troles, J., Coulombier, Q., Canat, G. et al. (2010). *Opt. Express* **18**: 26647.

46 Gao, W., Cheng, T., Xue, X. et al. (2016). *Opt. Express* **24**: 3278.

47 Yao, J., Zhang, B., Yin, K. et al. (2016). *Opt. Express* **24**: 14717.

48 Mairaj, A., Petrovich, M., West, Y. et al. (2001). *Fiber Optic Sensor Technol. II* **4204**: 278.

49 Troles, J., L. Brilland, F. Smektala et al. (2007). Chalcogenide photonic crystal fibers for near and middle infrared applications. *9th International Conference on Transparent Optical Networks*, Rome, Italy (1–5 July 2007), pp. 297–300.

50 Cordeiro, C., Chillcce, E., Barbosa, L., and Cruz, C. (2006). *J. Non-Cryst. Solids* **352**: 3423.

51 Kiang, K., Frampton, K., Monro, T. et al. (2002). *Electron. Lett.* **38**: 546.

52 Fuochi, M., Poli, F., Cucinotta, A., and Vincetti, L. (2003). *J. Lightwave Technol.* **21**: 2247.

53 Varshney, S., Fujisawa, T., Saito, K., and Koshiba, M. (2005). *Opt. Express* **13**: 9516.

54 Varshney, S., Saito, K., Iizawa, K. et al. (2008). *Opt. Lett.* **33**: 2431.

55 Beaud, P., Hodel, W., Zysset, B., and Weber, H. (1987). *IEEE J. Quantum Electron.* **23**: 1938.

56 Dudley, J. and Taylor, J. (ed.) (2010). *Supercontinuum Generation in Optical Fibers*. Cambridge, UK: Cambridge University Press.
57 Agrawal, G.P. (2007). *Nonlinear Fiber Optics*, 4ee. San Diego, CA: Academic Press.
58 Stolen, R., Gordon, J., Tomlinson, W., and Haus, H. (1989). *J. Opt. Soc. Am. B* **6**: 1159.
59 Blow, K. and Wood, D. (1989). *IEEE J. Quantum Electron.* **25**: 2665.
60 Lin, Q. and Agrawal, G. (2006). *Opt. Lett.* **31**: 3086.
61 Kodama, Y. and Hasegawa, A. (1987). *IEEE J. Quantum Electron.* **23**: 510.
62 Chen, C. and Kelley, P. (2002). *J. Opt. Soc. Am. B* **19**: 1961.
63 Dianov, E., Nikonova, Z., Prokhorov, A., and Serkin, V. (1986). *Sov. Tech. Phys. Lett.* **12**: 311.
64 Gordon, J. (1987). *Opt. Lett.* **11**: 659.
65 Mitschke, F. and Mollenauer, L. (1986). *Opt. Lett.* **11**: 659.
66 Liu, X., Xu, C., Knox, W. et al. (2001). *Opt. Lett.* **26**: 358.
67 Washburn, B., Ralph, S., Lacourt, P. et al. (2001). *Electron. Lett.* **37**: 1510.
68 Cormack, I., Reid, D., Wadsworth, W. et al. (2002). *Electron. Lett.* **38**: 167.
69 Ouzounov, D., Ahmad, F., Muller, D. et al. (2003). *Science* **301**: 1702.
70 Nishizawa, N., Ito, Y., and Goto, T. (2002). *IEEE Photon. Technol. Lett.* **14**: 986.
71 Efimov, A., Taylor, A., Omenetto, F., and Vanin, E. (2004). *Opt. Lett.* **29**: 271.
72 K. Abedin and F. Kubota, *IEEE J. Sel. Top. Quantum Electron.* **10**, 1203 (2004).
73 Lee, J.H., Howe, J., Xu, C., and Liu, X. (2008). *J. Sel. Top. Quantum Electron.* **14**: 713.
74 Nishizawa, N. and Goto, T. (2003). *Opt. Express* **11**: 359.
75 Kato, M., Fujiura, K., and Kurihara, T. (2004). *Electron. Lett.* **40**: 381.
76 Oda, S. and Maruta, A. (2006). *Opt. Express* **14**: 7895.
77 Husakou, A. and Herrmann, J. (2001). *Phys. Rev. Lett.* **87**: 203901.
78 Akhmediev, N. and Karlsson, M. (1995). *Phys. Rev. A* **51**: 2602.
79 Nishizawa, N. and Goto, T. (2001). *Opt. Express* **8**: 328.
80 Skryabin, D., Luan, F., Knight, J., and Russell, P. (2003). *Science* **301**: 1705.
81 Biancalana, F., Skryabin, D., and Yulin, A. (2004). *Phys. Rev. E* **70**: 016615.
82 Pitois, S. and Millot, G. (2003). *Opt. Commun.* **226**: 415.
83 Wadsworth, W., Joly, N., Knight, P. et al. (2004). *Opt. Express* **12**: 299.
84 Rarity, J., Fulconis, J., Duligall, J. et al. (2005). *Opt. Express* **13**: 534.
85 Fan, J., Migdall, A., and Wang, L. (2005). *Opt. Lett.* **30**: 3368.
86 Fulconis, J., Alibart, O., Wadsworth, W. et al. (2005). *Opt. Express* **13**: 7572.
87 Sharping, J. (2008). *J. Lightwave Technol.* **26**: 2184.
88 Sharping, J., Fiorentino, M., Kumar, P., and Windeler, R. (2002). *Opt. Lett.* **27**: 1675.
89 Deng, Y., Lin, Q., Lu, F. et al. (2005). *Opt. Lett.* **30**: 1234.
90 Sharping, J., Foster, M., Gaeta, A. et al. (2007). *Opt. Express* **15**: 1474.
91 Sharping, J., Sanborn, J., Foster, M. et al. (2008). *Opt. Express* **16**: 18050.
92 Liu, X., Yang, X., Lu, F. et al. (2005). *Opt. Express* **13**: 142.
93 Fok, M. and Shu, C. (2007). *Opt. Express* **15**: 5925.
94 Yu, C., Yan, L., Luo, T. et al. (2005). *Technol. Lett.* **17**: 636.
95 Fok, M. and Shu, C. (2006). *Proceeding of the Fiber Communication Conference*, March, Paper OWI 34.
96 Seeds, A. and Williams, K. (2006). *J. Lightwave Technol.* **24**: 4628.
97 Russell, P. (2003). *Science* **299**: 358.
98 Luan, F., Knight, J., Russell, P. et al. (2004). *Opt. Express* **12**: 835.
99 Roberts, P., Couny, F., Sabert, H. et al. (2005). *Opt. Express* **13**: 236.
100 Russell, P., Hölzer, P., Chang, W. et al. (2014). *Nat. Photonics* **8**: 278.

101 Couny, F., Benabid, F., and Light, P. (2006). *Opt. Lett.* **31**: 3574.
102 Pearce, G., Wiederhecker, G., Poulton, C. et al. (2007). *Opt. Express* **15**: 12680.
103 Travers, J., Chang, W., Nold, J. et al. (2011). *J Opt. Soc. Am. B* **28**: A11.
104 Ding, W. and Wang, Y. (2014). *Opt. Express* **22**: 27242.
105 Hartung, A., Kobelke, J., Schwuchow, A. et al. (2014). *Opt. Express* **22**: 19131.
106 Chen, L., Pearce, G., Birks, T., and Bird, D. (2011). *Opt. Express* **19**: 6945.
107 Mak, K., Travers, J., Hölzer, P. et al. (2013). *Opt. Express* **21**: 10942.
108 Azhar, M., Wong, G., Chang, W. et al. (2013). *Opt. Express* **21**: 4405.
109 Rodrigues, S., Facão, M., and Ferreira, M. (2015). *Fiber Int. Opt.* **34**: 76.
110 Rodrigues, S., Facão, M., and Ferreira, M. (2018). *Opt. Commun.* **412**: 102.
111 Rodrigues, S., Facão, M., Carvalho, M., and Ferreira, M. (2020). *Opt. Commun.* **468**, paper 125791.
112 Rodrigues, S., Facão, M., and Ferreira, M. (2020). *Fiber Int. Opt.* **39**: 215.
113 Im, S., Husakou, A., and Herrmann, J. (2009). *Opt. Express* **17**: 13050.
114 Börzsönyi, A., Heiner, Z., Kalashnikov, M. et al. (2008). *Appl. Opt.* **47**: 4856.
115 COMSOL. (2011). Multiphysics, user's guide, COMSOL AB, version 42, http://www.comsol.com.
116 Börzsönyi, Á., Heiner, Z., Kovács, A. et al. (2010). *Opt. Express* **18**: 25847.
117 Brée, C. (2012). Nonlinear optics in the filamentation regime. Springer theses. Humboldt University of Berlin, Germany.
118 Joly, N., Nold, J., Chang, W. et al. (2011). *Phys. Rev. Lett.* **106**: 203901.
119 Debord, B., Alharbi, M., Vincetti, L. et al. (2014). *Opt. Express* **22**: 10735.
120 Mousavi, S., Mulvad, H., Wheeler, N. et al. (2018). *Opt. Express* **26**: 8866.
121 Debord, B., Maurel, M., Gérôme, F. et al. (2019). *Photonics Res.* **7**: 1134.
122 Luan, J., Russell, P., and Novoa, D. (2021). *Optics Express* **29**: 13787.
123 Nibbering, E., Grillon, G., Franco, M. et al. (1997). *J. Opt. Soc. Am. B* **14**: 650.
124 Mlejnek, M., Wright, E.M., and Moloney, J.V. (1998). *Opt. Lett.* **23**: 382.
125 Guichard, F., Giree, A., Zaouter, Y. et al. (2015). *Opt. Express* **23**: 7416.
126 Mak, K., Travers, J., Joly, N. et al. (2013). *Opt. Lett.* **38**: 3592.
127 Bhagwat, A. and Gaeta, A. (2008). *Opt. Express* **16**: 5035.
128 Yatsenko, Y., Pleteneva, E., Okhrimchuk, A. et al. (2017). *Quantum Electron.* **47**: 553.
129 Rosenthal, E.W., Palastro, J.P., Jhajj, N. et al. (2015). *J. Phys. B* **48**: 094011.
130 Zahedpour, S., Wahlstrand, J., and Milchberg, H. (2015). *Opt. Lett.* **40**: 5794.
131 Belli, F., Abdolvand, A., Chang, W. et al. (2015). *Optica* **2**: 292.
132 Hosseini, P., Ermolov, A., Tani, F. et al. (2018). *ACS Photonics* **5**: 2426.

17

Supercontinuum Generation

One of the most impressive nonlinear phenomena that can be observed in highly nonlinear fibers is supercontinuum generation (SCG). It results generally from the synergy between several fundamental nonlinear processes, such as self-phase modulation (SPM), cross-phase modulation (XPM), stimulated Raman scattering (SRS), modulation instability (MI), and four-wave mixing (FWM) [1]. The spectral locations and powers of the pumps, as well as the nonlinear and dispersive characteristics of the medium, determine the relative importance and the interaction between these nonlinear processes. Soliton-related effects also play a significant role whenever light with sufficient power propagates in the anomalous dispersion regime [1]. The first observation of supercontinuum (SC) was realized in 1970 by Alfano and Shapiro in bulk borosilicate glass [2]. SC in fibers occurred for the first time in a 1976 experiment, when 10 ns pulses with more than 1-kW peak power were launched in a 20-m long fiber, producing a 180-nm-wide spectrum [3].

An SC source can find applications in the area of biomedical optics, where it allows the improvement of longitudinal resolution in optical coherence tomography by more than an order of magnitude, in optical frequency metrology, in all kinds of spectroscopy, in hyperspectral lidars, and as a multiwavelength source in the telecommunications area [4–13].

17.1 Pumping with Femtosecond Pulses

Pumping in the normal dispersion regime using femtosecond pulses, where soliton formation is not allowed, leads only to spectral broadening through SPM and optical wave-breaking [14–16]. As seen in Chapter 4, SPM-induced spectral broadening depends on the propagation distance, peak power, and shape of the input pulse. Its role becomes particularly significant at high input powers and for ultrashort pulses so that the rate of power variation with time, dP/dt, is large. If the pump wavelength is located near the zero-dispersion wavelength (ZDW), the spectrum can extend into the anomalous dispersion regime. In such a case, the soliton dynamics can have a significant contribution to spectral broadening [17].

Soliton dynamical effects play a prominent role when pumping in the anomalous dispersion regime using femtosecond pulses. If the power of the pump pulses is high enough, they can evolve as higher-order solitons. As seen in Chapter 6, these pulses begin experiencing spectral broadening and temporal compression which are typical of higher-order solitons [18, 19]. However, because of perturbations such as higher-order dispersion or intrapulse Raman scattering (IRS), the dynamics of such pulses departs from the behavior expected of ideal high-order solitons and the pulses break up. This process is known as soliton fission and determines the separation of each higher-order

Solitons in Optical Fiber Systems, First Edition. Mario F. S. Ferreira.

soliton pulse into N fundamental solitons whose widths and peak powers are given by Eqs. (16.8) and (16.9) [20].

The soliton fission process is affected if the input pulse wavelength is near a ZDW. As seen in Chapter 16, in such a case, the ejected solitons shed some of their energy to dispersive waves (DWs), also called *nonsoliton radiation* (NSR), generated on short-wavelength side of the ZDW in fibers with positive dispersion slopes ($\beta_3 > 0$) [21, 22]. Since the DWs have a lower group velocity, they will lag behind the solitons. However, the group velocity of the redshifting solitons decreases during the propagation and eventually they meet with the DWs. The temporal overlap between a soliton and a DW allows them to interact via the XPM effect, which determines a blueshift and a decrease in the group velocity of the DW. In this way, as long as this cycle continues, the blueshifted radiation cannot escape from the soliton, which is usually referred to as a "trapping effect" [23, 24]. The XPM effect determines the spectral broadening of the DWs and merging of different peaks, helping in the formation of a broadband SC [25–29].

The FWM effect also plays an important role in the process of SC formation by femtosecond pulse pumping in the anomalous dispersion regime. Provided an appropriate phase-matching condition is satisfied, a Raman soliton can act as a pump and interact through FWM with a dispersive wave, giving origin to an idler wave, as described in Section 4.3 [15, 30]. In one possible FWM process, corresponding to the conventional FWM, the phase-matching condition is given by

$$\beta_d(\omega_3) + \beta_d(\omega_4) = \beta_s(\omega_3) + \beta_s(\omega_4) \tag{17.1}$$

where $\beta_d(\omega)$ and $\beta_s(\omega)$ are the propagation constants of the DW and of the soliton, respectively. In this case, the idler frequency ω_4 lies far from the original DW. Considering a positive third-order dispersion, this FWM process contributes to the extension of the SC toward the infrared side of the input wavelength.

Another possibility corresponds to a Bragg-scattering-type FWM process, in which the phase-matching condition is given by

$$\beta_d(\omega_3) - \beta_d(\omega_4) = \beta_s(\omega_3) - \beta_s(\omega_4) \tag{17.2}$$

In this case, the idler frequency ω_4 lies close to the original DW. Considering a positive third-order dispersion, this FWM process plays an important role in extending the SC toward the blue side of the input wavelength.

Figure 17.1 shows the experimental results for the SC spectrum provided by a 1-m-long microstructured optical fiber (MOF) pumped with 140-fs pulses as a function of input power, for different degrees of anomalous dispersion [25]. The ZDW of the used fiber was 780 nm and the pumping wavelength was tuned at (a) 790, (b) 805, and (c) 850 nm. A pump power much higher than that needed to form a fundamental soliton excites higher-order solitons. For example, a bounded soliton around 860 nm pumped at 850 nm with 0.4-kW peak power corresponds to $N = 4.87$. A pronounced pulse splitting and the soliton self-frequency shift (SSFS) due to IRS of the ejected solitons can be observed. After the higher-order soliton fission, the blueshifted peak is generated as indicated by arrows in Figure 17.1a–c.

It was observed that pumping in the vicinity of the ZDW provided a smooth SC with relatively low input powers [25]. Peaks in the proximity of the ZDW serve as parametric pumps, which generate further peaks between blueshifted and SSFS peaks. At higher pump energies, all those peaks merge to give a continuous spectrum, whose bandwidth increases by increasing the input power. XPM followed by all the generated frequencies also made the resultant spectrum smooth. The generation of SC at deep anomalous dispersion exhibits substantial broad SC at relatively low powers, but the generated SC profile is less smooth [25].

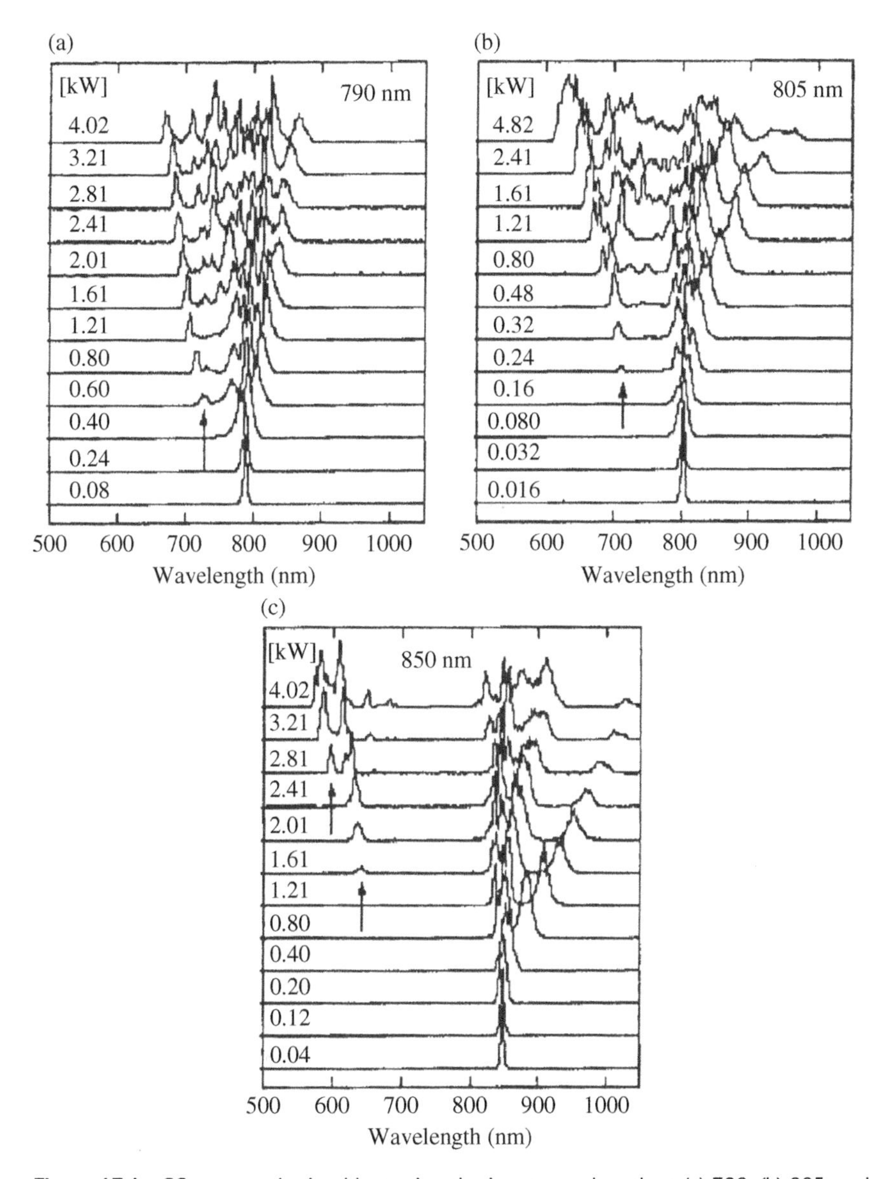

Figure 17.1 SC spectra obtained by tuning the input wavelength at (a) 790, (b) 805, and (c) 850 nm as a function of input power for a 1-m-long MOF. *Source:* After Ref. [25].

In another 2004 experiment, it was demonstrated that, besides contributing to a smooth SC, XPM also plays a crucial role to extend the SC toward shorter wavelengths [26]. Figure 17.2a shows the observed spectra at the output of a 1.5-m-long MOF with a core diameter of 1.67 μm when 27-fs pulses at 790 nm were coupled into it. The fiber exhibited two ZDWs: one at 700 nm and another at 1400 nm. For the highest average input power of ~120 mW, the spectrum extends from 450 to 1600 nm. For low input power, two distinct spectral peaks are observed. The peak shifted most to the red corresponds to a soliton experiencing SSFS as discussed in Chapter 16. As the input power is increased, the redshift of the soliton is enhanced with the simultaneous appearance of blue

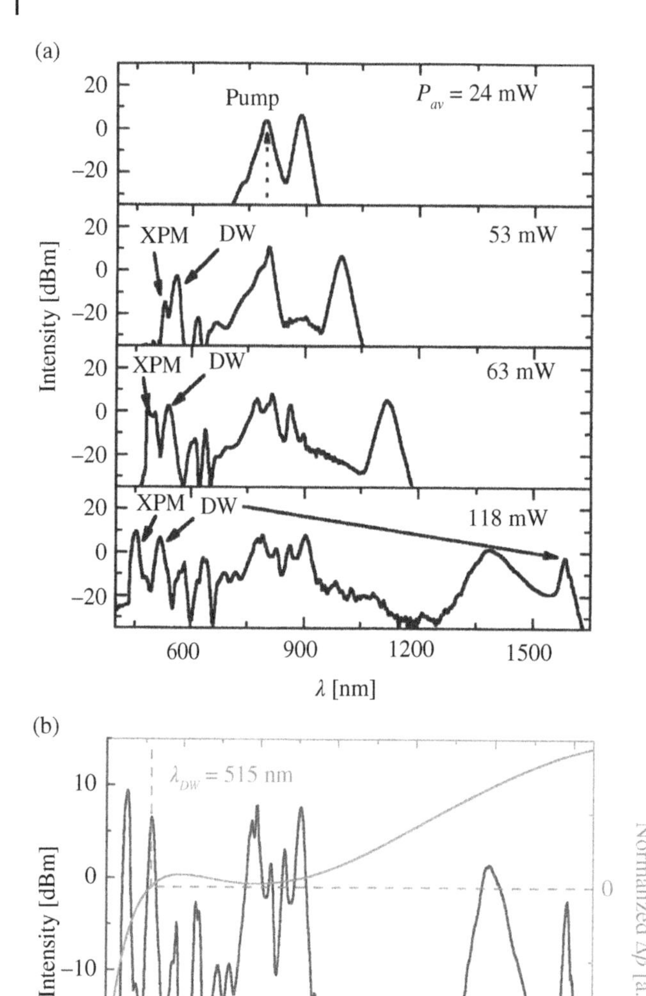

Figure 17.2 (a) Supercontinuum (SC) spectra at the output of the MOF pumped by 27-fs pulses at 790 nm for increasing average input power; (b) Phase-matching condition for the generation of the dispersive wave (DW). The SC spectrum for P_{av} = 118 mW is displayed. P_{av}: average power, XPM: cross-phase modulation, DW: dispersive wave. *Source:* After Ref. [26].

components in the spectrum, corresponding to a DW. Besides, a new spectral peak marked with XPM and located on the blue edge of the spectrum appears as the input power is increased. The peak near 1600 nm in the spectrum for P_{av} = 118 mW also represents a DW, which is generated when the Raman soliton approaches the second ZDW. In this case, we observe the suppression of SSFS due to the spectral recoil of the soliton into the anomalous-GVD regime, as discussed in Chapter 16 [31].

Figure 17.2b shows the spectrum for $P_{av} = 118$ mW together with the phase mismatch as a function of wavelength between the shortest soliton and DW [26]. The phase matching occurs at 515 nm and the peak located there corresponds to the DW. The peak located at the blue edge of the spectrum should not result from the generation of a DW since its location is detuned by more than 60 nm from the phase-matched wavelength at 515 nm. Numerical simulations based on the generalized nonlinear Schrödinger equation (GNLSE) show that the soliton slows down as its spectrum shifts toward the red and begins to overlap with the DW; the blueshift peak appears as a result of their interaction through XPM. Concerning the second DW created near 1600 nm, it travels faster than the soliton, which prevents the generation of any XPM-induced spectral peak.

17.2 Modeling the Supercontinuum

Modeling the SCG can be realized considering the generalized NLSE given by Eq. (16.5), which includes higher-order dispersive and nonlinear effects, [32–37]. It has proven successful in modeling most features of supercontinua observed experimentally. The split-step Fourier method described in Section 6.8 is generally used to solve Eq. (16.5), where the number of dispersive terms to be considered in each case is not obvious. Actually, dispersion to all orders can be included numerically since the dispersive step is carried out in the spectral domain of the pulse after neglecting the nonlinear terms. On the other hand, both the step size along the fiber length and the time resolution used for the temporal window also deserve careful consideration. In particular, the time window must be wide enough to include the entire SC.

The simulation of the SC was realized in Ref. [36] using fiber and laser parameters taken from Ref. [33]. Such parameters represent typical experimental conditions for SCG in MOFs employing femtosecond lasers and pumping in the anomalous dispersion regime. An input pulse, $U(0.\tau) = \sqrt{P_0}\text{sech}(\tau/t_0)$, was assumed, where $P_0 = 10$ kW and the width $t_0 = 28.4$ fs, which corresponds to an intensity full width at half maximum (FWHM) of 50 fs. The fiber has a length $L = 10$ cm and a nonlinear parameter length $\gamma = 45\ \text{W}^{-1}/\text{km}$. The pulse was initially centered at 850 nm and the following fiber parameters were also assumed: $\beta_2 = -12.76\ \text{ps}^2/\text{km}$, $\beta_3 = 8.119\times10^{-2}\ \text{ps}^3/\text{km}$, $\beta_4 = -1.321\times10^{-4}\ \text{ps}^4/\text{km}$, $\beta_5 = 3.032\times10^{-7}\ \text{ps}^5/\text{km}$, $\beta_6 = -4.196\times10^{-10}\ \text{ps}^6/\text{km}$, $\beta_7 = 2.570\times10^{-13}\ \text{ps}^7/\text{km}$. For these conditions, the soliton order is N ~5.

The spectral evolution of the SC at different positions along the MOF is shown in Figure 17.3, which reproduces very well the experimental observations reported in Ref. [33]. Dramatic spectral broadening is observed in the first centimeter of the fiber. The initial broadening, which is mainly due to the interaction between the SPM and GVD, is symmetric. Afterward, the spectrum becomes asymmetric and distinct spectral peaks develop on both sides of the pump wavelength as higher-order dispersive and nonlinear perturbations cause the fission of the higher-order soliton. The resulting fundamental solitons then undergo a continuous self-frequency shifting to longer wavelengths because of IRS. The emergence and self-shifting of one of these fundamental solitons on the long-wavelength side are clearly seen in Figure 17.3b. In this process, each Raman soliton sheds some of its energy in the form of non-soliton radiation on the short-wavelength side of the pump, which also leads to the appearance of discrete spectral components in this region of the spectrum.

Figure 17.4 shows the temporal and spectral evolution of an optical pulse propagating in a 15-cm-long silica MOF with a hexagonal pattern of holes of diameter $d = 1.4$ μm spaced by $\Lambda = 1.6$ μm. This result was obtained taking into account the full dispersion curve of such fiber, represented in

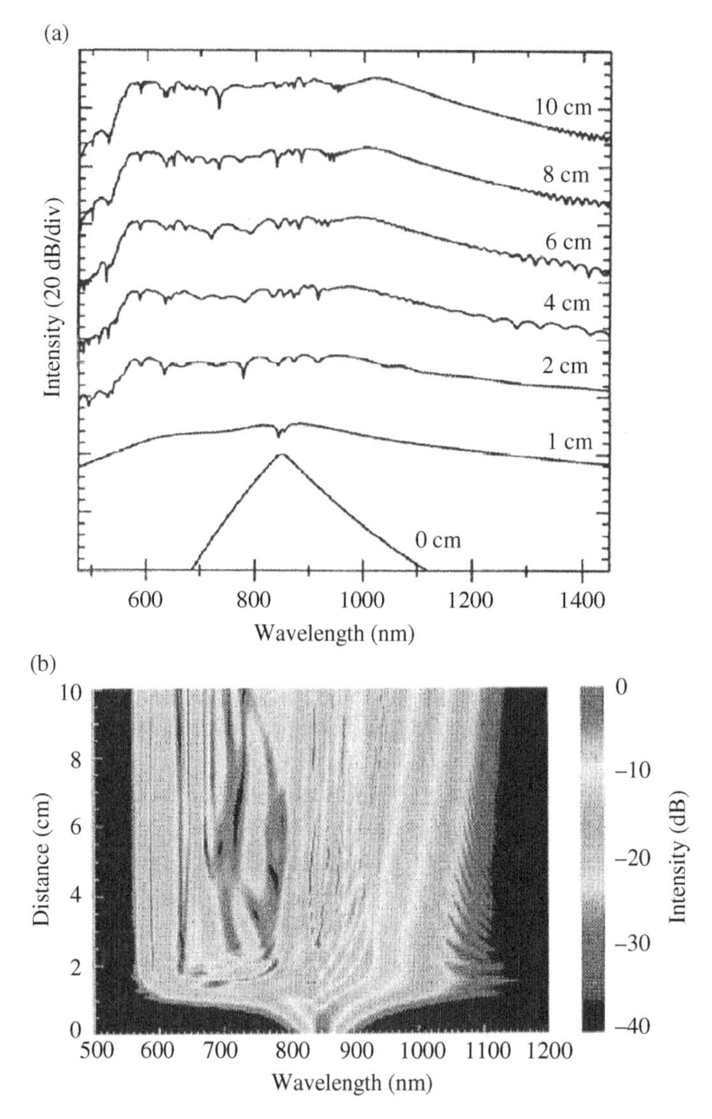

Figure 17.3 Results from a numerical simulation of SCG in an MOF. (a) Spectra at discrete locations along the fiber; and (b) density plot as a function of the propagation distance. *Source:* After Ref. [36].

Figure 17.4b. The fiber presents a single ZDW at 735 nm, at which the third-order dispersion parameter β_3 is positive. The pump is in the anomalous dispersion regime at 790 nm, where the nonlinear parameter is $\gamma = 117\mathrm{W}^{-1}\mathrm{km}^{-1}$. An input pulse, $U(0,\tau) = \sqrt{P_0}\mathrm{sech}(\tau/t_0)$, is assumed, where $P_0 = 5$ kW, and the width $t_0 = 14.2$ fs, which corresponds to an intensity FWHM of 25 fs. For the assumed pulse and fiber parameters, we have a soliton order $N = 3.01$.

For the pulse and fiber parameters assumed, we have a fission distance L_{fiss}~0.51 cm, which can be confirmed in Figure 17.4. As a consequence of the SSFS induced by IRS, the solitons arising from the fission process separate from each other. Since the SSFS is the largest for the shortest soliton, its spectrum shifts the most toward the red side in Figure 17.4a. The change of the soliton′s frequency determines a reduction in the soliton’s speed because of dispersion.

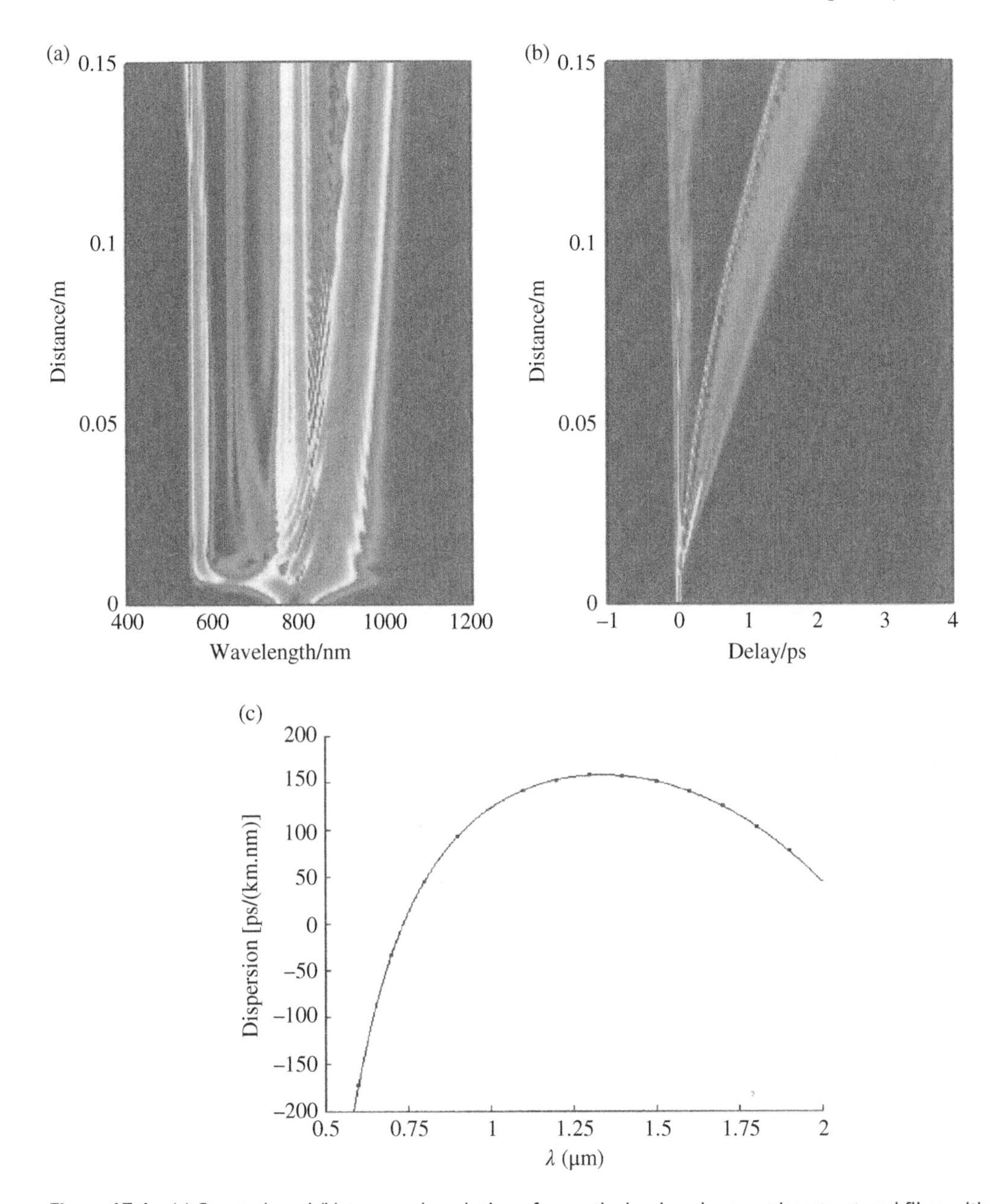

Figure 17.4 (a) Spectral; and (b) temporal evolution of an optical pulse along a microstructured fiber with a hexagonal pattern of holes of diameter d = 1.4 μm spaced by Λ = 1.6 μm; (c) Dispersion curve of the MOF, presenting a ZDW at 735 nm. The input pulse has an FWHM of 25 fs and is launched at 790 nm with a peak power of 5 kW.

This deceleration appears as a bending of the soliton trajectory in the time domain, as observed in Figure 17.4b. A clear signature of soliton fission in Figure 17.4a is the appearance of a new spectral peak at $z = L_{fiss}$. This spectral peak corresponds to the DWs, generated on the short-wavelength side of the ZDW in fibers with positive dispersion slopes ($\beta_3 > 0$) [21, 22].

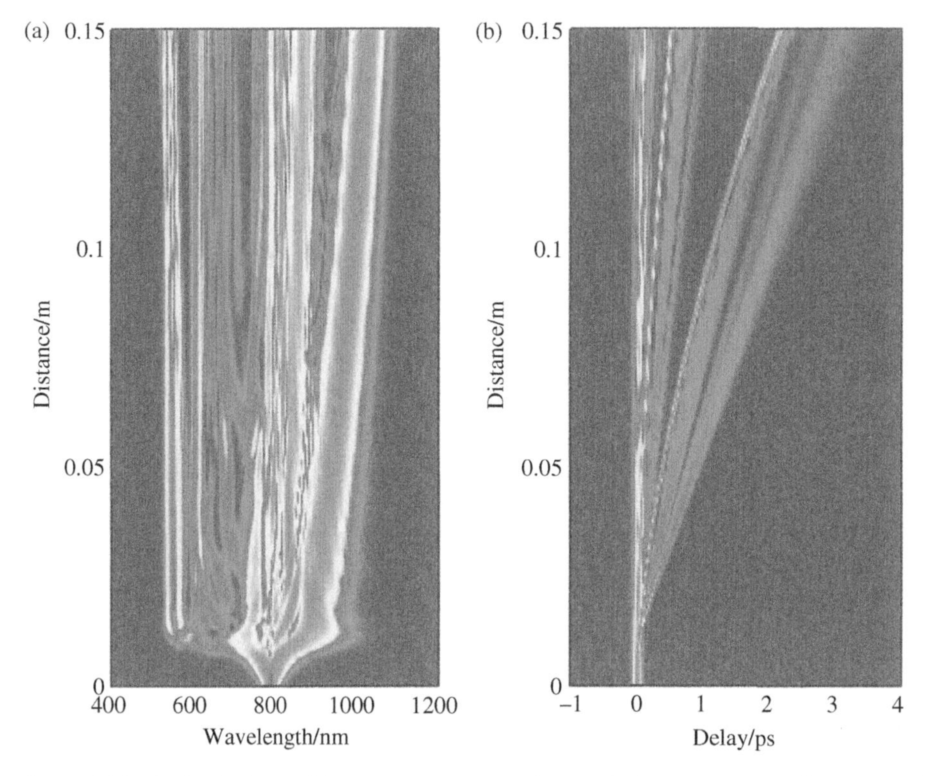

Figure 17.5 (a) Spectral; and (b) temporal evolution of an optical pulse along a MOF with d = 1.4 and Λ = 1.6 μm. The input pulse has a peak power of 5 kW and an FWHM of 50 fs.

Figure 17.5 shows the temporal and spectral evolution of a sech pulse with peak power $P_0 = 5$ kW, and $t_0 = 28.4$ (FWHM = 50 fs) propagating at 790 nm in the same fiber considered in Figure 17.4. Compared with this figure, we observe in Figure 17.5 a wider and much more uniform SC. Figure 17.6 shows the initial and final pulse profiles, both in frequency and time domains. Since the power of the first ejected soliton is greater than in the case of Figure 17.4, it experiences a larger SSFS, according to Eq. (16.10). At the same time, the corresponding DWs are generated at a lower wavelength in the normal region. In contrast to the continuous shift to longer wavelengths, the short-wavelength edge of the SC does not undergo further extension with propagation.

In spite of the significant progress in the development of SC sources and successful application of theoretical models to optimize the generation of broad radiation sources, the comprehension of the complex nonlinear mechanisms involved in SC generation was not fully achieved yet. Actually, a quantitative agreement between experimental and numerical simulation results of SC generation is not an easy task. Among other aspects, the polarization effects and the coherence properties of SC are two issues, which continue to attract considerable attention from researchers.

17.3 Pumping with Picosecond Pulses

The higher-order soliton evolution and soliton fission described in the previous section become progressively less important when pumping in the anomalous dispersion regime near the ZDW with long duration and high-peak-power pulses. This is because the fission distance L_{fiss} also increases

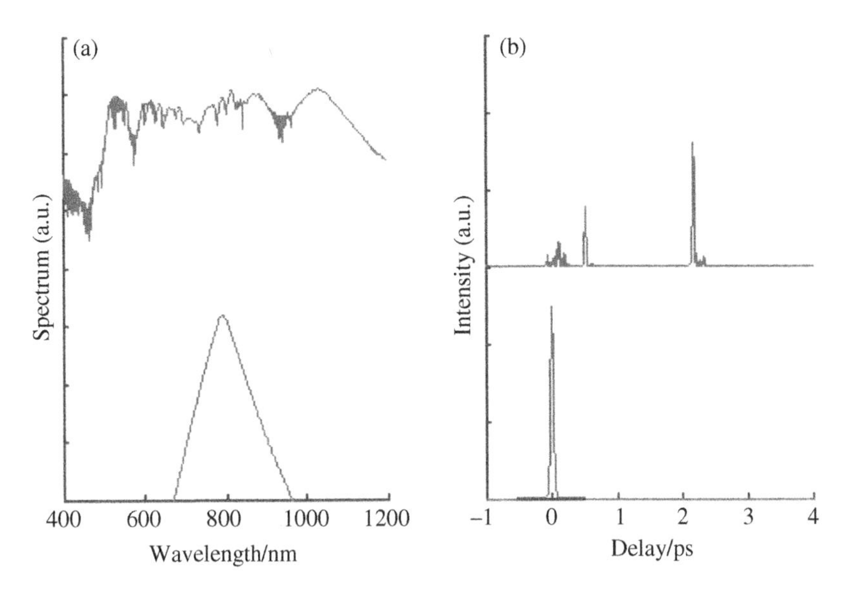

Figure 17.6 Initial and final pulse profiles in (a) frequency and (b) time domains for the case of Figure 17.5.

with the pump pulse duration. Under these conditions, the spontaneous modulation instability becomes the dominant process in the initial propagation phase. In the frequency domain, the modulation instability causes the spread of energy from the central wave frequency to sidebands with a frequency separation given by [18, 19]:

$$\Omega_s = \sqrt{\frac{2\gamma P_p}{|\beta_2|}} \tag{17.3}$$

where β_2 is the group velocity dispersion, γ is the nonlinear parameter, and P_p is the pump peak power. The MI period can be obtained from the above frequency separation as

$$T_s = \frac{2\pi}{\Omega_s} = \sqrt{\frac{2\pi^2|\beta_2|}{\gamma P_p}} \tag{17.4}$$

The modulated picoseconds pulse can then break up into a train of solitons with durations approximately given by T_s. The subsequent evolution of these solitons then leads to additional spectral broadening and SC formation through a variety of mechanisms. If such solitons are sufficiently short, they are redshifted due to the IRS effect [38, 39]. As seen in Chapter 16, the rate of frequency shift is inversely proportional to the fourth power of the soliton width (or proportional to the square of the soliton peak power). If the solitons resulting from the initial pulse breakup are not sufficiently short, they can undergo soliton collisions, which determine the transfer of energy among them due to Raman scattering. In this process, some solitons achieve higher peak power and then undergo a larger redshift. Optical rogue waves can be observed in association with redshifted solitons with uncommon probability distributions arising from picosecond SC generation [40–42].

As observed above, if the pump wavelength is near the ZDW, the redshift of the solitons is accompanied by the generation of DWs on the short-wavelength side of the ZDW in fibers with positive dispersion slopes ($\beta_3 > 0$). The temporal overlap between a soliton and a DW allows them to interact via the XPM effect. This process is responsible for the maximum blueshift of the SC, which is only limited by the ability of the solitons to redshift.

Actually, most of the common silica fiber SC sources are limited on the short-wavelength edge, normally around 450 nm [43]. However, a blue-enhanced SC can be achieved efficiently using tapered or specially designed MOFs. In a 2011 experiment, blue-extended SC spectra down to 372 nm in simply designed high air-filling fraction (>0.9) MOFs was reported [44]. Later, a flat ultraviolet-extended SC down to 352 nm pumped at 1064 nm in cascading MOF tapers has been reported [45]. More recently, an ultraviolet-enhanced SC, spanning from 350 to 2400 nm, was obtained using a specifically designed seven-core MOF pumped by a picosecond Yb-doped master oscillator power amplifier (MOPA) [46].

The solitonic effects are absent, at least in the initial propagation phase, if pumping is realized far into the normal dispersion regime. In this case, the spectral broadening occurs initially due to SRS. The contribution of the SRS process for the SCG is especially effective on the long-wavelength side of the spectrum, which tends to make the overall spectrum asymmetric.

The FWM effect becomes progressively more important if the pumping is closer to the ZDW, since the parametric gain is higher than the Raman gain [18]. Using pump pulses of sub-kilowatt peak power and duration of some tens of ps, a spatially single-mode SC with more than 600 nm width could be generated by the interplay of SRS and FWM [47, 48].

The dispersive properties of the fiber become especially important in the case of the FWM process. If the required phase-matching conditions are satisfied, FWM generates sidebands on the short- and long-wavelength sides of the pulse spectrum. This process can produce a wideband SC even in the normal-GVD region of an optical fiber. Actually, the phase-matching condition for the FWM process requires that dispersive effects compensate nonlinear ones, as given by Eq. (16.14).

In a 2008 experiment, a flat octave-spanning SC was generated from the visible to the near-infrared (NIR), by pumping a MOF at 1064.5 nm, corresponding to the normal dispersion regime, with 0.6-ns pulses with a 7 kHz repetition rate and ~60 mW average power [49]. Figure 17.7 shows

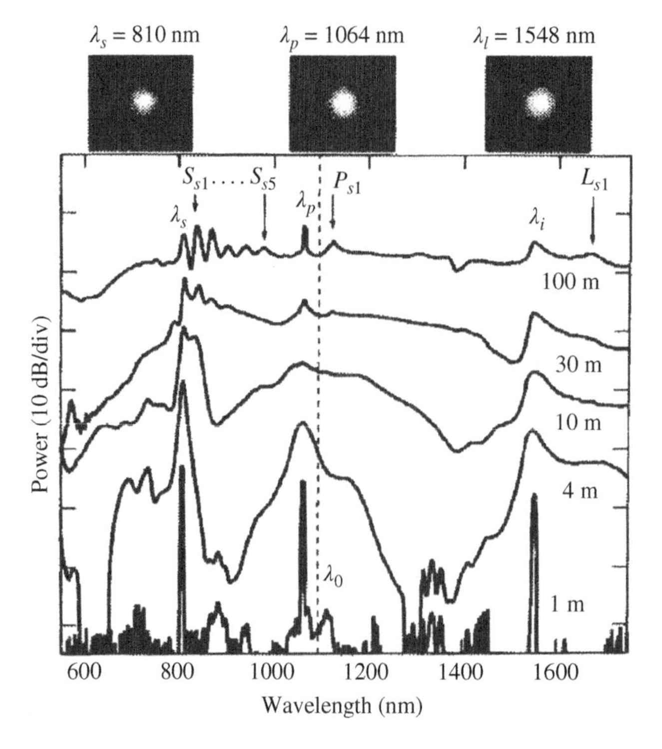

Figure 17.7 Evolution of the output spectrum as a function of the fiber length for a pump peak power of 2 kW launched into the MOF. The vertical line represents the ZDW. *Source:* After Ref. [49].

the output spectra for L increasing from 1 to 100 m, for a peak power of 2 kW launched into the fiber. For $L = 1$ m (bottom of Figure 17.7), the signal and idler waves arising from degenerate FWM are clearly visible. They are located, respectively, at 810 and 1548 nm, which is in excellent agreement with Eq. (16.14). By increasing the fiber length, the spectral width of these two waves and that of the pump increase. The broadening of the pump spectrum is expected since the injected peak power is much higher than the Raman threshold. As a consequence, the phase-matching relation of Eq. (16.14) can be satisfied between additional wavelengths located close to the pump and other wavelengths located around the signal and idler waves. For a fiber length of 30 m, the signal and idler waves start generating their own Raman cascade. For a fiber length of 100 m, the power of the signal wave becomes strong enough to generate up to five Raman Stokes orders located in the normal GVD region, very well defined and labelled S_{S1} to S_{S5} in Figure 17.7. In contrast, the Raman orders generated by the pump and idler waves fall in the anomalous regime and broaden rapidly, evolving into a continuum. This behavior can be attributed to MI combined with solitonic effects.

MOFs based on silica are particularly attractive candidates for high-power SC generation in the visible and NIR regions, which require low-loss materials in the wavelength range of interest, as well as appropriate nonlinear coefficient and controllable dispersion. SC sources in these regions have attracted considerable attention due to potential applications in optical frequency metrology, spectroscopy, hyperspectral lidar, medicine and biomedicine, and the optical communications area [4–13]. In a 2016 experiment, a picosecond ytterbium-doped fiber laser has been used to pump a piece of 5-m-long MOF to obtain 67.9 W SC with spectrum coverage of 500–1700 nm [50]. In 2018, an 80 W SC with spectra covering 350–2400 nm from multicore MOFs was demonstrated [51]. In the same year, a MOF SC source with 215 W of average power and spectra spanning from 480 nm to over 2000 nm was reported [52]. A linearly polarized SC with 93 W average output power and spectrum ranging from 520 to 2300 nm was demonstrated in a 2019 experiment using a piece of 2.6 m long polarization maintaining MOF and a picosecond pump source [53].

17.4 Continuous Wave Supercontinuum Generation

Compared with long pulse-pumped systems, continuous wave (CW) pumped systems allow a simple experimental realization. Typically, the experimental arrangement consists simply of a CW fiber laser spliced directly to a suitable fiber. This brings advantages in terms of robustness and stability. An Er-doped fiber laser is generally used for pumping around 1400–1600 nm, whereas a Yb-doped fiber laser is commonly used around 1060 nm. Typically, a fiber-coupled power of 5–50 W is used for CW continuum generation. These relatively high values for the coupled power lead to very high spectral powers in the resulting continuum, typically greater than 10 mW/nm.

The general features observed in the case of long-duration pulses can also be extended to the CW pumping in the anomalous dispersion regime. In this case, the SCG also starts with the MI effect, which evolves from background noise and is often stimulated by the intensity fluctuations which are typical of a CW pump [54–56]. The MI leads to the splitting of the CW into fundamental solitons.

In some circumstances, the fundamental solitons arising from MI can be short enough in order to enable the occurrence of IRS, leading to the SSFS. As the MI evolves from noise, there will be some jitter on the period and hence also on the duration of the generated solitons. In this case, a range of SSFSs occurs, creating a smooth redshifted continuum evolution.

Soliton collisions also constitute an important mechanism for the spectral expansion in CW-pumped supercontinua [57, 58]. In this case, the number of solitons can be particularly large and as redshifted solitons travel more slowly than those at higher frequencies, they can eventually collide while propagating through a fiber. During such collisions, some energy will be transferred to the most redshifted soliton, if it is within the spectral range of Raman scattering. The energy increase leads to a corresponding compression, which results in a further redshift for that soliton, according to Eq. (16.10). The increased redshift also slows down the soliton, leading it to pass by more solitons and hence experience more collision events. Soliton collisions can considerably increase the extent of an SC compared to Raman self-scattering alone.

The expansion of the SC to longer wavelengths can be limited by either the dispersion magnitude increasing or nonlinearity decreasing. Depending on the dispersion profile, the existence of a second ZDW can lead to the generation of DWs and the spectral recoil of the solitons [59, 60]. Concerning the extension to shorter wavelengths, it is determined by the excitation of phase-matched DWs and soliton trapping mechanisms, as discussed previously. Experimentally, this phenomenon has been combined with dispersion-engineered MOFs to further extend SCs to the UV in nanosecond and picosecond pumping schemes [61]. This idea consists of modifying the dispersion curve of the fiber so that group velocity-matching conditions for trapped DWs continuously evolve along propagation. This leads to the generation of new wavelengths as the ZDW decreases along propagation.

In a 2008 experiment, an SC ranging from 670 to 1350 nm with 9.55 W output power was achieved in the CW pumping regime, using a 200-m-long ZDW decreasing MOF pumped by a 20-W ytterbium fiber laser at 1.06 μm [62]. The fiber was composed by a 100-m-long section with a constant dispersion followed by a 100-m-long section with linearly decreasing ZDW. The initial ZDW was located at 1053 nm, just below the pump wavelength and decreased to 950 nm in a quasi-linear way. Figure 17.8 shows the output spectrum measured for three different input powers of 8.2, 11.3, and 13.5 W. As expected, the broadening of the output spectrum increases on both sides with increasing launch power. The spectrum is limited at long

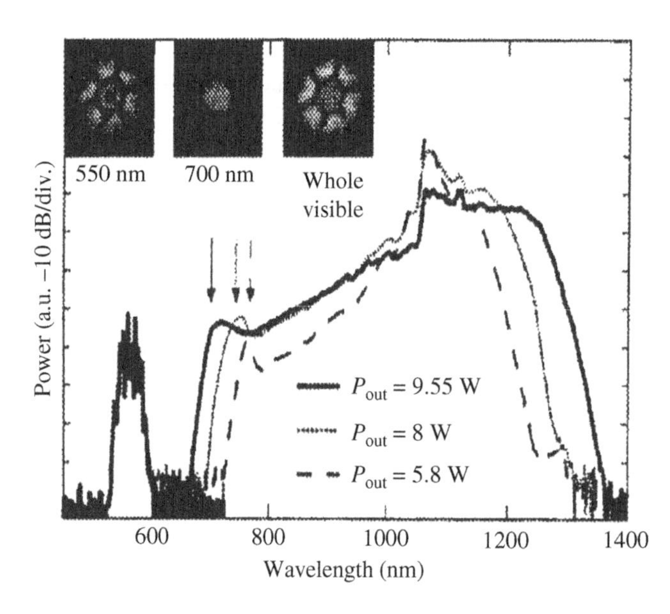

Figure 17.8 Output spectra for output powers of 5.8 W (dashed curve), 8 W (dotted curve), and 9.55 W (bold curve). The inset shows the far-field output spot at 550, 700 nm, and in the whole visible range. *Source:* After Ref. [62].

wavelengths by solitons redshifted by the SSFS effect. The trapped DW appears as a power increase located at 763, 751, and 720 nm, indicated by the arrows in Figure 17.8. The obtained results confirm that the extension of the SC toward short wavelengths is mainly due to the trapping of DWs by redshifted solitons [17].

In a 2018 experiment, a continuous-wave-pumped SC spanning over 1000 nm (>1 octave) was demonstrated using a 2-km-long standard telecom fiber and a high-power ytterbium-doped fiber laser [63]. By undergoing cascaded Raman conversion, the power from the 1117-nm pump was transferred to the anomalous dispersion region (at 1310 nm), after 3 Raman stokes conversions. The growth of higher-order Raman stokes in the normal dispersion region with an increase in pump power is shown in Figure 17.9a and b. Once the 1310-nm stokes starts growing, CW light breaks into pulses because of MI and this results in spectral broadening as shown in Figure 17.9c. When the power is increased further, other nonlinearities extend the spectrum both in shorter and longer wavelength sides, as shown in Figure 17.9d. While SRS and Raman-induced frequency shift (RIFS) extend the spectrum toward the longer wavelength side, FWM and DW generation mainly extends the spectrum toward the shorter wavelength region.

Figure 17.10a shows a spectrum spanning from 880 to 1900 nm within a 20-dB bandwidth for an input power of 76 W. The output power from the SC increases linearly with the pump power till 61 W of pump power, as shown in Figure 17.10b. A further increase in pump power determines a saturation of the SC since the spectrum has grown full in the longer wavelength side and further power transfer to longer wavelengths is attenuated by the silica fiber losses.

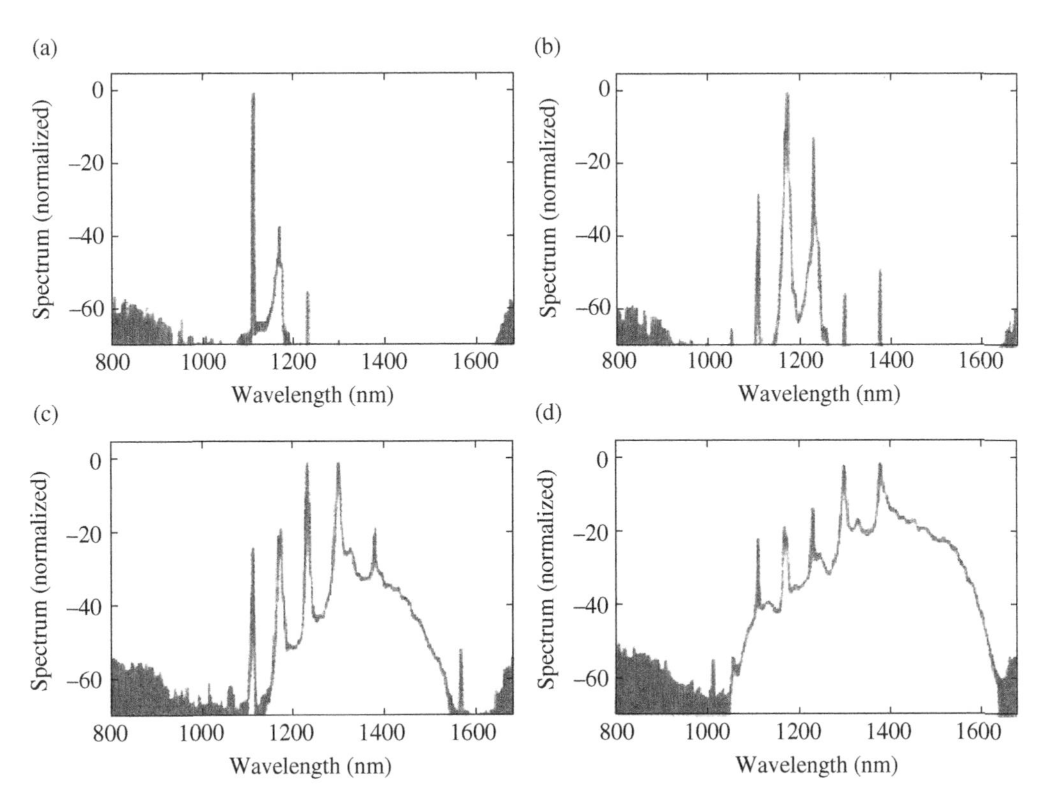

Figure 17.9 SC evolution at different output powers (a) 3 W; (b) 8 W; (c) 11 W; and (d) 15 W. *Source:* After Ref. [63].

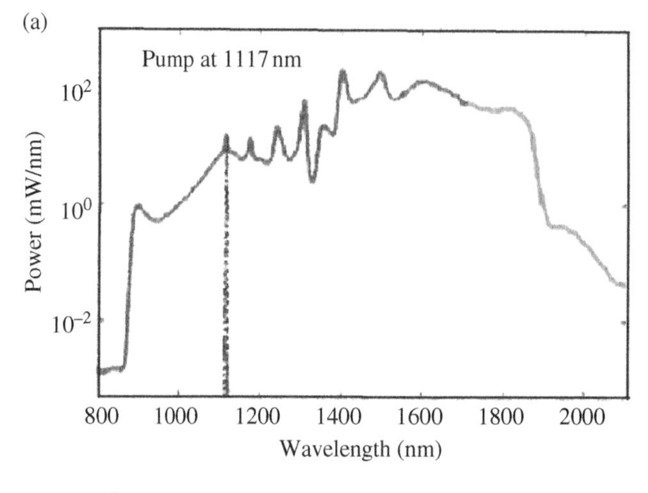

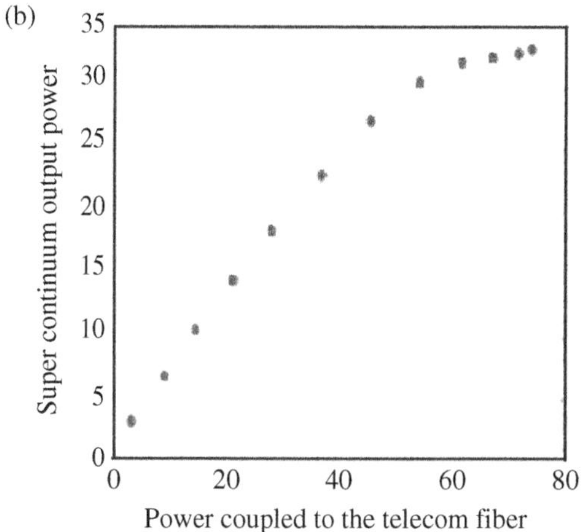

Figure 17.10 Full spectra for 1117 nm pumping at full power (pump location shown in dotted lines); (b) SC output power vs. input power coupled to the telecom fiber. *Source:* After Ref. [63]).

17.5 Mid-IR Supercontinuum Generation

As seen in previous Sections, silica fibers have shown an impressive performance for SCG in visible and NIR bands. However, the material loss in silica increases drastically at wavelengths beyond ~2 μm, effectively preventing the spectral evolution of SCG into the mid-infrared (MIR) region, where many molecular bonds have fundamental vibration frequencies [64]. MIR SC light sources show remarkable potential in various applications, such as tomography [65], food quality monitoring [66], disease diagnosis [67], and national defense [68]. Different types of glasses have been used to make fibers able to generate MIR SC, namely tellurite [69, 70], fluoride [71, 72], and chalcogenide (ChG) glasses [73, 74]. Compared with other glasses, ChG glass exhibits a wider MIR transmission window up to 20 μm and a higher optical nonlinearity up to three orders of magnitude higher than that of silica glasses [75], making them promising candidates for MIR SC generation.

In recent years, many ultrabroadband SCs in step-index ChG fibers have been generated by pumping in the anomalous dispersion regime [76]. In a 2014 experiment, an ultrabroad MIR SC spanning from 1.4 to 13.3 μm using an 85-mm-long step-index fiber made of ChG fiber pumped

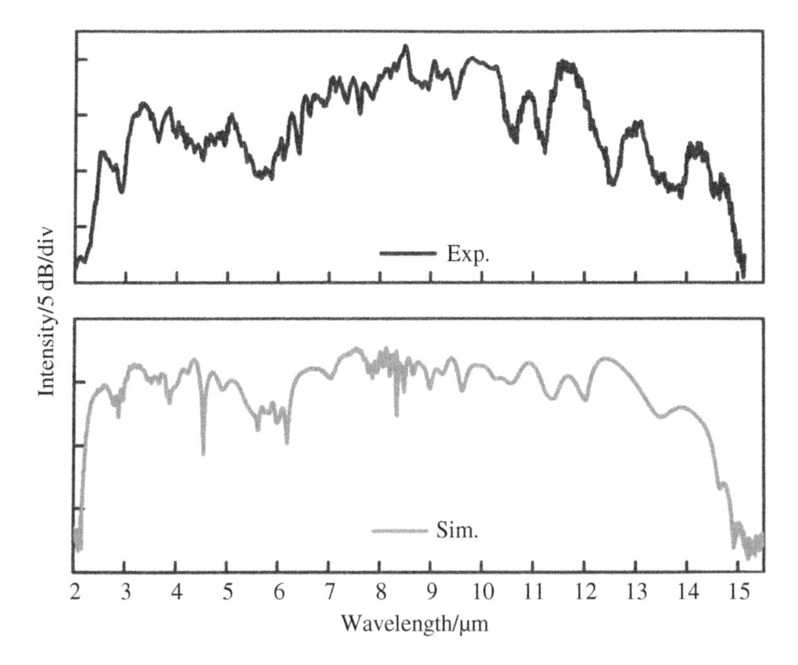

Figure 17.11 Measured (upper-curve) and simulated (lower-curve) MIR SC generation in a 3-cm-long chalcogenide (ChG) fiber pumped at 9.8 μm with 170-fs pulses with the peak power of ~2.89 MW. *Source:* After Ref. [79].

at 6.3 μm was demonstrated [77]. A broadband SC covering 2–12 μm spectral range in an 11-cm-long Ge-Sb-Se fiber pumped at 4.485 μm with 330-fs pulse was reported in 2016 [78]. In another 2016 experiment, a MIR SCG spanning ~2.0 to 15.1 μm in a 3-cm-long ChG step-index fiber pumped at 9.8 μm with 170-fs pulses has been demonstrated [79]. Figure 17.11 shows the SC spectrum obtained in such experiment (a), as well as the simulation result (b), obtained by solving the GNLSE given by Eq. (16.5). The nonlinear response function of the glass, $R(t)$, was assumed as in Eq. (16.6), whereas the delayed Raman response was approximated by Eq. (16.7). A good agreement is observed between both results.

One problem of the ChG fibers concerns their damage threshold, which is much lower than other fibers, namely fluoride fibers [80]. This indicates the unfitness of ChG fibers for high-power SC generation. Actually, the average power of the ChG-fiber-based SC laser sources have been limited to hundreds of milliwatts [81, 82]. An all-fiber MIR-SC laser source with a spectral coverage of 1.9–4.9 μm and a maximal average power of 11.8 W was demonstrated in a 2020 experiment [83]. The SC laser involved a silica-fiber-based SC laser as a pump source and a piece of fluoroindate (InF3) fiber as a nonlinear medium.

In all cases mentioned above, SC is generated by pumping in the anomalous dispersion regime and it is mainly due to soliton dynamics, such as soliton fission and self-frequency shift. However, this broadening process generates a spectrum with a complex temporal profile, which is sensitive to noise amplification, leading to different spectral components [84]. These features cause coherent degradation of SC and reduce resolution or precision in many applications. A common approach to avoid these limitations is to pump fibers with an all-normal dispersion profile by using ultrashort pulse lasers [85–87]. In this case, the fiber exhibits flattened normal dispersion over the entire spectral region of interest, and the SC spectral broadening is mainly a result of SPM and optical wave breaking [88], which can eliminate noise-sensitive soliton dynamics [89–91]. A fiber-based SC source delivering up to 825 mW of average output power between 2.5 and 5.0 μm generated in an all-normal dispersion regime has been reported in a 2020 experiment [82]. Femtosecond input pulses at 3.6 μm with an average output power exceeding the watt level were spectrally broadened through SPM using commercial ChG-based step-index fibers.

17.6 Supercontinuum Coherence

The coherence of the SC can be characterized by the degree of coherence associated with each spectral component, given by [16, 84]:

$$g_{12}(\omega,z) = \left| \frac{\left\langle \widetilde{U}_1^*(\omega,z)\widetilde{U}_2(\omega,z)\right\rangle}{\left[\left\langle \left|\widetilde{U}_1(\omega,z)\right|^2\right\rangle\left\langle \left|\widetilde{U}_2(\omega,z)\right|^2\right\rangle\right]^{1/2}} \right| \tag{17.5}$$

where $\widetilde{U}_1$ and $\widetilde{U}_2$ denote the Fourier transforms of two outputs resulting from two independently generated input pulses with random noise, and the angle brackets denote an average calculation over the entire ensemble of pulses. It has been shown that $g_{12}(\omega)$ depends on the average value of the input pulse parameters, namely on the pulse width [16]. The coherence is significantly degraded in the case of a high-energy short pulse propagating in the anomalous dispersion regime when a large number of fundamental solitons are created through the fission process. In fact, it has been shown that this process can be strongly perturbed by the amplified spontaneous emission present on higher-order solitons injected into conventional fibers [92]. This results in large fluctuations in the amplitude and duration of subsequently generated fundamental solitons, which are then converted into wavelength fluctuations through the SSFS effect. The coherence degradation effects become more dramatic in the case of highly nonlinear MOFs, where the pump laser shot noise replaces amplified spontaneous emission as the input noise seed.

Figure 17.12 illustrates the spectral, temporal and coherence evolution along a MOF with a length of 10 cm of (a) 100 fs and (b) 150 fs pulses with 10 kW peak power pulses at 835 nm pump wavelength. Since the ZDW of this fiber is at 780 nm, this corresponds to the anomalous dispersion regime. In spite of the fact that the spectral bandwidths are similar in both cases, the coherence evolution is significantly different. In particular, we observe that the initial spectral broadening in (a) is coherent, with near-unity coherence over the major part of the SC from 500 to 1200 nm. In this case, the pulse becomes temporally compressed so quickly that the soliton fission process dominates the initial dynamics.

In case (b) of Figure 17.12, a high degree of coherence is only maintained in the immediate vicinity of the broadened pump around 750–900 nm. In this case, the MI amplified noise is the dominant feature and the soliton fission occurs randomly. This process occurs even before other perturbations such as higher-order dispersive and nonlinear effects become significant. For typical experimental conditions, it has been found that the SC generated by higher-order input solitons presents a high coherence when $N < 20$, whereas it has a low coherence for $N > 40$ [16, 93].

Of course, the generated SC is highly coherent when the fiber is pumped in the normal GVD regime since MI and soliton-related effects do not occur in this case. However, the drawback is

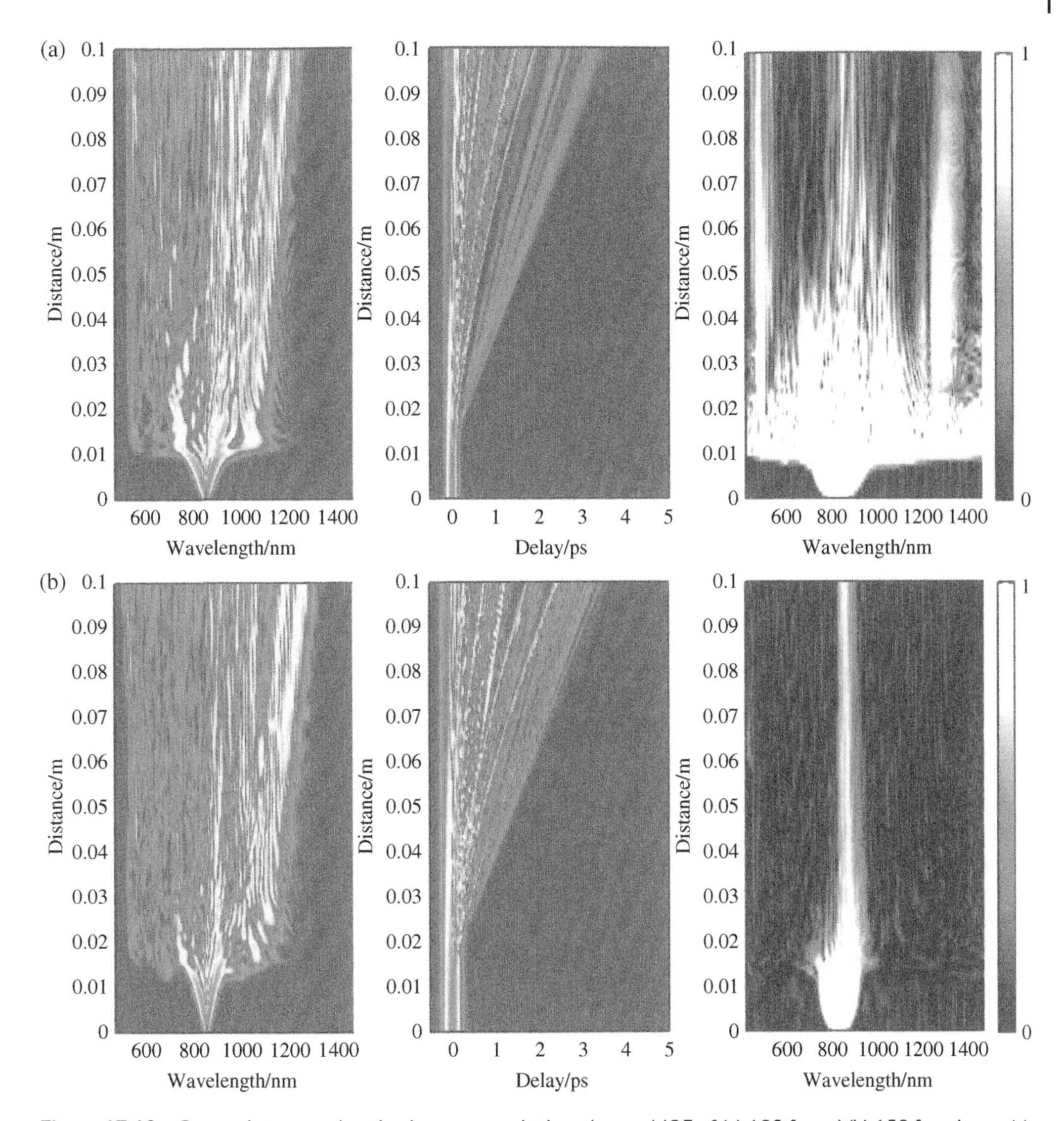

Figure 17.12 Spectral, temporal and coherence evolution along a MOF of (a) 100 fs and (b) 150 fs pulses with 10 kW peak power pulses at 835 nm pump wavelength. The fiber ZDW is at 780 nm.

that the SC spectral width (at the same peak power) is comparatively much smaller due to the rapid initial temporal spreading of the pump pulses [92].

Coherent SCG has been demonstrated using silica, tellurite and ChG fibers by pumping in the all-normal dispersion regime using short-wavelength pumps around 1.5 and 2 μm [94–101]. However, the long-wavelength edge could not reach 4 μm. Recent advances in fiber lasers have extended the central wavelength beyond 3 μm with average powers of a few tens Watts [102–104]. A coherent SC covering 1.7–12.7 μm was achieved in 2019, by pumping an all-normal dispersion 7 cm-long Te-based ChG tapered fiber at 5.5 μm [91]. In a 2020 experiment, a ChG all-solid hybrid MOF with an all-normal and flattened dispersion has been demonstrated [105]. A broad SC spectrum extending from 2.2 to 10 μm was obtained when the fiber was pumped at 5 μm with an input power of 3.9 mW.

17.6.1 Spectral Incoherent Solitons

As seen in Section 17.4, SCG does not involve the fission process when a CW pump is used in the anomalous dispersion region. In this case, SCG starts with the MI process, which leads to the splitting of the CW into a multitude of fundamental solitons. As the MI evolves from noise, there will be some jitter on the period and hence also on the duration of the generated solitons, leading to an incoherent SC. In the case of a highly nonlinear MOF, whose dispersion curve exhibits two ZDWs, it has been shown that rapid temporal fluctuations of the field prevent the formation of robust coherent structures and solitons do not play any significant role in the spectral broadening process. A thermodynamic interpretation of this incoherent regime of SC generation has been provided in [106] on the basis of wave turbulence theory [107–109]. According to this interpretation, once the field has reached its thermodynamic equilibrium state, the spectral broadening process saturates and the spectrum of the optical field no longer evolves during propagation [106]. Several studies have confirmed that the incoherence of the SC results through a thermalization process and leads to the formation of spectral incoherent solitons [106, 110–112]. These solitons were found numerically in 2008 and differ from those described in Chapter 6 in the sense that they are confined in the spectral domain and not in the temporal domain [110].

Spectral incoherent solitons were observed for the first time in a 2009 experiment, through SC generation inside a 40-m-long MOF exhibiting two ZDWs: one at 1033 nm and another at 1209 nm [111]. The MOF was pumped at 1064 nm using 660-ps pulses with a peak power of 3.5 kW delivered by passively Q-switched Nd:YAG laser. Figure 17.13 shows (a) the evolution of the spectrum along the fiber; and (c) its dependence on the input peak power (length 40 m). We observe in

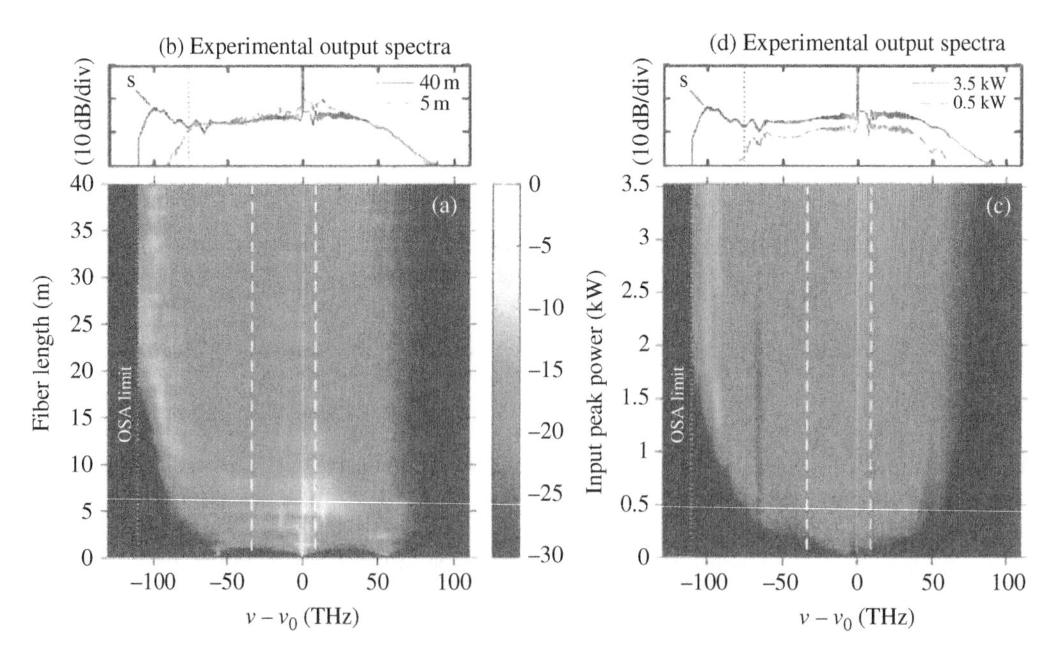

Figure 17.13 (a) Output spectra as a function of the fiber length (P_0 = 3.5 kW); and (c) as a function of the input peak power (length 40 m); (b) Experimental output spectra obtained after 5 and 40 m of propagation; (d) Experimental output spectra at P_0 = 0.5 kW and P_0 = 3.5 kW. The white dashed lines show the two ZDWs of the optical fiber. "S" indicates the position of the spectral incoherent soliton. *Source:* After Ref. [111].

Figure 17.13a an initial spectral broadening in the first meter of propagation in the MOF, which is due to the development of two sets of MI gain-bands: one in the anomalous dispersion regime and the other in the normal dispersion regime. After some few meters, the process of spectral broadening saturates in the high-frequency part of the spectrum and for large propagation distances the SC edge exhibits a slight depletion. Such depletion is due to the Raman effect, which leads to a transfer of power toward the low-frequency components. The saturation of spectral broadening may be regarded as a signature of the thermalization process of the optical field. The Raman effect is responsible for the generation of spectral incoherent solitons in the low-frequency part of the SC spectrum (normal dispersion regime), which is marked by S in Figure 17.13b and d. The spectral soliton moves away from the main SC spectrum, which reduces the power in the low-frequency edge of the SC spectrum. The Raman effect thus becomes almost inefficient, thus leading to a saturation of the spectral broadening in the low-frequency edge of the main SC spectrum.

Figure 17.14a shows the evolution of the spectrum of the field obtained by numerically solving Eq. (16.5) considering the fiber characteristics of Figure 17.13 [111]. The initial condition refers to a 60-ps (FWHM) Gaussian pulse with a peak power of 3.5 kW. Figure 17.14c,d illustrates the same numerical simulation, but starting with a noisy CW beam and neglecting the Raman and the self-steepening effects. In this case, the SC thermalization is confirmed since a clear saturation of spectral broadening occurs in both the high- and low-frequency edges of the SC spectrum. A comparison of Figure 17.14a and c shows that the Raman effect prevents the establishment of

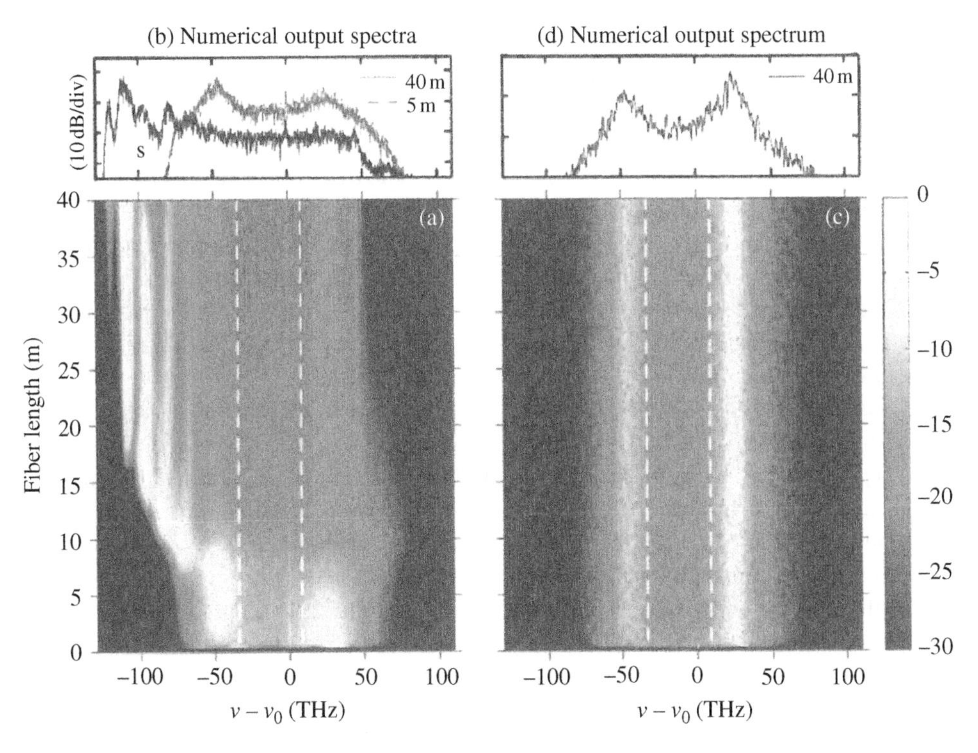

Figure 17.14 (a) Numerical simulations based on the generalized nonlinear Schrödinger equation (16.5), for an input power of 3.5 kW and an initial Gaussian pulse of 60 ps. (b) Numerical spectra for both propagation lengths 5 and 40 m. In (c, d), the simulation starts with a noisy CW beam and neglects the Raman and the self-steepening effects. The white dashed lines indicate the fiber ZDWs. "S" indicates the position of the spectral incoherent soliton. *Source:* After Ref. [111].

a truly stationary state in the high-frequency edge of the SC spectrum. Besides the saturation of the spectral broadening process, we observe in Figure 17.14c,d the development of a double-peak in the spectrum evolution, which is a specific signature of the wave thermalization process, as predicted by the kinetic wave theory [106].

17.7 Supercontinuum Generation in Hollow-Core Kagomé Fibers

SC sources based on solid-core conventional and microstructured optical fibers can provide broadband light from the mid-IR to the near-UV [1, 16]. However, they have not been demonstrated for wavelengths shorter than 280 nm [113], and are found to suffer from limited optical transparency and cumulative optical damage when delivering UV light. Actually, no solid-state or fiber-based SC source exists that can span the vacuum ultraviolet (VUV) (∼100–200 nm) and deep-ultraviolet (DUV) (∼200–300 nm) spectral ranges.

Hollow-core MOFs provide a means of eliminating the diffraction of laser light in gases while offering precise control of group velocity dispersion and long low loss interaction paths [114]. In particular, kagomé MOFs offer a broadband transmission and weak anomalous dispersion from the UV to the near-IR. These properties not only help support ultrafast soliton dynamics but also allow the guidance of any UV light that is subsequently generated at relatively low losses [115–117]. Actually, noble-gas-filled kagomé MOFs have been used to demonstrate the widely tunable and efficient generation of ∼5-nm-wide bands of UV light down to 176 nm [118–121].

Pulse propagation in a gas-filled kagomé MOF can be described by Eq. (16.17), where the Kerr coefficient depends linearly on the gas pressure. In the case of a noble gas, the Raman effect is ignored and the ZDW varies with pressure, as illustrated in Figure 16.16.

Figure 17.15a shows the propagation of a 600-fs pulse in a kagomé MOF with a 30 μm core-filled argon at a pressure of 25 bar. The ZDW of the fiber is 750 nm and the pulses have an energy of 10 μJ at 800 nm, corresponding to a soliton of order $N \sim 245$. Pulse propagation was modeled based on Eq. (16.17) by including the effect of quantum noise. Figure 17.15b shows the splitting of the input pulse into a large number of ultrashort solitons, that subsequently undergo multiple collisions. This leads to a smooth, flat, and high-energy SC spanning the range from 350 to 1500 nm, as observed from Figure 17.15c. We observe that, in the absence of Raman scattering, the SC shows a blue enhanced asymmetry, in contrast to what is seen in glass-core fibers.

Figures 17.16–17.18 show the spectral evolution of 40-fs pulses launched into a kagomé MOF with a core diameter of 40 μm filled with argon, xenon, and helium, respectively, at different pressures [120, 121]. The pulses are launched at $\lambda_{\text{pump}} = 790$ nm and have an energy of 10 μJ. These results can be understood considering the dependence of the ZDW, λ_{ZD}, with the pressure for each gas, as illustrated by Figure 16.17. In the case of Figure 17.16a, the pumping is made in the anomalous dispersion regime and the input soliton order is N~38. However, in Figure 17.16b–d, the pumping is realized in the normal dispersion regime and the spectral expansion becomes less significant. We may observe that by increasing the pressure, the spectral broadening occurs within shorter distances, which is due to the increased nonlinearity for higher pressure values. This is also observed in Figure 17.17, in which case the pumping is made in the normal dispersion regime for all the assumed pressure values, Nevertheless, the distance for spectral broadening in the xenon-filled fiber is one order of magnitude smaller compared with the corresponding distance for the argon-filled fiber, due to the higher nonlinearity of xenon.

In the case of helium-filled kagomé fiber (Figure 17.18), the pumping is always in the anomalous dispersion regime. By increasing the gas pressure, the ZDW becomes closer to the pumping

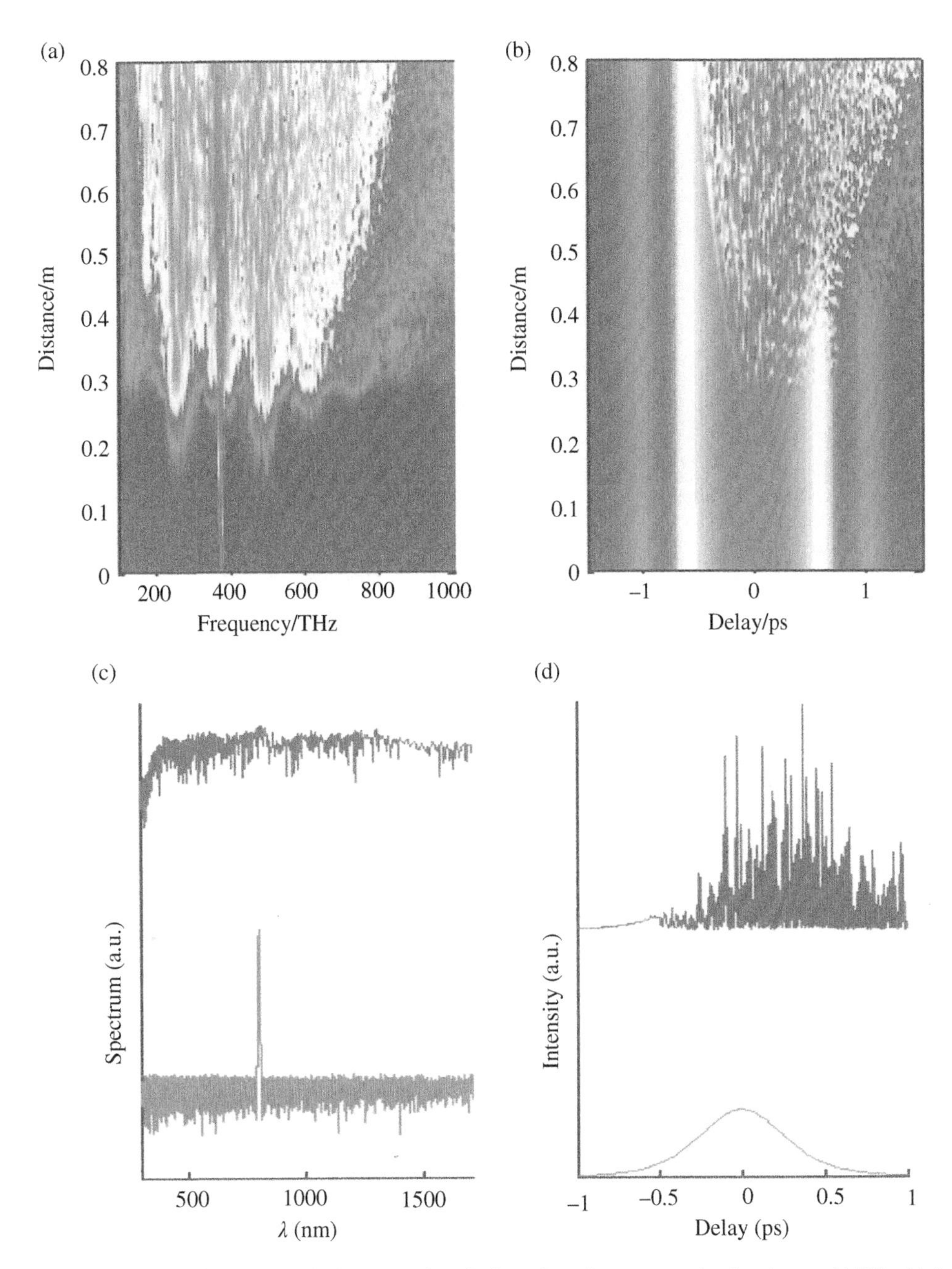

Figure 17.15 (a) Spectral; and (b) temporal evolution of a pulse propagating in a kagomé MOF with inclusion of quantum noise; (c) input (lower) and output (upper) pulses in spectral and (d) temporal domains.

wavelength and the fiber nonlinear parameter also increases so that the soliton order in Figure 17.18a–d is $N = 4.9$, 7.0, 10.3, and 15.8, respectively. Consequently, the soliton fission length decreases with pressure.

As a consequence of the higher-order soliton compression, its spectrum expands and overlaps with resonant dispersive-wave frequencies, which are consequently excited in the UV region.

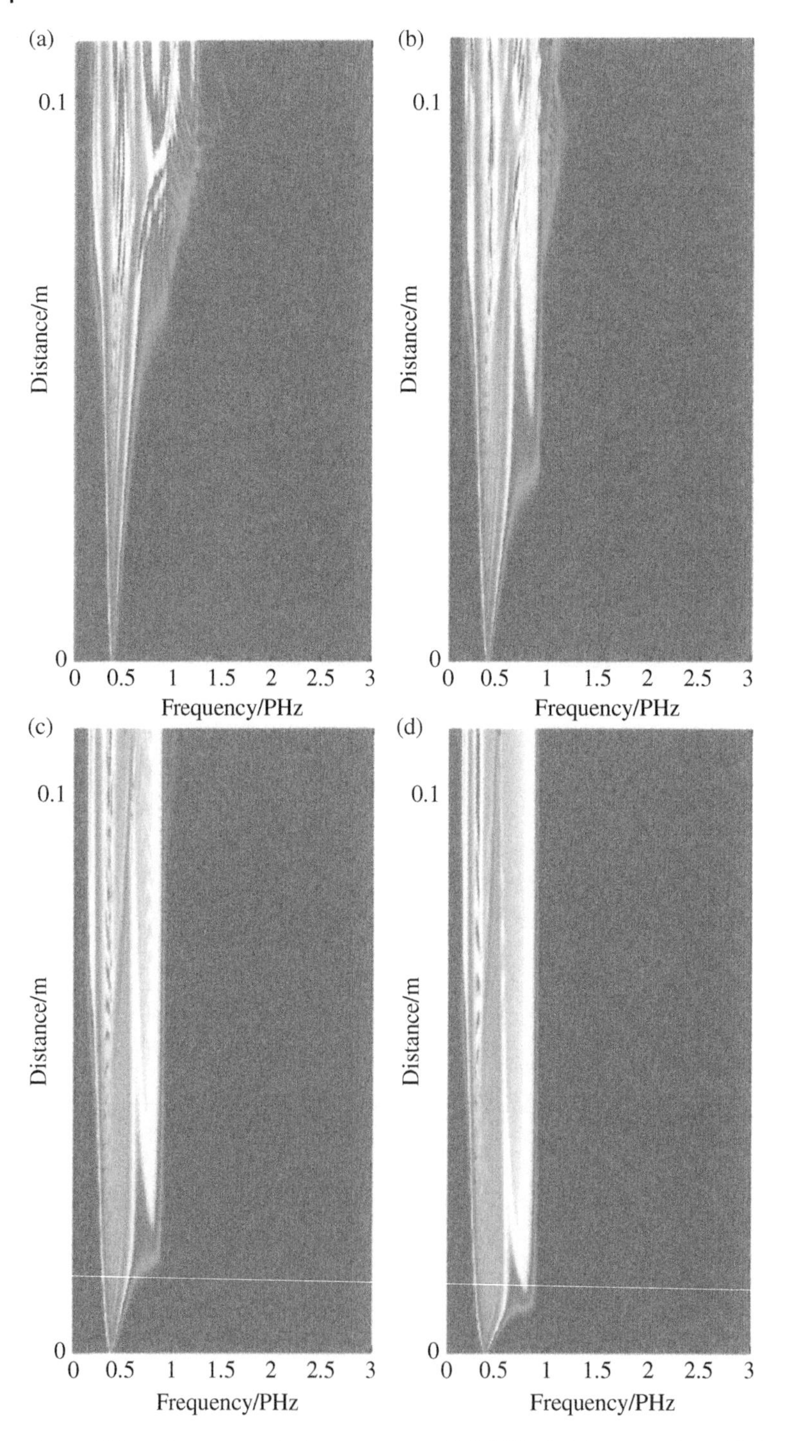

Figure 17.16 Spectral evolution in a kagomé MOF filled with argon of a 40-fs input pulse, with an energy of 10 μJ and $\lambda = 790$ nm, at pressures (a) 10 bar; (b) 20 bar; (c) 40 bar; and (d) 80 bar.

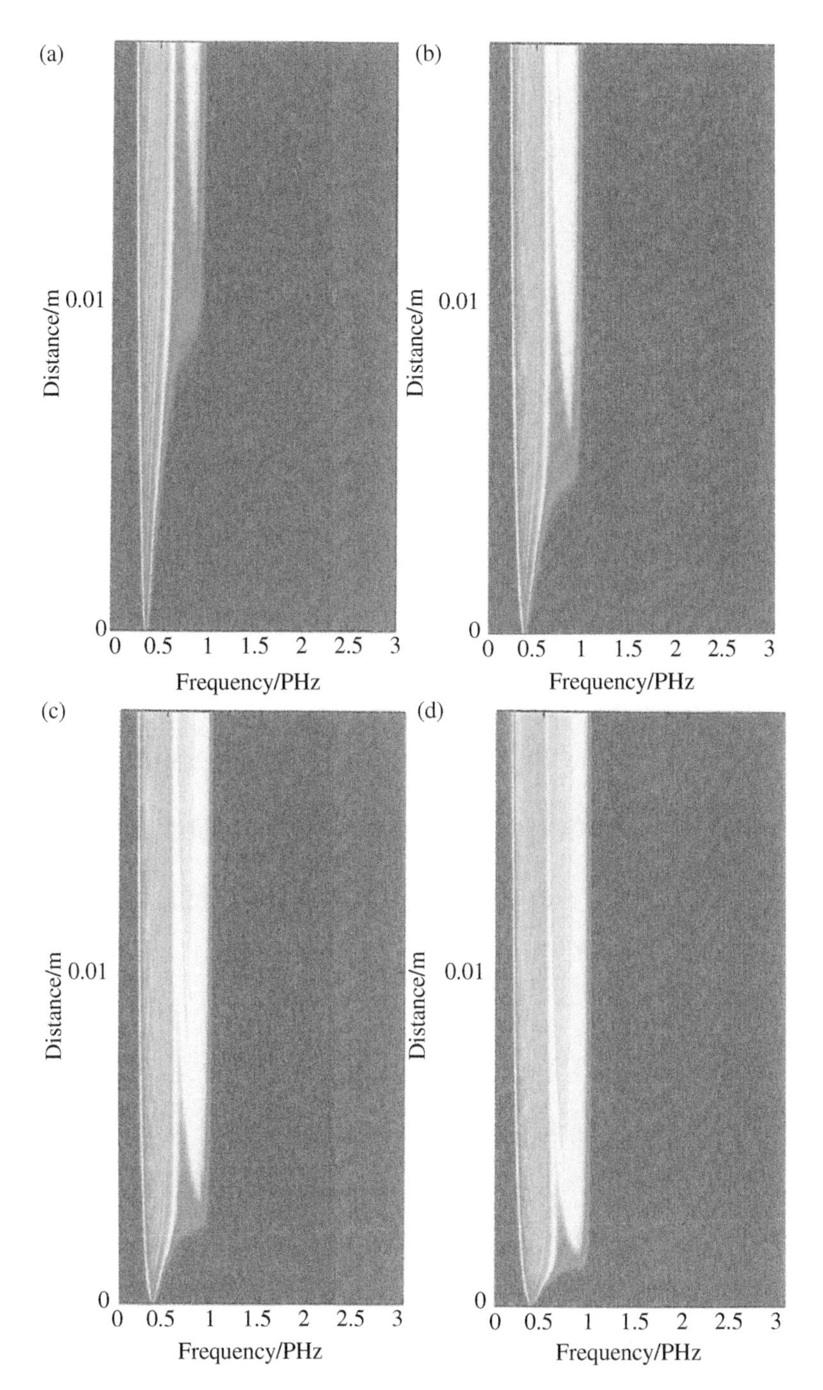

Figure 17.17 Spectral evolution in a kagomé MOF filled with xenon of a 40-fs input pulse, with an energy of 10 μJ and λ = 790 nm, at pressures (a) 10 bar; (b) 20 bar; (c) 40 bar; and (d) 80 bar.

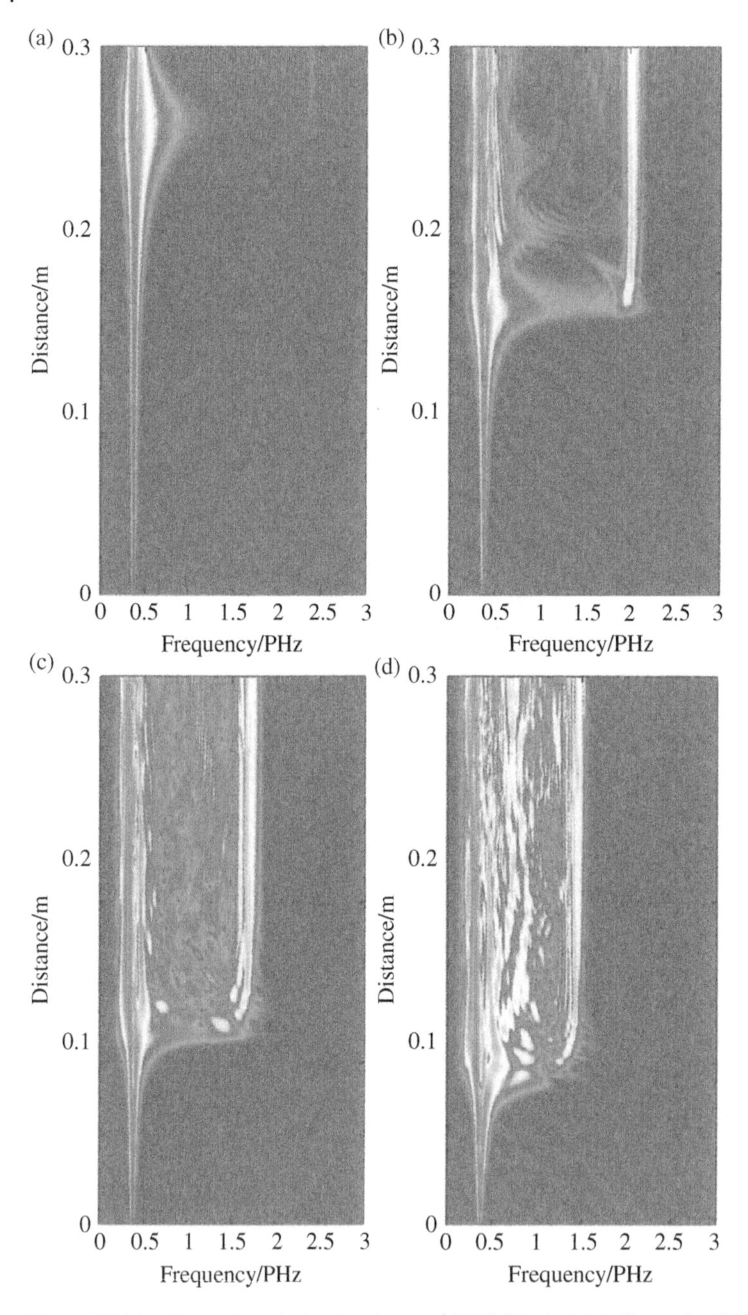

Figure 17.18 Spectral evolution in a kagomé MOF filled with helium of a 40-fs input pulse, with an energy of 10 μJ and λ = 790 nm, at pressures (a) 10 bar; (b) 20 bar; (c) 40 bar; and (d) 80 bar.

The resonance condition is given by Eq. (16.11), where the soliton peak power P_s must be substituted by the peak power of the compressed pulse. This condition provides the resonance frequencies $\omega \sim 2.4$, 2.0, 1.7, and 1.6 PHz for the cases of Figure 17.18a–d, respectively, which agrees reasonably with the central frequencies of the generated UV bands.

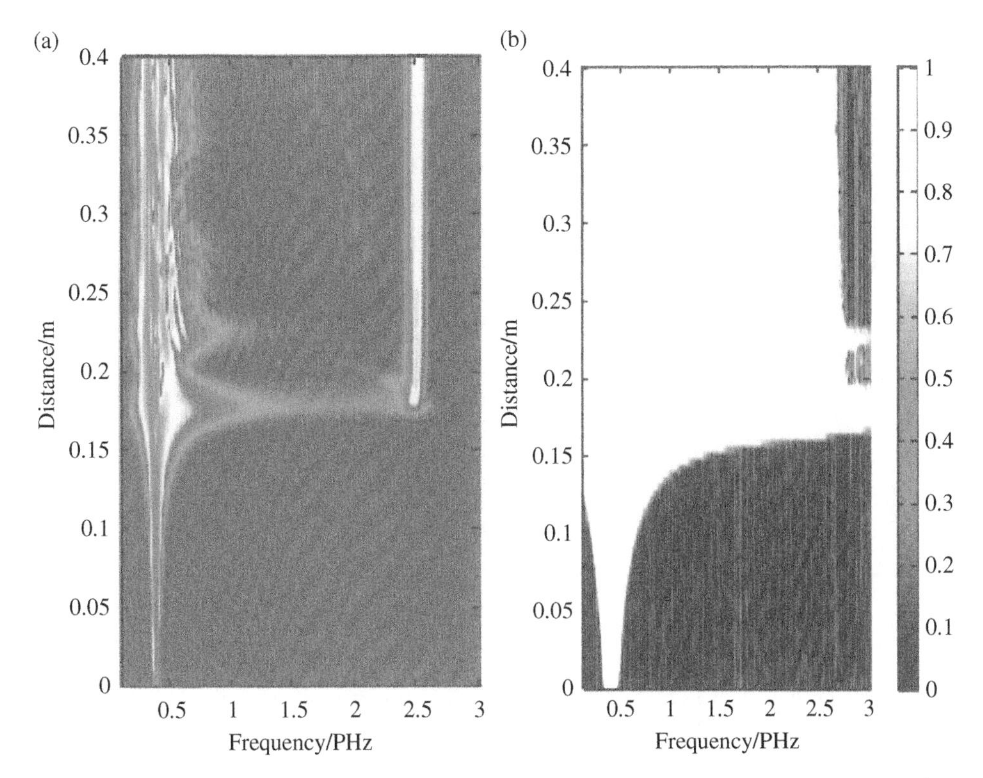

Figure 17.19 (a) Spectral; and (b) coherence evolution of ultrafast nonlinear UV-generation in a kagomé MOF with a core of diameter $d = 40.0\,\mu m$, filled with helium at a pressure $p = 10$ bar; the 40-fs pump pulses have a wavelength $\lambda_{pump} = 800$ nm and an energy of 20 μJ.

Figure 17.19 illustrates an optimized situation for the generation of UV light, in which 40-fs pulses with an energy of 20 μJ at $\lambda_{pump} = 800$ nm are launched into a kagomé HC-MOF with a core diameter $d = 40.0\,\mu m$, filled with helium at a pressure $p = 10$ bar. The UV light is generated when the pulse achieves the maximum compression. At this stage, as a consequence of the higher-order dispersion, the pulse suffers soliton fission. Figure 17.19b shows that the generated spectrum has a very high coherence, which may be useful in several applications.

Concerning the generation of UV light, the conversion efficiency can be defined as the fraction of the total output power at wavelengths shorter than 350 nm. In the case of Figure 17.19, we have a conversion efficiency of 5.2%. The quality of the UV emission can also be characterized through a factor $Q = {P_{UV}}^{FWHM}/P_{UV}$, where ${P_{UV}}^{FWHM}$ is the spectral power within the FWHM of the strongest UV peak and P_{UV} is the total spectral power in the UV region [120]. In the case of Figure 17.19, the UV output has a quality factor $Q = 85.9\%$.

The UV-generated light source can be tuned by varying the input pulse parameters or the gas pressure [120]. For example, Figure 17.20a shows that decreasing the pump wavelength from 800 to 700 nm, the UV center frequency is shifted to lower values. The overall profile of the output spectrum is similar to that of Figure 17.19; however, the quality factor is 66.1% and the conversion efficiency is 8.8%. On the other hand, reducing the pulse FWHM from 40 to 10 fs provides a slight increase of the UV center frequency, as illustrated by Figure 17.20b. Compared with the case of Figure 17.19, there is a reduction of the conversion efficiency to 1.7%, whereas the quality factor

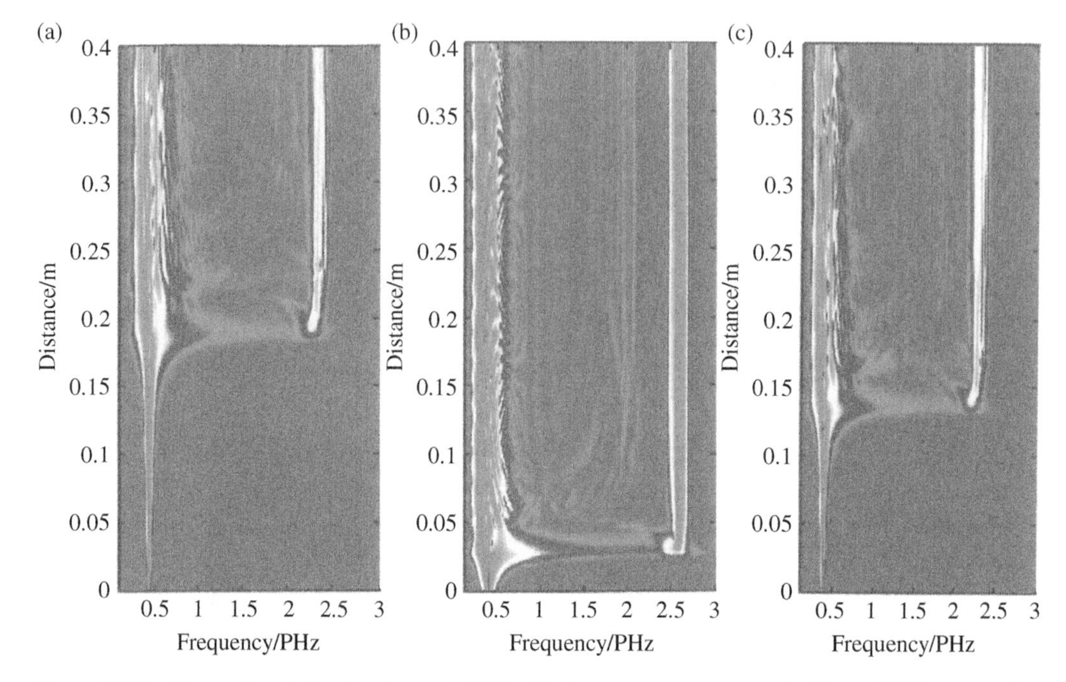

Figure 17.20 Spectral evolution in a kagomé MOF with the same parameters as in Figure 17.18, except that (a) λ_{pump} = 700 nm; (b) the input pulse width is 10 fs; and (c) the gas pressure is 15 bar.

of the UV radiation remains similar (82.6%). The fission distance decreases for smaller pulse widths, as expected. Increasing the gas pressure to $p = 15$ bar causes a shorter fission distance, as observed from Figure 17.20c. On the other hand, compared with the case of Figure 17.19, the quality factor is reduced to 37.7%, whereas de efficiency of conversion is increased to 9.6%.

The comparison of Figures 17.20c and 17.19a shows that by increasing the gas pressure, the central frequency of the UV light decreases. Figure 17.21 illustrates the possibility of tuning the frequency of the final UV pulse by changing the pressure of the gas in the fiber's core. The variation of the gas pressure between 7 and 16 bar provides a tuning of the UV light main peak frequency by ~0.4 PHz.

Conversion efficiencies of several percent to a UV wavelength band a few nm wide, tunable over 1 PHz from 176 to 550 nm, was obtained in a 2013 experiment using a kagomé MOF filled with different noble gases and few-μJ 38 fs pulses at 800 nm [119]. It was demonstrated that the tunability could be further controlled by introducing a gas pressure gradient.

The use of HC-MOFs filled with other gases or with ambient air has also been considered in recent years [122–127]. For example, the generation of a spatially coherent SC extending from the VUV (124 nm) to the NIR (beyond 1200 nm), in a hydrogen-filled kagomé MOF was reported in 2015 [124]. Few-microjoule, 30-fs pump pulses at a wavelength of 805 nm were launched into the fiber, where they undergo self-compression via the Raman-enhanced Kerr effect. Among the different molecular gases, hydrogen offers several advantages, such as a broad transmission range down to the extreme UV (~70 nm), a long dephasing time of

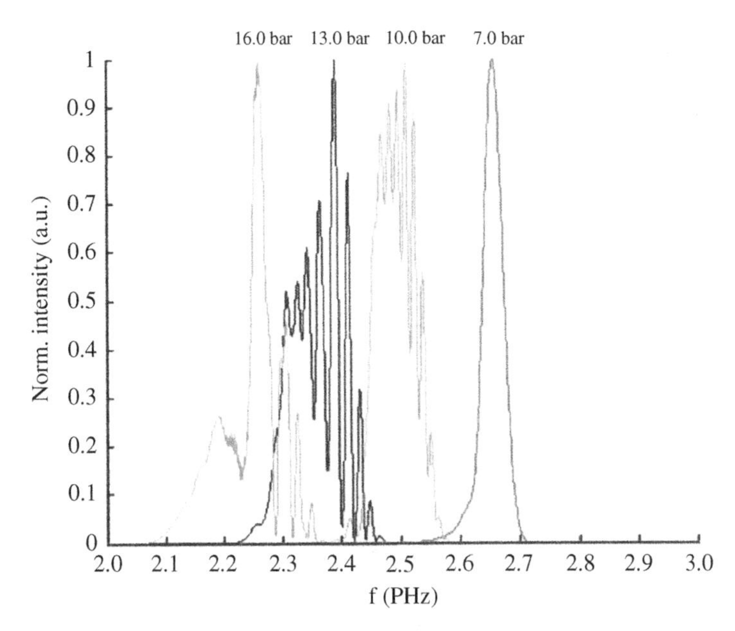

Figure 17.21 Output spectral profile of UV light generated in a kagomé MOF; all the parameters are the same as in Figure 17.18 except for the gas pressure, which is indicated.

its molecular oscillations, a large high-pressure Raman gain coefficient, a relatively high ionization potential, and the highest rotational and vibrational frequency shifts of all molecules. The combination of these factors was essential in achieving the observed super-broadening [124].

In a 2018 experiment, the spectral broadening of ~6-ps input pulses in a 13.8-m-long HC-MOF filled with air was investigated [126]. Figure 17.22a shows the experimental output spectra at different input powers. The origin of the initial pulse-broadening process was associated with rotational Raman scattering (RRS) at low power levels. Due to the broadband and low loss transmission window of the kagomé MOF, it was possible to observe the transition from this initial stage to a high spectral-power density SC spanning from 850 to 1600 nm.

To simulate the experimental results, a semi-quantum model (SQM) for air was developed that includes both the RRS and the vibrational Raman scattering (VRS) responses and that can be used in conjunction with the standard generalized nonlinear Schrödinger equation [126]. An excellent agreement between the experimental and simulation results can be observed in Figure 17.22, particularly at low input powers (~5 W), where the simulations reproduce both the RRS and VRS dynamics well. As the power increases the signature of higher-order RRS, alongside the roto-VRS is also well reproduced, as seen in Figure 17.22d. Increasing the input power not only stimulates higher-order Raman peaks but also produces noticeable broadenings from the Kerr nonlinearity, which plays a key role in the sideband expansion. This broadening process is well reproduced by simulations, as shown by the snapshots at which the input laser and VRS bands start to overlap (at ~1250 nm), as seen in Figure 17.22e,f.

The same model presented in [124] was used to describe the results obtained in a 2021 experiment, in which spatially filtered 54-fs pulses at 400 nm were launched into an 80-cm-long kagomé MOF filled with common air [127]. As the pump energy was increased, the spectrum at the fiber output broadened and shifted toward lower frequency, due to the IRS. It was verified that the rotational responses of N_2 and O_2 were dominant in the Raman response of air.

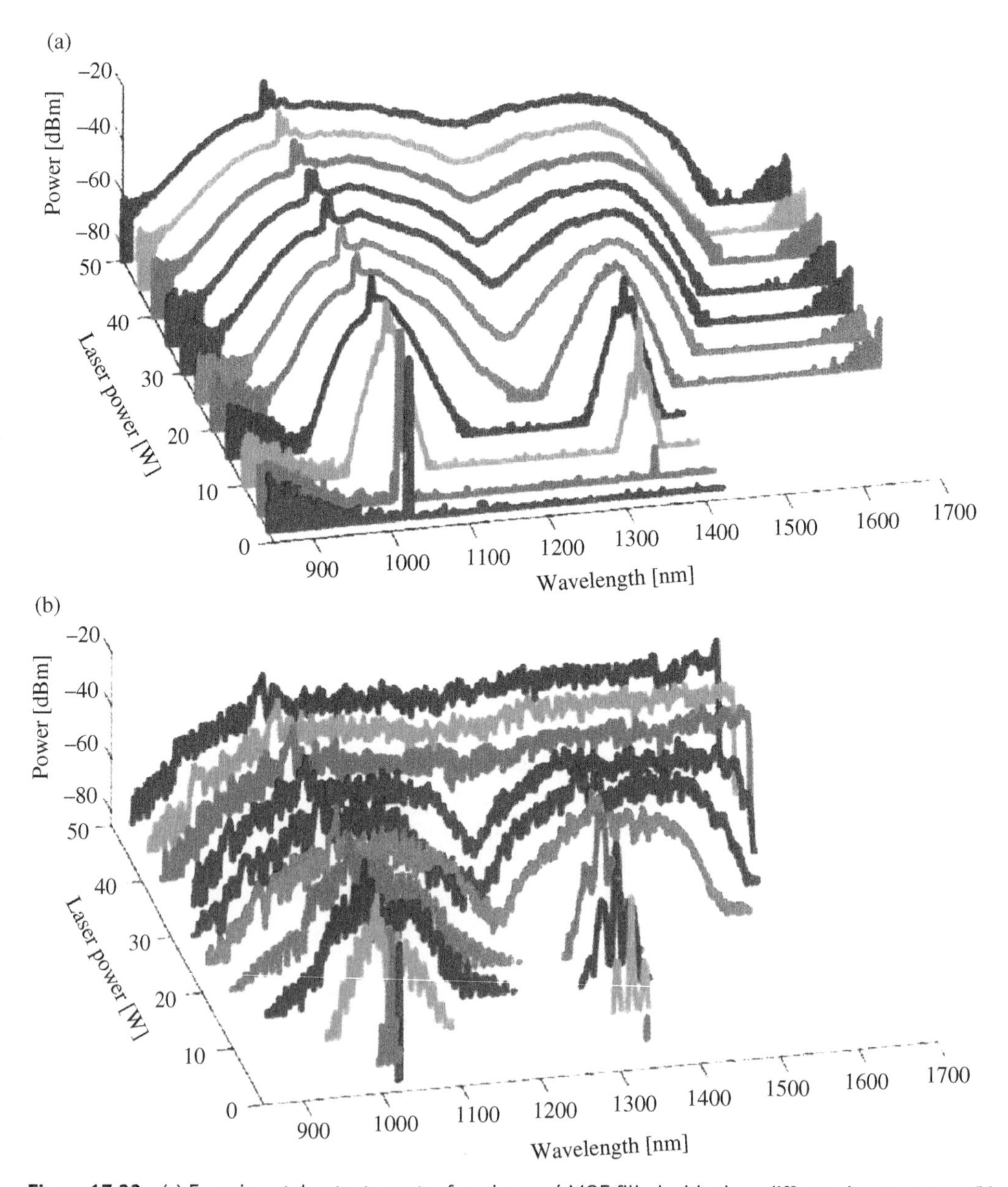

Figure 17.22 (a) Experimental output spectra for a kagomé MOF filled with air at different input powers; (b) simulation results using a semi-quantum model (SQM) in the GNLSE (averaged over 20 shots). The comparison of simulated and experimental results is presented for (c) 5 W; (d) 15 W; (e) 25 W; (f) 35 W; (g) 50 W. *Source:* After Ref. [126].

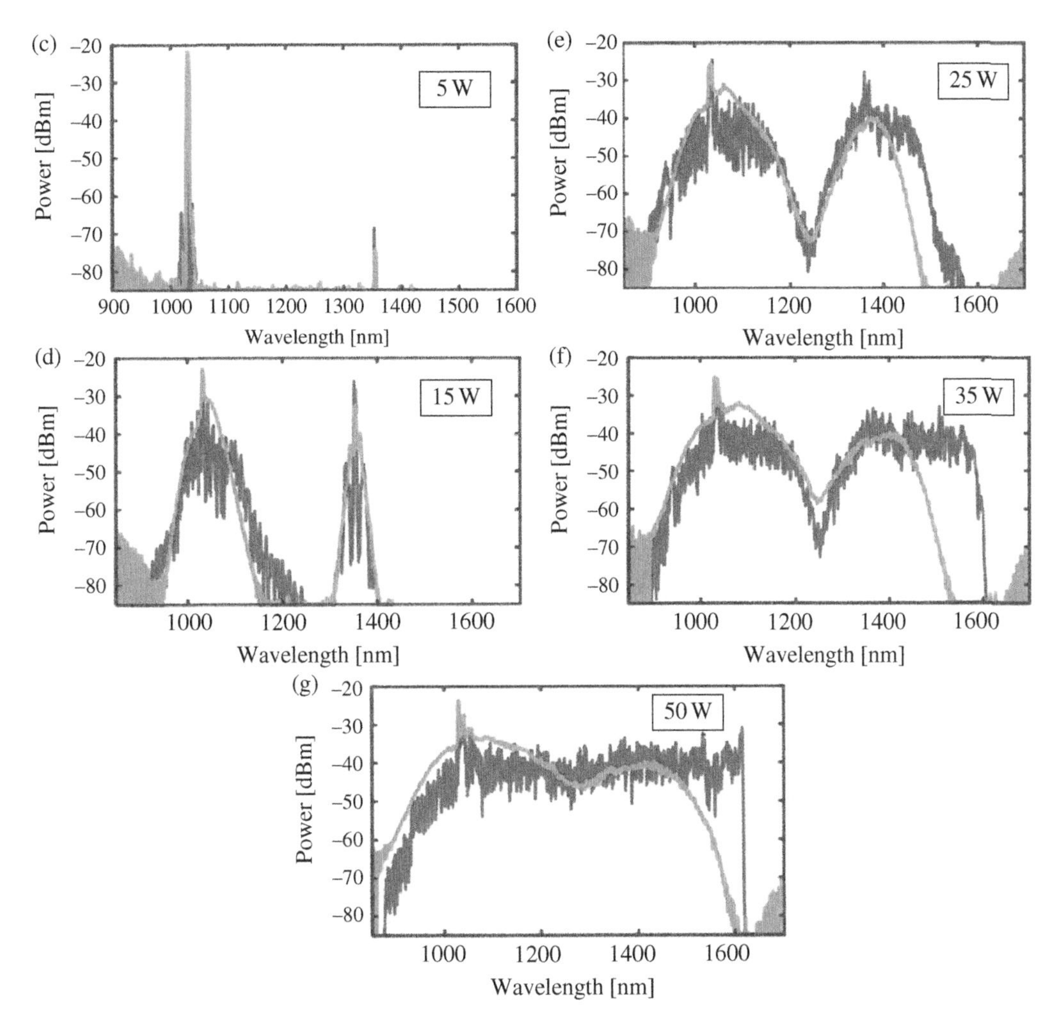

Figure 17.22 (Continued)

References

1 Dudley, J. and Taylor, J. (ed.) (2010). *Supercontinuum Generation in Optical Fibers*. Cambridge, UK: Cambridge University Press.

2 Alfano, R.R. and Shapiro, S.L. (1970). *Phys. Rev. Lett.* **24**: 584.

3 Lin, C. and Stolen, R. (1976). *Appl. Phys. Lett.* **28**: 216.

4 Bellini, M. and Hänsch, T. (2000). *Opt. Lett.* **25**: 1049.

5 Udem, T., Holzwarth, R., and Hänsch, T. (2002). *Nature* **416(** 233.

6 Fercher, A., Drexler, W., Hitzenberger, C., and Lasser, T. (2003). *Rep. Progr. Phys.* **66**: 239.

7 Lindfors, K., Kalkbrenner, T., Stoller, P., and Sandoghdar, V. (2004). *Phys. Rev. Lett.* **93**: 037401.

8 Hundertmark, H., Wandt, D., Fallnich, C. et al. (2004). *Opt. Express* **12**: 770.

9 Takara, H., Ohara, T., Yamamoto, T. et al. (2005). *Electron. Lett.* **41**: 270.

10 Humbert, G., Wadsworth, W., Leon-Saval, S. et al. (2006). *Opt. Express* **14**: 1596.

11 Labruyère, A., Tonello, A., Couderc, V. et al. (2012). *Opt. Fiber Technol.* **18**: 375.

12 Hakala, T., Suomalainen, J., Kaasalainen, S., and Chen, Y. (2012). *Opt. Express* **20**: 7119.

13 Kawagoe, H., Ishida, S., Aramaki, M. et al. (2014). *Biomed. Opt. Express* **5**: 932.

14 Ortigosa-Blanch, A., Knight, J., and Russel, P. (2002). *J. Opt. Soc. Am. B* **19**: 2567.

15 Genty, G., Lehtonen, M., Ludvigsen, H. et al. (2002). *Opt. Express* **10**: 1083.

16 Dudley, J., Genty, G., and Cohen, S. (2006). *Rev. Modern Phys.* **78**: 1135.

17 Gorbach, A., Skyabin, D., Stone, J., and Knight, J. (2006). *Opt. Express* **14**: 9854.

18 Agrawal, G.P. (2007). *Nonlinear Fiber Optics*, 4e. Burlington, MA: Academic Press.

19 Ferreira, M.F. (2011). *Nonlinear Effects in Optical Fibers*. Hoboken, NJ: John Wiley & Sons.

20 Kodama, Y. and Hasegawa, A. (1987). *IEEE J. Quant. Electron.* **23**: 510.

21 Akhmediev, N. and Karlsson, M. (1995). *Phys. Rev. A* **51**: 2602.

22 Husakou, A. and Herrmann, J. (2001). *Phys. Rev. Lett.* **87**: 203901.

23 Nishizawa, N. and Goto, T. (2002). *Opt. Express* **10**: 1151.

24 Gorbach, A. and Skryabin, D. (2007). *Nature Photonics* **1**: 653.

25 Sakamaki, S., Nakao, M., Naganuma, M., and Izutsu, M. (2004). *IEEE J. Sel. Top. Quantum Electron.* **10**: 876.

26 Genty, G., Lehtonen, M., and Ludvigsen, H. (2004). *Opt. Express* **12**: 4614.

27 Frosz, M., Falk, P., and Bang, O. (2005). *Opt. Express* **13**: 6181.

28 Judge, A., Bang, O., and Sterke, C. (2010). *J. Opt. Soc. Am. B* **27**: 2195.

29 Roy, S., Bhadra, S., Saitoh, K. et al. (2011). *Opt. Express* **19**: 10443.

30 Skryabin, D. and Yulin, A. (2005). *Phys. Rev. E* **72**: 016619.

31 Skryabin, D., Luan, F., Knight, J., and Russell, P. (2003). *Science* **301**: 1705.

32 Washburn, B., Ralph, S., and Windeler, R. (2002). *Opt. Express* **10**: 575.

33 Dudley, J. and Coen, S. (2002). *J. Sel. Top. Quantum Electron.* **8**: 651.

34 Gaeta, A. (2002). *Opt. Lett.* **27**: 924.

35 Husakou, A. and Herrmann, J. (2002). *J. Opt. Soc. Am. B* **19**: 2171.

36 Hult, J. (2007). *J. Lightwave Technol.* **25**: 3770.

37 Price, J., Monro, T., Ebendorff-Heidepriem, H. et al. (2007). *IEEE J. Sel. Top. Quantum Electron.* **23**: 738.

38 Mitschke, F. and Mollenauer, L. (1986). *Opt. Lett.* **11**: 659.

39 Gordon, J. (1986). *Opt. Lett.* **11**: 662.

40 Solli, D., Ropers, C., Koonath, P., and Jalali, B. (2007). *Nature* **450**: 1054.

41 Monfared, Y. and Ponomarenko, S. (2017). *Opt. Express* **25**: 5941.

42 Ankiewicz, A., Bokaeeyan, M., and Akhmediev, N. (2018). *J. Opt. Soc. Am. B* **35**: 899.

43 Stone, J. and Knight, J. (2008). *Opt. Express* **16**: 2670.

44 Ghosh, D., Roy, S., Pal, M. et al. (2011). *J. Lightwave Technol.* **29**: 146.

45 Chen, H., Chen, Z., Zhou, X., and Hou, J. (2013). *Laser Phys. Lett.* **10**: 085401.

46 Wang, N., Cai, J., Qi, X. et al. (2018). *Opt. Express* **26**: 1689.

47 Coen, S., Chau, A., Leonhardt, R. et al. (2002). *J. Opt. Soc. Am. B* **19**: 753.

48 Dudley, J., Provino, L., Grossard, N. et al. (2002). *J. Opt. Soc. Am. B* **19**: 765.

49 Kudlinski, A., Pureur, V., Bouwrnans, G., and Mussot, A. (2008). *Opt. Lett.* **33**: 2488.

50 Sun, C., Ge, T., Li, S. et al. (2016). *Appl. Opt.* **55**: 3746.

51 Qi, X., Chen, S., Li, Z. et al. (2018). *Opt. Lett.* **43**: 1019.

52 Zhao, L., Li, Y., Guo, C. et al. (2018). *Opt. Commun.* **425**: 118.

53 Tao, Y., Chen, S., and Xu, H. (2019). *Opt. Express* **27**: 26044.

54 Hasegawa, A. and Brinkman, W. (1980). *IEEE J. Quantum Electron.* **16**: 694.

55 Tai, K., Hasegawa, A., and Tomita, A. (1986). *Phys. Rev. Lett.* **56**: 135.

56 Mussot, A., Lantz, E., Maillotte, H. et al. (2004). *Opt. Express* **12**: 2838.

57 Frosz, M., Bang, O., and Bjarklev, A. (2006). *Opt. Express* **14**: 9391.

58 Korneev, N., Kuzin, E., Ibarra-Escamilla, B., and Flores-Rosas, M. (2008). *Opt. Express* **16**: 2636.

59 Biancalana, F., Skryabin, D., and Yulin, A. (2004). *Phys. Rev. E* **70**: 016615.

60 Cumberland, B., Travers, J., and Popov, S. (2008). *J. Taylor. Opt. Express* **16**: 5954.

61 Kudlinski, A., George, A., Knight, J. et al. (2006). *Opt. Express* **14**: 5715.

62 Kudlinski, A. and Mussot, A. (2008). *Opt. Lett.* **33**: 2407.

63 Arun, S., Choudhury, V., Balaswamy, V. et al. (2018). *Opt. Express* **26**: 7979.

64 Moller, U., Yu, Y., Kubat, I. et al. (2015). *Opt. Express* **23**: 3282.

65 Cimalla, P., Walther, J., Mittasch, M., and Koch, E. (2011). *J. Biomed. Opt.* **16**: 116020.

66 Ringsted, T., Siesler, H., and Engelsen, S. (2017). *J. Cereal Sci.* **75**: 92.

67 Petersen, C., Prtljaga, N., Farries, M. et al. (2018). *Opt. Lett.* **43**: 999.

68 Islam, M., Freeman, M., Peterson, L. et al. (2016). *Appl. Opt.* **55**: 1584.

69 Domachuk, P., Wolchover, N., Cronin-Golomb, M. et al. (2008). *Opt. Express* **16**: 7161.

70 Liao, M., Gao, W., Duan, Z. et al. (2012). *Opt. Lett.* **37**: 2127.

71 Qin, G., Yan, X., Kito, C. et al. (2009). *Appl. Phys. Lett.* **95**: 161103.

72 Wang, F., Wang, K., Yao, C. et al. (2016). *Opt. Lett.* **41**: 634.

73 Kubat, I., Agger, C., Møller, U. et al. (2014). *Opt. Express* **22**: 19169.

74 Saini, T., Kum, A., Zhao, Z. et al. (2015). *J. Lightwave Technol.* **33**: 3914.

75 Zhao, Z., Wang, X., Dai, S. et al. (2016). *Opt. Lett.* **41**: 5222.

76 Dai, S., Wang, Y., Peng, X. et al. (2018). *Appl. Sci. (Basel)* **8**: 707.

77 Petersen, C., Møller, U., Kubat, I. et al. (2014). *Nat. Photonics* **8**: 830.

78 Zhang, B., Yu, Y., Zhai, C. et al. (2016). *J. Am. Ceram. Soc.* **99**: 2565.

79 Cheng, T., Nagasaka, K., Tuan, T. et al. (2016). *Opt. Lett.* **41**: 2117.

80 Tao, G., Ebendorff-Heidepriem, H., Stolyarov, A. et al. (2015). *Adv. Opt. Photonics* **7**: 379.

81 Théberge, F., Bérubé, N., Poulain, S. et al. (2018). *Opt. Express* **26**: 13952.

82 Robichaud, L., Duval, S., Pleau, L. et al. (2020). *Opt. Express* **28**: 107.

83 Yang, L., Zhang, B., He, X. et al. (2020). *Opt. Express* **28**: 14973.

84 Dudley, J. and Coen, S. (2002). *Opt. Lett.* **27**: 1180.

85 Huang, C., Liao, M., Bi, W. et al. (2018). *Photon. Res.* **6**: 601.

86 Diouf, M., Cherif, R., Ben Salem, A., and Waguem, A. (2017). *J. Mod. Opt.* **64**: 1335.

87 Diouf, M., Salem, A., Cherif, R. et al. (2017). *Appl. Opt.* **56**: 163.

88 Heidt, A. (2010). *J. Opt. Soc. Am. B* **27**: 550.

89 Heidt, A., Hartung, A., Bosman, G. et al. (2011). *Opt. Express* **19**: 3775.

90 Hooper, L., Mosley, P., Muir, A. et al. (2011). *Opt. Express* **19**: 4902.

91 Zhang, N., Peng, X., Wang, Y. et al. (2019). *Opt. Express* **27**: 10311.

92 Nakazawa, M., Tamura, K., Kubota, H., and Yoshida, E. (1998). *Opt. Fiber Technol.* **4**: 215.

93 Genty, G., Coen, S., and Dudley, J. (2007). *J. Opt. Soc. Am. B* **24**: 1771.

94 Klimczak, M., Siwicki, B., Skibinski, P. et al. (2014). *Opt. Express* **22**: 18824.

95 Al-Kadry, A., Li, L., Amraoui, M. et al. (2015). *Opt. Lett.* **40**: 4687.

96 Klimczak, M., Siwicki, B., Zhou, B. et al. (2016). *Opt. Express* **24**: 29406.

97 Liu, L., Cheng, T., Nagasaka, K. et al. (2016). *Opt. Lett.* **41**: 392.

98 Froidevaux, P., Lemière, A., Kibler, B. et al. (2018). *Appl. Sci.* **8**: 1875.

99 Saini, T., Nguyen, P., Tuan, T. et al. (2019). *Appl. Opt.* **58**: 415.

100 Xing, S., Kharitonov, S., Hu, J., and Brès, C.-S. (2018). *Opt. Express* **26**: 19627.

101 Nguyen, H., Tong, T., Saini, T. et al. (2019). *Appl. Phys. Express* **12**: 042010.

102 Jackson, S. (2012). *Nat. Photonics* **6**: 423.

103 Fortin, V., Bernier, M., Bah, S.T., and Vallée, R. (2015). *Opt. Lett.* **40**: 2882.
104 Maes, F., Fortin, V., Bernier, M., and Vallée, R. (2017). *Opt. Lett.* **42**: 2054.
105 Nguyen, H., Tuan, T., Xing, L. et al. (2020). *Opt. Express* **28**: 17539.
106 Barviau, B., Kibler, B., Coen, S., and Picozzi, A. (2008). *Opt. Lett.* **33**: 2833.
107 Dyachenko, S., Newell, A., Pushkarev, A., and Zakharov, V. (1992). *Physica D* **57**: 96.
108 Newell, A., Nazarenko, S., and Biven, L. (2001). *Physica D* **152**: 520.
109 Zakharov, V., Dias, F., and Pushkarev, A. (2004). *Phys. Rep.* **398**: 1.
110 Picozzi, A., Pitois, S., and Millot, G. (2008). *Phys. Rev. Lett.* **101**: 093901.
111 Barviau, B., Kibler, B., Kudlinski, A. et al. (2009). *Opt. Express* **17**: 7392.
112 Michel, C., Kibler, N., and Picozzi, A. (2011). *Phys. Rev. A* **83**: 023806.
113 Stark, S., Travers, J., and Russell, P. (2012). *Opt. Lett.* **37**: 770.
114 Russell, P. (2006). *J. Lightwave Technol.* **24**: 4729.
115 Im, S., Husakou, A., and Herrmann, J. (2010). *Opt. Express* **18**: 5367.
116 Durfee, C. III, Backus, S., Kapteyn, H., and Murnane, M. (1999). *Opt. Lett.* **24**: 697.
117 Nold, J., Hoelzer, P., Joly, N. et al. (2010). *Opt. Lett.* **35**: 2922.
118 Joly, N., Nold, J., Chang, W. et al. (2011). *Phys. Rev. Lett.* **106**: 203901.
119 Mak, K., Travers, J., Hölzer, P. et al. (2013). *Opt. Express* **21**: 10942.
120 Rodrigues, S., Facão, M., and Ferreira, M. (2015). *Fiber Int. Opt.* **34**: 76.
121 Rodrigues, S., Facão, M., and Ferreira, M. (2018). *Opt. Commun.* **412**: 102.
122 Benabid, F., Knight, J., Antonopoulos, G., and Russell, P. (2002). *Science* **298**: 399.
123 Russell, P., Hölzer, P., Chang, W. et al. (2014). *Nat. Photonics* **8**: 278.
124 Belli, F., Abdolvand, A., Chang, W. et al. (2015). *Optica* **2**: 292.
125 Hosseini, P., Ermolov, A., Tani, F. et al. (2018). *ACS Photonics* **5**: 2426.
126 Mousavi, S., Mulvad, H., Wheeler, N. et al. (2018). *Opt. Express* **26**: 8866.
127 Luan, J., Russell, P., and Novoa, D. (2021). *Opt. Express* **29**: 13787.

Index

t

u

v

w

x

z

Printed and bound by CPI Group (UK) Ltd, Croydon, CR0 4YY
16/08/2022
03141867-0001